DISCOVERING GENOMICS, PROTEOMICS, AND BIOINFORMATICS

A. Malcolm Campbell
Davidson College

Laurie J. Heyer
Davidson College

San Francisco Boston New York
Cape Town Hong Kong London Madrid Mexico City
Montreal Munich Paris Singapore Sydney Tokyo Toronto

Senior Project Manager: Peggy Williams
Publishing Assistant: Michael McArdle
Production Editors: Larry Lazopoulos, Jamie Sue Brooks
Production and Composition Services: The Left Coast Group
Text Designer: Andrew Ogus
Cover Designer: Yvo Reizebos
Copyeditor: Gary Morris
Illustrations: Steve McEntee and Mary Ann Tenorio
Proofreader: Martha Ghent
Photo Research: Kathleen Olson, Maureen Spuhler
Manufacturing Supervisor: Vivian McDougal
Marketing Manager: Josh Frost
Cover Printer: Phoenix Color
Printer: Courier Book Companies

Library of Congress Cataloging-in-Publication Data

Campbell, A. Malcolm
    Discovering genomics, proteomics, and bioinformatics / A. Malcolm Campbell,
Laurie J. Heyer.
    p. cm.
    Includes index.
    ISBN 0-8053-4722-4
    1. Genomics. 2. Proteomics. 3. Bioinformatics. I. Heyer, Laurie, J. II. Title.

QH447 .C35 2002
572.8'6—dc21   2002067456

Text, photography, and illustration credits appear following the Glossary.

Many of the designations used by manufacturers and sellers to distinguish their products are claimed
as trademarks. Where those designations appear in this book, and the publisher was aware of a trademark claim, the designations have been printed in initial caps or all caps.

ISBN 0-8053-4722-4

2 3 4 5 6 7 8 9 10 — CRW — 06 05 04 03

www.aw.com/bc

CONTENTS

CHAPTER 11

Why Can't I Just Take a Pill to Lose Weight?   306

Hungry for Knowledge   307

CHAPTER 12

Why Can't We Cure More Diseases?   320

How to Develop a New Medication   321

FOREWORD

Stan Fields
University of Washington

New fields of biology often open up when new tools become available to analyze living creatures and their constituent parts. Developments in microscopy and cell fractionation led to the rise of cell biology; the availability of X-ray sources and later Nuclear Magnetic Resonance machines brought about the three-dimensional images of structural biology, and recombinant DNA technologies advanced the already spreading techniques of molecular biology. As these disciplines matured, their approaches began to affect the thinking that surrounds major questions in biology; at the same time, these new branches retained some of their distinctive styles and adherents.

The advent of the Human Genome Project captured the imagination of both scientists and the general public. On 26 June 2000, President Bill Clinton, at a press conference with the heads of the National Human Genome Research Institute and Celera Genomics Corporation, declared "We are here to celebrate the completion of the first survey of the entire human genome. Without a doubt, this is the most important, most wondrous map ever produced by humankind." For biologists, the Human Genome Project meant the arrival of billions of base pairs of DNA sequence, not just of humans but of model organisms such as *E. coli,* yeast, the nematode worm, the fruit fly, and the mustard weed. In conjunction with these sequences came a flurry of technologies to interpret the DNA sequence, the RNAs that are copied from it, and the proteins that are synthesized from these RNAs.

Discovering Genomics, Proteomics, and Bioinformatics gives students a detailed view of the revolutionary methodologies of the last few years and their impact on biological thinking. It provides a lucid explanation of technologies used for such tasks as DNA sequencing, detecting polymorphisms, arraying genes, making gene knockouts, and identifying protein interactions. The application of these technologies to specific biological problems provides a model for how to approach many other questions in biology—from development to human disease. The text also illustrates how the acquisition of large data sets has led to the generation of networks of genes and proteins, as well as models of cellular behavior. A series of *Discovery Questions* in each chapter guides the reader through scientific questions as well as ethical, legal, and social issues.

The disciplines of genomics and proteomics cannot be uncoupled from bioinformatics, the tools to handle and analyze the prodigious amounts of data that continue to emerge from large-scale DNA, RNA, and protein projects. The book draws on numerous databases and computational algorithms to make this coupling clear. The inclusion of *Math Minutes* provides the statistical and probability calculations that add rigor to the interpretations of these data sets.

Have these powerful approaches led to a new way of thinking? At one level, the answer is clearly "yes," as syntheses of genomic and proteomic data make possible inferences not previously apparent. Yet some might argue that biologists have always studied "systems biology," and what has changed is not the thinking but the experimental tools. When it was possible to detect only a single transcript, an RNA might be analyzed under various parameters to measure single gene responses. Now that it is feasible to employ DNA microarrays, scientists can detect virtually all the transcripts of an organism. Perhaps students beginning their studies now will enter a discipline in which these experimental and computational tools will constitute not "systems biology" but simply "biology."

PREFACE

The term "genomics" was derived from the term "genome," which means the complete (haploid) DNA content of an organism; genomics is the field of genome studies. Once genome and genomics became popular terms, a flurry of new terms that ended in "-omes" and "-omics" began to appear in publications. *Discovering Genomics, Proteomics, and Bioinformatics* is more than a "tome of -omes" because the field has expanded beyond a narrow definition of genomics. Genomics, as presented in this book, includes the interaction of molecules inside cells, including DNA, protein, lipids, and carbohydrates. In the spirit of discovery, we will explore the tools and questions behind the revolution that is changing the way biology is studied.

Discovering Genomics, Proteomics, and Bioinformatics is based on two pedagogical principles that have been successful for many teachers; teach in the context of an interesting question and on a need-to-know basis. This is how everyone learns new information, and it is the best way to help students learn so they will be motivated and more likely to retain the information. *Discovering Genomics, Proteomics, and Bioinformatics* is built on "stories" or case studies taken from scientific publications. In answering the many questions raised by these case studies, we will explore scientific content and process. The content includes all the major areas such as sequences (whole genomes and variations), microarrays, and proteomics.

This book is designed as an interactive resource to use when exploring topics in genomics and proteomics. The figures provide real data that you can mine to extract more information than is initially apparent. The online databases engage you in real-time discoveries using the same databases investigators are using for their own research. Discovery Questions focus your attention on critical information and urge you to think for yourself using the tools and information presented in text and figures. Traditional textbooks supply you with facts and details that you are inclined to memorize for tests. Genomics requires you to analyze, hypothesize, think, and formulate models; this book was designed to develop your critical thinking skills.

Studying biology in the twenty-first century offers you unique opportunities. Therefore, this textbook must also look and "feel" different. You will need to use the computer a lot to access the latest information. To fully understand genomics, proteomics, and bioinformatics, you will read about real and compelling cases that challenge and encourage you to learn. So, immerse yourself in the case studies and discover what genomics is all about.

Writing Style

The text is written in a style that is easy to read and comprehend. It avoids unnecessary jargon, yet new terms are included when essential to help you understand the material.

Discovery Questions

The process of critical thinking is enhanced by Discovery Questions which are imbedded within the case studies rather than saved for the end of each chapter. Discovery Questions focus your attention on key concepts as well as experimental design, interpretation of data, and the need to support your opinion with data. Analyzing real data reproduced from peer-reviewed publications will allow you to reach your own conclusions and the text will help guide you through the data. To answer some Discovery Questions (see the following examples), you will use online public databases, many of which are regularly updated.

DISCOVERY QUESTIONS

16. Go to *S. cerevisiae* Genomic View and find the location (base 1,236,754 of chromosome 4) of the NORF identified by this research. Find the annotated gene that overlaps with the insertion site.

17. Go to SGD Gene/Sequence Resource page, retrieve the nucleotides on chromosome 4 ranging from 1,236,454–1,237,054, and click on "6-Frame translation." The mTn insertion happened at base 300 of these 600 bases. How many potential proteins are located at this site?

All the Discovery Questions are available via the companion web site, to facilitate your interaction with online resources and permit you to submit your answers to your instructor via email.

Media Menu

Throughout every chapter are Media Menus to alert you to the resources contained on the book's companion web site. From this site, you can read in-depth descriptions of methods, access sequence information, view a 3D structure, and link to related web sites. These media tools allow you to participate in the interactive process of discovery that is at the core of genomics.

METHODS
QuickPDB

STRUCTURES
Cyclooxygenase

SEQUENCES
Uncharacterized Protein

LINKS
Conserved Domain
PDB
PREDATOR

Math Minutes

Most biologists in the cell/molecular field do not use much math in their work, but genomics, proteomics, and bioinformatics are changing this reality; these fields rely heavily on mathematics. To facilitate appreciation of how the data were analyzed and the role mathematics plays in understanding biology, we have produced Math Minutes as enrichment for those who want to discover the interaction of math and biology. Math Minutes (see sample below) use the case studies as foundations for concise lessons in statistical analysis, probability and computational methods.

MATH MINUTE 6.2 • IS Sup35 A CENTRAL PROTEIN IN THE NETWORK?

In Figure 6.16, Sup35 appears to be central to the network of interacting proteins. However, looks can be deceiving, particularly in such a complex network. We would like to quantify whether Sup35 interacts with an unusually large number of proteins, compared to other proteins in the network.

A mathematician would call the network of interacting proteins in Figure 6.16 a *graph*. In graph theory, the lines between nodes are called *arcs* or *edges*. A *directed* graph has arrows on the edges to indicate the direction of information flow between two nodes. In Figure 6.16 where the connections or relationships between nodes are important but there is no directional information, the arrows are left off and the graph is *undirected*. The number of edges touching a node is called the *degree* of the node. In graph theory terms, the question at hand is whether the Sup35 node has a significantly greater degree than is "typical" for this graph. If so, investigators are led to believe that Sup35 plays a central role in the function of the entire network of proteins.

One way to approach questions about graphs like that in Figure 6.16 is to build a probabilistic model of the graph, called a *random graph*. Under this model, an edge is drawn between each pair of nodes with a certain probability, each edge independent of the others. Advanced probability theory and graph theory provide precise ways to determine the probability that the maximum degree in an arbitrarily large graph exceeds a given number. However, under the simplifying assumption that the degree numbers are approximately normally distributed (see Math Minute 8.1), you can evaluate whether the degree of the Sup35 node is significantly greater than expected using the mean and standard deviation of the degree numbers. Since Figure 6.16 is too complex to illustrate this process, consider the following smaller network:

In this graph, nodes 4, 5, and 7 have degree 1; nodes 2, 3, and 6 have degree 2; nodes 1 and 8 have degree 3; and node 9 has degree 5. The mean degree is the average of the nine degree numbers (2.22). The standard deviation of the nine degree numbers (1.23) is used to determine if the degree of node 9 is unusual. Specifically, if node 9's degree is more than two standard deviations larger than the mean degree, it can be considered unusually large. Since $5 > 4.68 = 2.22 + 2 \times 1.23$, we conclude that node 9 has an unusually large degree. This same procedure could be applied to the graph in Figure 6.16 to help quantify whether Sup35 plays a central role in this network of proteins.

Art Program

Detailed and abundant illustrations are reproductions of original data and expand on the basic information provided in the text, as shown in the example below. Your understanding will be enhanced by analysis of the figures from which you can extract additional information.

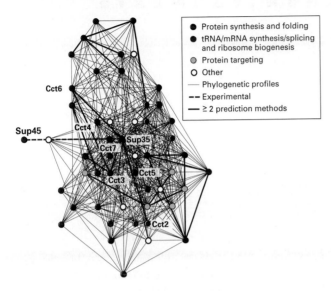

Figures on CD-ROM

All the illustrations in the text appear on the CD-ROM that accompanies every copy of the book. Some figures, which are best seen in full color and large format do not appear in the text but are available exclusively on the CD-ROM. These illustrations are designated in the text as follows:

> **FIGURE 4.19 • Stress-induced expression of 900 genes clustered into the ESR.**
> Go to the CD-ROM to view this figure.

These figures may be projected in class to enhance classroom lectures and discussions.

Transition from Genetics to Genomics

Typically, people think of genomics as high-volume genetics. However, a genome is more than the sum of its parts and you need to approach it with a new mindset. The last unit, Unit Four, contains three chapters to help ease your transition into genomics. Specifically, these three chapters confront the common misconception of "one gene, one protein, one phenotype." The case studies contain real data for you to interpret and discover the interconnectivity of the "cell web." The discovery approach to learning will foster scientific skills of analyzing data and formulating models to explain the data.

Media

The companion web site for *Discovering Genomics, Proteomics, and Bioinformatics* (www.geneticsplace.com) is a tool to enhance your study of genomics.

- *Methods:* These web pages explain how molecular and genomic methods are conducted and what type of data they produce. They are intended to supplement the textbook and provide background information if you have not learned them in previous courses.
- *Sequences:* In several Discovery Questions, you will use online bioinformatics tools to analyze protein or DNA sequences. To save you time and the potential problem of typos, all sequences are supplied in web pages (see the illustration below) that you can copy and paste for analysis.
- *Structures:* A significant aspect of any protein is its 3D structure. Periodically, web pages with chime tutorials have been created to illustrate structural features that are best understood when you interact with them.
- *Links:* Two types of links have been collected for each chapter. The first provides direct access to online databases and bioinformatics analysis tools such as the National Center for Biotechnology Information and the Protein Data Bank. The second facilitates easy access to investigators' laboratory web pages when you are particularly interested in a case study or area of research.

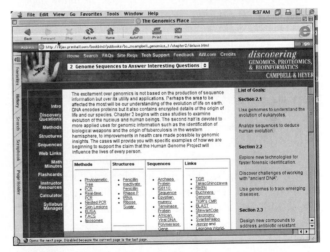

For the Instructor
Instructor's Guide ISBN 0-8053-4726-7

The printed version of the Instructor's Guide provides tips on presenting the material and answers to all the Discovery Questions. In addition to the written answers, the electronic version of the Instructor's Guide (www.geneticsplace.com) provides color figures to accompany some answers. For example, if students are asked to draw a graph or construct a circuit diagram, your electronic version will illustrate one possible answer. These illustrations, along with all illustrations on the CD-ROM, may be used for lecture presentation such as PowerPoint or web pages.

ACKNOWLEDGMENTS

Davidson College is home to many supportive people. The leadership and backing of Clark Ross, Bobby Vagt, and Verna Case provided the foundation for much of this work. Colleagues in many departments helped the authors in numerous ways: Karen Bernd, Karen Hales, Dave Wessner, Barbara Lom, Pam Hay, Julio Ramirez, David Brown, Jeanne O'Neill, Suzanne Churchill, Chris Paradise, Mur Muchane, Peggy Maiorano, and Betty Hartsell. The Waksman Foundation for Microbiology, the National Science Foundation, the Duke Endowment, and the Associated Colleges of the South made the funding possible for my sabbatical year in Seattle. Perhaps most importantly, I would like to thank the fourteen brave students who tested out a rough draft version of this book: Sean Burke, Amber Hartman, Ben Havard, Julie Hwang, Dennis Jones, Peter Lowry, Jennifer Madden, Emily Oldham, La Powell, Lisa Robinson, Elizabeth Sellars, Liz Shafer, J. D. Willson, and Marisa Wilson.

Much of the first draft was written while I was on sabbatical in Seattle. Mary Claire King, Maynard Olson, and Leroy Hood were very supportive hosts during this time. Several members of the University of Washington Genome Center provided technical and moral support. People at the Institute for Systems Biology (ISB) established a wonderfully supportive environment to learn genomics and proteomics. Eric Davidson's presentation at the ISB annual retreat was inspirational in many ways. And thanks to all the investigators around the world who freely shared their work and experience, and answered questions. It was heartwarming to be a part of a community where knowledge is shared and education valued. Personally, our friends and family in Seattle provided a wonderful home away from home, especially the Hope-Young family, the Hill family, Ginger Armbrust and Susan Francis, the Walker family, and the 20 families in the Meadowbrook Co-op Preschool.

Any major project requires a large number of people to create it. This book began as a vague idea and was quickly shaped into a reality thanks to the nurturing guidance of my friend Michele Sordi. Without her, it never would have come to fruition. Producing the actual book was largely due to the diligence and sweat equity of Peggy Williams. Her labors were critical to the successful invention of a new book in a new field. In addition, wisdom and editorial collaboration at Cold Spring Harbor Labratory Press was shepherded by John Inglis, who leads a substantial team of dedicated professionals. At Benjamin Cummings, many people played substantial roles in shaping this book, including Steve McEntee and Mary Ann Tenorio whose artistic skills were able to make the figures true to their original form while also enhancing their pedagogical value. Jamie Sue Brooks and Larry Lazopoulos at Benjamin Cummings, and the staff at The Left Coast Group admirably shaped the production work. Thanks to Andrew Ogus for creating the look of the book, and to Yvo Rezebios for the cover design. Finally, Michael McArdle as publishing assistant kept everyone in touch with each other.

Finally, the most important support came from those closest to home. Almost overnight a wonderful collaboration was born between a mathematician who knows a lot of biology and a biologist who tries his best with math; thanks, Laurie. Members of the Genome Consortium for Active Teaching (GCAT) provided a collegial support system that encouraged me when I needed it. Susan, Paulina, and Celeste crossed the country twice and provided me with the stability needed to produce this book. My extended family also supplied needed support along the way without whom I would still be making floor tiles in the factory.

—*A. Malcolm Campbell*

Colleagues in the mathematics department, especially chair Stephen Davis and mentor Donna Molinek, offered endless encouragement and support to me. Our students are a continuing source of energy and inspiration. I am particularly indebted to Computational Biology students Frank Chemotti, Amber Hartman, Soren Johnson, Jennifer Kawwass, Peter Leese, Rachel Patton McCord, Cecilia Mendiondo, Emily Oldham, La Powell, Talbot Presley, Stanley Prybe, Lang Robertson, Lisa Robinson, and Megan Shafer. This interdisciplinary team produced first-class Kyte Doolittle and hierarchical clustering web pages. Students in both Genomics and Computational Biology battle-tested numerous Math Minutes; special thanks to those whose questions led to Math Minute 6.1.

I continue to be inspired by the former teachers and mentors who led me to and through this field: Mike Waterman, Simon Tavaré, Gary Stormo, and John Williamson. They, along with friends and my family, made my contribution to this book a possibility. Malcolm, who was willing to share his vision for genomics education with a new colleague, made it a reality. Finally, I thank my husband Bill, whose patience and love never wear thin. He is my greatest support in all things.

—*Laurie J. Heyer*

REVIEWERS

UNIT ONE

Genome

Sequences

CHAPTER 1

Genome Sequence

Acquisition

and Analysis

With any new field of study such as genomics comes new technology and terminology. The first part of Chapter 1 concentrates on how DNA sequence information is obtained and analyzed. Along the way, we will learn the essential vocabulary. The second half of Chapter 1 focuses on early lessons from the draft sequence of the human genome. In short, Chapter 1 provides an overview of obtaining and analyzing genome sequences.

1.1 Defining Genomes

We begin this section with technical information about sequencing methodology so you will understand how data are produced. People often assume genomics includes only completed genomes, but we will examine rich databases dedicated to short segments of DNA. Periodically, you will be asked Discovery Questions, which are interactive opportunities to use public databases. Discovery Questions use specific examples to illustrate general principles of genome data analysis. Because databases are continuously updated, answers and web page layout will change; an unknown gene today may be a newly discovered cancer gene tomorrow.

What Is Genomics?

Genomics is an unusual scientific term because its definition varies from person to person. The root word "genome" is universally defined as the total DNA content of a haploid cell or half the DNA content of a diploid cell. You would think the discipline genomics would be the study of genomes, but this simple definition is too simplistic. In one sense, all of biology is related to the study of genomes because an organism is shaped by its genome. However, most biologists would agree that disciplines such as anatomy and zoology should not be lumped into the current usage of genomics. How should we define genomics?

For most people, genomics involves large data sets (about 3 billion base pairs for the human genome) and **high-throughput** methods (fast methods for collecting the data). Genomics includes sequencing DNA and collecting genome variations within a population as well as transcriptional control of genes. Once the terms genome and genomics gained popularity, a cascade of new terms was initiated so each new area of research became an "-omic" or the subject under investigation was an "-ome." The best examples are proteome and proteomics; a **proteome** is the complete protein content of a cell/organism at a given moment. Other terms include transcriptome, metabolome, glycome, and variome. Do these newest fields fit under the larger umbrella of "genomics," or are they distinct? It depends on whom you ask. Throughout this book, we will define terms as needed and not focus on minor points. For us, the field of genomics includes a wide range of topics, including proteomics.

LINKS

Fred Sanger

One last point to make about genomics. Although some people consider genomics to be nothing more than a collection of new methods, we will include one more component to our definition—a new perspective. With new methods come new types of questions and new ways to understand life. For many years, molecular methods were used as reductionist tools to dissect cells and understand how the parts work in isolation. The field of genomics asks "expansionist" questions to understand how all parts work together. How does a functioning genome respond to environmental changes? What proteins interact with each other? These questions bring new interpretations that have many names—for instance, networks, systems biology, circuits. In this book, we use the term **circuits** to refer to the holistic interactions of genomes and proteomes. Chapters 7, 8, and 9 consider three organizational levels of circuits—single genes, multiple genes, and whole genomes. Therefore, this book considers a broad definition of **genomics,** from DNA sequence analysis to an organism's response to environmental perturbations. The best place to start, however, is at the beginning, so let's learn about the sequencing of genomes.

How Are Whole Genomes Sequenced?

The most popular method to sequence DNA is the **dideoxy method,** sometimes referred to as the **Sanger method** in honor of Fred Sanger, who was awarded a Nobel Prize in 1980 for it. We begin by describing the original radioactive procedure, and then address subsequent modifications that have improved sequencing efficiency.

At the heart of sequencing is DNA replication, which takes place in every dividing cell (Figure 1.1). To sequence DNA, many copies of a double-stranded DNA are denatured into

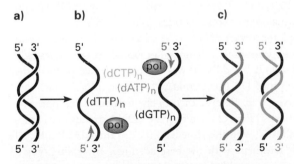

FIGURE 1.1 • DNA replication in three steps.
a) Double-stranded DNA (blue and black interwoven strands) is shown aligned in antiparallel orientation. **b)** The strands are separated and primers (light blue and gray arrows) bind to the 3' ends along with DNA polymerases (ovals). The dNTPs are available for incorporation into the new second strands. **c)** The final products are two copies of the original DNA with each older strand (dark blue and black) interwoven with the newer strands (gray and light blue).

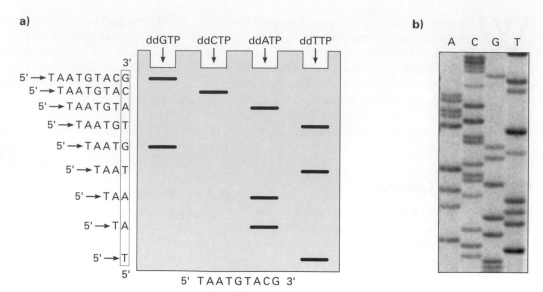

FIGURE 1.2 • How to read a DNA sequencing gel. a) The diagram shows a sequencing gel where each of the four reactions was added to separate wells of the gel. Charged DNA polymers migrate toward the positive pole at the bottom of the gel, with the smallest fragments moving fastest. On the left side of the gel is the DNA sequence in each band that contains millions of copies of identical DNA segments. In reality, you cannot see bands as short as shown here for illustration purposes. **b)** Part of an X-ray film is shown, and typically DNA gels were loaded at two different times to separate shorter from longer pieces of ddNTP-terminated DNA.

METHODS

X-ray film

STRUCTURES

ddNTPs

LINKS

BLASTn

Leroy Hood

single strands. The investigator mixes denatured DNA, DNA polymerase, a primer, and **deoxyribonucleotide triphosphates** of all four bases (**dNTPs** = dGTP, dTTP, dATP, and dCTP). One of the dNTPs includes a radioactive atom for reasons that will become clear. This mixture is aliquotted into four tubes, and to each tube is added a small dose of one of the four **dideoxyribonucleotide triphosphates** (**ddNTPs** = ddGTP, ddTTP, ddATP, and ddCTP). A ddNTP lacks the 3' hydroxyl group (−OH) found on normal deoxyribonucleotides. When DNA is being polymerized, the newest dNTP to be added forms a covalent bond onto the protruding 3' hydroxyl group of the last successfully added dNTP. If a ddNTP was the last one incorporated, the elongation of DNA is terminated because ddNTPs lack a 3' hydroxyl group. Because each of the four tubes has only one type of ddNTP, all of the terminated DNA strands in a particular tube end with the same base (i.e., in the ddATP tube, all strands terminate with the base adenine). By adding only a small amount of ddATP, when a growing strand incorporates an adenine, normally a dATP will be added to the growing strand of DNA. About 1% of all adenine-containing nucleotides are ddATP, so 99/100 times a normal dATP is incorporated and the strand continues to elongate until a ddATP is incorporated. After a few minutes, the DNA polymerization reaction is stopped. The contents of each of the four tubes is loaded onto four separate lanes of a gel, and the different length DNA molecules separate according to their sizes, with the

smallest molecules migrating the fastest (Figure 1.2). The DNA sequencing gel is exposed to X-ray film, which is developed and read from the bottom of the gel (5' end) to the top of the gel (3' end).

DISCOVERY QUESTIONS

1. Read the sequence from the real X-ray film in Figure 1.2. Record the sequence for both strands of DNA, with the top strand containing the sequence on the gel. Be sure to keep track of 5' and 3' ends for both strands.

2. Perform a **BLASTn** search with the top strand of DNA. For what gene did you just read the sequence? Now try a BLASTn search with the bottom strand (remember to enter it 5' to 3'). Do you get the same result?

Radioactive sequencing worked well, but it was labor intensive. You could collect only 500 bases of sequence on a gel that took about 24 hours to make, run, and expose the X-ray film. A faster method was needed—ideally one that did not require radioactivity. Working at Caltech, Leroy Hood and members of his lab developed a nonradioactive method that used ddNTPs with each base coupled to a different color fluorescent dye. With this innovation, labs could obtain more sequence information from each reaction and eliminate the time-consuming step of exposing the gel to X-ray film and manually reading the sequence. Only one tube was used to sequence the DNA; this was loaded onto a single lane of a gel in a dark cabinet where a laser scanned the gel as each band migrated down the gel. As each band flashed its color, indicating which ddNTP terminated that segment of DNA, the data were recorded into a computer. Fluorescence-based sequencing began the era of automated sequencing. Without this innovation, sequencing an entire genome, even small prokaryotic genomes, would have been too expensive and slow. Instead of black bars in four lanes for each piece of DNA, sequencers produced four-colored **chromatograms** (**chromat** for short), which depicted the base (one color for each base), the intensity of the light signal (height of the peak), and which base the automated software determined was at each position (printed above the graphs).

DISCOVERY QUESTIONS

3. Go to the Chromat 1 web page and look at the overall sequence. Don't bother trying to read the letters, but see if it is possible to tell which end is the 5' end.

4. Beginning at base 80, read 50 bases of the sequence and write down both strands of the DNA, with the top strand being the one on the chromat.

5. Perform a BLASTn search with this sequence, but only use 30 bases of your 50. What was your best match? Look to the right at the E-value and record this number.

Now submit all 50 bases and compare your results from the search with only 30 bases. Why did the E-value change even though you retrieved identical sequences?

METHODS

Automated Sequencing

Chromatograms

SEQUENCES

Chromat 1

MATH MINUTE 1.1 • WHAT IS AN E-VALUE?

A BLASTn search returns hits, sequences that produce "significant" alignments to the query sequence. The significance of a hit is measured by its **E-value,** or **expect value.** Each alignment has a bit score (S), a measure of similarity between the hit and the query, given in the column next to the E-value. The E-value of a hit is the number of alignments with bit score $\geq S$ that you expect to find by chance (i.e., with no evolutionary explanation). Biologically significant hits will tend to have E-values much less than 1.0. An E-value near 1.0, or even substantially larger than 1.0, does not necessarily mean that the corresponding hit is biologically irrelevant; however, the larger the E-value, the greater the chance that the similarity between the hit and the query is due to mere coincidence.

E-values are calculated from the following three factors:

1. *The bit score.* Since a larger bit score is less likely to be obtained by chance than is a smaller bit score, larger bit scores correspond to smaller E-values.

2. *Length of the query.* Since a particular bit score is more easily obtained by chance with a longer query than with a shorter query, longer queries correspond to larger E-values.

3. *Size of the database.* Since a larger database makes a particular bit score more easily obtained by chance, a larger database results in larger E-values.

To see how E-values are calculated, we revisit the BLASTn search made in Discovery Question 5. Modify the 50 base query sequence by inserting the letters ggg between positions 30 and 31. Do a BLASTn search with this modified query, but this time, under "Options for advanced blasting," set Expect to 50 (rather than the default value of 10). The higher threshold allows you to see hits with E-values as large as 50. Above each alignment in the BLASTn report is the associated bit score (S) and, in parentheses after the bit score, the raw score (R). For example, the first hit (E-value near 0.003) has $S = 46.1$ bits and $R = 23$. E-values are computed directly from bit scores, which are in turn computed directly from raw scores.

Raw Scores

The raw score is calculated by counting the number of identities, mismatches, gaps, and "–" characters in the alignment and setting $R = aI + bX - cO - dG$, where:

I is the number of identities in the alignment, and a is the reward for each identity

X is the number of mismatched nucleotides, and b is the "reward" for each mismatch

O is the number of gaps, and c is the penalty for opening a gap

G is the total number of "–" characters, and d is the penalty for each "–" in the gap

The values of the parameters *a, b, c,* and *d* are printed at the bottom of the BLASTn report. The default values are $a = 1$, $b = -3$, $c = 5$, and $d = 2$; these values can be changed on the "Other advanced" line. For example, in our first alignment,

```
Query: 1    atgctctggccacggcacttgcggatcccagggtgatctgtgcacctgcgata 53

            |||||||||||||  ||||  ||||||||   |||| |||||||||||||||||

Sbjct: 107  atgctctggccacggatcttgtgggatccca---tgatatgtgcacctgcgata 156
```

there are 46 identities, 4 mismatches, 1 gap, and 3 "-" characters, resulting in

$$R = 46 + (-3)(4) - (5)(1) - 3(2) = 23.$$

Bit Scores

The bit score is obtained from the raw score using the equation

$$S = \frac{\lambda R - \ln K}{\ln 2},$$

where λ and K (also printed at the bottom of the BLASTn report) are normalizing parameters. In our example, $\lambda = 1.37$ and $K = 0.711$, and the first alignment has a bit score of

$$S = \frac{(1.37)(23) - \ln(0.711)}{\ln 2} \approx 46.$$

As a result of the normalizing process, bit scores and E-values are independent of the scoring system, allowing those calculated with particular rewards and penalties to be compared directly to those calculated with different rewards and penalties.

E-values

Finally, the E-value is calculated as $E = mn2^{-S}$, where *m* is the effective length of the query, and *n* is the effective length (total number of bases) of the database. (Effective lengths are adjusted from the actual lengths to account for the fact that an alignment cannot start near the end of the sequence, the so-called edge effect.) In our example, $m = 34$ (19 nucleotides fewer than the 53 bases submitted) and $n = 5,854,611,841$. Plugging these numbers into the E-value equation gives $E = 0.003$, as shown in the BLASTn report.

METHODS

PCR

Cycle Sequencing

96-Well Plate

Capillary Electrophoresis

Arabidopsis

LINKS

Template DNA

Norm Dovichi

Colony Picking

NCBI database

HGP

With the advent of **polymerase chain reaction (PCR),** it did not take long for people to develop **cycle sequencing,** which combined the improvements of automated sequencing with the power of PCR. PCR works with very small amounts of template, and you don't have to produce single-stranded template. As cycle sequencing has grown in popularity, companies have created kits that perform 96 reactions simultaneously in a 96-well plate and analyze the results 96 samples at a time. With the increase in sequencing capacity, there was a need to improve DNA separation. Norm Dovichi and his colleagues helped invent **capillary electrophoresis,** which uses long, flexible, and very thin capillary tubes filled with a grainy matrix that resolves miniscule amounts of DNA. Using the 96-well sequencing format and capillary electrophorests enabled investigators to analyze simultaneously all 96 samples. Automated sequencing and capillary electrophoresis pushed the technology far enough that whole genome sequencing projects were completed ahead of schedule and under budget. In addition, colony picking from libraries and production of template DNA became automated. The number of species and the size of the **National Center for Biotechnology Information (NCBI)** database demonstrate the success of high-throughput DNA sequencing.

Why Do the Databases Contain So Many Partial Sequences?

Now that we can produce DNA sequences quickly and cheaply, we need to address the strategies for sequencing a complete genome. When the **Human Genome Project (HGP)** was proposed, the master plan included preliminary steps worth discussing. The HGP was not limited to the human genome; it also targeted yeast, fly, worm, mouse, and a flowering plant called *Arabidopsis.* By comparing different genomes, we should be able to better understand genomes in general, and the human genome in particular. The *E. coli* genome was being sequenced by a single lab at the University of Wisconsin, Madison, and its sequence would be helpful as well (see pages 64–66). The current list of sequenced genomes includes many species, predominantly prokaryotes. To understand the mapping strategy, we will talk about BACs, YACs, STSs, and ESTs. Molecular biologists love to create acronyms and then use them as if they are real words.

There is a significant difference between sequencing the genome of a prokaryote with only 2 **megabases (Mb)** (2 Mb = 2,000,000 bases) of DNA compared to a mammalian genome with roughly three orders of magnitude increase in size (2×10^3 Mb = 2 billion bp; the human genome is composed of about 3 billion bp). Big genomes contain repetitive DNA sequences from old viruses and **transposons** that have accumulated over the millennia. Piecing together the full genome sequence would be significantly easier with known markers along the way, a type of DNA bread crumb trail to indicate we were on the right path. A few genes and **cDNAs** had been sequenced, but these were too far apart to be useful. So the HGP decided to identify short segments of unique DNA sequence along every chromosome, which were called **sequence-tagged sites (STSs).** For the most part, STSs were defined by a pair of PCR primers that amplified only one segment of the genome. In conjunction with STSs, each chromosome was chopped into smaller fragments and inserted into vectors that could maintain the human DNA inside bacteria or yeast. Bacterial vectors that could carry large pieces of DNA (about 150 kb) were called **bacterial artificial chromosomes (BACs,** pronounced "backs") and replicate in *E. coli.* **Yeast artificial chromosomes (YACs,** pronounced like more than one of the animal yak) replicate in yeast and carry DNA inserts from 150 kb to 1.5 Mb). By using restriction maps, investigators could determine which BACs and YACs contained overlapping DNA, and assemble **contiguous** overlapping segments of DNA, often referred to as **contigs.** As contigs were generated, STSs were layered over the chromosomal DNA,

and the HGP was able to map the entire human genome. This genomic map would ensure that labs could place their DNA sequences at the correct location along each chromosome (Figure 1.3).

METHODS

cDNAs

LINKS

Sequenced Genomes

Prokaryotes

Human Map Viewer

Electronic PCR

DISCOVERY QUESTIONS

6. Search NCBI's Human Map Viewer using the term "obesity." You will get hits for every locus that has obesity associated with it. How many loci do you see? Are they clustered or distributed throughout the genome?

7. Click on the blue number "10" below the cartoon of chromosome 10. What gene did you identify?

8. Click on "Maps & Options" to modify the view. From the new window, you can choose from the list in the left window and your choices are displayed in the right window. Modify the display until only "Gene," "Morbid/Disease," and "Ideogram" will be displayed. Click on "Morbid/Disease" and then on the "Make Master" button followed by "Apply." The ideogram on the far left shows how much of the chromosome you are viewing. You can zoom in or out as needed.

   This database allows you to search for your favorite disease or condition and track down all this information. You could place an order for this DNA or amplify it yourself using PCR.

9. Go to Electronic PCR and enter this Accession number, "M18533," in the big open box to determine if there are any STSs in this sequence. What gene have you located, and how many STS markers are there? Click on one of the blue links and see how much information is there. Do you have all the information you need to isolate this STS? What else did you learn, other than sequences of the primers?

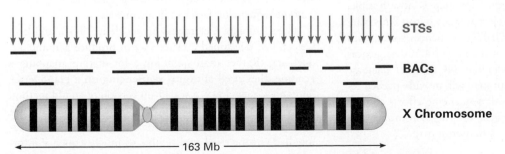

FIGURE 1.3 • Relationships of chromosomes to genome sequencing markers. The X chromosome is about 163 Mb in length. In this diagram, there are 16 overlapping BAC clones that span the entire length. In reality, 1,408 BACs were needed to span the X chromosome. Arrows (top) mark STSs scattered throughout the chromosome and on overlapping BACs.

METHODS
Whole Genome

SEQUENCES
Mystery Sequence

LINKS
EST Database
UniGene
UniGene Statistics
DOE
NIH
Wellcome Trust
TIGR
Celera

While some hurried to completely sequence the human genome, others continued to map the genome. One mapping technique tried to identify all the coding DNA to distinguish coding DNA from "**junk DNA**." To do this, several labs began sequencing very short segments (<500 bases) from either the 5' or 3' end of every cDNA they could clone. The short segments of cDNA were called **expressed sequence tags** (**ESTs**), and it did not matter if the investigators could determine what gene they came from or what the encoded protein did. Their goal was to compile every EST. In addition to identifying genes, ESTs hinted at the size of the genome and alternative ways to splice mRNA (see pages 15–16). ESTs have been helpful for labs interested in cloning particular genes. If investigators sequenced a piece of DNA that contained an EST, they would know the DNA was in a coding region and they had cloned a gene. The number of ESTs grew so quickly that a separate database was created for them, although ESTs are also integrated into larger databases such as GenBank and EMBL. To organize all these ESTs, UniGene was created. Here is how NCBI described UniGene in August 1996:

> The UniGene collection, now accessible through NCBI's Home Page, contains more than 48,000 clusters of sequences, each representing the transcription product of a distinct human gene. With current (1996) estimates of 80,000 to 100,000 genes in the human genome, this is close to the 50% mark. The clusters are largely based on EST sequences, so most of the sequences are not complete and most of the genes have still not been characterized. But one important use of the UniGene clusters is to identify novel, nonredundant mapping candidates for generating a transcript map that identifies all coding sequences in the genome.
>
> . . . Recent advancements in EST technology and the increased public availability of EST sequences have dramatically increased the numbers of genes in GenBank, so that developing a dense transcript map is now feasible. The Merck-funded EST project at Washington University alone has produced 320,000 EST sequences so far, with new data being submitted at the rate of 4,500 sequences per week. Mark Boguski, who leads NCBI's EST database project, says, "The transcript map will provide needed reality checks for the large-scale sequencing efforts ahead," and adds that "the disease gene hunting community has long had a desire to develop a transcript map."

By now, you have a sense of the data publicly available, thanks to a combination of your tax dollars (largely through the Department of Energy [DOE] and the **National Institutes of Health [NIH]**) and charitable contributions from

DISCOVERY QUESTIONS

10. BLASTn this mystery sequence and select "est_mouse" from the "Choose database" menu. What can you learn based on the hits you obtained? For example, what gene have you identified? Scroll down and see how many tissues are described in this search. Imagine you were studying obesity in mice; how might this help your efforts? (See Chapter 11 for details.)

11. Did the EST database provide you with more than just sequence identification information? With the completion of many genomes, is there any utility for the EST databases? Support your answer with specific examples.

12. Go to the UniGene Statistics page to read the latest information on human ESTs. Based on this information alone, guess how many genes are in the human genome. Don't bias your estimate by considering the published draft sequences, but use the UniGene data to determine your estimate. Write down the number you guessed so when you read Section 1.2, you can experience the same reaction as biologists all over the world when the total gene number was announced in February 2001.

foundations such as the Wellcome Trust of Britain, the world's largest medical research charity. Today, the hottest debate in science is about public vs. private genome sequences. In the mid-1990s, **The Institute for Genomics Research (TIGR)** was created as a private, nonprofit research institute. In July 1995, TIGR produced the world's first whole-genome sequence (1.8 Mb) from the bacterium *Haemophilus influenzae.* Shortly thereafter, TIGR was divided in two and Celera, a for-profit company, was created. Celera was led by TIGR cofounder Craig Venter until early 2002. Interestingly, Celera's name is derived from the Latin word *celere,* which means swift; their goal was to produce genome sequencing faster than anyone else.

The naming of Celera was no accident. TIGR, and later Celera, had developed a faster way to produce genomic sequencing. Rather than spending a lot of time mapping a genome, they used **shotgun sequencing** on a grand scale. Shotgun sequencing is a scaled-up version of an old method where you cut all the DNA into random pieces and sequence them all. The new twist for Celera was the scale. Never before had anyone tried to assemble so much sequence information. To test their ability to assemble whole genomes, Celera started with prokaryote genomes. Later they collaborated with a consortium of academic labs to shotgun-sequence the *Drosophila* genome with impressive results that are available at Flybase and Berkeley *Drosophila* Genome Project. The shotgun approach has been validated, and the HGP has adopted it as needed.

Science never rests, and there are more genomes to be sequenced. Many disease-causing species such as malaria and model species such as zebra fish and dogs will soon join the list of completed genomes (see the **Joint Genome Institute [JGI]**). Sequencing technology is improving, too. Many labs and companies are trying to invent new and improved ways to bring genome sequencing to you and your life. Some believe DNA sequencing will be routine in clinical settings in the near future.

A final word under the heading of "Tools of the Trade." The human genome sequence published in February 2001 is only a draft. On the same day of the big press conference announcing the completion of the genome, HGP labs were still sequencing their respective BACs and YACs to fill in the gaps and double-check accuracy. As the data become more complete, we will find new stories and questions that will require additional research from investigators in a variety of disciplines.

How Do We Make Sense of All These Bases?

Let's imagine you are a student learning how to analyze DNA sequences. The first thing you must do is start with what is already known. The best way to do that is to search the literature, starting with the mother of all collections; NCBI contains all available biomedical information. Enter the phrase "homo cyclooxygenase," to search cyclooxygenases of *Homo sapiens*. Select PubMed from the search menu and hit "Go." You will get a list of all the research and review papers published about "homo" *and* "cyclooxygenase." Now

select the "Nucleotide" database and hit "Go." You will see the sequences related to the terms "homo" *and* "cyclooxygenase." You can see on the right side there are terms that will allow you to search PubMed, protein sequences, related sequences, [3D] structures **Online Mendelian Inheritance in Man (OMIM)**—a sexist name but a comprehensive list of everything known about human biology and diseases—and taxonomy in case you don't recognize an organism's Latin name.

For the sake of completeness, you should know there are two other powerful ways to find out about **annotated** (characterized) genes. Try GeneCard, which is based at the Weizmann Institute in Israel; GeneCard is mirrored all over the world. Choose the closest site, enter the term "obese," and click on "Go." You will get several intermediate hits. Find leptin receptor, and click on the link to the left, which says "Display the complete GeneCard for this gene." You will see a treasure trove of genomic, proteomic, and phenotypic information. GeneCard is limited to human sequences, so if you want one-stop shopping for other species, try LocusLink. Type in the word "prion" and you will get a list of loci, an abbreviation for the source species (e.g., *Hs* for *Homo sapiens*, *Mm* for *Mus musculus*, *Dm* for *Drosophila melanogaster*) as well as a colorful key for the linked databases.

From all these databases linked to each other, you begin to see the power behind public genome projects. If you are conducting research on a particular gene, phenotype,

LINKS

Flybase

Berkeley
Drosophila
Genome Project

JGI

Near Future

NCBI

PubMed

OMIM

LocusLink

George Church

GeneCard Mirror

BOX 1.1 • *Which Draft Sequence Is Better?*

Much has been written about the public vs. private sequencing of the human genome. The two experimental designs for whole-genome sequencing were different, and each presented its own problems. But what we, the end users, want to know is how good are they and can we use them? Access to the public database is obviously much better and cheaper (i.e., free). Celera's complete database works on a pay-per-view system and is unavailable to all but the richest. As for quality, both draft sequences have gaps and some typos, too. One troublesome consequence of Celera's shotgun approach is that duplicated genes may

be combined into a single locus when in fact they should be two separate loci. When George Church at Harvard University compared the two draft versions, his group focused on two paralogs with 99.9% sequence identity. The HGP placed these two genes on different chromosomes, whereas Celera placed them at a single locus. Although this is only one example, it does point out the potential problems of complete shotgun sequencing. Luckily for Celera, they can use (at no cost) the public database to improve their own data, though HGP did not benefit from the Celera data. HGP and Celera were unable to agree on a

collaboration, so each group produced their own human genome sequences. There are many stories about the acrimony surrounding the private vs. public effort, and much of this is detailed in Kevin Davies' book *Cracking the Genome: Inside the Race to Unlock Human DNA*. Politics, egos, money, fame . . . science does not work in a social vacuum. However, in a few years, the differences between the two versions will asymptotically approach zero, and all we will remember is that on 15 February 2001, the human genome was first published in draft form, twice.

or disease, you do not have to start from scratch, clone any genes, or do much sequencing. These steps have been done for you and compiled in searchable databases. Now that this "grunt work" is completed, we are free to ask interesting and complex questions.

ORFs and Translation Before we start mining databases, let's run through some basic tools to analyze information. Let's start by finding **open reading frames** (**ORFs,** pronounced like the first four letters in the word *orph*an), which is a shorthand way of describing a segment of codons, ranging from a few bases to thousands, that begin with a start codon and end with a stop codon. For most people, ORFs are the interesting part of any genome, but later we will see there are other human genome gems hiding outside of the 1–2% of coding DNA.

Perform a BLASTn search on the ORF sequence. Record the **accession number** of the first hit. You have located a complete cDNA, so it contains all the **coding sequence** (**CDS**) as well as the 5' and 3' untranslated regions.

Go to ORF Finder, enter the accession number into the appropriate box, and click on "OrfFind." The cDNA will be translated in all six reading frames with the results shown in graphical format. Remember, DNA is double stranded, so there are three reading frames for the top strand and three more for the bottom strand. ORFs are indicated by the colored boxes, and you are looking for the largest ORF. The software ranks the ORFs from largest to smallest and indicates the reading frame for each ORF. For example, +3 means the ORF used the third reading frame (3) of the top strand (+). Click on the colored box next to the +3 from the list to see the deduced amino acid sequence.

DISCOVERY QUESTIONS

13. What gene have you identified? What species? How big is the encoded protein?

14. Did your ORF +3 translation match the sequence of the protein listed by BLASTn?

15. Go to Entrez, and use the term "cyclooxygenase 1" to search for the human sequence from the "nucleotide" database. Record the accession number for human cyclooxygenase 1 (ignore all patent-related sequences). You will need both cyclooxygenase accession numbers for the next section.

BLAST2 Sequences allows you to compare two nucleotide or amino acid sequences. Enter your two cyclooxy-

genase accession numbers (in the small, not the large boxes), and click on "Align." The genes have three sections that are well conserved at the nucleotide level, but the rest of these two ORFs are much more divergent. To see the bp alignment, click on the small boxes to the left of the diagonal graph.

DISCOVERY QUESTIONS

16. Calculate the average percent nucleotide identity for these three regions.

17. Go back to BLAST2 Sequences and now enter these two protein accession numbers for COX2, "NP_000954," and COX1, "NP_000953." Be sure to change the search from BLASTn (nucleotide) to BLASTp (protein). The protein you enter in the top blank will be on top of the resulting page.
    a. What is the overall amino acid identity? Is this higher or lower than the *overall* nucleotide identity?
    b. Notice that there is a separate percentage calculated for similarities (called "Positives") that takes into account similar structures of some amino acids. What is the percent similarity?
    c. Which parts of the proteins are not well conserved? Look at the sequence alignment that uses the single amino acid code and find where one protein has several Xs in a row, where a protein lacks amino acids present in the other protein.
    d. Use your browser's "Find" function to locate the amino acid sequence GAPFS. Serine (S) is the amino acid modified by aspirin. Is GAPFS in a region of high sequence identity or similarity? (See pages 328–329 for details.)

Can We Predict Protein Functions?

A good predictive tool worth learning is the **hydropathy plot,** or **Kyte-Doolittle plot,** which predicts whether a protein is an integral membrane protein or not. In a Kyte-Doolittle plot, every peak at 1.8 or higher indicates the potential for a transmembrane domain. Copy the leptin amino acid sequence and paste it in the appropriate box on the Kyte-Doolittle Hydropathy plot page. When the plot is visible, print or save it. Repeat this process to produce a plot for the leptin receptor. Compare the hydropathy plot predictions with what we know about these two proteins. Leptin is secreted and has no transmembrane domains, but it does contain a **signal sequence** (first 20 amino acids that target it to be synthesized on the rough ER) that has a peak close to 2. The leptin receptor is an integral membrane protein, so it should have produced one peak of 1.8 or higher. How well did the computer predict the proteins' structures?

The 3D shape of a protein is probably its most important characteristic. If you are lucky, the 3D structure has been determined for your favorite protein and you know its true shape. For most proteins however, you will have to make some reasonable guesses about 3D structure based on amino acid sequence similarity to proteins with known structures. Let's look at some methods for predicting 3D structures. Perform a **conserved domain** (**CD**) accession number search (from the pull-down menu) for the protein "AAC83646" (type it in the large box).

DISCOVERY QUESTIONS

18. Mouse over the domain boxes to determine the number of different CDs. Don't just count the number of boxes, but determine the types of domains revealed when you mouse over each box. The red and blue colors indicate alternatively spliced mRNAs.

19. Click on the "Show Domain" button (top left) and see what hits you get. At what protein have you been looking?

20. Go back and click on "gnl/CDD/7333," to the left of "smart00291,ZnF_ZZ, . . ." Read the text at the top of your screen. Does this domain have an important function? Explain your answer.

21. Analyze this uncharacterized protein amino acid sequence. It has not been submitted to GenBank, so you will want to copy and paste its sequence. Use the various web pages outlined above to determine as much as you can about its structure and function.

3D Structures Finally, if you want to know if the 3D structure of your favorite protein has been determined, go to either **Protein Data Bank** (**PDB**) or Entrez Structure and search by sequence (nucleotide or protein) or key word. Try "utrophin" and follow the links to view the structure (QuickPDB contains detailed instructions for 3D views). How many polypeptides do you see, and how do they interact? How does its structure compare with dystrophin's?

If you have the good fortune to discover a protein with no known ortholog, you might want to predict its 3D structure.

Currently, there are no reliable computer programs to perform this task. However, you can use programs such as PREDATOR to predict the *secondary* structure of your protein. Submit the amino acid sequence for human cyclooxygenase 1, and compare the PREDATOR prediction to the real structure. How close is the prediction to reality? The holy grail of computational structural biology is to predict the 3D structure of a protein based on its primary amino acid sequence. We are not there yet, but many mathematicians and computer scientists are working on this problem.

METHODS
QuickPDB
STRUCTURES
Cyclooxygenase
SEQUENCES
Uncharacterized Protein
LINKS
Conserved Domain
PDB
PREDATOR
Entrez Structure
Gene Ontology

Structure–Function Relationships Throughout this book, we will be looking at the structure of proteins and discussing their functions. However, the term "function" has become a bit outdated because it is too vague. When we say function, do we mean that the protein plays a role in signal transduction, or that it is a kinase, or that it spans the membrane? All three descriptors could be categorized as "functions," and yet they are not synonymous. As more genomes become sequenced, we need unified terminology to describe the role each gene/protein plays inside the cell. A consortium of investigators who work with different model organisms decided to create a more coherent nomenclature to describe genes and their products. The consortium, **Gene Ontology,** decided that three hierarchical terms were needed to describe different aspects of every protein: why? what? and where?

1. **Biological process** is the *why*—the overall objective toward which this protein contributes.

2. **Molecular function** is the *what*—the biochemical activity the protein accomplishes.

3. **Cellular component** is the *where*—the location of protein activity.

Let's look at one simple example. The protein aquaporin (see pages 174–175 for more information) is an integral membrane protein in red blood cells, kidney cells, and many others. Aquaporin's *cellular component* is the plasma membrane. Its *molecular function* is a channel that permits water to cross the phospholipid bilayer, which otherwise would be impossible since water is polar and cannot pass through the plasma membrane. Maintaining osmotic balance is aquaporin's *biological process* since it permits osmotic pressure to draw water (but not protons) back and forth across the membrane. Many proteins will have more than one answer to the why? what? and where? questions. Nevertheless, Gene Ontology's unification of protein roles will help us communicate more effectively as we determine how genomes produce multifunctional cells.

METHODS

Dendrogram

Phylogenetic Tree

LINKS

WormBase

COG

Enzyme
Commission

Swiss-Prot

Joe Felsenstein

DISCOVERY QUESTIONS

22. How can a single protein have more than one answer to the why? what? and where? questions to describe its roles in cells?

23. Go to WormBase, search for the gene *"pmr-1"* and learn its biological process, molecular function, and cellular components. Does this protein have more than one biological process, molecular function, and cellular component?

How Well Are Genes Conserved in Diverse Species?

Now that you know how to use a lot of the databases, let's try one more useful tool. Imagine you have just cloned a gene from a green alga and you want to know how well your protein is conserved relative to other species. We will evaluate this tool using the model enzyme **isocitrate dehydrogenase (IDH),** which is found in every organism. To get an overview of the role IDH plays in cellular metabolism, go to OMIM and search for "IDH." Click on the option for IDH3A (gene 3, alpha subunit) and read the short summary.

With a basic understanding of why IDH is found in every organism, let's see what sort of diversity there is among IDH genes. Go to the **Clusters of Orthologous Groups (COG),** and scroll down a bit. On the right side, under the heading of "Functional Categories," click on the letter C, which is one of several letters in a patch of colored boxes. When you get to this page, you will see we are looking for proteins under the *molecular function* (Gene Ontology usage) of "Energy production and conversion." Do a "Find" function in your web browser for "isocitrate," and then click on the "COGO538" hyperlink next to isocitrate dehydrogenase. You will see a table at the top left that says there are 29 proteins in this database (as of May 2002), and this enzyme is involved in the TCA cycle. Below is a list of the species abbreviations and a hyperlink for each of the proteins for each species. At the bottom, you see a **dendrogram** (a branching diagram that puts the most similar sequences next to each other). Print this page for later use. Clicking on the dendrogram brings up a figure with a circle and some colored diamonds. Selecting "centered" and clicking on the word "ROTATE" rotates the 3D dendogram and can be stopped by circling "STOP." Notice the two major parts, indicating there are two **isoforms** (distinct genes encoding similar proteins with similar roles but no assumptions implied about evolutionary relationships) of IDH

proteins. As you read at OMIM, IDH is split into two major groups.

Since COG is limited to certain species, let's try a less restrictive web site. The Enzyme Commission established a numbering system, **Enzyme Commission (EC) numbers,** so everyone could use the same descriptor for an enzyme that may have many different names. The two EC numbers for IDH, 1.1.1.42 and 1.1.1.41, reveal the split you saw in the dendrogram at COG. Go to **Swiss-Prot** and enter one of the two EC numbers for IDH. Swiss-Prot will display more than 70 1.1.1.42 orthologs and more than 30 1.1.1.41 orthologs.

Go to the Phylogenetic Tree web page where you will see 24 IDH proteins listed for 11 species. Notice that these IDHs are a mixture of 1.1.1.42 and 1.1.1.41 forms. Click on "Align Sequences" and wait for the results. All 24 amino acid sequences have been aligned with dashes inserted to fill gaps and to maximize alignments. From this alignment you can detect regions of all IDHs that are highly conserved and other regions that are conserved in a subset of IDHs. Once the sequences have been aligned, we can create two types of **phylogenetic trees.** Phylogenetic trees are similar to the dendrogram you saw at COG, but they assume that sequence conservation reveals evolutionary relationships. This assumption is based on the idea that all genes mutate at a constant rate, which is not necessarily true, so be careful when interpreting a "phylogenetic" tree.

Click on the "Rooted Tree" button, print the results and then repeat the same process using the "Unrooted Tree" button. These trees were generated by a program created in Joe Felsenstein's lab at the University of Washington. In our example, we wanted to know the relationships for these 24 IDHs from a wide range of species. Phylogenetic trees display each form of IDH on a different branch of the tree. The closer two particular branches are, the more highly conserved their sequences are. Compare the rooted and unrooted trees to see how they differ.

DISCOVERY QUESTIONS

24. Do the NADP⁺- and NAD⁺-dependent IDHs cluster into the two halves of each tree? Which NADP⁺-dependent IDHs are clustered with the NAD⁺-dependent IDHs?

25. Do all forms of IDH from one species cluster together or are they clustered with different species? Compare this trend for both the NADP⁺- and NAD⁺-dependent IDHs.

26. Find the two mycobacterial IDHs on the trees. Are they clustered near each other? Similarly, find the two Archaeal IDHs (M.jannNADP and

T.aquaNADP). Are they clustered near each other? Explain the implications of these two pairs of IDHs.

27. Locate the "IDH-like" protein from *Arabidopsis* (A.thaNADP?). The genome annotation indicates it may or may not be a true IDH enzyme. What functional roles can you predict based on your phylogenetic trees?

Most people use phylogenetic trees to illustrate either an evolutionary relationship or differences in protein structures. Similarity in proteins raises the problem of three terms that are often used loosely and can be confusing. In the case of IDH, eukaryotes have from three to five different IDH genes, sometimes called *isoforms* as noted above. If these genes arose from a common ancestral gene within one species, they are called **paralogs,** or are said to be *paralogous.* If the same IDH genes in two organisms evolved from a common ancestral gene in another species, then they are **orthologs,** or described as *orthologous.* Finally, the terms synteny and homology are confusing because their definitions have been blurred by usage that is not in keeping with their historical origins. Originally, **synteny** described genetic loci located on the same chromosome within a species, even if they were separated by a great distance. With the completion of many genome sequences, synteny has shifted its meaning to describe multiple genetic loci from different species being located on a chromosomal region of common evolutionary ancestry. When "synteny" appears in this book, we use it the way it was published in the original case study to be consistent with its source. Floating the definition to match the publications recognizes that English is a living language; just as the terms "cool" and "gay" have changed meanings in their popular usage over the years, so too have some scientific terms. **Homology** is losing its popularity and also has two meanings. Originally, two sequences were described as homologous if their sequences were similar because of a common evolutionary origin. Currently, homology is used simply to denote sequence similarity, whereas the terms orthologs and paralogs are used when discussing evolutionary relationships. The phrases "sequence similarity" or "sequence identity" are used when discussing the order of bases or amino acids with no assumptions made about evolutionary relationships. Nevertheless, you will still find many scientific publications that use these terms vaguely, as you will see in our next set of Discovery Questions.

DISCOVERY QUESTIONS

28. Go to the Human-Mouse Homology Map (Flash graphics works better on Netscape than Internet Explorer). Choose NCBI vs. MGP and then "Human" from the "Master" menu and "9" from the "Chromosome" menu. Look at the cytogenetic region 9q32–q33.3 and find *Cox1* (called *PTGS1* for prostaglandinendoperoxide synthase-1). On which chromosome is the mouse *Cox1* ortholog? Notice where it is located relative to the break point in the recombination of chromosomes.

29. Now try human chromosome 1 and locate *PTGS2 (Cox 2)* at 1q25.2–25.3. On which mouse chromosome is *Cox2* located?

30. Describe the chromosomal relationships between the human and mouse *Cox1* and *Cox2* genes.

SEQUENCES

Fly Period

LINKS

Homology Map

BLAST

Can We Search for Distantly Related Proteins in Different Species? Eukaryotic genomes are huge, so to find distantly related orthologs, we need computer programs. PSI-BLAST allows us to search outward in a spiraling pattern from a central starting point of a particular gene or protein. As an example, start on the BLAST page and click on the "PSI-and PHI-BLAST" link under the Protein BLAST heading. From the Fly Period page, copy and paste the amino acid sequence for *Drosophila* period into the PSI-BLAST input window. Follow the links, set the limit to 250, "Format" your request, and get your first **iteration** (first result) of this search. Your browser will open a new window, but don't close the old one yet. On the results page, you will get a list of the top 250 proteins that contain similar sequences. A graphic will show you which parts line up with your original fly protein sequence. Now there are two options from which you can choose, as follows.

Option 1 Take the default settings and submit your PSI-BLAST for a *second* iteration. This will reactivate the original PSI-BLAST window. You will need to press "Format" again to collect your results, but be sure to first limit your hits to 250.

While you wait, the program is computing the frequency of amino acids at each position and then using this to select proteins with the most similar amino acid sequences. Highly conserved amino acids are emphasized in the selection of proteins for the new list. To broaden the original BLAST results, variations from the original hits are incorporated in a new query of the database. If you want to narrow your search, you can de-select the proteins you want to exclude from your amino acid frequency calculation. When the results finally appear, you will see a list of 250 proteins. Most of them will have green dots by their names, which means they were in the first iteration. Scroll down to find some that have "NEW" next to the names. Can you find the human ortholog for the fly protein?

Try PSI-BLAST for a *third* iteration. Do you begin to pull out new proteins, or are they all old hits? By selecting which proteins you want to use, you can gradually work your way from one particular sequence in one species to orthologs and paralogs in a wide range of species with less and less sequence conservation.

Option 2 After your first PSI-BLAST, go to the human sequence (perform a "Find" function with your browser for "human") and perform a new PSI-BLAST using the human sequence as a starting place for a first iteration. If you limit your results to the top 250, you will see all three human orthologs. Select only these three and submit a second iteration. What do you find in the top 250 hits of the second iteration?

These two PSI-BLAST exercises illustrate how computer programs can search in an ever-widening circle, and with enough time, effort, and patience, you can annotate ORFs from a newly sequenced genome.

You now have the background necessary to analyze whole genomes. We have been working with genomes that were annotated. Imagine what it would be like to annotate a list of millions or even billions of bases—to predict where the genes are, what each protein does, and how the entire organism uses this information to live, reproduce, and respond to environmental fluctuations. This work takes many people with different academic backgrounds working together in teams. As you can imagine, software development for genome analysis is a very hot research area in computer science, mathematics, engineering, and biology. Since few people can master more than one or two of these areas, collaborations have become standard, though increasingly biologists are learning more math to improve their training and career options. If you can do both, you have many career opportunities ahead of you, but that's another story best told by your advisor.

Take a few minutes to draw a flowchart illustrating the steps you would take to annotate (define the genes; describe the proteins' biological process, molecular function, and cellular component; as well as summarize the major metabolic pathways the organism needs to survive and evolve) a newly sequenced genome.

How Do You Know Which Bases Form a Gene?

Let's imagine you have completed the sequencing and compiled it into a full-length chromosome, or chromosomes. Now what? The amount of information you have at this point is overwhelming. You don't know anything about the gene content of your species, its history, or which metabolic pathways are functional. Where do you start?

A good place to start is the beginning—*what is a gene?* Back in the old days (before genomes were sequenced), most everyone agreed that a gene was a segment of DNA that encodes a protein or RNA molecule that performs some cellular role. Today, that "written in stone" rule is being challenged by people who think any piece of DNA that performs a function, even if it is not transcribed, should be considered a gene. For example, telomeres and centromeres are critical for chromosome function, but they are not transcribed. Should we consider them as functional units, and thus genes? Finally, on the extreme end are those who say that an obvious cellular role is not necessary, and that any piece of DNA that has survived long periods without recombination should be considered a gene because recombination was selected against and thus its sequence remains intact for some reason.

However, most people define a gene as a piece of DNA of which some is transcribed and includes a promoter, coding sequences, and a signal for the RNA polymerase to stop. For prokaryotes, genes contain ORFs with no introns. Their genomes do not contain much **intergenic sequence** (technical term for DNA lacking genes), and they have smaller genes. Eukaryotes have more complexity in gene and genome structures. Promoters often contain specific sequences (e.g., **TATA** [pronounced ta-ta with short "a" sounds] and **CAAT** [pronounced cat] **boxes**) that are recognized by RNA polymerase and transcription factors. Further **upstream** are enhancers with complex sequence patterns that are recognized by transcription factors. The start codon often has a consensus sequence called a Kozak signal (named for Marilyn Kozak, who discovered this pattern), which consists of gccrccATGg, where the start codon is ATG and r stands for either purine (A or G). Eukaryotic genes may contain **introns** as well as the coding **exons.** The size of exons and introns can vary, with the smallest intron being about 70 bases while the largest are over 30,000 bp. The average human gene is 30,000 bp long, but the dystrophin gene is over 2 million bp. Combined with the complexity of real genes, the mammalian genome also contains 225 bp (averaged size) ORFs for every 1 kb that are not transcribed. These ORFs are called **pseudogenes** since mutation has rendered them nonfunctional. So finding a "real" gene amidst all this complexity is a very difficult task.

Tools for Gene Hunting Obviously, we cannot read a page or two of DNA sequence and pick out the genes. We need computers to help us sort through the data. DNA sequences are being collected faster, so we need programs that can process large sequences. When analyzing entire genomes, there are a few tricks that make the task easier. For organisms that have been studied genetically, there are genetic and **cytogenetic markers** (chromosome banding patterns) to narrow down where a gene might be located. The gene *white,* the first mutant gene of *Drosophila,* was genetically mapped to a small region of the X chromosome, and its cytogenetic location is visible on polytene chromosomes. ESTs, STSs, BACs, YACs, and cDNAs have improved the utility of cytogenetic maps and facilitated genome annotation.

Prokaryote and eukaryote genomes are significantly different and thus require different tools. Many commercial software packages exist, but luckily there are many free programs available, too. Most of the free programs run on your hard drive, but a few are available online through your web browser. GeneMark was originally created to locate genes in prokaryotes, but Mark Borodovsky and his research group at Georgia Tech have expanded the options to include a significant number of model eukaryotes. MIT hosts another online program to locate genes within a genome. GenScan will accept up to 1 million bp of sequence, but if you want to search multiple sequences, you can download the program and run it on a Unix-based computer. If you want to download and use a Perl (computer language) program you might consider Glimmer or GlimmerM from TIGR, which locate genes within whole genomes for prokaryotes and eukaryotes, respectively. There is a GlimmerM web server, but only a few species are available with a limit of 200 kb. With time, more gene-finding web sites will become available and more flexible.

Annotated Genomes Online   Once a genome has been sequenced and annotated, we can learn a lot about it. Let's look at the X chromosome using the HGP Genome Browser based at the University of California, Santa Cruz (Ensemble Genome Browser in the U.K. is an alternative site). At the HGP Browser, click on "Genome Browser" at the top left side. Enter the expression "Xq22.1" and hit "Submit." You will get a rather numbing picture that we will modify. Scroll to the bottom to change the settings. We want base position "on"; "Known Genes" and "Human mRNAs" set to "full"; "Chromosome Band," "STS Markers," and "Human ESTs" set to "dense"; all others set to "hide." Hit "refresh."

Now you will have zoomed in on one band of X chromosome. By choosing "dense" or "hide" for most of these options, we have compacted a lot of information into single lines of visual information. The three "full" settings produced a new line for each entry in the database and facilitated reading information about a particular gene. Read the list of known genes (blue text), and find *Nox1*. Notice that there are three separate entries for *Nox1*. Click on one of the three blue "*Nox1*" names and then click on the GeneCards link to find out more about the gene *Nox1*. Pay special attention to the "alternative forms" portion of *Nox1*.

link "sv." You should see a graphical version of this gene.
   a. How many different mRNAs (listed as CDS for coding sequences) are produced by this one gene? The color code is just below the expanded view of this gene. Notice that only the first exon is shown in the sequence (as denoted by the red bracket in the cartoon above).
   b. Use the navigation button icon to zoom out by clicking on the top blue line. How many genes are in this region of the X chromosome? Do any other genes produce more than one mRNA?
   c. Do you think this is a complete gene count for the 800,000-plus bp?

How Many Proteins Can One Gene Make?

A group of investigators in Switzerland and Hungary were interested in ion channels, specifically voltage-gated H^+ ion channels. They used partial cDNA sequences to locate *Nox1* as the source of an H^+ ion channel. To their surprise, *Nox1* encoded the NADPH oxidase and not an H^+ ion channel. When they searched the EST database, the investigators quickly realized that *Nox1* encodes three different mRNAs. There are two long versions and one short version (Figure 1.4). All three appear to be integral membrane proteins, but the short form does not contain the NADPH binding site. When expressed in tissue culture cells, this short form was able to transport H^+ ions. The three mRNAs are produced by alternative splicing that occurs in a tissue-specific manner: "Nox-1 long" was produced in the colon and uterus, and "Nox-1 short" was produced only in the colon. The third mRNA produces a slight variation of the long form—no one knows whether its role is unique or not. The fact that one gene produces three mRNAs illustrates the challenge in defining a gene. If one piece of DNA can produce three proteins, should we consider that piece of DNA to be three genes? This case study illustrates the utility of the EST database and how its value will increase as more ESTs are added for different species.

DISCOVERY QUESTIONS

31. What does GeneCards say about *Nox1*? Is it an NADPH oxidase?

32. Go to the MapViewer for the human genome, enter "*Nox1*," and hit "Find." You will see a red dash next to the X chromosome. Click on the blue "X" under the ideogram and notice the other genes in the area. Next to the gene "*NOX1*," click on the

DISCOVERY QUESTIONS

33. Analyze the dystrophin gene, with the HGP Genome Browser page; enter the name "dystrophin" and click the "Submit" button. You will get a list of hits. Click on the complete CDS link and then on the top link of the next page. You will

get a graphic view of the human dystrophin gene. Use the button to zoom out 1.5X.

a. Are there any STS markers in this region?

b. Look at the Gap and Coverage lines. Has the public HGP sequenced all of the chromosome in this region? (white = no coverage; light gray = predraft of less than 4X shotgun coverage; medium gray = draft of at least 4X shotgun coverage; dark gray = multiple draft coverage with overlapping pieces of the same DNA region; black = finished sequence)

c. How many mRNAs are there that use more than one exon? What does this indicate?

34. In the "position" box, enter "7p15.2" and then hit the "jump" button. At the bottom of the page, hide all features except set to full "Human mRNAs" and "RefSeq Genes" and then set to dense "Mouse Blat" and "Fish Blat."

 You want to focus on the possible evolutionary conservation of this region. We are looking at some **Hox genes,** which are critical to embryonic body development.

a. Does this region appear well conserved in the mouse? Support your answer with data.

b. Does puffer fish express Hox genes too?

c. Set repeat masker to "dense" and refresh this

diagram. Do Hox genes contain a lot of repeats (black indicates repeat sequences) compared to portions outside the Hox genes? Do you think this is significant? Explain your answer.

d. Set the two **SNPs (single nucleotide polymorphisms,** which can be thought of as point mutations) options to dense. Are there many polymorphisms in the Hox genes? Do you think this is significant? Explain your answer. For a comparison, click on "Move <<<" at the top left.

Summary 1.1

In Section 1.1, you learned how to use public databases that are funded by government and private organizations in the United States, Europe, and Japan. With these public domain tools, we can use small segments of sequence to retrieve full-length coding regions, deduce protein sequences, and make educated guesses about possible cellular roles for proteins. We can take advantage of evolutionary relationships and compare sequences from different species to examine the degree of conserved domains. By using dynamic databases, you have become familiar with research tools used by investigators all over the world. In the next section, we examine what has been gleaned from the human genome draft sequences.

FIGURE 1.4 • Two proteins produced by alternative splicing of RNA. a) The full-length mRNA was translated to produce the larger protein (NOX-1L) that is an NADPH oxidase. **b)** Several exons were excised to produce a smaller protein (NOX-1S) that functions as a H^+ channel.

a)

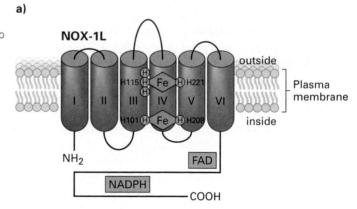

b)

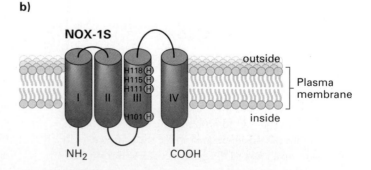

1.2 What Have We Learned from the Human Genome Draft Sequences?

In Section 1.2, we will study the public domain draft version of the human genome sequence, which was published in February 2001. We begin with a list of findings that outline what was learned when 3 billion base pairs were examined for the most obvious traits. Following this overview list, we examine some exceptions to popular understanding of what our genomes are. How can we be sure our initial analysis is correct? Are some genes not what they appear to be? Can the activity of our genomes be modified without changing the DNA sequence? A number of online databases and analysis tools will provide information to answer Discovery Questions that immerse you in current research projects and ongoing debates over defining our genome.

Overview of Human Genome First Draft

Of course, it is impossible to fully summarize the human genome, in the sense that new discoveries will continue for

many years. Computer analysis of the genome will never extract *all* the meaningful information. Therefore, investigators will need to perform laboratory experiments to glean more information than is apparent from computer searches. For example, alternative splicing adds a significant degree of complexity in defining the proteome but we cannot detect it without bench work, although ESTs are helpful. We will briefly survey some of the initial findings and look in depth at others later in the chapter.

LINKS

Genome Hub
Funding Statement
Caltech
BACPAC

Summary Statements

- Publication of draft sequence on 15 February 2001 was based on data "frozen" on 7 October 2000. Humans have approximately 35,000 genes. The exact number of genes will change with the discovery of new genes and the realization that some were incorrectly annotated. The Genome Hub is a centralized web site for all HGP links. **Draft sequence** means the DNA was sequenced on average four times, with **finished sequence** having eightfold

---

BOX 1.2 • *Whose DNA Did We Sequence?*

One area of controversy that arose early in the planning of the Human Genome Project was whose DNA will be sequenced. In the end, it turns out that nine different sources of DNA were used, eight of them identified as male. Interestingly, only three of the nine DNA samples were taken from germ line cells (sperm); the other six were somatic cells, and at least five of these were male.

The issue of gender has raised a number of questions. A female donor will have two copies of X and no Y chromosome. Conversely, male donors will have only one copy of X, but they also have a Y chromosome. Since some genes (e.g., immunoglobulin and T-cell receptor genes) rearrange in somatic cells, there is an interest in sequencing nonrearranged genes, and thus germ line cells. However, obtaining eggs from women is much more difficult than obtaining sperm from men. In addition to difficulties in

obtaining eggs, the number of sperm cells that can be readily obtained is advantageous as well.

For many years, females have been underrepresented in research—both as investigators and as participants in clinical trials. Therefore, there was an appeal to include DNA from women in the Human Genome Project. Of course, every male donor received half of his DNA from his mother; nevertheless, some of the DNA libraries used for sequencing were derived from anonymous women donors.

It might seem that the easiest DNA to obtain would be from members of the laboratories where the DNA libraries were constructed. However, personal involvement with the research may lead to difficult ethical issues. Perhaps a technician might feel compelled to donate DNA, or confidentiality of the donors could be breached. How would *you* feel if a mutation were

discovered as a part of the HGP and you knew your DNA was the one being sequenced? The final consideration was that the HGP might be perceived as elitist if only the top scientists were able to have their DNA sequenced. For these reasons, many donors were recruited from different locations and only a subset of the collected tissue was used for sequencing. The identity of the DNA source has been hidden to minimize potential ethical dilemmas. You can read more about what steps were taken to choose DNA samples at the Human Genome Funding Statement, Caltech, and the BACPAC Resource Center in Oakland, CA. In April 2002, Craig Venter disclosed that his DNA was used for Celera's sequencing project. This revelation contradicts Celera's public statements in February 2001. Venter's DNA sequence accounts for three-fifths of the total Celera sequence.

coverage and errors estimated to be one in 10,000 bp. Gaps, "typos," and contig assembly errors in the draft sequence will be corrected during the finishing process. There are three types of gaps: (1) gaps within unfinished sequenced clones; (2) gaps between sequenced BAC clones; and (3) gaps between mapped BACs. Sizes of gaps had to be estimated.

- Two software packages were critical to the success of HGP: PHRED and PHRAP both developed by Phil Green at the University of Washington, Seattle. PHRED analyzes the quality of each sequenced base and assigns a quality score to each base, enabling genome centers to detect "bad calls." PHRAP takes the PHRED data and uses the quality assessment to assemble the overlapping sequences into larger contigs.

- The mutation rate in males is twice as high as in females. Recombination occurs more frequently nearer the telomeres and on the "petite" arms of chromosomes. Long chromosomes have a recombination rate of 1 cM per Mb, whereas short arms have about twice that rate. The extreme examples are the short (2.6 Mb) pseudoautosomal regions of the petite arms of the sex chromosomes, Xp and Yp, at about 20 cM per Mb. Crossing over is believed to be necessary for normal separation of chromosome pairs during anaphase I of meiosis.

- Some human genes initially appeared to have entered our genome via horizontal transfer from bacteria, but this interpretation has been refuted by several labs (see page 22 for details).

- CpG dinucleotides form "CpG islands," and the cytosine base is often methylated. When the cytosine spontaneously deaminate (lose an amine group), the cytosine is converted into thymidine base. Therefore, over time, CpG dinucleotides gradually get converted to TpG dinucleotides. If the cytosine is not methylated, then the deamination produces uracil, which is fixed by the cell's quality control machinery. The distribution of deamination is not random.

MATH MINUTE 1.2 • HOW DO YOU FIT A LINE TO DATA?

Figure 1.5a depicts the relationship between CpG island density and gene density for each chromosome. The researchers report that chromosomes 16, 17, 19, and 22 are outliers because they fail to follow the same linear trend as the remaining chromosomes. To explore how this conclusion might have been drawn from the data in Figure 1.5a, try to draw straight lines on the graph in the figure supporting each of the following statements:

a. Chromosomes 16, 17, 19, and 22 are outliers.

b. Chromosomes 16 and 19 are outliers, but chromosomes 17 and 22 are not.

c. None of the chromosomes are outliers.

Which line do you think fits the data best?

Since you drew these three lines somewhat subjectively, your lines may look different from someone else's lines. However, there is a mathematical approach, called simple linear regression, by which the exact same lines can be drawn by everyone, every time. Regression analysis is routinely used in scientific studies to model linear relationships between variables. (The term regression comes from the work by the founder of eugenics, Sir Francis Galton, who described the tendency for offspring to be closer to average size than their parents as "regression towards mediocrity.")

You can quantify how well a particular line fits a set of data points by measuring the *vertical* distance (i.e., absolute value of the difference in Y-coordinates) from each point to the line. If you seek a line that fits all 23 data points, you might try to find the line that minimizes the sum of these 23 distances. Mathematically, however, it is more convenient to determine the slope and intercept of the line that minimizes the sum of *squared* distances. Since the quantity to be minimized (sum of squared distances) is a function of two variables (slope and intercept), partial derivatives are key to this process. Squared distances are more convenient because it is easier to work with derivatives of squared quantities than derivatives of absolute values.

For the data in Figure 1.5a, the investigators did not offer a rationale for their choice of line (a) over lines (b) and (c). However, it appears that the regression line was constrained to go through the origin. In this case, the intercept is 0 and only the slope needs to be determined. With the additional restriction on statements (a), (b), and (c) that the line must go through the origin, do you agree with the investigators that line (a) is the best fit?

FIGURE 1.5 • GC content of the genome. a) For each chromosome, the number of CpG islands per Mb is plotted on the X-axis, genes per Mb on the Y-axis. Most of the chromosomes fit along a diagonal line, but four do not. **b)** The ratio of GC content is plotted on the X-axis with frequency of each ratio on the Y-axis. GC content for the whole genome and individual genes is not identical. To produce the ratios, gene content was calculated using sections of DNA 20 kb long and centered on the 9,315 known genes, while genome content was determined by 20 kb of nonoverlapping adjacent DNA.

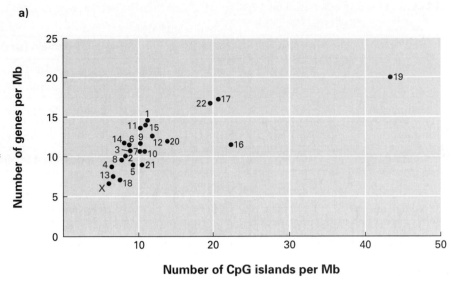

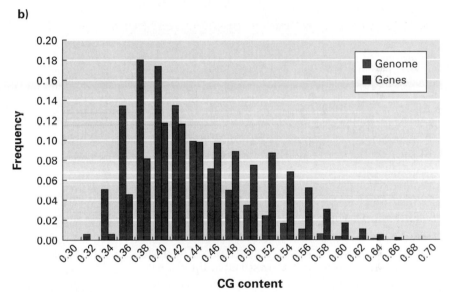

- There are five classes of repeated sequences: (1) transposon-derived; (2) pseudogenes; (3) simple repeats, sometimes called VNTR (variable number of tandem repeats, e.g., $(CA)_n$; (4) segmental duplications where one chromosomal region gets copied onto another chromosome; and (5) regions near telomeres and centromeres and ribosomal genes.

- Segmental duplications can occur on the same chromosome (syntenic) and lead to localized regions of deletion, which has been cited as the cause for several genetic diseases.

- Transposable elements fall into four major categories: (1) SINEs (short interspersed elements), including Alu, which collectively comprise about 13% of the entire human genome and may confer a functional advantage to us; (2) LINEs (long interspersed elements), which collectively comprise about 21% of the entire genome; (3) LTR retrotransposons, which comprise about 8% of the genome;

and (4) DNA **transposons,** which are about 3% of the genome. All together, these elements comprise about 45% of the human genome! One of the great treasures hidden in these repeat elements is evolutionary data. By measuring the degree of sequence divergence within the transposable elements, we can begin to measure time of divergence if care is taken to calibrate the "molecular clock," though some investigators think calibration is impossible.

- The X chromosome has the highest concentration of transposable elements, with a 525 kb section being 89% transposon, of which a 200 kb subset is 98% transposon. Conversely, there are some regions, such as the Hox region we examined above, where the density of transposons is approximately 2%.

- The Y chromosome is the site of greatest LINE transposition, and thus the Y chromosome looks the "youngest" of all the chromosomes due to the high degree of recent

TABLE 1.1 • Average based on 1,804 genes chosen from RefSeq and unambiguously aligned over their entire length with finished sequence.

	Median	Mean	Sample (size)
Internal exon	122 bp	145 bp	RefSeq alignments to draft genome sequence with confirmed intron boundaries (43,317 exons)
Exon number	7	8.8	RefSeq alignments to finished sequence (3,501 genes)
Introns	1,023 bp	3,365 bp	RefSeq alignments to finished sequence (27,238 introns)
3' URT	400 bp	770 bp	Confirmed by mRNA or EST on chromosome 22 (689)
5' UTR	240 bp	300 bp	Confirmed by mRNA or EST on chromosome 22 (463)
Coding Sequence	1,100 bp	1,340 bp	Selected RefSeq entries (1,804)
(CDS)	367 aa	447 aa	
Genomic extent	14 kb	27 kb	Selected RefSeq entries (1,804)

Reprinted by permission from *Nature* 409: 896. Copyright 2001, Macmillan Magazines Ltd.

DNA insertions. The density of LINEs is probably a consequence of the relative paucity of genes on the Y chromosome, which can tolerate insertions more readily than gene-dense chromosomes.

- A hypothesis has been proposed for a function for SINEs. SINEs are transcribed under conditions of stress, and the resulting RNAs hybridize to a particular protein kinase that normally inhibits translation. Thus, SINEs would enable a cell to produce more proteins when stressed, which is exactly what happens in yeast (see pages 176–178). The hypothesis predicts that SINEs would provide a selective advantage if they enabled the cells to survive the stress better. Interestingly, Alu sequences are concentrated around genes, which may be selected for as predicted by this hypothesis.

- It appears about 50 human genes were derived from transposons. Included in this list are RAG1 and RAG2, which play a significant role in producing the high degree of diversity in antibody binding capacity.

Can We Describe a Typical Human Gene?

Let's look at an average human gene (Table 1.1). Human genes tend to be composed of several exons (average number is about nine, though some genes have only one). Exons are small (on average about 50 codons) interspersed by large introns (on average about 3.3 kb). The average 5' **UTR (untranslated region)** is 300 bp, and the 3' UTR averages 770 bp, with a total coding region of 1,340 bp to produce an average protein size of 447 amino acids. For

the average gene, all of this genetic material is distributed over 27,000 bp. You can see that the human gene is not a model of efficiency of space, which is one of the major reasons why deciphering the human genome is difficult (Figure 1.6).

DISCOVERY QUESTIONS

35. Let's do some quick estimations about our DNA using these numbers: haploid genome of 3,289,000,000 bp, 35,000 genes, and the numbers from Table 1.1.
    a. What percentage of your genome is spent on genes? exons? introns?
    b. What percentage of your genes is spent on exons? introns?

36. Are any chromosomes missing from Figure 1.5a?

A nagging question has been how can humans be more complex at the protein level if our genes are only twice as numerous as those of worms with 1,000 cells? Our complexity is hidden in our genome and expressed in our proteome. One source of greater proteome complexity than genome complexity is alternative splicing, as we saw with Nox1-long vs. -short. The *Nox1* gene produced three mRNAs and at least two functionally different proteins. Using a sampling method, the HGP found about 60% of all human genes produce more than one mRNA due to alternative splicing, with an average of 2.6 transcripts per gene. *C. elegans* is not as

a)

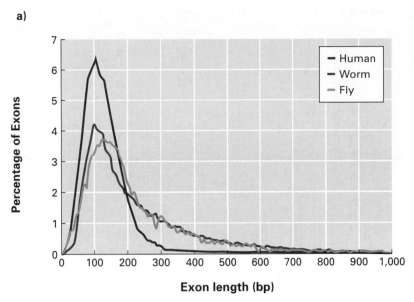

b)

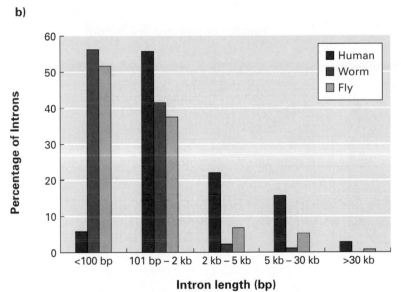

FIGURE 1.6 • **Size distribution of exons and introns. a)** Species comparison of exon sizes in human, *C. elegans,* and *Drosophila* genomes. **b)** Species comparison of intron sizes in human, *C. elegans,* and *Drosophila* genomes. Note the inverse correlation between intron and exon sizes.

complex, with about 22% of its genes producing more than one mRNA (average of 1.34 transcripts per gene). However, fewer *C. elegans* ESTs are available for analysis, so estimates of alternative splicing may be artificially low, and thus alternative splicing may not be the main source of greater human proteome complexity.

In addition to the number of genes that produce multiple mRNAs, there is another issue. If each gene can be either on or off, then the combination of active genes in any given cell is 2^x with x being the number of genes in the genome. Therefore, flies have $2^{18,000}$ possible combinations of on and off genes, while humans have $2^{35,000}$ possible combinations. Humans have more cells with specialized functions than yeast, flies, or worms, and therefore we need

to regulate gene expression very carefully. This leads to a prediction that humans need more transcription factors than these other organisms, supported by comparison of the genomes (Figure 1.7). Proteome complexity is regulated by gene expression. The subset of proteins in each cell, tissue, and organism is unique, and thus worms need more gene regulation than yeast, flies more than worms, and humans more than flies.

What makes humans unique? This is a difficult question to answer, but we do have some relevant information (Figure 1.8 on page 23). You can see that each species has a unique combination of proteins for the different biological processes. This is not too surprising, but what is rather humbling is the pie chart in Figure 1.8b where we see that only

FIGURE 1.7 • **Genome-wide comparisons of transcription factor families.** Listed are nine different transcription factor families (X-axis), with abundance in each family on the Y-axis.

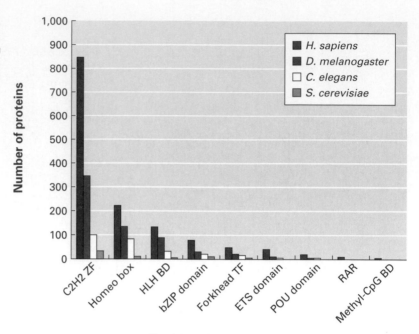

Families of transcription factors

LINKS

TIGR Investigators

1% of our genes have not been found in other species and that we share one-fifth of our genes with all other organisms, including bacteria.

The field of genomics is a combination of hypothesis testing and discovery science. **Discovery science** is what Leeuwenhoek performed after inventing the microscope. "I wonder what I'll find if I look at. . . ." While mining data from genome sequences, we can make unexpected discoveries. One such discovery focused on 223 human genes found only in vertebrates and which appeared to be orthologs of bacterial genes. At least 113 are found in many diverse bacteria, which led investigators to conclude that these genes entered some point in the vertebrate lineage by **horizontal transfer** (nonsexual transmission of DNA from one species to another) from bacteria. In case you think this small number of genes is not physiologically important, let's look at one example. We have two monoamine oxidases (MAOs), enzymes that deactivate neurotransmitters such as norepinephrine and serotonin. MAO inhibitors are used to treat depression and Parkinson's disease and cannot be taken with painkillers such as Tylenol.

The hypothesis of horizontal transfer of bacterial genes into the human genome is not universally accepted. Contradictory publications claiming the impression of horizontally transferred bacterial genes was based on the limited number of genome sequences. TIGR investigators reported that only about 41 genes may have been horizontally transferred, but they predict this number will gradually reach zero as more genome sequences are analyzed. The TIGR group proposes that all eukaryotes once had these genes but many have lost them over time, thus making the genes appear to be horizontally transferred from bacteria to mammals. Although the initial report of horizontal transfer of bacterial genes into the human genome appears to be incorrect, it is a good example of discovery science producing an unexpected hypothesis that was later tested experimentally.

DISCOVERY QUESTIONS

37. Given the information in Figure 1.8, what is one aspect that makes humans different from other species?

38. Given that other species have orthologs of 99% of our genes, how can we be different from all other species?

Where will the complete sequence of the human genome lead us? In a couple of years, we will have a *finished* human genome sequence and many current uncertainties will be settled. But don't expect all the questions to be answered. Science advances when data produce more questions than answers. We will continue to utilize the genome sequence to determine the molecular causes of diseases. In addition to single-gene diseases, we will discover new ways to track the molecular causes of complex phenotypes such as Alzheimer's, cancer, and schizophrenia (see Chapters 10–12). As disease-causing genes are identified, new drugs can be developed to modify their cellular roles. For example, two proteins are known to play a role in amyloid

a)

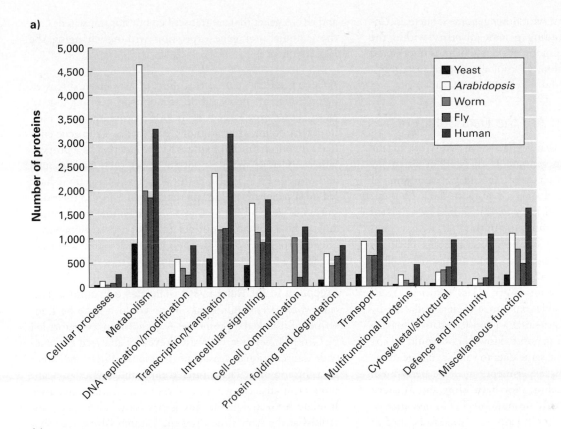

b)

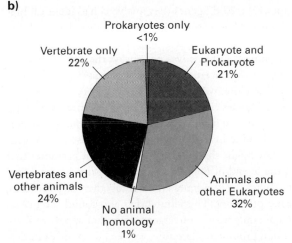

FIGURE 1.8 • **Functional categories of proteins. a)** Comparison of 12 functional categories in eukaryotic genomes. The number of proteins is plotted on the Y-axis, and the 12 categories are on the X-axis. **b)** Distribution of orthologs for the predicted human proteins. These data are based on genome sequences available February 2001.

plaque formation in the brains of Alzheimer's patients. One is called APP (amyloid precursor protein), which is cleaved by the protease called BACE (β-site APP-cleaving enzyme). BACE was identified in 1999, and is located on chromosome 11. When the investigators searched the human genome, they identified BACE2, which shares a 52% amino acid identity with BACE1. Interestingly, BACE2 is located in the Down's syndrome region of chro-

mosome 21. Perhaps the amyloid plaque formation found in individuals with Down's syndrome (caused by three copies of chromosome 21) is caused by excessive amounts of BACE2, and both Down's syndrome and Alzheimer's might be treatable with a common drug.

Eventually, we want to identify all the regulatory regions for all the genes to better understand what governs transcription (see Chapter 7). We'd like to compare the human

METHODS

Northern Blots
Western Blots
RT-PCR
Immuno-
precipitations
Enzyme Assays

LINKS

Stanley Prusiner
Nobel Prize

genome with other genome sequences. Understanding genetic diversity within the human population will help us understand evolution, physiology, disease susceptibility, and drug efficacy (see Chapter 3).

When Are the Data Sufficient?

Now that we have an overview of the human genome as analyzed **in silico** (by computer), you should be wondering if anyone has experimental data to support these findings. Whenever a new method is created, it should be validated using more traditional techniques. In this case, we have analyzed genomic data before but never on this scale. There are larger unsequenced genomes, especially in the plant kingdom, but there is something special about the human genome that makes us want to get it 100% correct. Years' worth of publicly available experimental data collected by labs around the world provided a good foundation for understanding the human genome. However, it would be circular reasoning to use the same data to validate our genomic analysis. Therefore, new experiments must be performed. Traditional methods such as **Northern blots** and **Western blots,** PCR and **RT-PCR,** immunoprecipitations, enzyme assays, and cell culture experiments, etc., need to be used to test predictions made from the genome annotations. But these methods are very specific and work best when a particular molecule or enzymatic pathway is studied in isolation. Are there any genome-wide analysis tools available?

What we all want to know is where the real genes are, which exons are used, and when. This is a straightforward goal, and yet accomplishing it will take many, many years. The question at hand is at the heart of science—when can you be sure of your conclusions? Science makes progress through reproducible results, peer review, and hypothesis formulation, with everyone else in the world trying to shoot down your hypothesis. When a hypothesis has withstood all challenges, it is believed to be correct (for the moment at least). The scientific process breaks down when everyone begins to believe that there are no alternative explanations left to consider. This is what happened when Stanley Prusiner proposed that the cause of mad cow disease was 100% protein and lacked any DNA or RNA genome. Everyone *knew* that nucleic acids were the only way for an infectious disease to spread, and thus the world was unprepared to listen to alternative hypotheses. Eventually, the scientific process did prevail as more and more data have supported the protein-only hypothesis that eventually led to a Nobel Prize for Prusiner.

Each of us has to use our own criteria for evaluating the quality of data. The human genome contains many more hidden stories that are waiting to be uncovered. Plus, there is a world of other genomes that have their own stories to tell. Before we look at some of those stories, we need to answer a couple more questions. Are there genes other than tRNA

and rRNA genes that are transcribed but not translated? Can the genome alter gene expression without changing the DNA sequence?

A Gene Is a Gene Is a Gene. . . Sort of As we have discussed before, there are many subtle variations in defining a gene. Most people expect genes to include a promoter, some coding DNA, followed by a bit of DNA that says the gene ends here. If your definition requires the production of a protein, you would miss many important genes such as ribosomal and transfer RNA genes. Essential genes such as these have led most people to include all transcribed DNA in their definition of genes. Let's look at some exceptions to the rules and see how this affects our definition of a gene. Remember, science progresses when exceptions are discovered and rules are modified.

Exception to the Rule: Every Gene Has a Promoter This rule would seem immutable. How could a gene be transcribed if it lacked a promoter? A group of investigators led by George Church at Harvard University analyzed genome-wide transcription of yeast genes that located near each other on a chromosome. The Church team found that a significant number of adjacent genes were transcribed simultaneously. It might seem strange that two genes should always be transcribed at the same times, but the Church lab noticed that 387 out of the 2,087 genes they analyzed had identical biological roles. How can two genes be regulated identically? The first hypothesis would be that the two genes have identical promoters, but this was not always the case. When the genome was sequenced, the upstream regions were often dissimilar. Therefore, we need more information.

Adjacent genes can be oriented in one of three ways (Figure 1.9). They can lie tail to tail (divergent), head to tail (tandem), or head to head (convergent). In the case of divergent genes, they might share a single bidirectional promoter that lies at the 5' end of both coding regions. This is rather surprising, but if the two genes are close together, they can share the same promoter. The Church lab also found several examples of tandem and convergent genes that appear to share a common promoter. Therefore, we need to modify our definition of a gene to allow genes to lack a unique promoter as

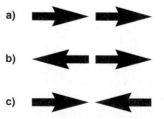

FIGURE 1.9 • Gene orientation. Three ways that two genes can be oriented relative to each other are **a)** tandem, **b)** divergent, and **c)** convergent. The arrows indicate the orientation of each ORF, with the point of the arrow representing the downstream portion of the gene that is transcribed last.

long as they are near another gene whose promoter can initiate transcription for both genes.

Exception to the Rule: Other Than rRNA and tRNA, All Genes Produce Proteins

You are familiar with tRNA and rRNA, and life would not exist without these untranslated genes. There are other, lesser-known RNA-only genes: examples include **small nucleolar RNAs (snoRNAs)** and **small nuclear RNAs (snRNAs)**. RNA splicing requires several of the snRNAs, while rRNA is post-transcriptionally modified by snoRNAs. Untranslated genes or **noncoding RNAs (ncRNAs)** usually do not look like ORFs, nor do they become modified by polyadenylation so new ncRNAs are easily missed by traditional computer and laboratory methods such as EST sequencing. Are there any undiscovered human ncRNAs, or have all of them been identified?

How would you go about finding a new ncRNA-encoding gene? You could not make an antibody to it. You could not purify the protein and sequence it, and it would not show up as an EST since ESTs contain poly-A$^+$ tails. About the only way you could find such a gene would be through a genetic analysis. For example, Amy Pasquinelli and Gary Ruvkun at the Massachusetts General Hospital, isolated a mutant strain of *C. elegans* that did not develop properly into an adult worm. The mutated gene is called *let-7,* and the active gene product is a 21-base RNA molecule that binds to the 3' UTR region of an mRNA encoded by a different gene that normally suppresses early development. When the **wild-type** *let-7* RNA binds to the complementary mRNA target, translation of the targeted mRNA is blocked, permitting development to continue normally. In the mutant worms, *let-7* does not bind the target mRNA and development is suppressed. Other small ncRNAs block protein production; they represent another gene category that requires us to modify our traditional definition of a gene. Are there other nontranslated gene categories still hiding in your genome? This question can only be answered with more benchwork.

Can the Genome Alter Gene Expression Without Changing the DNA Sequence?

There are two genetic diseases you probably have not heard of, and yet they have provided us with a valuable insight into the way our genomes work. Prader-Willi Syndrome (PWS) is caused by a megabase-sized deletion in the region of 15q11–q13. A different genetic condition called Angelman Syndrome (AS) is caused by a large deletion in the region of 15q11–q13. As you can see, these two different conditions are caused by the loss of the same section of our genomes, and yet the symptoms are different. Whether you have AS or PWS depends on which parent gave you the mutated chromosome. If your father passed on the deletion you will develop PWS, but if your mother passed on the deletion, you will develop AS. Children with PWS have weakened muscles, fail to suckle, will not grow very tall, have short hands and feet, and will develop obesity and mental retardation as they mature. Children with AS have problems with balance and motor skills, appear hyperactive and overly happy (excessive laughter), and are often more severely retarded than PWS patients. These conditions are not sex linked since they occur in boys and girls with equal frequency, but the sex of the parent who passed on the mutated chromosome always determines whether the child will have AS or PWS.

This situation is an example of **imprinting,** a process mammals use to mark a small set of genes during gametogenesis so that only the paternal *or* maternal copy will be transcribed and the other allele at the same locus will remain silent. As of 2002, there are more than 20 mammalian genes in the mammalian genome that are known to be imprinted. Silencing one allele depending on which parent passed it on presents an exception to the rule of gene expression. Monotremes and nonmammalian vertebrates do not imprint their genes, which indicates that placental mammals are unique in their requirement of imprinting. Looking at the list of imprinted genes shows that many of them play a role in embryo and placental development. When you think about it, the placenta and developing embryos are similar to parasites that drain resources from the mother. In species where the female mates with more than one male, the two parents have opposing interests in the newly formed zygotes. A male will prefer that his offspring extract the maximum amount of nutrients from the mother even if it hurts the mother's ability to reproduce later with other males. Conversely, females would prefer to ration resources so that each of her offspring has the potential to survive and she can continue to reproduce with any male she chooses. This opposing interest creates a genetic tug-of-war, which may have been the selection pressure that led to imprinting. Paternally expressed genes should promote growth, while maternally expressed genes should reduce the amount of growth as long as the mother is the sole source of nutrients (pregnancy and nursing).

This hypothesis sounds logically consistent, but we need some data to support it. To examine the predictions of this genetic tug-of-war hypothesis, many investigators have focused their labs on the mystery that surrounds genetic imprinting. One of the world leaders in this field has been Shirley Tilghman from Princeton University, whose lab has made significant contributions toward our understanding. One paternally expressed gene, insulin-like growth factor 2 *(Igf2),* promotes the growth of the embryo. Only the paternal allele is expressed in the developing embryo. The Igf2 receptor (Igf2r) is expressed only by the maternally inherited allele. By controlling the embryonic expression of the receptor, the mother can maintain control of the paternally driven ligand encoded by *Igf2.*

Tilghman's lab demonstrated imprinting in a series of matings between two closely related strains of mice abbreviated LS and PO. Because they are closely related, the

LINKS

Gary Ruvkun

PWS

AS

20 Mammalian Genes

Shirley Tilghman

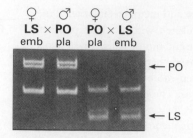

FIGURE 1.10 • **Expression of *Igf2* alleles.** RNA was isolated from placenta and embryo from two *Peromyscus polionotus* mouse strains called LS and PO. The RNA was used as template for reverse-transcriptase PCR, digested with a restriction enzyme and analyzed by agarose gel electrophoresis. Depending on which strain was the mother or father, two different banding patterns are observed due to slight differences in the sequences of the *Igf2* alleles in PO and LS.

STRUCTURES
Dnmt1p

LINKS
Human Epigenome Consortium

mice can mate successfully, but their DNA sequences are different enough that each species' *Igf2* alleles can be distinguished (Figure 1.10). *Igf2* from the PO strain produces three bands with two of similar molecular weight near the top of the gel. LS *Igf2* produces two major bands including one of small molecular weight. If the PO *Igf2* is expressed, the top two bands will be detected, while LS *Igf2* expression results in a band near the bottom of the gel. When the female LS mated with a male PO, the PO *Igf2* allele was expressed in the embryo and placenta and the maternal LS allele was completely silenced.

DISCOVERY QUESTIONS

In these three questions, we are working with only one strain of mice, not two.

39. If the *Igf2* gene were deleted in the sperm, predict the phenotype in the offspring. What would the phenotype be if the egg's *Igf2* gene were deleted?

40. If the maternally expressed gene *Igf2r* were deleted in eggs, predict the phenotype in the offspring. What would the phenotype be if the sperm lacked an *Igf2r* gene?

41. What would you predict for the offspring if the sperm's *Igf2* gene and the egg's *Igf2r* gene were deleted?

Many more experiments have contradicted the widely held dogma that both alleles are equal. Let's look at one more gene before we try to understand part of the mechanism that controls imprinting. When the promoter of another paternally expressed gene *(Snrpn)* was deleted, *Snrpn* and at least

three other genes that are normally transcribed from the paternal alleles were silent. The *Snrpn* mutation produced symptoms similar to several PWS phenotypes including failure to suckle. This is consistent with the hypothesis that paternally expressed genes promote the maximum utilization of maternal nutrients and thus promote growth and eating. Surprisingly, the coding region of *Snrpn* was not the critical component but its promoter was. Similarly, the coding portion of the maternally expressed gene *H19* can be deleted with no ill effects, but when its promoter was mutated, normal imprinting was disrupted for both *H19* and the paternally expressed *Igf2*. There are many more hard-to-explain imprinting stories, but let's turn our attention to possible mechanisms.

What Is the Fifth Base in DNA? Methyl-Cytosine To understand imprinting, we need to discuss methylated DNA. The genomes of many species contain a gene called *DNA methyltransferase 1 (Dnmt1)*. Dnmt1p is an enzyme that adds a methyl (CH_3) group onto hundreds of thousands—but not all—cytosines throughout the genome. Since the 1960s, methylated cytosines have been associated with genes that are unable to be transcribed. Natural loss of methylation is seen in some cancers and noncancerous aging cells. Regulation of gene expression without altering the DNA sequence is called **epigenetic regulation.** In December 1999, an international team created the Human Epigenome Consortium to map the human **methylome,** a complete description of the methylation status of the genome. If there are about 400,000 methylated cytosines in any given cell, that alone would take a lot of work to study. However, each cell type has its own unique methylome, and since there are about 100 unique cell types in the human body, the team will need to catalog 40 million methylation events, and perhaps twice that many since gender-specific imprinting utilizes methylation.

The precise mechanism for imprinting is unknown, but there are a few models that help us understand how a particular gene could be differentially expressed depending on whether it reached the embryo inside an egg or a sperm. During gametogenesis, DNA methyltransferase is activated, and it modifies gamete-specific sites that determine which alleles will be expressed in the embryo. Direct methylation may block the transcription of some genes (Figure 1.11a), but this mechanism alone is not sufficient to explain all the data. In some cases, it appears that methylation of the promoter allows a distant enhancer sequence to activate an alternative gene on the same chromosome (Figure 1.11b). Still other data indicate that there are insulator regions of chromatin that can block the activity of a downstream enhancer, but the insulator is neutralized by methylation (Figure 1.11c). Another model predicts that chromatin structure (e.g., heterochromatin) might work over great distances to silence a region, and this structure is regulated by methylation (Figure 1.11d). Tilghman's lab demonstrated that a

FIGURE 1.11 • Four proposed models to explain imprinted gene silencing mechanism. Each panel contains two lines that represent two different copies of the same locus. The active gene is depicted with filled boxes and the silenced genes with open boxes and an X. Methylated DNA is denoted by a series of vertical blue lines and CH₃; an enhancer by a filled circle; a boundary or insulator by a hatched box; closed heterochromatin by overlapping open circles. Arrows coming from enhancers indicated gene activation.

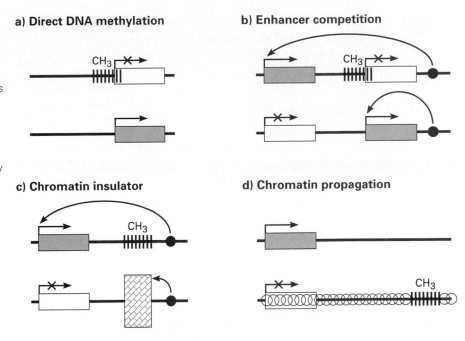

DNA-binding protein called CTCF recognizes the methylation status of imprint-controlled regions on chromatin, and CTCF only binds to unmethylated DNA. The site of CTCF binding is between imprinted loci and their enhancers, which puts it in the right location to be a key player in Figures 1.11b and 1.11c. The complex regulation of imprinting illustrates that the DNA between genes (junk DNA) may be more important than we had imagined. In addition, imprinting is even more complex since at least one maternally expressed gene, *Mash,* is unaffected by the loss of methylation, which indicates that other epigenetic mechanisms may help control imprinted genes.

Imprinting, Methylation, and Cancer In January 2001, the laboratories of Albert de la Chapelle from Ohio State University and Yusuke Nakamura from the University of Tokyo reported that loss of imprinting of *Igf2* might predispose individuals to colorectal cancer. This team studied the insulator region near *Igf2,* where a cluster of CpG islands were hypermethylated and CTCF binds. Loss of imprinting of *Igf2* has been associated with a variety of childhood and adult tumors, as well as Beckwith-Wiedemann Syndrome, a congenital overgrowth disorder that predisposes the individual to tumors. With the loss of imprinting, the amount of *Igf2* transcription might be twice as high as normal, and excess growth factor might contribute to Wiedemann Syndrome and tumor formation. It has been shown previously that loss of *Igf2* does affect the growth of intestinal cancers.

Nakamura and de la Chapelle took tumorous and nontumorous biopsies from cancer patients who had particular DNA repeat units that were unstable and determined the degree of methylation in the *Igf2* insulator region. These samples were compared to other cancer patients who did not exhibit unstable DNA repeat units. In 20 tumor samples taken from patients with unstable repeat units, 17 were hypermethylated in the CTCF-binding region. In addition, 14 of the 20 were hypermethylated in their nontumorous colonic mucosa. This positive correlation indicates that the hypermethylation of imprinted genes might predispose some individuals to colorectal cancer.

Laurie Jackson-Grusby conducted a more comprehensive genomic response to a perturbation of the metabolome when she led a team from Eric Lander's and Rudolf Jaenisch's labs at MIT and Christopher Wilson from the University of Washington, Seattle. Typically *hypo*methylation of the genome is associated with senescence and cell death. Jackson-Grusby wanted to determine how the genome would respond to the deletion of DNA methyltransferase. Using skin cells grown in culture, she measured the change in transcription for about 6,000 mouse genes with and without *Dnmt1.* (See Chapters 4 and 5 to learn how this many genes can be monitored.) Approximately 10% of the genes showed altered expression levels; most of these were induced, but a few were repressed. The affected genes included imprinted genes, cell-cycle control genes, growth factor ligands, and receptors. The complete list of *Dnmt1*-affected genes is freely available online.

Jackson-Grusby's results verify that the methylome is a significant component of gene regulation and thus the

LINKS

Albert de la Chapelle

Yusuke Nakamura

Beckwith-Wiedemann Syndrome

Eric Lander

Rudolf Jaenisch

Christopher Wilson

Dnmt1-affected Genes

proteome. In order to understand a genome, we will also need to understand the methylome and recognize that, functionally, the genome has five bases: G, C, A, T, and methyl-C.

DISCOVERY QUESTIONS

42. Do you think methylation is an ancient mechanism or one limited to vertebrates? How could you answer this question?

43. Go to GenBank and search "methyltransferase." Can you find any DNA methyltransferases in organisms other than vertebrates?

44. Jackson-Grusby et al. found that most of the genes with altered expression increased their expression levels when the methyltransferase was deleted, as you would expect since methylation normally silences genes. Explain how some genes could be repressed when *hypo*methylated.

Summary 1.2

Understanding the human genome will take many years of continued research. Although the quality of the DNA sequence is fast approaching "finished" status, completion of the sequencing project is not the end but the beginning of our understanding. Human, fly, worm, *Arabidopsis,* yeast, and *E. coli* genome sequences will help us determine what makes us human. One major lesson from the HGP has been that we are more than the sum of our genes. Human cells regulate gene expression in more complex patterns since we have more transcription factors. Our genome produces more proteins than there are genes by alternatively splicing mRNAs. We use epigenetic regulation, including methylation, to provide our genomes with increased capacity for control of transcription. Over time, we will discover more complexity within our genome of "only" 35,000 genes.

Chapter 1 Conclusions

Chapter 1 provided you with a broad perspective. You know how genome sequences are produced and annotated. Online bioinformatics tools allow you to mine data from sequences in the public domain. But the apparent simplicity of a genome can be deceptive. Not long ago, the human genome was described as 98% junk DNA. Given the capacity of our genome to evolve and utilize nontraditional genes, we know that the field of genomics has begun a long and exciting series of discoveries, insights, and surprises. The remaining chapters of this book will enable you to use additional bioinformatics tools to explore genomic information, in many different contexts.

References

Tools

Bánfi, Botond, Andrés Maturana, et al. 2000. A mammalian H⁺ channel generated through alternative splicing of the NADPH oxidase homolog of *NOH-1. Science.* 287: 138–142.

Cohen, Barak A., Robi D. Mitra, et al. 2000. A computational analysis of whole-genome expression data reveals chromosomal domains of gene expression. *Nature Genetics.* 26: 183–186.

Gene Ontology Consortium. 2000. Gene onotology: Tool for the unification of biology. *Nature Genetics.* 25: 25–29.

International Human Genome Sequencing Consortium. 2001. Initial sequencing and analysis of the human genome. *Nature.* 409: 860–921.

Jackson-Grusby, Laurie, Caroline Beard, et al. 2001. *Nature Genetics.* 27: 31–39.

Larsen, Frank, Glenn Gundersen, et al. 1992. CpG islands as gene markers in the human genome. *Genomics.* 13: 1095–1107.

Nakagawa, Hidewaki, Robert B. Chadwick, et al. 2001. Loss of imprinting of the insulin-like growth factor II gene occurs by biallelic methylation in a core region of *H19*-associated CTCF-binding sites in colorectal cancer. *PNAS.* 98(2): 591–596.

Natale, Darren A., Uma T. Shankavaram, et al. 2000. Towards understanding the first genome sequence of a crenarchaeon by genome annotation using clusters of orthologous groups of proteins (COGs). *Genome Biology.* 1(5): research0009.1–0009.19.

Overbeek, Ross, Niels Larsen, et al. 2000. WIT: Integrated system for high-throughput genome sequence analysis and metabolic reconstruction. *Nucleic Acids Research.* 28(1): 123–125.

Pasquinelli, A. E., et al. 2000. Conservation of the sequence and temporal expression of *let-7* heterochronic regulatory RNA. *Nature.* 408: 86–89.

Pruitt, Kim D., and Donna R. Maglott. 2001. RefSeq and LocusLink: NCBI gene-centered resources. *Nucleic Acids Research.* 29(1): 137–140.

Shoemaker, D. D., et al. 2001. Experimental annotation of the human genome using microarray technology. *Nature.* 409: 922–927.

Tatusov, Roman L., Eugene V. Koonin, and David J. Lipman. 1997. A genomic perspective on protein families. *Science.* 278: 631–637.

Tilghman, Shirley M. 1999. The sins of the fathers and mothers: Genomic imprinting in mammalian development. *Cell.* 96: 185–193.

Vrana, Paul B., John A. Fossella, et al. 2000. Genetic and epigenetic incompatibilities underlie hybrid dysgenesis in Peromyscus. *Nature Genetics.* 25: 120–124.

Vrana, Paul B., Xiao-Juan Guan, et al. 1998. Genomic imprinting is disrupted in interspecific Peromyscus hybrids. *Nature Genetics.* 20: 362–365.

Review and Summary Sources
Aach, John, Martha L. Bulyk, et al. 2001. Computational comparison of two draft sequences of the human genome. *Nature.* 409: 856–859.

Adoutte, André. 2000. Small but mighty timekeepers. *Nature.* 408: 37–38.

Birney, Ewan, Alex Bateman, et al. 2001. Mining the draft human genome. *Nature.* 409: 827–828.

Feinberg, Andrew P. 2001. Methylation meets genomics. *Nature Genetics.* 27: 9–10.

Hagmann, Michael. 2000. Mapping a subtext in our genetic book (from News of the Week). *Science.* 288: 945–46.

Hillis, David M., and Mark T. Holder. 2000. Reconstructing the tree of life. New technologies for the life sciences. *A Trends Guide.* December. 47–50.

Littlejohn, Tim. 2000. Phylogenetics—a web of trees. *BioTechniques.* 29(3): 482–483.

Passarge, Eberhard, Bernhard Horsthemke, and Rosann A. Farber. 1999. Incorrect use of the term synteny. *Nature Genetics.* 23: 387.

Salzberg, Steven L., Owen White, et al. 2001. Microbial genes in the human genome: Lateral transfer or gene loss? *Science.* 292: 1848–1850.

Tupler, Rossella, Giovanni Perini, and Michael R. Green. 2001. Expressing the human genome. *Nature.* 409: 832–833.

Weinberg, Robert A., and Eric S. Lander. Genomics: Journey to the center of biology. *Science.* 287: 1777–1782.

Wickware, Potter. 2000. Next-generation biologists must straddle computation and biology. *Nature.* 404: 683–684.

Wolfsberg, Tyra G., Johanna McEntyre, and Gregory D. Schuler. 2001. Guide to the draft human genome. *Nature.* 409: 824–826.

University of California at Santa Cruz. 2001. Human genome browser user guide. <http://genome.ucsc.edu/goldenPath/help/hgTracksHelp.html>. Accessed 19 March 2002.

Genome Sequences Answer Interesting Questions

The excitement over genomics is not founded on the production of sequence information but on its utility and applications. Perhaps the area to be affected the most will be our understanding of the evolution of life on earth. DNA encodes proteins, but it also contains encrypted details of the origin of life and our species. Chapter 2 begins with case studies to examine the evolution of the nucleus and human beings. The second half is devoted to more applied uses for genomic information, from the identification of biological weapons and the origin of tuberculosis in the western hemisphere, to improvements in health care made possible by genomic insights. The cases provide specific examples to support the claim that the Human Genome Project will influence the life of every person.

2.1 Evolution of Genomes

Perhaps the area of biology to benefit most from genome sequences will be our understanding of evolution. Designing nongenomic experiments or finding fossils that provide new insight is extremely difficult. However, organisms alive today carry in their genomes information that has been shaped by evolution. Two fundamental questions have persisted since evolution was accepted as the shaping principle of biology: How did the nucleus of eukaryotic cells evolve? What is the origin of human beings? In Section 2.1, genomes can be viewed as missing links that enable us to answer these two difficult questions.

How Did Eukaryotes Evolve?

Evolution is the unifying theme of biology, but it has left us with many questions. How could a nucleus evolve inside what we now call eukaryotes? It seems impossible for a nucleus to have evolved inside a cell that already had its own genome. That evolution only takes place with small changes over time is a misconception. Sometimes drastic changes happen very quickly that radically change the gene pool, and selection favors individuals with advantageous capabilities. A second misconception is that genomes are relatively stable. Genomes change much more dramatically than we had realized before genomic sequences became available. Let's look at a series of cases that use genomic sequences to understand the origins of the eukaryotic nucleus.

Which Is the Chicken and Which Is the Egg? Evolution has been substantiated so many times that calling it a theory can be misleading to the general public. They often think the word implies that there are still other viable options, that it is one possible explanation among many. No, evolution is more like a principle in the same way that gravity is a princi-

ple. The challenge for us is to design experiments and analyze data to reveal how life has changed over time.

Most of you learned that the nucleus, chloroplast, and mitochondrion came into existence through **endosymbiotic** events. By this, we mean that one cell engulfed another and now the two former cells rely on each other for survival as a single eukaryotic cell. Although it is difficult to imagine this event, it does sound plausible given that these three organelles all contain their own genomes, and all are encased in double membranes. But the nuclear genome is very different from bacterial, chloroplastic, and mitochondrial genomes. How can one endosymbiotic genome look so different from the other two? The answer has been found in extreme places.

When endosymbiosis was first proposed as the origin of the nucleus, there were only two kinds of cells—bacteria and eukaryotes. Since then, we have discovered a third type of cell that is so different from either of the other two that new terminology had to be invented. The highest classification is now the **domain,** of which there are three: **Eubacteria** (true bacteria), **Eukaryotes** (true nuclei), and **Archaea,** the "**extremophiles**"—lovers of extreme environments. Archaea are not Eubacteria, because their cell walls, membrane composition, transcription mechanisms, and rRNA sequences all differ and they often live where Eubacteria cannot. They live in boiling springs (e.g., *Thermus aquaticus,* whose DNA polymerase is used for **PCR**), eat methane for food, thrive in five molar salt, and live in the plumes of hot vents under tremendous pressure at the bottom of oceans. These are habitats where textbooks used to say life was impossible. Nature abhors a vacuum, and it appears that life-forms fill all niches.

Prokaryotic genomes are easier to sequence than eukaryotic genomes because they are smaller. Several very dedicated labs around the world, including The Institute for Genomics Research (TIGR), have sequenced many Eubacteria and Archaea genomes. With this database, a group of four Japanese investigators used whole genome sequence comparisons to determine the origin of the nucleus. Chloroplast and mitochondrial genomes look very much like Eubacteria, so that debate has been settled already. But the nucleus has remained enigmatic, so their goal was challenging and interesting. Their strategy was to mine the genomic data in a novel way, so all their work was done on computers.

The team of four worked in Takao Shinozawa's lab at Gumma University in Kiryu, Japan. Shinozawa and his colleagues used the BLAST program to determine the number of orthologous yeast genes in six Archaea and nine Eubacteria genomes. They performed this BLAST at different **expect values (E-values),** which is a statistical measurement of the likelihood of similar sequences occurring randomly (see Math Minute 1.1 on page 5). They used a statistical test

LINKS

TIGR

Takao Shinozawa

BLAST

31

FIGURE 2.1 • Nuclear genes. Hit numbers **a)** of 2,688 yeast "nuclear organization" genes and *t*-statistic values **b)** at each E-value threshold. E-values (−log E-value scale) are shown on the X-axis. Gray lines correspond to Archaea, and blue lines correspond to Eubacteria. *t*-statistic values are shown in the range of E-values over which the hit number (number of matching sequences in the database) was ≥ 5 in at least one bacterium (bars) or < 5 (double-headed arrow).

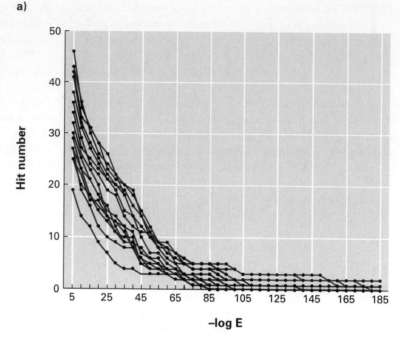

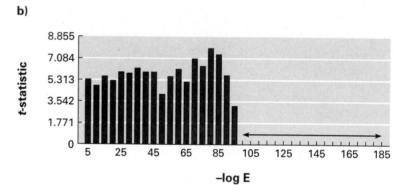

called a *t*-test to infer the prokaryotic origin for each of four categories of yeast genes.

The first group of yeast genes they tested were those involved in "nuclear organization," and they compared the number of orthologs to yeast genes (i.e, the hit number) in each prokaryotic genome (Figure 2.1). Subcategories of nuclear organization include meiosis, DNA replication, cell-cycle control, mitosis, transcription, ribosomal proteins, translation, nuclear structure, and the endoplasmic reticulum (ER) for a total of 2,688 genes. To read Figure 2.1a, compare the number of Archaea orthologs (gray lines) to the number of Eubacteria orthologs (blue lines). Archaea genomes contain more nuclear organization orthologs as denoted by higher hit numbers for the six Archaea lines compared to the nine Eubacteria lines in the graph. The X-axis is a sliding scale of stringency or E-values. If the E-value is set to exclude potential orthologs with low sequence similarity (moving to the right on the X-axis), the

hit numbers decline for each genome. As the stringency is lowered to include potential orthologs with lower sequence similarity (moving to the left on the X-axis), the hit numbers for each genome increases. Therefore, with −log E-values 1 through 95, the greatest number of yeast orthologs are found in Archaea genomes. The *t*-test (Figure 2.1b) validated the impression that nuclear organization genes originated in Archaea and not Eubacteria (see Math Minute 2.1 for details).

The next two categories of yeast genes Shinozawa's group analyzed were 601 mitochondria-related **open reading frames (ORFs),** which included tricarboxylic acid cycle (TCA) and electron transport genes (Figure 2.2a and b) and 1,266 genes related to energy/metabolism (Figure 2.2c and d). In both cases, the yeast orthologs are more numerous in Eubacteria genomes (with *t*-statistic values less than −1.771) than Archaea genomes over a wide range of E-values. The Eubacterial origin of mitochondria and energy

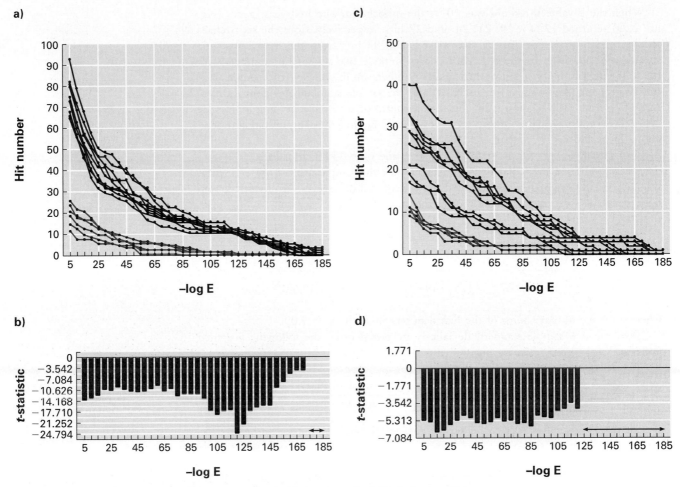

FIGURE 2.2 • Mitochondrial and energy genes. Hit numbers **a)** of 601 "mitochondrial genes" and *t*-statistic values **b)** at each E-value threshold. Hit numbers **c)** of 1,266 "energy genes" and *t*-statistic values **d)** at each E-value threshold. Gray lines correspond to Archaea, and blue lines correspond to Eubacteria.

MATH MINUTE 2.1 • ARE THE HIT NUMBERS SIGNIFICANTLY DIFFERENT?

The analysis in Figures 2.1–2.3 is based on BLAST searches for several sets of yeast ORFs at progressively smaller E-values (horizontal axes in the figures), limited to six Archaea and nine Eubacteria genomes that are fully sequenced. Since the input sequence length and database size are fixed, BLAST hits must have increasingly larger bit scores as the E-value gets smaller (see Math Minute 1.1). In other words, a lower E-value threshold corresponds to a smaller number of hits. Therefore, every line in Figures 2.1–2.3 either remains flat or decreases as you read from left to right.

What the investigators looked for in these data is a significant separation between the number of Archaea hits and the number of Eubacteria hits. A separation can be detected graphically if the gray lines bunch together, the blue lines bunch together, and there is not much overlap between gray and blue lines. But how much bunching is necessary? how little overlap? To help answer these questions and determine whether a separation between gray and blue lines is statistically significant, the researchers used the *t*-test. Let's walk through the steps of a *t*-test on these data.

When the E-value threshold was 10^{-5}, the mitochondria-related yeast genes (Figure 2.2a) returned 12, 15, 19, 21, 24, and 27 hits, respectively, from the six Archaea species, and 65, 66, 68, 73, 75, 80, 81, 82, and 93 hits from the nine Eubacteria species. The first step in the *t*-test is to describe these two sets of hit numbers with the following summary statistics: how many numbers are in each set (the sample size), the average (or sample mean) of each set, and the sample standard deviation (a measure of "spread," or variation from the sample mean) of each set. The sample sizes are denoted by n_1 and n_2, the sample means by $\bar{y}_1$ and $\bar{y}_2$, and the sample standard deviations by s_1 and s_2.

In this example, $n_1 = 6$, $\bar{y}_1 \approx 19.67$, $n_2 = 9$, and $\bar{y}_2 \approx 75.89$. To calculate the standard deviation for a particular data set, (1) subtract the sample mean from each measurement, (2) square each difference, (3) sum the squared differences, (4) divide the sum by $n - 1$, and (5) take the square root of the quotient. Thus,

$$s_1 = \sqrt{\frac{1}{n_1 - 1} \sum_{i=1}^{n_1} (y_{1i} = y_1)^2}$$

$$\approx \sqrt{\frac{1}{5}\left[(12 - 19.67)^2 + (15 - 19.67)^2 + \cdots + (27 - 19.67)^2\right]} \approx 5.6$$

where y_{1i} is the *i*th data point of the first data set. Similarly, $s_2 \approx 9.09$.

Next, the two sample standard deviations are *pooled* using the following weighted average:

$$s_p = \sqrt{\frac{(n_1 - 1)\,s_1^2 + (n_2 - 1)\,s_2^2}{n_1 + n_2 - 2}} \approx \sqrt{\frac{(5)(5.6)^2 + (8)(9.09)^2}{6 + 9 - 2}} \approx 7.93$$

Finally, the *t*-statistic for testing whether the Archaea hits and the Eubacteria hits are significantly different is

$$t = \frac{\bar{y}_1 - \bar{y}_2}{s_p \sqrt{\dfrac{1}{n_1} + \dfrac{1}{n_2}}} \approx \frac{19.67 - 75.89}{7.93 \sqrt{\dfrac{1}{6} + \dfrac{1}{9}}} \approx -13.45$$

This is the value shown in Figure 2.2b under the E-value threshold 10^{-5} ($-\log E = 5$). Note that large positive values of *t* correspond to cases in which $\bar{y}_1$ is significantly larger than $\bar{y}_2$, and large negative values of *t* correspond to cases in which $\bar{y}_2$ is significantly larger than $\bar{y}_1$.

The final determination of statistical significance is made by comparing each *t*-statistic to a cutoff value from a table of the *t*-distribution. For this study, the investigators chose the cutoff values of ± 1.771. That is, *t*-statistic values greater than 1.771 favor an Archaeal origin, and *t*-statistic values less than -1.771 favor a Eubacterial origin.

---

metabolism was expected, and validated Shinozawa's conclusion that nuclear organization genes originated in Archaea.

Shinozawa's team analyzed a fourth group of genes, ones associated with "organization of the cytoplasm," which included subcategories such as cytoskeleton, Golgi, vacuoles, plasma membrane, protein folding, sorting, and modifications. This collection of genes resulted in an interesting finding that further validated their method of considering many different E-values (Figure 2.3). When the E-value was low (e.g., 10^{-5} to 10^{-15}, or $-\log E$ of 5–15), the origin of the yeast genes appeared to be Archaea. However, as the stringency of the E-value cutoff was increased, the *t*-statistic values changed. With E-values of 10^{-50} to 10^{-170}, the yeast orthologs appeared to be more prevalent in Eubacteria. Therefore, the "cytoplasmic" genes were not classified as either Archaea or Eubacteria, since the interpretation depended on an arbitrary selection of an E-value. By examining the prokaryotic origins of yeast genes, Shinozawa utilized public domain databases and the power of seeing data from a genomic perspective to gain new insight. His group synthesized their

a)

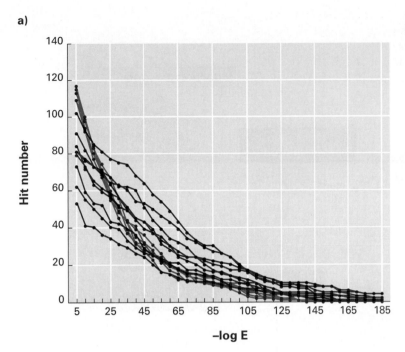

b)

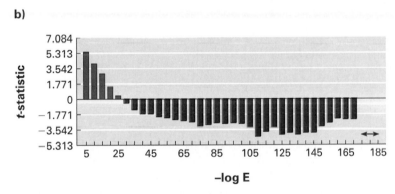

FIGURE 2.3 • Cytoplasmic genes. Hit numbers **a)** of 2,995 yeast "cytoplasmic genes" and *t*-statistic values **b)** at each E-value threshold. The critical point to note is that the gray (Archaea) and blue (Eubacteria) lines cross at about −log E of 25, and the *t*-statistic value changes with different E-values.

findings into a comprehensive model, allowing us to make new predictions that can be experimentally tested (Figure 2.4).

DISCOVERY QUESTIONS

1. Based on Shinozawa's findings, do you think Archaea transcription factors and DNA polymerases more resemble Eubacterial or human orthologs? Test your prediction by using BLASTp and the Archaea protein sequence.

   What type of protein is this? What species is it from? Are there any human orthologs on the results page (perform a "Find" function with your browser for the word "human")?

2. If you wanted to use an antibiotic to kill an Archaea but not a eukaryote, what cellular processes would make the best targets? Which would be the worst? Explain your answers.

3. Why does it make sense for proteins in the ER and protein translation to be among those originating in Archaea rather than the "cytoplasm" category, which has a less clear origin?

SEQUENCES

Archaea Protein

Is There Evidence of Intermediate Stages in Genomic Evolution? Data reveal the laws of nature, so errors in science usually arise from poor experimental design, interpretations, and assumptions. With this in mind, you should be wanting additional data to support the endosymbiotic origin of the nucleus. When we talk about the fossil record and its evidence supporting the principle of evolution, we often think of the

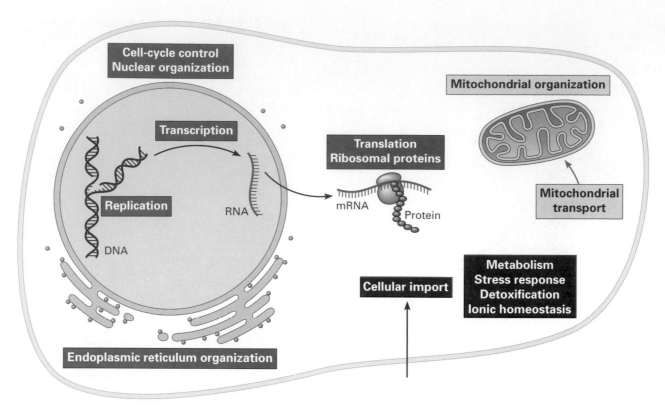

FIGURE 2.4 • **Summary of the prokaryotic origins for eukaryotic genes.** Yeast genes of functional groups are colored gray if they originated from Archaea, dark blue if they originated from Eubacteria; mitochondrial genome of bacterial origin is shown in light blue.

LINKS

RIKEN

"missing link." A missing link is a hypothetical fossil that shows a halfway transitional organism that will fill in the missing data point to complete the picture. Wouldn't it be nice if nature provided us with current examples of evolution where prokaryotes are in various stages of symbiosis and genomic degeneration (Figure 2.5)? Genomes sequenced in 2000 and 2001 provide us with examples of missing links that illustrate what might have happened when the first Archaea was engulfed by a bacterium and eukaryotes were poised to evolve.

Are You Going to Eat That? This story focuses on a biological curiosity that at first is difficult to explain. Did you know that aphids, the tiny insects that draw their food from the phloem of plants, do not excrete nitrogenous waste? This might explain why ants can drink what comes out of the back end of an aphid since it lacks toxic nitrogenous waste. One rule of biology is that all animals produce nitrogenous waste. Mammals urinate; birds excrete uric acid in their droppings, which is why they are so white. We accumulate nitrogenous waste when we break down proteins and amino acids, and we rid ourselves of this waste. Aphids are curious creatures since they are exceptions to this rule. Any time we observe an exception to a rule, we are faced with a few

choices; accept these apparent oddities as "exceptions that prove the rule," ignore these exceptions, or modify our rules. It seems unwise to say nitrogen waste is either not produced or is not toxic. Therefore, we need to find another reason to keep our rule and still accept the observation that aphids don't excrete nitrogenous waste.

Most aphid species have 60–80 large cells in their abdomens called bacteriocytes, which are home to a round bacterium called *Buchnera*. These bacteria are transmitted to the next generation of aphids from the mother, and the mutualism between the insect and the bacterium is so complete (established about 225 million years ago) that neither species can reproduce in the absence of the other. Shuji Shigenobu and colleagues from the University of Tokyo and the **RIKEN** Genome Sciences Centre sequenced the genome of *Buchnera*. The circular chromosome is 640,681 bp, and the genome includes two smaller plasmids of about 8 kb each. This is the second smallest genome sequenced (*Mycoplasma genitalium* has only 580,070 bp), which indicates this once free-living bacterium might have lost some of its original DNA.

When Hajime Ishikawa and his collaborators looked for ORFs, they found only 583, with an average gene size of 988 bp. Using BLAST to search the nonredundant database,

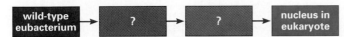

FIGURE 2.5 • Schematic of missing links in the evolution of the eukaryotic nucleus. We know a lot about bacterial and nuclear genomes, but we would like to discover intermediate steps in this transition.

biological roles were assigned for 500 of the genes. Most of the genes were very similar to those found in *E. coli* (which has 4,288 genes), and many of the *Buchnera* genes are found in the same order as seen in *E. coli*. Seventy-nine of the remaining ORFs were similar to hypothetical genes annotated for other bacterial genomes. Only four genes were found to be unique to the *Buchnera* genome. Based on this, Ishikawa's team concluded that *Buchnera* evolved from a close relative of modern *E. coli*, although *Buchnera* has lost about 75% of its original genome.

What makes the aphid/*Buchnera* symbiosis so interesting is the degree of interdependence. At the creation of eukaryotes, when an Archaea was first engulfed by a bacterium, presumably each cell retained its full complement of genes from its original genome. Current eukaryotic cells contain portions of both genomes, but many genes have been lost over time. The genes found in eukaryotes today are the result of a "compromise," with many of the redundancies between Eubacterial and Archaeal genomes deleted in order to be more efficient and advantageous. Aphids, like all animals, can synthesize some amino acids, but the remaining essential ones must be supplied through diet. *Buchnera* has lost many genes necessary to synthesize amino acids except the ones that aphids cannot synthesize. Therefore, *Buchnera* needs the amino acids aphids produce, and aphids need the amino acids *Buchnera* produces. Through this mutualistic relationship, the two are collectively self-sufficient, and they do not need any amino acids to be supplied in their diet.

DISCOVERY QUESTIONS

4. Search *Buchnera* for genes in the list below, where you will see an amino acid is followed by one gene necessary for its production. Deduce which amino acids are supplied by *Buchnera* and which are supplied by the aphid.

Arginine—*ArgA*	Histidine—*HisA*
Valine—*IvHI*	Tyrosine—*TyrA*
Isoleucine—*IlvC*	Proline—*ProA*
Leucine—*LeuA*	Serine—*SerA*
Phenylalanine—*PheA*	Glycine—*GlyA*
Tryptophan—*TrpAB*	Cysteine—*CysE*
Threonine—*ThrA*	Glutamate—*GdhA*
Methionine—*MetC*	Glutamine—*GlnA*
Lysine—*LysA*	Aspartate—*AspC*
Asparagine—*AsnA*	Alanine—*AlaB*

LINKS

Buchnera Genome

Search *Buchnera*

5. Are there any amino acids that are considered nonessential for most animals that *Buchnera* can produce? The nonessential amino acids for most animals are tyrosine, proline, serine, glycine, cysteine, glutamate, glutamine, aspartate, asparagine, and alanine.

6. Based on your findings for Discovery Questions 4 and 5, can you make any predictions about what might happen to the *Buchnera* genome over the next 100 million years?

Given that *Buchnera* has a very accommodating host, you might suspect other "essential genes" for free-living life outside aphids might have been lost as well. Indeed, *Buchnera* would not survive for long if it were "set free," since it lacks many DNA repair enzymes and cell wall synthesizing enzymes, and has no recognizable phospholipid synthesis capacity. It also lacks many signaling pathways most bacteria use to sense and respond to changes in their environment. Therefore, *Buchnera* is dependent on a stable environment where DNA damage is rare and many metabolites are supplied for its use. In return, *Buchnera* produces a few amino acids and vitamins as a type of rent to its aphid host. Given that aphids reproduce rapidly and thrive globally, this symbiotic arrangement appears to work very nicely from the host's perspective as well. But no landlord likes to be taken advantage of, and *Buchnera* seems to "know" this rule for subletting space. *Buchnera* retains all the genes necessary to carry out ATP synthesis including ATP synthase and the electron transport system, although it lacks the genes necessary for fermentation and anaerobic metabolism. So, *Buchnera* carries its own weight by aerobically producing ATP in addition to a few amino acids and vitamins (Figure 2.6).

In some ways, *Buchnera* is like a mitochondrion. *Buchnera* produces ATP, contains a small but essential genome, and requires many metabolites produced indirectly by the nuclear genome including the membrane phospholipids that separate it from the cytoplasm. In this comparison, the aphid supplies the nuclear genome, and *Buchnera* behaves as if it were a newly evolving organelle—"aminoacid-plast." In this way, *Buchnera* looks very much like an intracellular missing link as it makes the transition from endosymbiont to organelle. Ishikawa recognized this similarity when his lab produced the complex integrated **circuit diagram** (see Figure 2.6) showing how host and symbiont mutually benefit each other. By taking a genome-wide look at *Buchnera*, we developed new connections and insights.

We began this story with the dilemma that either aphids retain their nitrogenous waste or we must revise our rules for animal metabolism. With the full genome sequence of *Buchnera*, the explanation becomes evident and we realize that aphids are not an exception to the rule. Aphids rarely

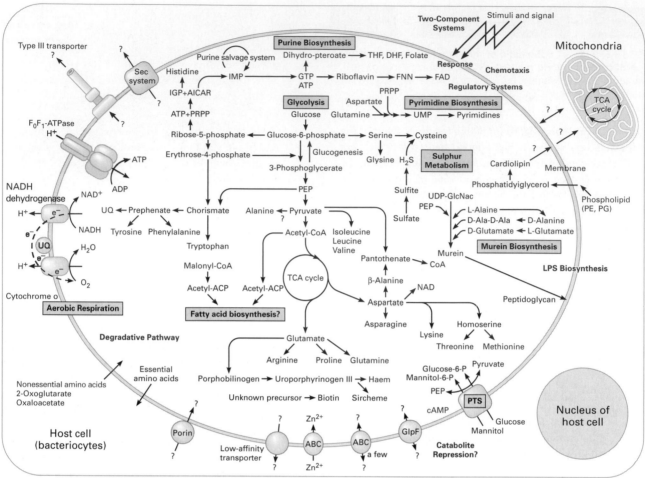

FIGURE 2.6 • **Whole-cell view of *Buchnera* metabolism as deduced from its genome.**
Buchnera is shown as an oval and inside a rectangular aphid bacteriocyte. Pathways or
steps for which no enzymes were identified are shown in blue, as are metabolites that lack
de novo synthetic pathways. Question marks indicate uncertainties or pathways that are
poorly understood.

LINKS

TIGR's CMR

secrete nitrogenous waste because they re-
cycle the amino groups that all animals
produce with protein catalysis. Aphids
produce excess glutamine that *Buchnera* uses as a substrate
for the synthesis of the aphid's essential amino acids. Waste
is a relative term in biology since every species produces it
and inevitably another will consume it. By engulfing *Buchnera,* aphids have streamlined the nutrient cycle and improved their efficiency. Evolution favors those with a
selective advantage, and converting waste nitrogen into essential amino acids has to be near the top of the list of adaptive solutions to selection pressures.

DISCOVERY QUESTIONS

7. Aphids can destroy houseplants very quickly. Based
on what you have learned, can you design a way to
kill aphids but not harm other insects, animals, or
plants?

8. Ishikawa and his colleagues propose that *Buchnera*
evolved from *E. coli.* Go to TIGR's Comprehensive
Microbial Resource (CMR), where you can analyze
whole genomes.

a. Scroll down and click on the link that says
"Align Whole Genomes." Choose *Buchnera* for
the Y-axis and *E. coli K-12* for the X-axis, and
then click on "Launch Alignment." You will see
a picture that aligns the two genomes with a
color code as described below the graph. Determine the degree of genome alignment for these
two species.

b. Go back to the CMR home page and choose
"Genome v/s Genome Protein Hits." First,
choose *E. coli K-12* as the reference genome.

Second, choose *Buchnera* from the pop-up menu and then click on "Add Molecule." Choose *E. coli K-12* and "Add Molecule." This will allow you to compare the encoded proteins (as defined by the published annotations). Click on "Generate Display." The results are provided graphically (left side) and in text format (right). If you mouse over the graphic displays of the genomes, you can see where the similarities lie within the reference genome. Knowing that *Buchnera* has 583 ORFs and *E. coli* has 4,288, does this comparison match your expectations?

We have seen our first example of a missing link where *Buchnera* lost some of its genome. But again, you might think this is unsubstantiated speculation since you cannot prove a negative. In other words, it is impossible to prove that genes have been lost when they are not present currently. It would be comforting if we could find a free-living bacterium whose genome is halfway between a typical bacterium and a symbiont that has lost a lot of genes already (Figure 2.7).

A Missing Link of Biblical Proportions A human pathogen may provide the missing data to complete our understanding of the transitional stages that led to the eukaryotic nucleus. For many centuries, one bacterium was the scourge of the world—*Mycobacterium leprae*. *M. leprae* is the causative agent for **leprosy,** which was highly contagious, disfiguring, lethal, and resulted in patients being quarantined in leper colonies. The mode of *M. leprae* transmission is uncertain, although it is believed to enter through the nasal passages. Once inside your body, it lives in the most unlikely places, inside your phagocytotic white blood cells (called **macrophages**), which normally engulf and kill pathogenic bacteria. Eventually, it also infects Schwann cells, which provide the fatty insulation for your nerves (called **myelin**). Infected Schwann cells lead to the loss of myelin, nerve damage, loss of sensation, and eventually loss of extremities due to reduced blood circulation. Although we understand leprosy better, we are unable to eliminate it. The World Health Organization reported that 690,000 new cases of leprosy were diagnosed in 1998. The last remaining leper

LINKS
Stewart Cole
M. leprae
M. tuberculosis

colony in the United States was located in Carville, Louisiana, and closed in the late 1990s.

M. leprae was first isolated in 1873, but during the last 100+ years, no one has ever been able to grow it in a culture medium. The best way to grow *M. leprae* is to use the nine-banded armadillo as a surrogate host. In this biological incubator, the bacteria reproduce by fission only once every 14 days, compared to *E. coli,* which can reproduce two to three times an hour. The inability of *M. leprae* to grow in a defined culture medium and its slow in vivo growth rates have baffled the research community, until now. In February 2001, one week after the draft human genome sequence was published, a British and French team led by Stewart Cole reported the complete genome sequence of *M. leprae*. Although prokaryotic genomes are easier to decipher than the human genome, it is informative to compare genomes of similar species. Fortunately, in 1998 the complete genome of *Mycobacterium tuberculosis* (the causative agent for tuberculosis) also was published by Cole's team.

The genome of *M. leprae* contains 3,268,203 bp, while *M. tuberculosis* contains 4,411,532 bp. When Cole's lab compared the content of the two related species, the contrast was striking (Table 2.1). Only 50% of the leprosy genome encodes 1,604 proteins, compared to tuberculosis's 91% coding DNA, which produces 3,959 proteins. The average gene lengths are nearly identical, but the number of **pseudogenes** (DNA that looks like genes but is unable to produce functional protein because of accumulated mutations) is very different, with *M. leprae* containing over 1,000 pseudogenes and *M. tuberculosis* a scant 6. Therefore, it appears that since the time of their evolutionary divergence, the *M. leprae* genome has lost over 2,000 functional genes and over 1 million base pairs of DNA. It is likely that *M. leprae* has evolved as a *Buchnera*-like intracellular recipient of host metabolites.

Let's look at a couple of examples that indicate the leprosy genome has undergone significant changes, including recombinations, rearrangements, and deletions. When the genomes of leprosy and tuberculosis are compared, there are about 65 segments that are syntenic. However, the relative order and orientation of the genes in the regions differ, and some of the genes are missing in *M. leprae* (Figure 2.8). When we look at the gene called *proS*, which encodes prolyl-tRNA synthetase (the enzyme that covalently joins the amino acid proline to its tRNA), its sequence is very different from *M. tuberculosis*. Instead, *proS* from *M. leprae* resembles the bacterial pathogen that causes Lyme disease, *Borrelia burgdorferi,* as well as *proS* from eukaryotes such as humans, flies, and yeast. *M. leprae* may have acquired a second *proS* ortholog and later lost its original *proS* gene. Consistent with the recent acquisition of *proS*, the *M. leprae proS* gene is out of order and in the opposite orientation compared to *M. tuberculosis*.

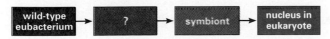

FIGURE 2.7 • Updated schematic of missing links in evolution of eukaryotic nucleus. Current symbionts (e.g., *Buchnera*) that have lost some of their genome may resemble what happened billions of years ago as an internalized prokaryote became a nucleus.

TABLE 2.1 • Comparison of *Mycobacterium* genomes.

Feature	*M. leprae*	*M. tuberculosis*
Genome size (bp)	3,268,203	4,411,532
G+C(%)	57.79	65.61
Protein coding (%)	49.5	90.8
Protein-coding genes (no.)	1,604	3,959
Pseudogenes (no.)	1,116	6
Gene density (bp per gene)	2,037	1,114
Average gene length (bp)	1,011	1,012
Average unknown gene length (bp)	338	653

Used with permission from Stewart Cole.

What Else Is *M. leprae* Missing? The leprosy genome is lacking many genes compared to its more vigorous relative *M. tuberculosis* (Figure 2.9). In particular, *M. leprae* is barely able to synthesize its membrane lipids. Surprisingly, *M. leprae* has one lipid-synthesizing gene not present in *M. tuberculosis,* and this particular lipid is abundant in the *M. leprae* membrane. It has evolved a compensatory mechanism for the loss of more common lipids. *M. leprae* derives most of its energy from lipid catalysis, and its pathway for carbon metabolism is lacking redundant genes found in most other organisms. For example, it lacks malic enzyme, and has only one isocitrate lyase gene as well as many other genes in the TCA and glycolysis pathways. Interestingly, it is unable to synthesize the amino acid methionine, which is not usually an essential amino acid but appears to be for *M. leprae.*

DISCOVERY QUESTIONS

9. Explain how the genome sequence could be used to develop a growth medium for *M. leprae.*

10. Given the limited metabolic capacity of *M. leprae,* how could investigators develop new drugs to kill *M. leprae*? Support your answer with data provided in this section.

With the *M. leprae* genome, we can see another possible step along the way toward creation of nuclei from prokaryotes with reduced genomes (Figure 2.10). Rarely do we have data that represent all the steps in a chain of evolutionary events. The fossil evidence for the reduction of toes on horses is unusually complete for data supporting evolution. Now we are able to examine fossils of a different sort—living bacterial genomes that display different degrees of evolutionary reduction in complexity, size, and ability to live independently. The original Archaea that evolved into the first nucleus has had billions of years to adapt to its new environment. Only through genomic analysis were we able to understand how intracellular genomes were formed from Eubacteria and Archaea to produce the third and newest domain of life—eukaryotes.

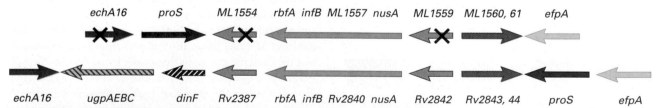

FIGURE 2.8 • Differences between *Mycobacterium* genomes. Comparison of *proS* genes (solid black arrow) from *M. leprae* (above) and *M. tuberculosis* (below). Genes or operons are depicted by arrows with the point representing the downstream end. X denotes pseudogenes.

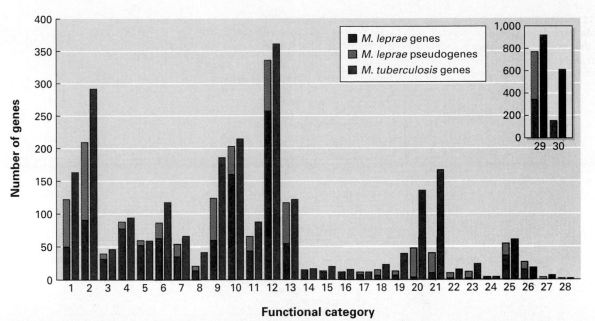

FIGURE 2.9 • **Distribution of genes by functional category.** The number of *M. leprae* genes (dark blue), pseudogenes (light blue), and *M. tuberculosis* genes (gray) are listed for each category. Functional categories: 1, small-molecule catabolism; 2, energy metabolism; 3, central intermediary metabolism; 4, amino acid biosynthesis; 5, nucleoside and nucleotide biosynthesis and metabolism; 6, biosynthesis of cofactors, prosthetic groups, and carriers; 7, lipid biosynthesis; 8, polyketide and nonribosomal peptide synthesis; 9, proteins performing regulatory functions; 10, synthesis and modification of macromolecules; 11, degradation of macromolecules; 12, cell-envelope constituents; 13, transport/binding proteins; 14, chaperones/heat-shock proteins; 15, cell division proteins; 16, protein and peptide secretion; 17, adaptations and atypical conditions; 18, detoxification; 19, virulence determinants; 20, IS elements and phage-derived proteins; 21, PE and PPE families; 22, antibiotic production and resistance; 23, cytochrome P450 enzymes; 24, coenzyme F420-dependent enzymes; 25, miscellaneous transferases; 26, miscellaneous phosphatases, lyases, and hydrolases; 27, cyclases; 28, chelatases. *Inset:* Y-axis shows the number of genes within each functional category; X-axis shows the functional categories: 29, conserved hypothetical proteins; 30, hypothetical proteins that share no significant similarity with any protein currently in the databases.

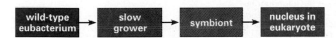

FIGURE 2.10 • **Updated schematic of missing links in evolution of eukaryotic nucleus.** Perhaps "genomically weak" prokaryotes were engulfed by other prokaryotes, which sparked a symbiotic relationship that eventually led to the formation of organelles as we recognize them today. This diagram is not intended to suggest that these steps necessarily happened, but it does reveal that intermediate missing links are alive today.

What Is the Origin of Our Species?

Since Darwin published his historic work outlining evolution, we have wondered how humans evolved. Of course, we did not evolve from present-day monkeys, but we do have common ancestors. We also share common ancestors with flies, worms, yeast, and even Archaea and the first prokaryotic cells. But that was too long ago for most of us to have any sense of kinship. What did *Homo sapiens*'s closest relatives look like, and where did they live 100,000 years ago?

This also leads us to wonder how closely related we are to all the other 6 billion humans alive today.

LINKS

Mitochondrial DNA

Placental Mammals and Insectivores Let's start our story with the evolution of mammals. A multinational team including investigators from The Netherlands; the University of California, Riverside; Belfast, Ireland; the University of Cincinnati; the University of Massachusetts in Amherst; and SmithKline Beecham Pharmaceuticals coordinated their efforts to deduce the evolutionary relationships among all taxonomic orders of placental mammals. To accomplish this hairy project, they sequenced some mitochondrial DNA as well as some nuclear DNA totaling over 8,000 bases for each species. When they compared the changes in DNA sequences for these two different genomes, they were able to piece together the mammalian family tree (Figure 2.11). It came as no surprise that our nearest relatives are other primates, specifically the great apes.

In agreement with others, this team concluded that the base from which all mammals arose was an insectivore-like mammal, which is supported by fossil evidence. They also concluded that different insectivore-like lineages have

FIGURE 2.11 • **Phylogenetic tree of placental mammals with a marsupial as the root.** This tree used the 2,947 bp nuclear sequences, which were available for a wider range of species than the longer 5,708 bp sequences (mixture of nuclear and mitochondrial sequences). The letters at each branch point indicate a decreasing likelihood score with "a" being the most likely rating. Blue arrow highlights the location of the human branch.

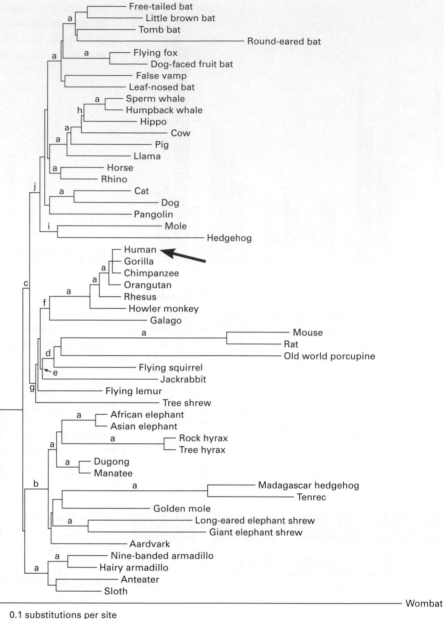

0.1 substitutions per site

independently given rise to diverse placental mammalian taxa in different geographical settings.

DISCOVERY QUESTIONS

11. Look at the mammalian tree and find humans and Galago. Go to NCBI Entrez Taxonomy home page and find out what a Galago is by using the search.

12. Although it is hard to see the exact branching, what types of mammals are most closely related to primates?

13. Find mice and rats, the two most common research animals. What is the scientific justification for using rodents as model organisms rather than cats, dogs, or other domesticated farm animals such as pigs?

Apes, Humans, and Diversity A related question to our evolutionary origin is our evolutionary divergence. How much genetic diversity is present in the human species? When we look at people from different regions of the world, we detect morphologically distinct traits. These traits are determined by our genomes, and so we are left with the impression that the human gene pool is extremely diverse. Is the human gene pool more diverse than the gene pool of other species? What about the genetic diversity of our nearest living relatives, the great apes? Do these primates really have less genetic diversity than we do?

To determine the answer, a German group led by Svante Pääbo examined about 10,000 bp of noncoding DNA from the X chromosomes of chimpanzees, lowland and mountain gorillas, and Bornean and Sumatran orangutans. By choosing noncoding DNA, they made the assumption that this section of the DNA would not be under selective pressure and thus would mutate at a constant rate. They examined the number of variable base pairs while taking into account the sample size for each species. In conducting this study, the investigators first acknowledged that data from mitochondrial and nuclear DNA do not always agree. For example, mitochondrial DNA indicates each subspecies is distinct, whereas nuclear DNA indicates all the great apes have exchanged DNA. Given this caveat, they move forward with their analysis to provide us with some interesting conclusions.

The time since the **most recent common ancestor (MRCA)** for all **hominid** species is about 540,000 years ago. Divergence of hominids took place more recently than did the other great apes since the MRCA for all gorilla species lived about 1.2 million years ago, chimpanzee's MRCA lived about 1.9 million years ago, and orangutan's MRCA lived about 2.1 million years ago. In general terms, this agrees with the mammalian tree shown in Figure 2.11. However, the German team concluded that humans are the least diverse of the primates they studied. The great apes have two to three times more diversity (depending on the species) than humans do. They speculated that if the now extinct hominid Neanderthal were still alive and considered human, then perhaps humans would have as much genetic diversity as the great apes. Since the Neanderthals became extinct about 30,000 years ago, we are left as the only surviving hominid. Our gene pool is less diverse than the great apes', even though there are many more humans alive than great apes. Furthermore, the data indicated that humans experienced a population expansion approximately 160,000–190,000 years ago. At this time, the size of the human population that migrated out of Africa was estimated to be about 3,700. Their estimate for the date of population expansion does not agree with the mitochondrial estimation of about 45,000 years ago. Pääbo's team recognized that their estimate does not have a confidence interval associated with it, and therefore their date may be estimating the same population expansion event as the mi-

tochondrial data predict, or perhaps a different and earlier expansion. Nevertheless, the overall finding is striking. Humans, with all our morphological diversity, have less genetic diversity than do the great apes among whom we cannot distinguish individuals. Our limited gene pool is due to a population that was very small and has expanded rapidly over the last 100,000 years.

Out of Africa to Your Home Town Finally, we are ready to determine where and when modern humans evolved.

There are two competing hypotheses to explain the evolution of humans. That *Homo erectus* evolved in Africa about 2 million years ago is widely accepted. The "recent Africa hypothesis" predicts that modern humans originated in Africa 100,000–200,000 years ago and a few migrated to cover the rest of the world. The more ancient *H. erectus* became extinct, and there was little or no genetic mixing between the African *H. sapiens* and *H. erectus*. The "multiregional hypothesis" predicts that ancient humans gradually evolved into modern humans at sites around the world. The main evidence for this hypothesis is the fossil evidence, which is updated periodically with new findings. For example, the 2001 discovery of a new hominid near Lake Turkana in Kenya might place the famous "Lucy" skeleton off the main line toward modern humans. The recent Africa and multiregional hypotheses about human evolution are diametrically opposed to each other and, therefore, it should be possible to distinguish which is correct.

A pair of labs from Sweden and Germany decided to stop looking at short pieces of genomes and they sequenced the entire mitochondrial genome (~16,500 bp long) of 53 people from diverse geographical origins. Mitochondria are transmitted from mother to child and therefore are not recombined with paternal DNA. They used the chimpanzee mitochondrial genome as their outgroup to root the phylogenetic tree containing the 53 human sequences. Based on genetic and palaeontological evidence, chimps and humans diverged about 5 million years ago. With this as a starting point, they calculated that humans have accumulated 1.7×10^{-8} nucleotide changes per site per year.

First, the genomic anthropologists looked at mitochondrial genome diversity among current Africans compared to all non-Africans and found that Africans have twice as much diversity among themselves (3.7×10^{-3} nucleotide changes per site) as do non-Africans (1.7×10^{-3} nucleotide changes per site). The greater African diversity could be the result of either a larger population size or a significantly longer genetic history. The diversity among Africans is reflected in the phylogenetic tree (Figure 2.12). Notice how the African half of the tree has many long branches compared to the non-African half with its short branches. The non-African

METHODS

Phylogenetic Tree

LINKS

Svante Pääbo

Extinct

FIGURE 2.12 • Rooted phylogenetic tree based on the complete mtDNA genomes. Bootstrap values (for 1,000 replicates) are shown at the branch points, and source of the DNA is listed at the leaves. Arrow highlights the branch point for the MRCA for branches that contain both African and non-African individuals. The "D loop" of mitochondrial DNA was excluded due to irregular mutation rates.

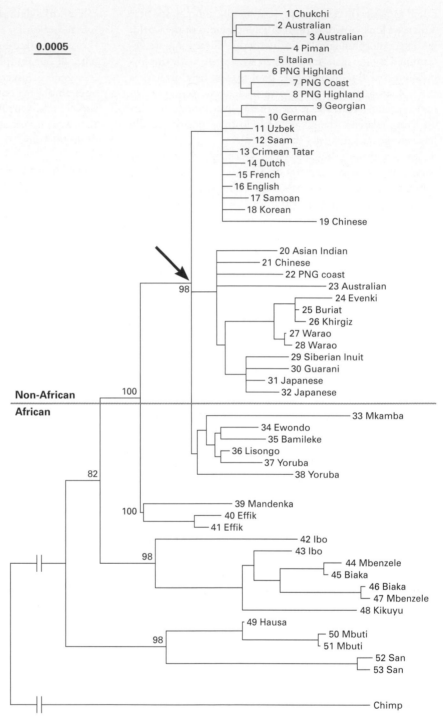

portion indicates a population bottleneck with a rapid radiation of diversity. Based on the distribution of the nucleotide differences within the mitochondrial genome, the investigators concluded that Africans have experienced a constant population size. Conversely, non-Africans exhibit a more uniform distribution of nucleotide differences that indicate a recent population expansion. The time of expansion was estimated to be 1,925 generations ago. With a generation time of 20 years, they calculated diversification occurred about 38,500 years ago.

MATH MINUTE 2.2 • HOW DO YOU KNOW IF THE TREE IS RIGHT?

The mitochondrial genome sequences used to form the tree in Figure 2.12 provide evidence of the evolutionary relationships among these individuals. However, several different trees might be consistent with the sequence evidence, depending on how you define consistent. Researchers typically form phylogenetic trees using one of three standard techniques (distance-based methods, maximum parsimony, or maximum likelihood), each of which implements slightly different criteria for joining individuals (or species, as in Figure 2.11) in the tree. For example, one distance-based method uses the distances between each pair of sequences to form a hierarchical clustering of the sequences (see Math Minute 4.3). No matter which method is used, the sum of the horizontal lengths of the tree branches from the root to a leaf is scaled to represent the difference between the root sequence and the leaf sequence.

One way to test the validity of a tree formed using one of these three techniques is to use one of the other techniques, and compare the results. Another way to evaluate a tree's validity is with **bootstrap analysis.** The idea behind bootstrapping is to repeat the tree-building process several times, each time using different subsequences of the original full-length sequences. If a large percentage of the resulting trees agree at a particular branch of the tree, you can place more faith in that branch point than if there were little agreement among the trees.

The bootstrap procedure works as follows. Suppose you have aligned all 54 sequences used to form the tree in Figure 2.12, using a multiple sequence alignment algorithm. Randomly select columns from the alignment, until you have selected as many columns as were in the original alignment. Column selection is performed "with replacement," which means you may choose the same column more than once and some columns not at all. Take the nucleotides from each sequence of the selected columns, resulting in 54 modified sequences. Build a new tree using these 54 modified sequences. Replicate this selection and tree-building process 1,000 times. Now go back to the tree built with the original sequences and choose a particular branch point, e.g., where the African and non-African people first joined (individuals 1–32 join with individuals 33–38). Label this branch point with the percent of the 1,000 trees in which this same branching structure (i.e., separating 1–32 from 33–38) occurs. In Figure 2.12, this particular branching structure occurred 98% of the time in the 1,000 replicates, as shown by the bootstrap value of 98. Branches with bootstrap values of 70 or greater are normally considered reliable. Thus the tree in Figure 2.12 is a very reliable representation of the relationship among these mitochondrial genome sequences.

Further analysis allowed a prediction for two different MRCA. If you use their molecular clock (1.7×10^{-8} nucleotide changes per site per year), it is possible to calculate the time of the MRCA for the two most distantly related human mitochondrial sequences to be $171,500 \pm 50,000$ years ago. The MRCA for Africans and non-Africans (arrow in Figure 2.12) is estimated to have existed $52,000 \pm 27,500$ years ago. Based on these findings, we are left with only one viable hypothesis—the recent Africa hypothesis. The multiregional hypothesis had predicted that modern humans evolved from ancient hominids who left Africa about 2 million years ago. This hypothesis predicted that *H. neanderthalensis* eventually became modern Europeans and *H. erectus* became modern Asians. Although we know that *H. neanderthalensis*

coexisted with modern humans for about 10,000 years, the data indicate that all non-Africans share common ancestors who originated in Africa and left there approximately 130,000 years ago (Figure 2.13). Based on a limited amount of Neanderthal DNA sequence, it has been estimated that the MRCA for all *Homo* species lived about 465,000 years ago and *H. sapiens* arose approximately 130,000–465,000 years ago with a compromise date of 200,000 years ago. Furthermore, it appears that all 6 billion of us modern humans owe our limited genetic diversity to a small number of breeding individuals (fewer than 10,000 combined) that included two main branches. One branch led to many African populations, while the other branch led to a mixture of African and non-African populations.

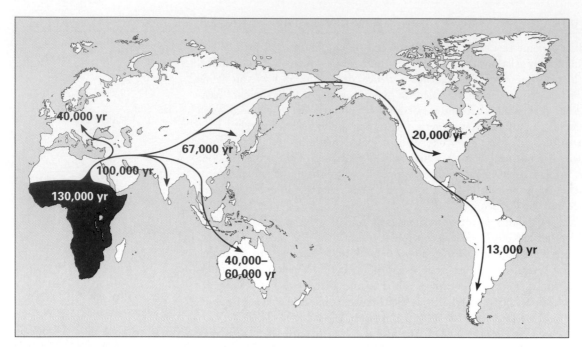

FIGURE 2.13 • **Map showing the origin and dispersal of modern humans.** The origin of *H. sapiens* may have been about 200,000 years ago. The earliest known fossil and archaeological evidence on each continent is shown on the map and is consistent with the interpretation presented. Used with permission from Blair Hedges.

DISCOVERY QUESTIONS

14. What two populations are the most distantly related based on mitochondrial sequence? How many mutations accumulated between the two most distantly related human sequences?

15. The Swedish and German team took great pains to demonstrate that the DNA mutation rates in the mitochondrial genome occur at a constant rate. There is one portion of the mitochondrial genome (called the D loop) that they exclude because it demonstrated irregular mutation rates. Why is this a critical point for their assumption that a "molecular clock" can be measured?

16. Find the three listings for Australians in Figure 2.12. Do any of these three branches represent the British colonization from the 19th century? Propose a hypothesis to explain the separation of Australians on this tree.

Although it is the most complete study to date (May 2002), other investigators will produce and analyze more data and the recent Africa hypothesis will be tested further. If it is repeatedly supported, then the hypothesis will stay substantially unchanged. If conflicting data are reported, the research community will modify the hypothesis until there is a consensus on our origins. For now, the data are clear that humans alive in the 21st century are all cousins to varying degrees.

Summary 2.1

Section 2.1 illustrates how genomic sequences can provide us with a more complete understanding of evolution. As more genomes are completely sequenced, we will fill other gaps in our knowledge. By analyzing complete genomes from different taxa, variations among individuals, the role of parasites and symbionts, and the genomes of organelles including chloroplasts, we will appreciate more fully the gradual population of the earth by organisms with diverse life histories and origins. The unifying principle in biology is evolution, and if genomes could talk, we'd be able to hear the story of the creation of life. We are learning how to listen, and soon we will be ready to hear what must be one of the most fantastic stories you could ever imagine. Stay tuned for more details about the pathway of evolution.

2.2 Genomic Identifications

Each species can be defined by its genome, and an individual's DNA sequence can be used to determine its species. However, we cannot easily sequence the entire genomes of every species, much less every individual of a species. Therefore, we rely on particular segments of DNA as molecular barcodes that can be scanned to reveal useful information. In this section, we examine four case studies where short segments of genomic DNA have been used to help identify their origins. The first case study examines how biological weapons can be identified with the aid of genome sequences. Two case studies examine DNA extracted from individuals 1,000 to 250 million years old. Finally, we will see how DNA is the ideal tool for diagnosing emerging diseases, which can lead to new treatments. With these cases, you should discover the diverse uses for genomic information.

How Can We Identify Biological Weapons?

On 22 March 2001, two downtown Toronto office buildings were sealed, with the workers still inside. Simultaneously, two different offices received packages that contained a "gray granular substance" and notes warning that the boxes contained anthrax. The recipients of the boxes called police, who restricted access in and out of the buildings for several hours. After three hours, the contents were determined to be harmless and the occupants were allowed to come and go freely. When anthrax spores were delivered to various people in the United States during the fall of 2001, identification times ranged from a few minutes to several days. Why were there such differences in the amount of time for determining whether anthrax spores were present? Can genomes lead to fast and reliable methods for biological weapon identification?

Anthrax is a spore-forming bacterium, *Bacillus anthracis,* that occurs in wild and domestic mammals (cattle, sheep, goats, camels, antelopes, and other herbivores) but can infect humans as well. Anthrax can be used as a biological weapon, which has prompted the U.S. Department of Defense to vaccinate all active-duty military personnel at risk for exposure. Anthrax spores can survive for many years and can lead to infection of the skin or lungs. Initial symptoms of respiratory infection resemble a common cold, but within days the symptoms become severe breathing problems, and can lead to shock and sometimes death. Fortunately, person-to-person contact is unlikely to spread *B. anthracis.* (For more information, you can consult the Centers for Disease Control and Prevention [CDC].)

How Did They Determine It Was Not Anthrax in Less Than Three Hours? Bacterial spores are similar to seeds, so you cannot quickly grow them up into rapidly dividing bacteria that can be analyzed by traditional microbiology methods. Even the fastest cells grow too slowly, and you might not know which growth medium would be best for supporting the unknown samples. Therefore, the only identifiable material available on short notice is DNA.

Bacterial chromosomes can be isolated in less than an hour, but raw DNA is not very helpful. We need to sequence the DNA to identify the species. Unfortunately, you would need species-specific primers, sequencing reactions, and sequencing machines, which are not easily transported for use outside the lab. **PCR** is a robust procedure that performs well in many

settings, but it typically takes 2–3 hours to amplify the DNA, and then the PCR products need to be analyzed by gel electrophoresis, which takes more time and processing. We need a faster process. The following case studies highlight recent advances in species identification using DNA-based methods.

METHODS
PCR
LINKS
CDC
Andreas Manz
David Burke

In 1998, Andreas Manz and his lab published a new and very fast method for PCR called chemical amplification continuous-flow PCR on a chip. The key to this technology is that PCR reactions do not need long incubation times, since the heat exchange takes place in about 100 milliseconds due to the design of the chip (Figure 2.14). DNA extracted from a biological sample is mixed with PCR reagents and, once combined, they rapidly travel through different temperature zones to produce PCR products that are analyzed later.

The Manz lab provided a great tool to speed up the PCR process, but detection of the product was still slow and required too much lab equipment for use in remote sites. What was needed was a self-contained unit that would perform the PCR and gel analysis in one step. David Burke and his colleagues at the University of Michigan have developed an integrated nanoliter DNA analysis device that is a good

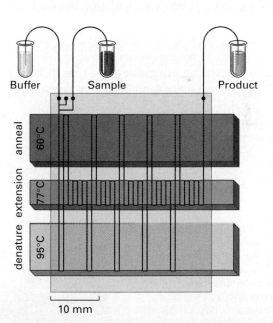

FIGURE 2.14 • Schematic of continuous-flow PCR. Three well-defined temperature zones are maintained at 60°C for annealing, 77°C for extension, and 95°C for denaturing. Notice that the pathway of the fluid, which is pumped through the channel, keeps the sample at 77°C longer by zigzagging back and forth at this temperature. The duration of time at each temperature is determined by the path length of the channel in each zone. The reagents are mixed just prior to entering the 60°C zone.

a)

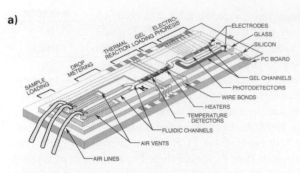

b)

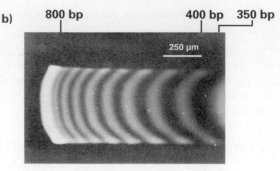

FIGURE 2.15 • **Integrated nanoliter DNA analysis. a)** Schematic of integrated device with two liquid samples and electrophoresis gel. The device is able to add set volumes of liquids in the "drop metering" section using a combination of hydrophobic regions and air intakes. **b)** Optical micrograph of a 50 bp DNA ladder in a 0.5 × 1.0 mm polyacrylamide gel. The DNA was loaded onto the gel and electrophoresed for 15 minutes. The band at the far right edge is the 350 bp band. The remaining bands are between 400 and 800 bp.

METHODS

Real-Time PCR

LINKS

Phillip Belgrader

solution (Figure 2.15a). The entire chip is produced as a single piece, and the user supplies the standard PCR reagents and template DNA—in our case unknown bacterial DNA. With this system, the PCR product could be sent directly onto a miniature gel that takes about 30 minutes to resolve the bands (Figure 2.15b). This device presents several advantages. First, the person in the field only needs to load DNA from a biological sample; the chip does everything else, including recording the results onto a computer. The chips are small, portable, require very little electricity, and do not require much training to operate compared to sequencing or agarose gels. This PCR and mini-gel chip is a workable solution, but even faster results would be helpful.

As an alternative to Burke's "lab on a chip," Phillip Belgrader from the Lawrence Livermore National Laboratory helped develop a very different machine that can identify samples in seven minutes. Belgrader's team built a device called real-time PCR that performs PCR in microfuge tubes

with volumes of 25 μL (Figure 2.16). This device has resulted in a significant improvement in our ability to identify unknown biological samples. **Real-time PCR** monitors the production of the amplified product at each cycle in the reaction so that as product accumulates, its production is monitored in "real time." The output is not a gel but a graph that displays the amount of PCR product. The user-friendly output converts a green flat line for no PCR product detected into a red graph to indicate successful identification of the unknown DNA.

Since the specific source of the DNA is unknown, the sample may be one or more pathogens. Many species-specific PCR primer pairs could be tested in a **multiplex** format (using more than one pair of PCR primers in a single reaction tube) where each species' PCR product emits a unique fluorescent color. Real-time PCR is extremely rapid, with a detection time of seven minutes (Table 2.2). Belgrader's team can measure the number of cells in the original template and the time it took to detect the PCR product (Figure 2.17). The device is easy to use and about the size of a big lunch box.

FIGURE 2.16 • **The portable Advanced Nucleic Acid Analyzer (ANAA) a)** An ANAA consists of an array of ten reaction modules and a laptop computer. **b)** Screen shot of laptop after running a reaction with bacteria (top left trace) and a negative control (top right trace). The detection of a signal produces a warning display.

a)

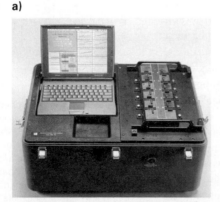

b)

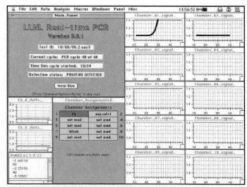

TABLE 2.2 • **Thermal cycling settings and detection times for bacterial cells.** Threshold value is the number of cycles needed to detect a positive signal.

Cycle Time (sec)	Denature Time (sec)	Anneal/Extend Time (sec)	Threshold Value (cycle number)	Detection Time (min)
38	4	19	23	14.6
38	4	10	23	10.7
24	4	5	24	9.6
17	1	1	25	7.0

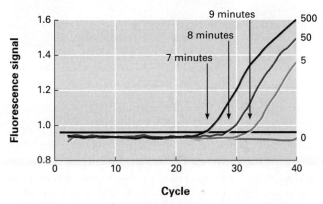

FIGURE 2.17 • **Quantitative real-time PCR performed with a 17-second cycle time.** The number of bacteria cells is listed to the right of each trace. The time it took before a positive signal was detected, triggering the warning signal, is shown above the arrows pointing to each trace crossing the threshold value (black line).

DISCOVERY QUESTIONS

17. The most important part of any experiment is the design of careful controls. Given that some errors might occur, what controls would be necessary each time a test is performed?

18. The ANAA device can run only ten reactions at a time. How could you address this limitation if you needed to test an unknown sample?

The advances in biological detection and identification allowed the workers in Toronto to learn in three hours that they had not been exposed to a lethal biological weapon. They were able to return to their normal lives quickly, and the buildings were reopened for business. For the cases in the United States, the quickest identifications used antibodies that bound to the bacterial cells and formed a line to indicate the presence of anthrax. This technology is similar to home pregnancy tests, requires many spores before detection is possible, and can lead to false positives when the antibodies

bind to harmless but similar bacteria. The most accurate identification technology is the classic microbiology method of growing cells and performing tests on them. This method is very accurate but results may take several days.

It seems likely that the integration of real-time PCR or the integrated nanoliter DNA analysis device will result in improvements in our ability to identify trace amounts of DNA. Since each organism has unique sequences, we will be able to utilize genomic information to identify more species as more genomes are sequenced. The U.S. Food and Drug Administration (FDA) uses a handheld ANAA for food inspections. Improvements for DNA extraction for PCR analysis should improve rapid and accurate identification of biological weapons.

How Long Can DNA Survive?

We often think of DNA as a fragile compound that is easily damaged. However, DNA is actually very stable and resists degradation much better than RNA and most proteins. Its ability to withstand extreme conditions has led to the discovery of "ancient DNA" synthesized thousands or even millions of years ago. To anyone familiar with *Jurassic Park* this concept is not new, but sometimes fact is more amazing than fiction.

Bacteria Survives in 250-Million-Year-Old Salt Crystal In 1995, a team of investigators reported the isolation of a previously unknown ancient species of bacteria that had been encapsulated in 25-million-year-old amber (along with an extinct bee species). In October 2000, biologists William Rosenzweig and Russell Vreeland, and geologist Dennis Powers made a discovery that was ten times as amazing. From deep in the earth, they isolated salt crystals that were dated at about 250 million years old (Figure 2.18). Inside the crystals were small, fluid-filled chambers that occasionally contain bacteria. The team was very careful not to use any crystals that indicated they had been penetrated by water or exhibited

FIGURE 2.18 • **Salt crystals from 1,850 feet down an air intake shaft in Carlsbad, New Mexico. a)** This crystal appeared undisturbed due to the clarity and shape of the fluid-filled chamber. This crystal contained the strain 2-9-3. The drill hole used to obtain the sample (shown above the arrow) permitted access to the inclusion (I), or chamber, containing the bacteria. **b)** This crystal was rejected since it contained cracks (arrow points to a large vertical crack) and the inclusion (I) is irregular in shape.

a)

b)

LINKS

2-9-3

NCBI's Entrez

a crack in the crystal structure. It is possible to identify pristine chambers by their rectangular shape, whereas disturbed chambers have irregular shapes and salt deposits on the back wall of the chamber.

Vreeland's team took great care to sterilize the outside of the crystal and all their tools and equipment, and they worked in sterile environments. They isolated a few microliters of fluid from 66 chambers in 56 different crystals. These small volumes were inoculated into different liquid media and allowed to incubate for three months in sealed tubes. Only 2 of the 56 crystals contained viable bacteria. The growing cells were spread onto solid media where only *Bacillus* colonies were identified. Three additional isolates have been obtained, but only the one they called 2-9-3 was characterized.

Chromosomal DNA was isolated from 2-9-3, the complete rDNA region was sequenced and compared with other sequences, and a phylogenetic tree was produced (Figure 2.19). The three species that form a lineage along with 2-9-3 are halophilic (salt-loving) bacteria. Because 2-9-3 was able to form spores, the researchers tested their ability to survive the sterilization techniques used to treat the outside of the crystals and collection tools. When 10^8 cells and 10^8 spores of 2-9-3 were exposed to the sterilizing procedure, none survived, supporting the conclusion that the isolated strain 2-9-3 is not a modern contamination from the outside of the crystals.

How could 2-9-3 survive inside salt crystals? Like modern species, 2-9-3 formed spores when cells were exposed to higher salt conditions as water evaporated. The spores became trapped in salt crystals, which protected them from environmental hazards that otherwise might have killed them. If this strain truly is ancient, it is the best

example we have of the stability of DNA over long periods of time. Many believe ancient DNA is only modern DNA contamination. Given the nature of ancient DNA research, it may be impossible to thoroughly demonstrate the true origin of the 2-9-3 DNA, and thus each of us must reach our own conclusions.

DISCOVERY QUESTIONS

19. Go to NCBI's Entrez and search the *nucleotide* database using the accession number AF166093 (2-9-3 is now called *Bacillus permians*). Perform a BLASTn search with the sequence, and identify which species are returned in your search. Are you able to reproduce results similar to those shown in Figure 2.19?

20. What is the greatest potential weakness with recovering an untainted ancient sample?

21. Based on DNA sequence information from one locus, is it possible to conclude whether this is a prehistoric strain or a modern contamination? What improvements in methodology would you suggest for future studies?

How Did Tuberculosis Reach North America?

Whenever we want to eliminate a disease-causing pathogen, we need to understand how the organism is spread. Does it spread by ticks (Lyme disease), mosquitoes (malaria), worms (elephantiasis), or by aerosolized bacteria (leprosy)? Knowing the transmission mechanism enables us to better combat the causative agent. Knowing how the disease was first intro-

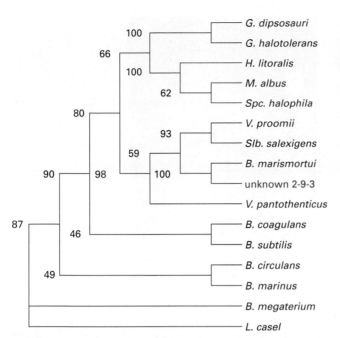

FIGURE 2.19 • **Rooted phylogenetic tree using ribosomal DNA sequence from 2-9-3.** Several species were compared with 2-9-3 to produce this tree. Bootstrap values from 100 repetitions are shown at each node.

FIGURE 2.20 • **Tuberculosis-like lesions on the right lung and hilar lymph region.** The lesions are designated N1, L, and N2, referring to the upper lymph node (N1), lung nodule (L), and lower lymph node (N2).

duced into humans would provide a more complete understanding of its life history. Unfortunately, most diseases predate modern medicine, so our knowledge is limited. However, with our ability to isolate ancient DNA, we can now retroactively diagnose ancient human remains. Can we use archaeological DNA remains to determine when and where a disease originated and spread?

One disease that is increasingly problematic is tuberculosis, since drug-resistant forms are becoming more prevalent. A long-standing belief holds that tuberculosis originated in Europe and was brought to North America by European explorers. However, in 1994 Wilmar Salo, the University of Minnesota, Duluth, Minnesota, reported the identification of *M. tuberculosis* DNA in a Peruvian mummy that predates the arrival of Christopher Columbus. The team located a spontaneously mummified body of a 40- to 45-year-old Chiribaya woman. The Chiribaya were an agricultural population that occupied this area of Peru about 1000–1300 A.D. Carbon dating from the mummy's remains placed the time of death at 934 ± 44 years A.D. Here is the description of the tissue sample isolation.

> Removal of the chest wall revealed intact, collapsed lungs bilaterally. The left lung and the middle and lower lobes of the right lung were not abnormal. The right upper lobe was diffusely adherent to the chest wall, and a largely calcified, discrete nodule 1.2 cm in diameter was present in a lateral, subpleural position [L in Figure 2.20]. At its point of adhesion to the parietal pleura overlaying the lateral part of the forth rib, a 1-mm aperture was apparent

> in the calcified shell, communicating with a central 3-mm diameter cavity within the nodule. In the right hilum, two partly calcified masses, each about 1.5 × 0.8 × 0.5 cm, were found in peribronchial positions [N1 and N2 in Figure 2.20]. No evidence of skeletal change could be identified in the spine, ribs, or elsewhere, and no abnormalities were found in the liver, heart, bowel, skin, trachea, or other soft tissue. No acid-fast bacilli were demonstrable in any of the lesions.

LINKS
Wilmar Salo

In short, they cut open the chest and found three bony nodules where the upper part of the right lung adhered to the inside lining of the chest wall. They also cut open and examined most of the soft tissue in the abdominal and thoracic cavities. They saw no bone deformities often associated with tuberculosis, nor did they identify any *M. tuberculosis* bacilli using the classic test of acid-fast staining. They later extracted the DNA from the three nodules and performed PCR on the DNA. The primers were designed to target "a segment of DNA unique to *Mycobacterium tuberculosis*" called IS6110. They amplified a 97 bp fragment and an overlapping fragment of 232 bp (Figure 2.21). Only tissue sample N2 was "clearly shown to contain the organism's DNA. The bacteria in the other lesions may have ceased to exist before the individual died." The team sequenced 9 of the 97 bp PCR products and verified that they had amplified IS6110. Based on these results, the team concluded that *M. tuberculosis* was present in Peruvian people prior to the arrival of Columbus.

Three years later, a team led by Andreas Nerlich from the Ludwig-Maximilian University in Munich, Germany, reported finding evidence of tuberculosis in a 3,000-year-old Egyptian mummy. Their work was published in 1997 as a one-page note in *The Lancet*:

> We were able to examine the torso from a male mummy aged less than 35 years. The head was missing and the

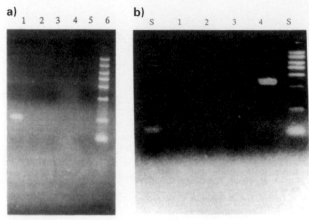

FIGURE 2.21 • **Nested PCR for IS6110. a)** Second-round amplification of the 97 bp target. This product was amplified by two rounds of PCR, where one microliter of the first PCR product (123 bp long) was used as template for the second round, which included nested primers. Lane 1 = DNA extracted from nodule N2; lane 2 = DNA extracted from lung nodule L; lane 3 = control reaction; 4 was empty; lane 5 = reaction control; lane 6 = molecular weight standards. From top to bottom: 1,000, 700, 500, 400, 300, 200, 100, and 50 bp. **b)** Nested PCR of the 232 bp target of IS6110. Two rounds of PCR were performed with the first product (247 bp long) used as the template for the second round of PCR. Lanes labeled S contain molecular weight standards; lane 1 = no first-round PCR template added; lane 2 = control reaction; lane 3 = control extract; lane 4 = N2 DNA extract.

FIGURE 2.22 • **Opened thorax of Egyptian mummy.** Large arrow points to pleural adhesions of right lung; small arrow points to destruction of bony elements of lumbar vertebral bodies L4 and L5.

METHODS

Nested Primers

SEQUENCES

IS6110

Egyptian Mummy

LINKS

Kathleen
Eisenbach

mummy had been broken in two parts just above the pelvis. The mummy was loosely wrapped in linen bindings and did not have evidence of evisceration. . . . After careful dissection of the intact anterior body wall, we noted residues of the lung with extensive pleural adhesions to the chest wall in the right thoracic cavity [Figure 2.22]. The left lung was collapsed and macroscopically unremarkable. In addition, there was a severe anterior destruction of two lumbar vertebral bodies, irregular osteolysis, and extensive reactive new bone formation was detectable on radiographs.

Tissue samples from both lungs, which we had removed immediately after opening the chest wall under sterile conditions, were subjected to DNA extraction and purification.

Using PCR, they amplified a 133 bp fragment that was not a part of IS6110 but encodes a surface protein expressed in several mycobacterial species including *M. tuberculosis* and *M. leprae.* They sequenced their PCR product and reported "the sequences showed homology to

the DNA of *M. tuberculosis,* thereby confirming that the mycobacterial DNA found in our samples was that of *M. tuberculosis.*" They did not publish the sequences from their PCR products, nor the percent similarity in their BLAST search.

DISCOVERY QUESTIONS

22. Perform a BLASTn search using the IS6110 sequence. Does this piece of DNA match anything besides *M. tuberculosis?*

23. Go to BLAST again and search with the 133 bp fragment sequence amplified from the Egyptian mummy. Does this piece of DNA match anything besides *M. tuberculosis?*

24. When IS6110 was first published, the authors clearly stated IS6110 was identical in all the *M. tuberculosis* complex, which includes three additional species: *M. bovis, M. microti,* and *M. africanum.* Does this information affect your interpretations of the two *M. tuberculosis* case studies?

Addendum In 1995, Joe Bates, Kathleen Eisenbach, and their colleagues at the University of Arkansas Medical School published a commentary on the 1994 Salo paper claiming to find *M. tuberculosis* DNA in a pre-Columbian Peruvian mummy. Bates and Eisenbach were two of the investigators

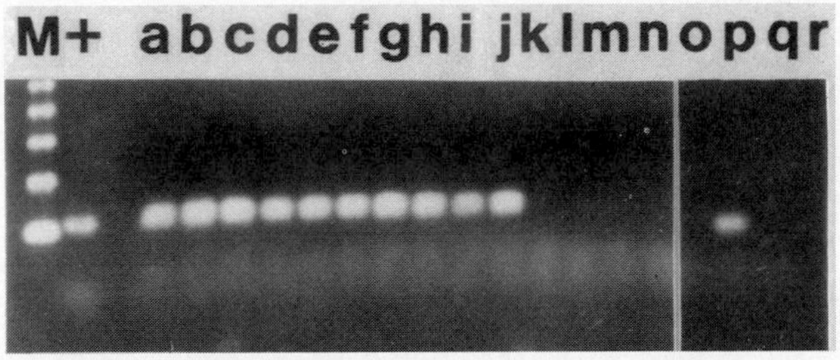

FIGURE 2.23 • **Amplification of DNA by PCR using various sources of *Mycobacterium* DNA.** Input DNA was 625 pg of DNA per reaction with 5 μl of the reaction product loaded on an agarose gel for analysis. Lanes: M = molecular weight marker; + = positive control; lanes a–g are different isolates of *M. tuberculosis;* lanes h, i, and j are different strains of *M. bovis,* with j being the BCG vaccine from Glaxo; lanes k–n are *M. kansasii* strains; o = *M. fortuitum;* p = *M. simiae;* q = *M. gordonae;* r = *M. chelonei.*

who initially characterized IS6110, and their commentary offered a different interpretation of the Peruvian results. They maintain that IS6110 is unable to distinguish among the four closely related species in the *M. tuberculosis* complex (Figure 2.23). They speculate that the DNA found in the Peruvian mummy was in fact *M. bovis* and not *M. tuberculosis.* They base their conclusions in part on the distribution of host species for *M. bovis,* which includes many wild and domesticated animals such as llamas and seals with which the Chiribaya people were known to have contact.

Eisenbach's group hypothesize that *M. bovis* predates *M. tuberculosis,* and as humans domesticated animals, they became infected with *M. bovis,* which later mutated into the current *M. tuberculosis* species and was spread by humans as we began to live in more densely populated centers in Europe and the Middle East around 1600 A.D. At this time, virtually 100% of European urban populations became infected with *M. tuberculosis,* and 25% of infected people died from it. High mortality would have created a strong selection pressure on genotypes that were able to survive the *M. tuberculosis* infection. As humans adapted, tuberculosis was less often a rapidly lethal disease but became a chronic, but nonlethal, pulmonary infection. Included among the chronically infected would have been European sailors who traveled to North America and came into contact with indigenous peoples. Consistent with this model are modern populations who were first exposed to Europeans in the 20th century. For example, people in central Africa in 1910, the highlands of New Guinea in 1950, and the high Amazon in the 1990s have all experienced the rapid and highly lethal response to their initial exposure to *M. tuberculosis* in the 20th century.

The hypothesis proposed by Bates et al. that *M. bovis* is more primitive than *M. tuberculosis* is not universally accepted. However, it is clear that the two cases of *M. tuberculosis* DNA are not as conclusive as they first appeared. The

regions of DNA amplified by PCR are not unique to *M. tuberculosis,* and thus there are alternative feasible explanations. It is possible the mummified individuals were infected with *M. bovis,* or that the samples were contaminated by *M. bovis,* since the mummified remains were exposed to the elements for hundreds or thousands of years. Ancient DNA offers an exciting potential to look into the past. For example, DNA has been extracted from petrified dinosaur droppings, permitting us to determine what species of plants were in the environment at the time. But great care must be taken when interpreting such findings since we have only partial sequences for some species and they may not be unique to a single species.

LINKS
New Diseases
Gary Hayward

DISCOVERY QUESTIONS
25. What additional information would you like to know if you were a peer reviewer for the paper where 3,000-year-old Egyptian mummies were determined to be positive for *M. tuberculosis?*
26. What controls would you design for similar studies?

How Are Newly Emerging Diseases Identified?

Most of this book has focused on genomes inside cells, but this is only a subset of the sequenced genomes. Many viral genomes have been sequenced, enabling a better understanding of viral diseases and perhaps better treatments, too. Initially, however, the greatest benefit is in the area of diagnosis. Every year, new diseases are identified in humans that have not been characterized before. Epidemiologists and pathologists are often the first brave investigators that (willingly) rush into an area of infestation to determine the causative agent. They collect samples from victims, animals, food sources, water, air, etc. Imagine being the first public health-care worker in a region that has suffered a new outbreak of Ebola. You don't know where it hides outside humans nor how it infects us. You take samples from a wide assortment of sites and go to the lab. The microscope is usually the first tool since many viruses have distinctive shapes that aid in diagnosis. At some point, DNA will be extracted and you will sequence it in hopes of determining the exact cause of the new outbreak. In this section, we will examine two outbreaks that occurred in 1999. The first took place in North American zoos, and the second in New York.

Case 1: Elephants Dying in Zoos As a graduate student working with Gary Hayward at Johns Hopkins Medical School, Laura Richman led a team to determine why so many young elephants were dying in captivity. Since zoos no longer capture wild elephants, the only source for zoos are elephants

a)

b)

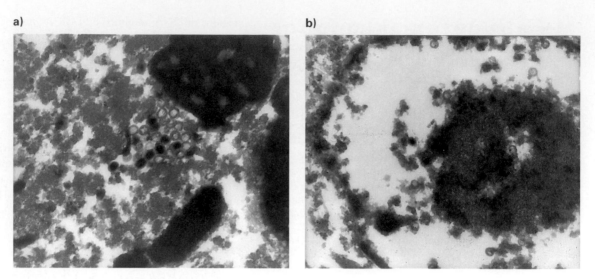

FIGURE 2.24 • **Elephant viruses. a)** and **b)** Transmission electron micrographs of herpesvirus "inclusion bodies" inside the nucleus of endothelial cells. These photomicrographs were produced from cardiac tissue obtained in an autopsy of an Asian elephant.

METHODS

Skin Lesions

SEQUENCES

Terminase Protein

born and raised in captivity. Between 1983 and 1996, 34 Asian elephants were born in North American zoos, of which 7 died with similar symptoms. During the same period, only seven African elephants were born, two of which died prematurely. Soon, most of the adult elephants in zoos will be too old to breed, and there will be too few young adults to sustain the desired populations, especially for African elephants. Therefore, we need to understand the cause of so many premature deaths.

The disease has a sudden onset characterized by fluid retention in the skin of the head and proboscis and internal hemorrhaging. Histological samples revealed abnormalities in the heart, liver, and tongue, with inclusions inside the nucleus of endothelial cells of the smallest blood vessels in these organs (Figure 2.24). Upon electron micrographic analysis, a diagnosis was made of herpesviruses that showed a preference for endothelial cells, which is unusual for herpesviruses. The high mortality rate was due to heart failure from capillary damage and leakage of fluid. Interestingly, some healthy adults had skin lesions that contained the same viral particles, and yet they seemed unharmed by these infections. The investigators were unable to culture the virions on a wide range of cells, so they decided to use DNA sequence to determine which strain of herpes was killing the elephants.

Based on the diagnosis of herpesvirus, PCR primers were designed to amplify two genes: terminase and DNA polymerase. PCR products of both genes isolated from African and Asian elephants were sequenced to produce unrooted phylogenetic trees (Figure 2.25). The virion sequences isolated from the elephants were highly conserved but unlike those from the alpha-, beta-, or gamma-herpesviruses. Based on the sequence information, it appears that zoo elephants

are infected with a previously unknown type of herpesvirus that was also detected in biopsies taken from noncaptive healthy adult African elephants (samples from healthy adult Asian elephants were unavailable).

One young elephant became sick and was successfully treated with the anti-herpesvirus medication famciclovir immediately after becoming ill. The success of quick treatment for infected elephants indicates that veterinary intervention can save the lives of many newborn elephants. In facilities that have experienced herpes-induced elephant deaths, African and Asian elephants have direct or indirect contact (e.g., from common feeding or bathing areas). Richman and her colleagues speculated that the two elephant species each have evolved resistance to their own variants of this herpesvirus but not to the virus found in the other species. This is a reasonable hypothesis since in the wild, the two elephant species never come into contact with each other and each harbors its own virus with no ill effects. But in captivity, the virus has been given the chance to "hop species" and infect a new host whose offspring have no genomic resistance. With the aid of DNA diagnosis, a mystery was solved and an effective treatment devised. This research may have a dramatic impact on the survivability of elephants born in captivity and may also be a lesson for us about mixing in zoos species that have not evolved in close proximity.

DISCOVERY QUESTIONS

27. Perform a BLASTp search using this region of the terminase protein.
    a. Which virus has the second highest percent *identity* after the elephants' virus?

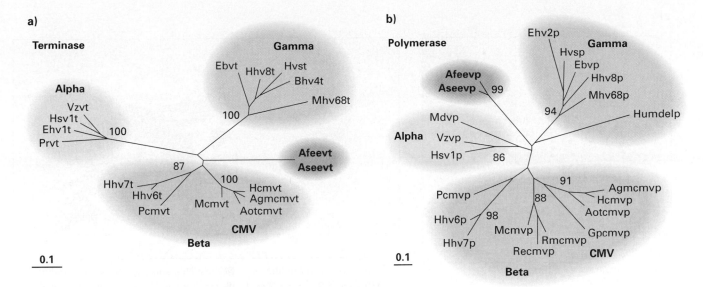

FIGURE 2.25 • **New herpesviruses.** Unrooted phylogenetic trees of herpesvirus **a)** terminase protein sequences and **b)** DNA polymerase protein sequences. The scale bars indicate the distance of a 10% change in amino acid sequence at any given position between any two proteins; numbers on branches indicate bootstrap values.

 b. Which isolate has the highest percent sequence *similarity* (listed as positives on the results page)? You should be able to find some with at least 80% positive.
 c. Predict which one of these two (answers to a and b above) might be the next most likely source of lethal infection for zoo elephants. Explain your answer.

28. Perform a BLASTn search using part of an African viral DNA polymerase gene. Compare these hits with the hits you get from the same region of an Asian viral DNA polymerase gene.
 a. What is the source of the hit you received for the Asian elephant virus?
 b. Did either of your searches return hits for both types of virus?
 c. Given that the authors hypothesized that each species infected the other, why didn't you get hits for both types of viruses?
 d. Did you get any hits for any other herpesviruses? Explain why this result is not surprising.

Case 2: New Human Encephalitis Outbreak in Northeastern United States

In August and September 1999, New York (Figure 2.26) experienced an outbreak of human encephalitis in which two people died. At the same time, there were a number of cases of both captive and wild birds dying. In a matter of weeks, the cause was diagnosed as a form of West Nile virus after a part of the 11,000 kb single-stranded RNA genome was sequenced. This was useful information since it told us that fatalities would be the exception.

Based on sampling methods in October 1999, approximately 1,256 people were infected with West Nile virus during the epidemic. Approximately 239 people (19%) may have had a mild clinical illness associated with their infection. However, prior to summer 1999, West Nile virus had never been reported in North America. These viruses are spread by insects—in this particular case mosquitoes. West Nile virus can live in a number of animal hosts, including birds that migrate huge distances each year. Where did this strain originate? How had it arrived? Did migrating birds bring the virus to New York? These questions needed to be answered so we can better understand how to prevent the arrival and spread of more West Nile virus.

The complete 11,029 nucleotide genome of the 1999 New York West Nile virus (WN-NY99) was sequenced. The source of the sequenced genome was a dead Chilean flamingo at the Bronx Zoo in New York. To determine the virus's origin, Robert Lanciotti from the National Center for Infectious Diseases in Fort Collins, Colorado, led an

SEQUENCES

African Viral DNA Polymerase Gene

Asian Viral DNA Polymerase Gene

LINKS

Robert Lanciotti

a)

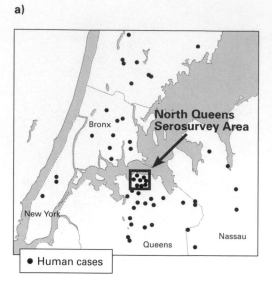

b)

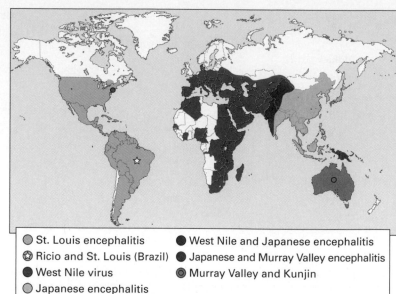

FIGURE 2.26 • **Distribution of Nile viruses. a)** A map of the New York City area with dots representing the locations where people lived who were confirmed to be infected with West Nile virus during the fall of 1999. **b)** Global distribution of different strains of Nile viruses. Note that prior to 1999, West Nile virus had not reached North America.

LINKS

West Nile Virus

Flash Animation Map

BLAST2

international team to use genome sequences to track its origin. They compared the genomes of 41 different strains of related viruses and produced a rooted phylogenetic tree (Figure 2.27). The analysis placed WN-NY99 in a cluster of similar viruses, so it did not appear to be new. Hundreds of birds and mosquito samples were collected; all the isolated viruses had 99.8% sequence identity, which indicated a single introduction of the West Nile virus rather than multiple separate introductions. Furthermore, when compared to its nearest neighbor isolated from a dead goose in Israel in 1998, the two isolates shared greater than 99.8% sequence identity.

Contamination of the 1998 Israeli sample with the WN-NY99 sample could not be the source of sequence similarity since the two genomes were sequenced on different continents. Therefore, it seems highly probable that the WN-NY99 virus was brought to North America from Israel. It is impossible to determine if the virus was in a human, an animal such as a pet or an illegally smuggled exotic animal, or an infected mosquito trapped on a plane. This case illustrates the rapidity with which pathogens can spread between geographically distant locations now that people travel rapidly around the world. The CDC was given several million dollars by the U.S. government to maintain a surveillance program designed to monitor the possible spread of West Nile virus. Samples will be taken from insects, birds, and mammals where investigators will search for RNA genomes

of the West Nile virus. As of fall 2001, West Nile virus has been reported as far south as Florida and as far west as Ohio. It appears the virus will become established in North America, and possibly spread throughout the continent.

DISCOVERY QUESTIONS

29. Go to the Flash animation map and hypothesize which arrived first, the mosquito vector, infected horses, infected birds, or human patients. To read this map, mouse over the icons in the lower left-hand corner and observe the map. You can change the time line by clicking the back arrow. Support your hypothesis with data.

30. Perform a BLAST2 alignment to compare partial coding sequences from an envelope glycoprotein gene of the Israeli 1998 virus (accession number af205882) with the Israeli 1952 virus (accession number af205881). Then compare the Israeli 1998 virus with the New York 1999 virus (accession number af196835). Finally, compare the Israeli 1998 virus with the Central African Republic virus from 1967 (accession number af001566). What hypotheses can you draw from these alignments? To see a very different virus, look at one isolated from the Central African Republic in 1972 (af001563).

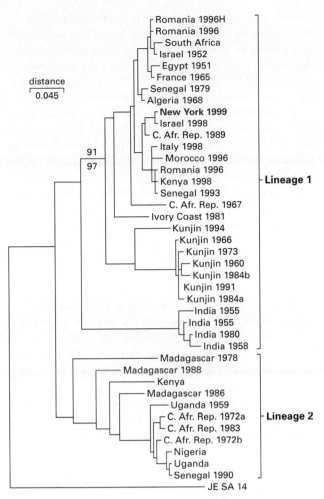

distance
0.045

91
97

Romania 1996H
Romania 1996
South Africa
Israel 1952
Egypt 1951
France 1965
Senegal 1979
Algeria 1968
New York 1999
Israel 1998
C. Afr. Rep. 1989
Italy 1998
Morocco 1996
Romania 1996
Kenya 1998
Senegal 1993
C. Afr. Rep. 1967
Ivory Coast 1981
Kunjin 1994
Kunjin 1966
Kunjin 1973
Kunjin 1960
Kunjin 1984b
Kunjin 1991
Kunjin 1984a
India 1955
India 1955
India 1980
India 1958
Madagascar 1978
Madagascar 1988
Kenya
Madagascar 1986
Uganda 1959
C. Afr. Rep. 1972a
C. Afr. Rep. 1983
C. Afr. Rep. 1972b
Nigeria
Uganda
Senegal 1990
JE SA 14

- Lineage 1

- Lineage 2

FIGURE 2.27 • Rooted phylogenetic tree of Nile viruses. Bootstrap value with 500 replicates (above) and confidence probability value (below) based on standard error are located at the branch between West Nile and Kunjin viruses.

If you are not from the New York area, this case study may seem too remote to be real. Public health officials took the West Nile virus very seriously and canceled summer concerts in Central Park during 1999 because they believed public gatherings might exacerbate the problem. They sprayed extensively to kill the mosquitoes, but bad news came when adult mosquitoes that had lived through the winter of 1999–2000 were found to contain the West Nile virus. In summer 2000, there were 20 human cases in New York and New Jersey, with one fatality. What precautions would you advise if you were in charge of public health?

Summary 2.2

DNA is a very stable polymer, which allows us to extract it from a wide range of sources. Soon, biological weapon identification will become reliable and quick, which will enable us to respond appropriately to hoaxes as well as real terror-

ism. At times, it seems that new diseases are evolving faster, but it may be that improving methods help us detect them better. Virus genomes can reveal the cause of diseases and potential treatments. Ancient DNA is a perplexing issue that is so fascinating we want to extract every ancient artifact to discover hidden information. The two cases in Section 2.2 illustrate how special caution must be used when working with ancient DNA. Biomedical research will lead to new methods for diagnosis, and as the technology improves, look for miniaturized DNA labs in your physician's office.

2.3 Biomedical Genome Research

"The Human Genome Project will change the way we practice medicine." This statement has been repeated in various ways since the HGP first captivated a small number of investigators as a realistic possibility. Now the human draft sequence is available, but where are the changes? Changes in medical treatment will take many years to develop, test, and bring to market. In most cases, however, genome sequences of pathogens, not humans, will prove to be easier to convert into medical improvements. This section relates two case studies about using pathogens to develop new treatments. A third case study shows how basic research, molecular methods, and knowledge of genome sequences might lead to one of the first examples of a new type of medication. Finally, we compare two *E. coli* genome sequences to gain insights into this pathogen. The *E. coli* case study illustrates how applied and basic research are intertwined and benefit one another.

Can We Use Genomic Sequences to Make New Vaccines?

Meningitis and sepsis involve inflammation of the outer lining of the brain (meningitis) or the blood (sepsis). If left untreated, they can rapidly progress to cause permanent nerve damage or death. The causative agents are often bacterial, so treatment is usually successful with antibiotics unless the pathogen is antibiotic resistant, in which case the infection is often fatal (about 10% of all cases are fatal, and about 25% of survivors have neurological damage). However, the initial symptoms are flu-like, and thus the severity of the infection usually is not appreciated until the immune response is vigorous and harmful. Therefore, vaccines for these types of infection are highly desirable.

A large percentage of meningitis and sepsis cases are caused by *Neisseria meningitidis,* which is a gram-negative bacterium classified into five major types based on antibody binding: **serotypes** A, B, C, Y, and W135. In the 1960s, vaccines were produced for serotypes A, C, Y, and W135 that work well in adults but have been less effective in young children and infants. Modern vaccines are under development that appear to be more effective in younger people, but no vaccine is available for serotype B. Serotype B is responsible for 45–80% of the *N. meningitidis* cases, depending on the country. The outer capsule of serotype B is made of a polysaccharide that is very similar to sugars found on human cells, which means the outer capsule polysaccharide would not be a good vaccine candidate. The outer surface of *N. meningitidis* contains a protein known to elicit a strong therapeutic immune response, but this protein is highly variable among different strains of serotype B and thus would not be a good candidate for vaccine production.

To address the need for an *N. meningitidis* vaccine, a group from TIGR and IRIS Chiron in Italy sequenced the 2,272,351 bp genome of *N. meningitidis* strain MC58 (serotype B) isolated from an infected patient. Using standard ORF-finding software and whole-genome comparisons with other bacteria, TIGR identified 570 ORFs that potentially encoded novel surface-exposed or secreted proteins (of 2,158 total ORFs). They isolated these ORFs and introduced them into *E. coli* for **overexpression.** Of these 570 ORFs, 350 produced protein, with 70 predicted to be lipoproteins, 96 periplasmic proteins, 87 inner membrane proteins, 45 outer membrane proteins, and 52 with uncertain location. Antibodies were produced that bound to the *E. coli*-produced candidate proteins. Enzyme-linked immunoabsorbent assay (ELISA) and fluorescence-activated cell sorting (FACS) were used to detect surface expression on a panel of diverse serotype B strains. In addition, a bacteriocidal test using serum was used to determine which proteins

elicited an immune response that might kill serotype B. From this series of assays, the group found 85 proteins that were positive in more than one assay. They selected for further testing 7 proteins (termed genome-derived Neisseria antigens [GNAs]) that were positive in all assays (representing 1.2% of the original 570 candidate proteins).

To determine the degree of conservation for the seven GNAs, they sequenced their orthologs in 31 *N. meningitidis* strains: 22 serotype B strains, 3 serotype A strains, 2 serotype C strains, and 1 strain each of serotypes X, Y, Z, and W135. In addition, they also sequenced orthologs of the seven GNAs in three different isolates of the closely related species that causes gonorrhoea, *N. gonorrhoeae* (Figure 2.28). From this analysis, they determined that GNA33, GNA1162, GNA1220, GNA1946, and GNA2001 had greater than 99% sequence conservation with the 31 *N. meningitides* strains, and all but GNA2001 showed greater than 95% sequence conservation with *N. gonorrhoeae.*

Of the seven GNAs, two were especially strong in FACS analysis when tested against both the original strain and a mixture of serotype B strains (Figure 2.29). Binding of the GNA33 and GNA1946 antisera were especially impressive when compared to the prototypical surface protein (called OMV), which works well against its original strain but not against a diverse mixture of strains. GNA33 is an enzyme critical in the formation of the bacterial cell wall, while GNA1946 is a lipoprotein in the outer membrane.

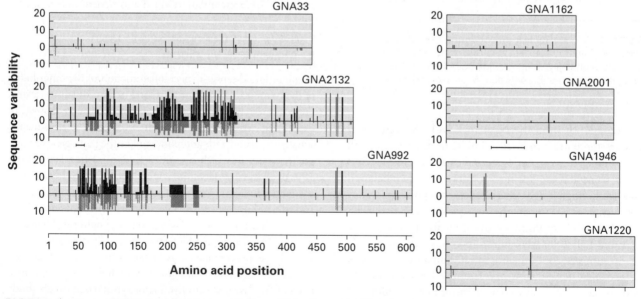

FIGURE 2.28 • **Amino acid sequence variation within *N. meningitidis* for the seven potential vaccine antigens.** X-axis is amino acid position and Y-axis is the number of strains analyzed. Line at zero represents the reference genome sequence. Amino acid differences within serotype B are indicated by blue bars above zero. Amino acid differences with the other serotypes are indicated by gray bars below zero. The height of the bars indicates the number of strains with differences. Horizontal bars below GNA2001 and GNA2132 represent DNA segments that are missing in some strains.

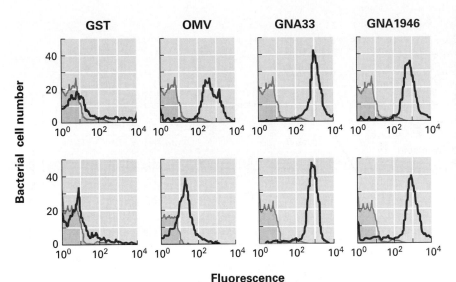

Fluorescence

FIGURE 2.29 • Two vaccine candidates. FACS analysis showing binding of polyclonal OMV, GNA33, and GNA1946 antisera to a uniform antigen (top row) and to a heterogeneous mixture of antigens (bottom row). Gray profiles show binding of negative control antisera; blue profiles show binding with experimental antisera. GST was used as the negative control antigen.

DISCOVERY QUESTIONS

31. The polysaccharide outer capsule of *N. meningitidis* looks like some of the polysaccharides on human cells. Why would this make the outer capsule a bad candidate for a vaccine?

32. Go to the Protein Data Bank site and search for the PDB ID code 1D0K (that is a zero, not the letter O). This is an *E. coli* lytic transglycosylase and closely related to GNA33. Click on "View Structure" and then on the next page, "Protein Explorer." Choose the version from the PDB server. You will see a rotating view and a page with three frames. In the top right frame, click on the "Explore 1D0K" link. Follow these directions:

   • Turn off the waters by clicking the "Water" button twice.
   • Click on the "Contact Surfaces" link.
   • Type in "protein" for the receptor and "ligand" for the ligand. Click on the "Zoom" button once and then the "Make" button once. It will take awhile, but don't click a second time.
   • Click and drag the structure around to find the five different ligands (N-acetylmuramic acid, D-glutamic acid, glycerol, and N-aceytl-D-glucosamine in the active site and a calcium ion on the opposite side). (*Advanced hint:* If you want to use the slab command, type "slab on" in the command line, which is the top box in the lower-right frame, to help you see inside the active site. You can turn off slab mode by typing "slab off" in the command line box.)

Here is the question. If you were to purify this type of protein for use as a vaccine, do you see any aspects that might prove problematic? Explain your answer.

LINKS
Protein Data Bank

The investigators started with whole-genome sequence information (2,158 total annotated ORFs) and selected two highly promising proteins to be used as vaccines against *N. meningitidis* serotype B that had resisted previous attempts. Of course, several years of testing and clinical trials will be required to confirm their utility and safety, but this work is a milestone. Genome sequence provided the key insight to a rational design of new vaccine candidates. As more pathogen genomes are sequenced, this approach will be repeated and refined until new and more protective vaccines are available to the public. However, the production of vaccines is relatively easy compared to the development of new antibiotics, which have additional design constraints, as we will see in the next section.

Can We Make New Types of Antibiotics?

The goal of a genomics approach to new antibiotics is to discover drugs that inactivate new targets. An ideal candidate is the outer surface, since this helps hide the bacteria from the immune system, blocks the entrance of many antibiotics, and is vital for bacterial survival. We need to understand a little about bacterial cell membranes and walls before we discuss the specifics of this case study. We will focus on **gram-negative bacteria,** which is the target of one new type of antibiotic.

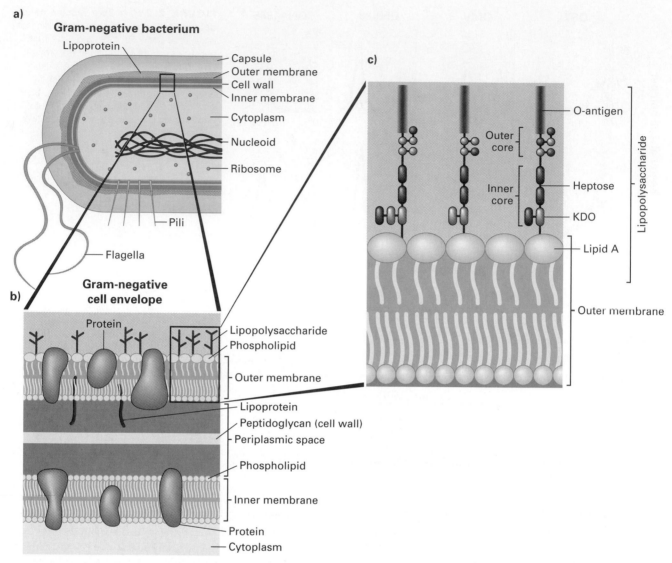

FIGURE 2.30 • **Schematics of gram-negative cell membranes and wall. a)** Half of a bacterium is shown for orientation. **b)** Magnification of one portion of the cell wall and membranes. **c)** Higher magnification for greater detail of the LPS with the different parts labeled.

LINKS

Christian Raetz

Gram-negative bacteria do not stain with a dye called Gram and thus they are called gram-negative. Gram-negative bacteria have several layers that make up their outer structure, as shown in Figure 2.30. There are two phospholipid bilayer membranes with a cell wall in between. On the outside of the outer membrane is a lipoprotein layer that contains proteins and lipids coated with different combinations of sugars (Figure 2.30c). Our story focuses on the lipid A portion of a lipopolysaccharide (LPS), which is a component of the outer membrane. The LPS layer is the key to the impermeability of gram-negative bacteria, and cells with poorly organized LPS are susceptible to immune system attacks. Excessive LPS is toxic when it overstimu-

lates the immune system and leads to septic shock and death. Furthermore, mutant cells that could not make LPS were not viable, so LPS is an ideal target for a new class of antibiotics.

Working with the gram-negative *E. coli* genome, a team led by Christian Raetz identified critical enzymes in the biosynthetic pathway of LPS. Raetz moved from the University of Wisconsin to the pharmaceutical giant Merck, where he continued his efforts to inactivate the pathway for LPS synthesis. Merck has millions of compounds in its chemical library, so they began a high throughput screening procedure to identify candidate drugs that could block LPS synthesis. In November 1996, Raetz and his colleagues published the results of their hunt. They identified a compound they call

a)

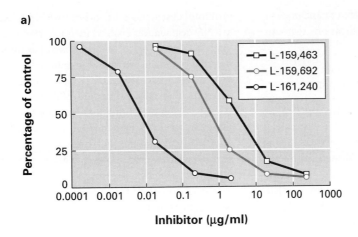

b)

L-161,240   L-159,692   L-159,463

FIGURE 2.31 • **Structures and effectiveness of new class of antibiotic. a)** Inhibition of
E. coli deacetylase in an in vitro assay. Activity of the enzyme is plotted on the Y-axis and
concentration of inhibitor on the X-axis. **b)** Three candidate drugs with differences highlighted
in blue.

L-161,240 and a variation of it called L-159,692, which
were able to block LPS production and cure mice infected
with *E. coli* (Figure 2.31). As predicted, these compounds
did not affect gram-positive bacteria or eukaryotic cells. Unfortunately, their compounds were not effective against two
other species of gram-negative bacteria.

The compound L-161,240 was about as effective as
ampicillin, rifampicin, and erythromycin against **wild-type**
E. coli. L-161,240 blocks a novel pathway and kills *E. coli*
cells in a matter of hours, so it looks like the ideal drug.
Many bacteria have evolved resistance to commonly used antibiotics and it is hoped L-161,240 might avoid neutraliza-
tion by the bacteria. Unfortunately, resistance to L-161,240
occurred at a frequency of about 1×10^{-9}, so L-161,240 is
not perfect. Cells that were resistant were able to synthesize
lipid A, so the target may still be viable. The candidate drug
may need some additional modifications to improve its ability to avoid resistance.

DISCOVERY QUESTIONS

33. What information would you like to collect in order to rationally design a variation of L-161,240
that could avoid resistance by *E. coli*? How might
you collect this information?

STRUCTURES
Penicillin
Inactivate Penicillin

34. Look at the structure of penicillin. It has a ring structure called a
β-lactam ring consisting of three
carbons and one nitrogen (click
on the button to highlight the ring). Two links
from the penicillin page show enzymes that inactivate penicillin. One cleaves the β-lactam
ring; the other adds a deactivating group onto
penicillin.
   a. Why do you think these two bacteria have
   evolved two different enzymes to inactivate
   penicillin?
   b. The fact that there are at least two different
   ways bacteria have evolved to inactivate penicillin tells us something about the selective
   pressure exerted by penicillin. What does this
   suggest to you about the possibility of creating
   an antibiotic that is resistance proof?

Raetz and his team have made a good start, and no doubt
they are testing more drugs that block LPS synthesis and are
not neutralized by antibiotic resistance. This case study illustrates how genomic information can stimulate the rational
design of new classes of antibiotics. By determining all the
genes in the lipid-A synthesis pathway, we should be able to

develop a series of antibiotics that block different enzymes in the pathway and produce a cocktail of medications that can kill even the most resistant bacteria.

Can We Invent New Types of Medication?

STRUCTURES
RNase P
tRNA

LINKS
Sidney Altman
Ribozyme
Pharmaceuticals

A new type of drug being developed by a small number of investigators is called **External Guide Sequence (EGS).** EGS was first described by Sidney Altman from Yale University, who was awarded a Nobel Prize for his research into RNA enzymes called **ribozymes.** Ribozymes are RNA molecules that can catalyze chemical reactions just as protein enzymes can. EGS is an offshoot of the ribozyme work and was first published in 1990. It has the potential to become a completely different type of pharmaceutical reagent.

EGSs are short oligoribonucleotides that are designed to base-pair with a target mRNA and lead to the destruction of the mRNA. This sounds similar to **antisense technology,** but it is significantly different for two reasons. Antisense technology relies upon RNase H to degrade the target mRNA, but EGS uses **RNase P** instead. EGS is more specific in its ability to initiate destruction of only the target mRNA and does not promote destruction of other mRNAs. RNase H–based mRNA degradation evolved in eukaryotes in response to viral infections, and thus its precision is not as important as killing the viral pathogen. RNase P is required to process precursor tRNAs, and its cleavage of tRNA precursors is much more precise than an antiviral assault. EGS works by binding to target mRNAs and forming a shape that resembles a precursor tRNA structure. When the cell detects an EGS-mRNA hybrid, RNase P cuts and inactivates the mRNA as if it were the normal substrate (Figure 2.32). This mechanism allows you to silence any gene by destroying all of its encoded mRNA.

Shaji George and his colleagues at Innovir Laboratories in New York (purchased in 2000 by Ribozyme Pharmaceuticals) have been experimenting with the minimum size for an effective EGS and possible chemical modifications for it. They demonstrated that an effective EGS could be reduced from the original 68 nucleotides to less than half that size. At 30 nucleotides, their shorter version was equally effective at degrading mRNA (Figure 2.33).

The next step, and probably the more important one, was to produce an EGS that could survive the onslaught of RNases present in every cell, in our blood, in our sweat, even dust. To address this problem, the group focused their attention on the sugar backbone of the EGS. DNA is much more stable than

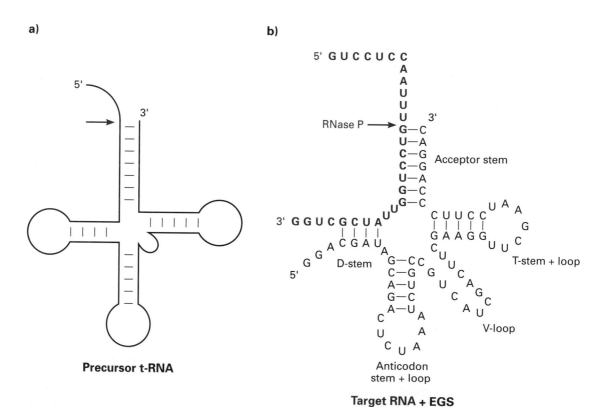

a) Precursor t-RNA

b) Target RNA + EGS

FIGURE 2.32 • **Two-dimensional representations of tRNA-like EGS structure.**
a) Schematic diagram of a precursor tRNA that is cleaved by RNase P (blue arrow) in vivo to produce a mature tRNA. **b)** EGS (black sequence) hybridized to RNA target (blue sequence). The EGS/RNA hybrid structure resembles *E. coli* tRNA^tyr. The RNase P cleavage site is indicated by the blue arrow.

a)

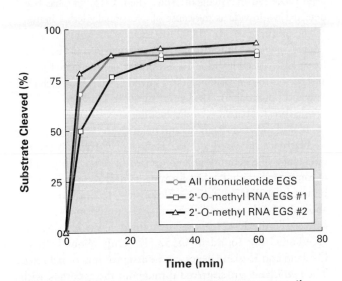

b)

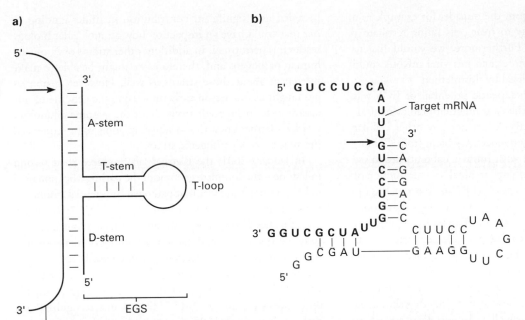

FIGURE 2.33 • The minimum EGS. a) Schematic showing the key parts in a minimized EGS, only 30 ribonucleotides long. **b)** 2D representations of the minimal EGS (black) bound to the same target sequence (blue sequence) as in Figure 2.32, which is cleaved about 61% as efficiently compared to full-length EGS-induced cleavage.

FIGURE 2.34 • A time course assay to measure the effectiveness of modified EGSs. The all-ribonucleotide EGS (gray line) was compared to two different EGSs that contained 2' O-methyl modified riboses outside the T-loop. The two experimental EGSs varied by one nucleotide in the loop; dUMP (blue line) and dTMP (black line).

RNA, as we saw in our study of ancient DNA. However, EGS made of DNA is ineffective, so they needed a modification that was easily synthesized by standard methods. They chose to add a methyl group to the 2' carbon of the ribose sugar. By replacing the hydrogen with a CH_3 on all of the EGS ribonucleotides except the loop portion, they produced an EGS that was stable for 24 hours in 50% human serum. Once they had a stable EGS, they tested its ability to cleave mRNA and found

it worked almost as well as the normal RNA version (Figure 2.34). By changing the purines to pyrimadines in the loop, they further increased the stability with no loss of activity. Unlike traditional antisense technology, EGSs are catalytic, which means they are not consumed in the cleavage reaction. The new RNase-resistant EGSs were able to catalyze about three cleavage reactions per molecule in vitro, which is only slightly worse than the normal RNA EGSs. In short, they had created a smaller and more stable EGS that worked almost as well as the original EGS. Now we need to determine if an EGS is capable of targeting only the mRNAs we want destroyed.

In 1998, Deborah Plehn-Dujowich and Sidney Altman wanted to determine if they could block the replication of influenza virus using EGS. They designed EGSs to bind to two different influenza transcripts, one coded for a viral RNA polymerase and the other for a viral nucleocapsid protein. First they tested the two targeted viral mRNAs to see which portions were accessible to RNase digestion. Then they designed several EGSs that would bind to the RNase-sensitive regions of the two viral mRNAs, and they tested them in vitro to see which worked best. They chose the best candidates and cloned the corresponding DNA into plasmids that would produce high levels of EGS RNA. These plasmids were incorporated into cultured cells that were later infected with influenza. Plehn-Dujowich and Altman were able to block viral replication by using two EGSs in a single cell. This demonstrated for the first time that EGSs could work on more than one mRNA in a single cell. The drawback was that the EGSs were produced inside the cells, not ideal for use in clinical settings. Ideally, we would add EGSs to cells

STRUCTURES
Ribose Sugar

METHODS
Liposomes

LINKS
Cy Stein
Fred Blattner
Pathogenic Strain

directly from the outside, for example as a nasal spray to treat cells lining respiratory passages. Furthermore, we would like to know if more than just viral mRNA could be neutralized by this method.

The therapeutic potential of EGS was assessed when a team attempted to neutralize one specific cellular mRNA but not any other cellular mRNAs with very similar sequences. Cy Stein from Columbia University collaborated with Innovir Laboratories to see if they could use EGS technology to block the production of the alpha form of protein kinase C (PKC-α) but not PKC-β or PKC-ζ. They designed several candidate EGSs that would bind to the 3' untranslated region of the PKC-α mRNA. Using **liposomes** to ferry the EGS inside cells, they compared a couple of different EGSs and determined that they had eliminated PKC-α but not PKC-β or PKC-ζ.

EGS therapy looks feasible and ideally suited for converting genome sequences into a new type of drug that should be able to combat pathogens as well as human diseases such as cancer. Many more years of research are needed, but EGS technology illustrates how basic research and genomics synergistically may provide new types of medications that could alleviate a lot of suffering.

DISCOVERY QUESTIONS

35. Go to NCBI's Entrez and search the protein database for "RNase P." Has this protein been found in many different species or just a few?

36. Antibiotics often fail in the long run because pathogens evolve mechanisms to evade them. What do you think about the ability of pathogens to evade EGSs?

37. If you wanted to generate an EGS that was specific to a particular mRNA, what aspect of the gene pool might make this more difficult? How could you address this problem so each person could still benefit from EGS therapy?

How Can *E. coli* Be Lethal and in Our Intestines at the Same Time?

It has been said that every molecular biologist is interested in two organisms, the one he or she studies and *E. coli*. Laboratory strains of *E. coli* enable researchers to clone DNA, express proteins, and isolate genes of special interest; without it labs cannot function. In 1997, the genome of a nonpathogenic laboratory strain of *E. coli* (called K-12) was published. The 4.6 Mb genome was sequenced in the laboratory of Fred Blattner at the University of Wisconsin. This milestone was

heralded as a significant contribution in understanding *E. coli* that could have an impact on how all molecular biology research is performed. In addition, other strains of *E. coli* are human pathogens and, therefore, we might be able to make inferences about those strains as well. However, genomics has taught us one lesson early on—using the genome of one organism to make predictions about the genome of another is risky. Blattner knew this as well as anyone, so he sequenced the genome of a pathogenic strain.

In January 2001, the Blattner lab produced their second milestone—the complete genome of the **pathogenic** (potential to harm its host) and **virulent** pathogen (substantial amount of harm or death to host) *E. coli* strain called **O157:H7.** Every one of us has *E. coli* in our intestinal tracts, and our intestinal strain occupies space and consumes limiting nutrients such as iron that help us resist infection by pathogenic bacteria. If you consume some undercooked beef that is contaminated with O157:H7, you might have the misfortune of experiencing the differences in genomes. O157:H7 was first recognized as a pathogenic strain in 1983 after an outbreak the previous year that originated from undercooked hamburger meat that had been processed in Michigan. Since then, O157:H7 has been detected worldwide as the cause of occasionally lethal hemorrhagic colitis, of which 75,000 new cases occur annually in the United States. In addition to transmission via infected beef, raw fruits and vegetables (due to fertilizers), contaminated water and person-to-person contact can spread O157:H7, which creates the potential for epidemics in some locations such as flooded areas. Therefore, understanding the O157:H7 genome could lead to better diagnosis and treatment.

Blattner's lab used a 1982 isolate of O157:H7 for sequencing. The genome was composed of a 5.5 Mb circular chromosome, with 4.1 Mb forming a "backbone" conserved between O157:H7 and K-12. However, O157:H7 contained about 1.34 Mb termed "O-islands" not found in K-12, and conversely, the K-12 genome contained about 0.53 Mb of "K-islands" not found in O157:H7. Only about half the O-island and K-island genes can be assigned functional roles. The two islands are scattered throughout the genomes, with O-islands containing 1,387 genes (out of 5,416 total genes) not found in K-12, and K-islands containing 528 genes not found in O157:H7. Many of the O157:H7-unique genes are predicted to be virulence genes since they encode toxins, metabolic pathways, transporters, and adhesion molecules. Ironically, K-12 also contains genes that could be predicted to be virulence genes, and yet it is not a human pathogen. This may indicate that K-12 could be pathogenic in a nonhuman host, that K-12 is evolving into a human pathogen, or that these genes may have no functional relationship with pathogenicity. To distinguish between these possibilities will require many years of laboratory research. Deciphering what these genes do will help us understand which ones enable O157:H7 to be virulent and K-12 to be a docile lab strain.

MATH MINUTE 2.3 • HOW CAN YOU TELL IF BASE COMPOSITIONS ARE DIFFERENT?

To say that the ratios of the four bases in the O-islands and K-islands are different from the backbone ratio, you must do some statistical analysis. You really want to be able to say they are *significantly* different. The standard tool for testing whether two sets of ratios, or frequencies, are significantly different is the chi-square test of homogeneity.

To illustrate how the chi-square test is performed, suppose that you had a 4 kb sequence containing 1,000 bases of each type, and a 3 kb sequence containing 600 A's, 800 C's, 700 G's, and 900 T's. Those data are summarized in the following table.

Base	Sequence 1	Sequence 2	Total
A	1,000	600	1,600
C	1,000	800	1,800
G	1,000	700	1,700
T	1,000	900	1,900
Total	4,000	3,000	7,000

For each of the eight cells in the interior of the table, compute the expected frequency of that cell as the cell's row total times the cell's column total divided by the grand total. For example, the expected frequency for the cell in the A row and the Sequence 1 column is $1,600 \times 4,000/7,000 \approx 914.29$. Subtract the expected frequency from the actual frequency ($1,000 - 914.29 \approx 85.71$), square the result ($85.71^2 \approx 7,346.2$), and divide this number by the expected frequency ($7,346.2/914.29 \approx 8.03$). Repeat this process for each of the eight cells, and sum the results. If this sum is larger than a cutoff value from the chi-square distribution, then the frequencies are said to be significantly different.

This is the math behind the scenes of the simple statement that the ratios of the four bases in most O-islands and K-islands are different from the ratios in the backbone. The investigators performed the chi-square test for 108 O-islands longer than 1 kb, and found 101 of them to have a significantly different base composition than the backbone.

One of the most striking features of O-islands and K-islands is their base compositions. Most of them do not have the same ratios of the four bases as found in the backbone. Furthermore, many of the genes have orthologs found in other species and in viruses. Therefore, it appears that large portions of these genomes may be the result of horizontal transfer from other organisms. Significant differences in these two genomes begs the question, How do we define "species" and "strain"? Genome sequences are forcing us to reconsider old definitions that seemed clear just a few years ago.

Since about 75% of O157:H7 and K-12 genomes are conserved in the backbone, you might predict that most of the genes in the backbone would encode identical proteins. Interestingly, only about 25% (911/3,574) of the backbone genes produce identical proteins, with the remaining proteins varying by at least one amino acid. This may sound like a trivial difference, but remember that *wt* versions of hemoglobin and cystic fibrosis protein (CFTR) differ from disease-causing versions by single amino acids. The difference in pathogenicity between O157:H7 and K-12 may be due to these subtle nucleotide variations.

The genome for O157:H7 had two gaps in the sequence database (4 kb and 54 kb) that were filled in after the publication of the genome sequence. Gap filling is often perceived

LINKS

NCBI's Microbial Genome Site

Genome of HIV

as uninteresting and not worthy of funding, but genomes can reveal information that can be understood only in the context of a completed sequence. Although these gaps are filled, the task is not completed. Various isolates of O157:H7 have been shown to have different restriction maps, and thus different genomes. These variations might correlate with disease phenotypes that could help us characterize the many genes with unknown functions. Nevertheless, the O157:H7 genome sequence is a good starting place for clinical uses such as diagnosis, vaccine production, and treatments. Since both *E. coli* genomes are in the public domain, anyone who wants to work in this area of biomedical research may do so.

DISCOVERY QUESTIONS

38. Go to TIGR's CMR and align K-12 and O157:H7 whole genomes. Are the backbones colinear their entire lengths?

39. Go to NCBI's microbial genome site and click on a name for one of the *E. coli* O157:H7 variants. Use the gene search field to determine if these pathogenic strains synthesize lipid A (the membrane protein targeted by a new class of antibiotics; see p. 59–62). Here is a list of genes that encode enzymes involved in lipid A synthesis: *envA, lpxA, lpxB, lpxD, lpC, firA, ssc,* and *omsA.* Could the lipid A antibiotic kill the strain you chose?

40. Since we need some *E. coli* to keep our intestines protected from virulent strains, traditional antibiotics are not ideal therapies. Would it be possible to devise a new class of antibiotics to target only O157:H7 and not K-12? Explain how you might develop this type of drug.

Now that we have sequenced the genomes of two strains of *E. coli,* we are hopeful that biomedical research will lead to better human health. However, easy fixes are unlikely. For example, the HIV genome was sequenced in 1985, and it contains only nine genes. With such a simple organism and a simplistic genome, you would think we could use genomic methods to develop a cure and vaccine very quickly. As you know, there is no cure for AIDS, and the only prevention is through behavioral modification. Look at the genome of HIV and you will see that it does not look challenging compared to the genomes we have considered in this chapter. Size can be intimidating, but the complexity of biology is difficult to understand fully.

Summary 2.3

Biomedical research may be the best-funded area of all the sciences. Many of the genomes that have been sequenced have direct medical utility. We can use genomes to accelerate the development of new vaccines and antibiotics. However, insights gained through basic research using molecular methods have led to the development of a new type of drug—EGS. When combined with genome sequences, we hope EGSs will target a wide range of diseases caused by pathogenic infections and genetic variations in our own genomes. The specificity of EGSs may allow individualized medications to target specific alleles within the human gene pool. As with all research, unexpected discoveries found in genome sequences can lead to new questions as well as immediate answers. We have seen genomic differences among different strains of *E. coli,* but unexpectedly we discovered that the differences between two strains within a species may vary more than the genomes of two different species. The entire phylogenetic tree of life may need substantial revisions as we refine definitions for "species" and "strains."

Chapter 2 Conclusions

In addition to new methods and vast amounts of data, whole-genome perspectives allow us to detect new patterns and connections that had been impossible to see. Genes come and go from genomes, and the origins of some genes may be from evolutionarily distant species. We hope to use sequences to develop better ways to diagnose, prevent, and treat diseases. Learning how to use this information will take a lot of time, money, and effort from people with diverse skills and perspectives. We can use genome sequences to look into the past or make predictions about what roles are performed by different parts of the genome. The completion of a genome sequencing project is no more the end of the story than birth is the end of a person's development. Billions of years of evolution are exposed by completely sequencing a **reference genome** (prototype sequence for a species). In Chapter 3, we examine case studies that focus on variation found among the genomes of individuals within every species. Deviations from a reference genome are the raw material for variations we see in any population, including humans. Why do individuals respond differently to the same pathogen or medical treatment? Variation must be taken into consideration when genomic insights are translated into medical treatments.

References

Evolution

Adcock, Gregory J., Elizabeth S. Dennis, et al. 2001. Mitochondrial DNA sequences in ancient Australians: Implications for modern human origins. *PNAS.* 98: 537–542.

Baliga, N. S., Y. A. Goo, et al. 2000. Is gene expression in Halobacterium NRC-1 regulated by multiple TBP and TFB transcription factors? *Molecular Microbiology.* 36(5): 1184–1185.

Cole, Stewart T., et al. 2001. Massive gene decay in the leprosy bacillus. *Nature.* 409: 1007–1011.

Horiike, Tokumassa, Kazuo Hamada, et al. 2001. Origin of eukaryotic cell nuclei by symbiosis of Archaea in bacteria is revealed by homology-hit analysis. *Nature Cell Biology.* 3: 210–214.

Hutchison, Clyde A. III, Scott N. Peterson, et al. 1999. Global transposon mutagenesis and a minimal *Mycoplasma* genome. *Science.* 286: 2165–2169.

Ingman, Max, Henrik Kaessmann, et al. 2000. Mitochondrial genome variation and the origin of modern humans. *Nature.* 408: 708–713.

Kaessmann, Henrik, Victor Wiebe, et al. 2001. Great ape DNA sequences reveal a reduced diversity and an expansion in humans. *Nature Genetics.* 27: 155–156.

Leakey, M. G., et al. 2001. New hominid genus from eastern Africa shows diverse middle Pliocene lineages. *Nature.* 410: 433–440.

Madsen, Ole, Mark Scally, et al. 2001. Parallel adaptive radiations in two major clades of placental mammals. *Nature.* 409: 610–614.

Nester, Eugene W., Denise G. Anderson, et al. 2001. *Microbiology: A Human Perspective.* Boston: McGraw-Hill. 504.

Ng, W. V., et al. 2000. Genome sequence of *Halobacterium* species NRC-1. *PNAS USA.* 97: 12176–12181.

Shigenobu, Shuji, Hidemi Watanabe, et al. 2000. Genome sequence of the endocellular bacterial symbiont of aphids Buchnera sp. APS. *Nature.* 407: 81–86.

Underhill, Peter A., et al. 2000. Y chromosome sequence variation and the history of human populations. 26: 358–361.

Review and Summary Sources
Clarke, Tom. 2001. Biotechnology: A genetic outcast. *Nature Science Update* (online) <http://www.nature.com/nsu/010301/010301-2.html>. Accessed 27 February 2002.

Hagmann, Michael. 2000. Leprosy's dying genome. *Science.* 288: 800–801.

Hedges, S. Blair. 2000. A start for population genomics. *Nature.* 408: 652–653.

Renfrew, Colin, Peter Forster, and Matthew Hurles. 2000. The past within us. *Nature Genetics.* 26: 253–254.

Identifications

Associated Press. 2001. Canadian office workers quarantined. New York Times on the web. 22 March 2001. <www.nytimes.com/aponline/world/AP-Canada-Biological Scare.html>.

Belgrader, Phillip, William Bennett, et al. 1999. PCR detection of bacteria in seven minutes. *Science.* 284: 449–450.

Burns, Mark A., Brian N. Johnson, et al. 1998. An integrated nanoliter DNA analysis device. *Science.* 282: 484–487.

Cole, Stewart. T., et al. 1998. Deciphering the biology of *Mycobacterium tuberculosis* from the complete genome sequence. *Nature.* 393: 537–544.

Cook, Suzanne M., Rene E. Bartos, et al. 1994. Detecting and characterization of atypical Mycobacteria by the polymerase chain reaction. *Diagnostic Molecular Pathology.* 3(1): 53–58.

Enserink, Martin. 2001. Biodefense hampered by inadequate tests. *Science.* 294: 1266–1267.

Koop, Martin U., Andrew J. de Mello, and Andreas Manz. 1998. Chemical amplification: Continuous-flow PCR on a chip. *Science.* 280: 1046–1048.

Nerlich, Andreas G., Christian J. Haas, et al. 1997. *The Lancet.* 350:1404.

Salo, Wilmar L., Arthur C. Aufderheide, et al. 1994. Identification of *Mycobacterium tuberculosis* DNA in a pre-Columbian Peruvian mummy. *PNAS.* 91: 2091–2094.

Stead, William W., Kathleen D. Eisenbach, et al. 1995. When did *Mycobacterium tuberculosis* infection first occur in the New World? *American Journal of Respiration and Critical Care Medicine.* 151: 1267–1268.

Thierry, Dominique, Corinne Chureau, et al. 1992. The detection of *Mycobacterium tuberculosis* in uncultured clinical specimens using the polymerase chain reaction and a non-radioactive DNA probe. *Molecular and Cellular Probes.* 6: 181–191.

Vreeland, Russell H., William D. Rosenzweig, and Dennis W. Powers. 2000. Isolation of a 250 million-year-old halotolerant bacterium from a primary salt crystal. *Nature.* 407: 897–900.

Williams, S. A., M. R. Lizotte-Waniewski, et al. 2000. The filarial genome project: Analysis of the nuclear, mitochondrial and endosymbiont genomes of *Brugia malayi. International Journal for Parasitology.* 30: 411–419.

Review and Summary Sources
Pennisi, Elizabeth. 1999. Bacterial partners for Filaria (News Focus section). *Science.* 283: 1105.

Biomedical Research

Anderson, John F., et al. 1999. Isolation of West Nile Virus from mosquitoes, crows, and a Cooper's hawk in Connecticut. *Science.* 286: 2331–2332.

Blattner, Frederick R., et al. The complete genome sequence of Escherichia coli K-12. *Science.* 277: 1453–1474.

Cann, Alan J. 2001. Malaria. <http://www.wehi.edu.au/MalDB www/intro.html>. Accessed 5 April 2002.

Fraser, Claire M., Jonathan Eisen, et al. 2000. Comparative genomics and understanding microbial biology. *Genomics.* 6(5): 505–512.

Glass, John I., Elliot J. Lefkowitz, et al. 2000. The complete sequence of the mucosal pathogen Ureaplasma urealyticum. *Nature.* 407: 757–762.

Lanciotti, Robert S., et al. 1999. Origin of the West Nile Virus responsible for an outbreak of encephalitis in the northeastern United States. *Science.* 286: 2333–2336.

Ma, Michael Y.-X., Biji Jacob-Samuel, et al. 1998. Nuclease-resistant external guide sequence-induced cleavage of target RNA by human ribonuclease P. *Antisense and Nucleic Acid Drug Development.* 8: 415–426.

Ma, Michael, Lyuba Benimetskaya, et al. 2000. Intracellular mRNA cleavage induced through activation of RNase P by nuclease-resistant external guide sequences. *Nature Biotechnology.* 18: 58–61.

Plehn-Dujowich, Debora and Sidney Altman. 1998. Effective inhibition of influenza virus production in cultured cells by external guide sequences and ribonuclease P. *PNAS USA.* 95: 7327–7332.

Muesing M. A., D. H. Smith, et al. 1985. Nucleic acid structure and expression of the human AIDS/lymphadenopathy retrovirus. *Nature.* 313: 450–458.

Onishi, H. Russell, Barbara A. Pelak, et al. 1996. Antibacterial agents that inhibit lipid A biosynthesis. *Science.* 274: 980–982.

Perna, Nicole T., et al. 2001. Genome sequence of enterohaemorrhagic Escherichia coli O157:H7. *Nature.* 409: 529–533.

Pizza, Mariagrazia, et al. 2000. Identification of vaccine candidates against serogroup B Meningococcus by whole-genome sequencing. *Science.* 287: 1816–1820.

Richman, Laura K., et al. 1999. Novel endotheliotropic herpesviruses fatal for Asian and African elephants. *Science.* 283: 1171–1176.

Salton, Milton R. J., and Kwang-Shin Kim. 2001. Structure Chapter 2 from Medmicro online textbook. <http://gsbs.utmb.edu/microbook/ch002.htm>. Accessed 5 April 2002.

Tettelin, Hervé, et al. 2000. Complete genome sequence of Neisseria meningitidis serogroup B strain MC58. *Science.* 287: 1809–1815.

The Institute for Genomic Research. 2001. *The Comprehensive Microbial Resource (CMR).* <http://www.tigr.org/CMR2/New Users.shtml>. Accessed 6 April 2002.

Weinstock, George M. 2000. Genomics and bacterial pathogenesis. *Genomics.* 6(5): 496–504.

Werner, Martina, Eddie Rosa, et al. 1999. Targeted cleavage of RNA molecules by human RNase P using minimized external guide sequences. *Antisense and Nucleic Acid Drug Development.* 9: 81–88.

Summary and Review Articles
Eisen, Jonathan. 2001. Gastrogenomics. *Nature.* Vol. 409: 463–466.

Goodman, Billy. Genomic strategies target bacteria. *The Scientist.* May 1, 2000. P 1, 14.

Rosamond, John, and Aileen Allsop. 2000. Harnessing the power of the genome in the search for new antibiotics. *Science.* 287: 1973–1976.

Vaara, Martii. 1996. Lipid A: Target for antibacterial drugs. *Science.* 274: 939–940.

CHAPTER 3

Genomic

Variations

Goals for Chapter 3

3.1   Environmental Case Study

Integrate genomic variations with ecological adaptations.

3.2   Human Genomic Variation

Understand human single nucleotide polymorphisms (SNPs) and mine SNP databases.

Discover how SNPs can cause diseases.

Examine how SNPs can affect medical therapies.

3.3   The Ultimate Genomic Phenotype—Death?

Explore the role of genomic variations in the aging process.

Recognize the trade-offs associated with genomic variations.

3.4   Ethical Consequences of Genomic Variations

Debate the potential harm vs. the potential benefit of genetically modified organisms.

Consider when genomic testing is beneficial.

Question the appropriate limits for genomic manipulations.

In previous chapters, we discussed genomic sequences as if there were one for each species. "*The* genome sequence" for any species is a **reference sequence;** the consequences of variation in a species' genome are often overlooked. Chapter 3 focuses on variations within genomes and how they affect species within complex ecosystems as well as an individual's life span. Finally, as we begin to understand the role of genomic variations, it is important to consider the ethical consequences of variations, including those engineered by humans.

METHODS

Microsatellites

PCR

Capillary
Electrophoresis

LINKS

David Mann

420,000 Years

Ginger Armbrust

3.1 Environmental Case Study

In this section, we examine one of many ecological examples of genomic variations in a species with an unsequenced genome. The case involves some of the ocean's smallest photosynthetic organisms. How can a mitotically dividing population maintain genomic variation, and what impact might this have on the accumulation of global CO_2 levels and global warming? The data that result from exploring these questions offer a cautionary warning about our ability to manage global climate.

Can Genomic Diversity Affect Global Warming?

In 2001, two groups attempted to catalog the number of species of small eukaryotes (< 5 μm in diameter). Based on ribosomal gene sequences, many new species from diverse taxa were present in samples taken at a single location on a single day. One type of **phytoplankton** (photosynthetic plankton), the diatom, includes as many as 200,000 different species that can reproduce asexually and sexually. Diatoms are found in all natural collections of water, fresh and salt. David Mann, from the Royal Botanic Garden in Edinburgh, Scotland, estimated that diatoms are responsible for fixing more CO_2 than all the rain forests combined; others estimate diatoms are responsible for as much as 50% of the global carbon fixation. Given that CO_2 levels are higher than they have been for at least 420,000 years, it will be important that we understand the full carbon cycle.

In the mid-1990s, oceanographers seeded a small patch of equatorial Pacific Ocean with iron, a limiting nutrient. The phytoplankton community responded with increased density and a dramatic change in species composition. Similar experiments were initiated in 2001 near Antarctica. This work has spawned serious proposals to use phytoplankton iron-seeding to increase fish production and reduce global levels of CO_2. Since we understand so little of the complexity in the ocean, there is a critical need to know more about diatoms and other phytoplankton. Surprising to most people, genomic tools and insights may prove to be critical to understanding global warming.

For many years, it had been assumed that diatoms succeed in a wide range of ecological niches because of their inter- and intraspecific physiological diversity. Individuals isolated from a single species can vary as much as 60-fold in their nutrient uptake rates, photosynthetic capabilities, and overall growth rates. We assumed this phenotypic diversity was due to a diversity in genotypes, but the supporting data were lacking. Ecologically important species are rarely considered for whole-genome sequencing so other methods are required to detect genomic diversity. To address this need, Tatiana Rynearson and her thesis advisor, Ginger Armbrust, at the University of Washington's School of Oceanography, used **microsatellite** loci in one species of diatom *(Ditylum brightwellii)* to document genome variation. Microsatellites are genomic regions of variability between individuals in a species and can be detected using **polymerase chain reaction (PCR)** and **capillary electrophoresis.**

Rynearson and Armbrust collected samples at three places in the Hood Canal near Puget Sound (Figure 3.1). This body of water is relatively stable and therefore should have a minimum of mixing (vertical and horizontal) of different water masses and introduction of new genotypes. All samples were taken from the same depth on 21 November 1997 using a 20 μm mesh net. Individual diatom cells were isolated within 24 hours of the initial collection, maintained sterilely in culture, and grown under constant laboratory conditions for DNA isolation and physiological analysis.

The first step was to define microsatellites for *D. Brightwellii.* Nine were cloned and sequenced, with locus Dbr4 used extensively (Figure 3.2). Dbr4 is a compound repeat (a mixture of GA and GT repeats) interrupted by nonrepeat nucleotides. Using one pair of PCR primers, Rynearson and Armbrust defined the genotypes of 23 different clonal isolates from their samples.

Analyzing the PCR products, the investigators identified ten unique genotypes that contained eight different Dbr4 alleles (Table 3.1, page 72). Allele 261 was the most abundant, and there were nine independent isolates that were homozygous for it. As often happens in research, they had some unusual but beneficial results. PCR of Dbr4 often produced three major peaks instead of the expected two (Figure 3.3, page 72). The PCR products were cloned and sequenced, revealing that the primers were amplifying two loci simultaneously. By slightly redesigning the primers, they were able to distinguish the two loci, increasing their ability to resolve genomic differences among the different isolates.

Using the two microsatellite loci, they identified 23 unique genotypes from 24 different isolates and calculated the observed heterozygosity (H_o; number of

a)

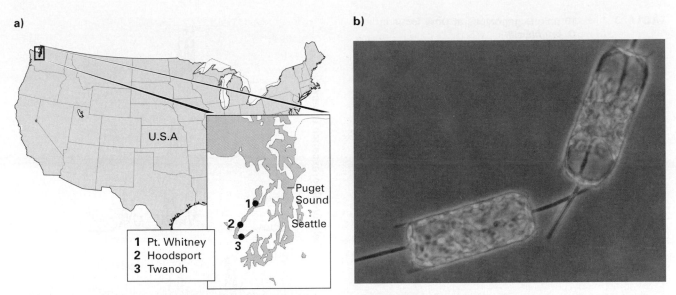

FIGURE 3.1 • Diatom study organism. a) Map of Washington state and Puget Sound in expanded view. Three sample collection sites are noted by numbers 1, 2, and 3. **b)** Light micrographs of two *Ditylum brightwellii* cells.

a)

```
  1   aaaacacaat  atgcattatt  gcgcaataac  acatatcaaa  tattgtccct  ataacaaaca
 61   tttttacaag  ctgtataatg  ctttaaattg  taacatatgt  ttacaaagag  acgcaatcta
121   acaaatgaac  atagctcctg  aGTGTGTGTG  TGTGTGTGTG CGCGCGCGCA  CGTGTGTGTG
181   TCTGTTTGTC  TGTGTGTCTG  TGCGTCTGTG  TCTGTGCGTG  TGTGTGTGCG  TGTGTGTCTG
241   TGTCTGTGTC  TGTGTGTGTG  TGTGTCTGTG  TCTGTGTCTG  TGTCTGTGTG  TGTGTCTGTA
301   TGTGTATGTC  TGTGTCTGTC  TGTGTCTGTG  CGTGTTTGTG  TGTCTGTGTT  TGTGTGTatc
361   cttgggaaaa  tt
```

b)

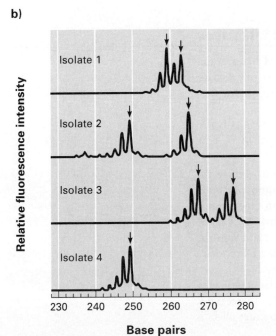

FIGURE 3.2 • Microsatellite Dbr4. a) Dbr4 sequence; location of PCR primer sequences are in blue; GT dinucleotide is underlined once; GC dinucleotide is underlined twice; (CT(GT)₂)₂ is underlined as shown; ((CTGTCT(GT)₂)₂ is underlined as shown. The reference repeat unit is 216 bp long and shown in all capital letters. Accession number AF263004. **b)** Electropherograms of Dbr4 alleles amplified from four different isolates using specific primers. Arrows indicate the allele sizes with smaller stutter bands also visible.

TABLE 3.1 • **10 unique genotypes at Dbr4 locus in *D. brightwellii*.**

Genotype*		Number of Isolates
243	243	1
255	259	1
257	257	1
257	259	1
259	259	1
261	261	9
261	263	3
261	267	1
263	263	4
263	271	1

*Allele numbers refer to bp lengths of microsatellite.
Source: Rynearson and Armbrust. 2000. *Limnology and Oceanography.* 45(6): 1334.

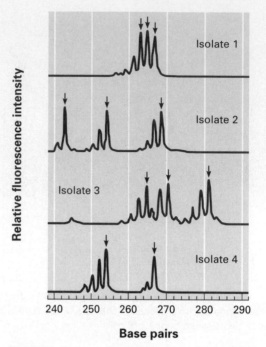

FIGURE 3.3 • **Electropherograms of microsatellite alleles from two loci amplified with less specific PCR primers.** Arrows indicate the allele size with smaller stutter bands also visible.

heterozygotes divided by the total number of samples) and the gene diversity (H_e; reflects the extent of diversity at individual loci). If all individuals in a population were mitotic siblings, H_e would be equal to zero; if all were unrelated, H_e would be very close to one. When examining only the original Dbr4 locus, H_o was 0.30 and H_e was 0.69. When the pair of loci were analyzed collectively, H_e was 0.88. A high H_e indicated that the population was very heterogeneous and that most individuals were not mitotic siblings even though the isolates were reproducing by mitosis. The Dbr4 locus H_o and H_e values were significantly different from each other (*p-value* < 0.0001), indicating that *D. brightwellii* were reproducing asexually, since there were more homozygotes than one would expect if these individuals were reproducing sexually. Consequently, it appears that the 24 diatom

isolates were products of mitotic reproduction, and yet the two-locus H_e value indicated most of the isolates were unrelated to each other. The data revealed a surprisingly high degree of genomic variation in a mitotically dividing population, prompting the question, "How can a population of asexual organisms maintain a high degree of genomic diversity?"

MATH MINUTE 3.1 • HOW DO YOU MEASURE GENETIC VARIATION?

Observed heterozygosity (H_o) and gene diversity (H_e; also called expected heterozygosity) are two measures of genetic variation within a population. H_o is easier to calculate, but H_e is more appropriate for inbred or asexually reproducing populations such as *D. brightwellii*.

To calculate H_o for the Dbr4 locus from the data in Table 3.1, divide the number of heterozygous isolates (7) by the total number of isolates (23) to get $H_o = 7/23 \approx 0.30$. This means that 70% of all isolates are homozygotes.

To calculate H_e for the Dbr4 locus, you must first find the frequency of each of the eight unique alleles that appear in Table 3.2. An allele frequency is the number of times the allele occurs in the 23 isolates, divided by the total number of chromosomes among the 23 isolates (46). For example, allele 261 has a frequency of $[(9 \times 2) + (3 \times 1) + (1 \times 1)]/46 = 22/46 \approx 0.48$. Square each of the eight allele frequencies, and sum the resulting eight numbers. Subtract this sum from 1 to get $H_e \approx 0.69$ for the Dbr4 locus. A similar calculation leads to the two-locus (Dbr4 and Dbr9) H_e of 0.88.

If only one allele is represented in the population, $H_e = 0$. If two alleles appear with equal frequency, $H_e = \frac{1}{2}$. This is the largest H_e can be for two alleles. To see why, suppose that the frequency of one allele is p. Then the frequency of the other allele is

$1 - p$, and using the formula described above, $H_e = 1 - [p^2 + (1 - p)^2]$. To find the maximum value of H_e, first take the derivative of this function of p, set it equal to 0, and solve for p:

$$-2p - 2(1 - p)(-1) = 0,$$

giving $p = \frac{1}{2}$. Then plug $p = \frac{1}{2}$ into H_e:

$$1 - [{\tfrac{1}{2}}^2 + (1 - \tfrac{1}{2})^2] = 1 - [\tfrac{1}{4} + \tfrac{1}{4}] = \tfrac{1}{2}.$$

More advanced calculus shows that the maximum value of H_e for n alleles occurs when all alleles have the same frequency (i.e., $1/n$). In this case, the maximum value of H_e is $(n - 1)/n$, which asymptotically approaches 1 as n increases.

By calculating H_o (0.3), the investigators determined the cells were reproducing asexually. However, calculating H_e (0.88) revealed that the diatom population was genetically diverse. These two calculations encouraged the investigators to test experimentally how diatoms maintain genomic diversity while reproducing asexually.

Trying to understand the genotype variation found in these diatoms brings us back to phenotypic diversity. It has been known for many years that individual cells had different physiological capacities, and now we see the genomic variability is high as well. Rynearson and Armbrust decided to test eight of their isolates for their ability to grow at three different light levels while maintained in otherwise equal growth conditions (Figure 3.4). They chose to test high, medium, and low light intensities since individual cells would experience similar differences as they drifted up and down in the water column. At first glance, the isolates may appear to be very similar, but notice the differences in growth rates and the small error bars for each light treatment.

As you would expect, all the isolates grew faster with more light. However, there are significant differences among them.

For example, at 166 μmol, isolates 2 and 8 grew about 20% slower than the others (p-value < 0.01). At 66 μmol, isolate 1 grew the fastest and isolate 2 grew the slowest with a difference of about 26% (p-value < 0.001). Three different growth rates (p-value < 0.01) were measured at 33 μmol. Isolates 1, 6, and 8 were the fastest; 3, 4, and 7 were intermediate; and 2 and 5 were the slowest, with a 33% slower rate compared to the fastest growers. What was especially interesting was that the growth rate at one light level was not a predictor for the growth rate at the other two light levels. Therefore, genomic diversity appears to be the cause of the difference in physiology, and the response to light intensity is probably a multilocus trait since there was no correlation in growth rates among the three light levels. This finding led to the testable prediction that the population must be under changing selection pressures (more than

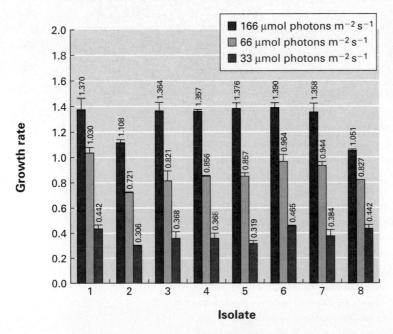

FIGURE 3.4 • Growth rates of genetically distinct isolates. Cells were grown at 166, 66, or 33 micromoles of photons per square meter per second in culture flasks maintained inside growth chambers. The numbers at the top of the bars are the mean growth rates. Error bars indicate the standard error of the mean growth rate.

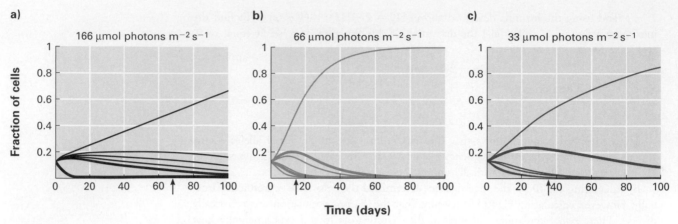

FIGURE 3.5 • **Diatom population simulations.** Computer simulations of changes in genomic diversity over time based on the mean growth rates of isolates maintained at 166, 66, and 33 μmol photons m^{-2} s^{-1}. Individual curves represent the relative abundance of each isolate. Arrows indicate the times required for the dominant strain to represent 50% of the population.

METHODS

Modeling

just changes in light intensities), which would explain the genomic variation observed in the 24 isolates.

The power of models is that they allow us to make predictions. Using their laboratory growth rates, Rynearson and Armbrust simulated the growth rates for all eight isolates if an equal number of cells were incubated in the same flask under three different light conditions (Figure 3.5). In each condition, a single isolate dominated the population. The time it took the dominant isolate to represent 50% of all the cells varied, as you would expect since the "losing" isolates had different relative growth rates for each light treatment. If the natural selection taking place in Hood Canal were due to a single and constant factor, you would predict that a population would decrease in genomic diversity and become more homogeneous. Based on

the investigators' analysis of the samples collected in Hood Canal, there must be multiple dynamic selection pressures that maintain genotypic diversity in the population. The 23 different genotypes isolated are each optimized to grow in certain conditions that must occur at specific locations or time periods in the Hood Canal. Otherwise, the genomic diversity would be reduced due to the competition for resources.

Diatoms often exhibit "blooms" where the number of individuals in a body of water increases rapidly. No one knows if the bloom is caused by the rapid growth of all genotypes or a subset of genotypes. Likewise, no one knows what the total genomic diversity is in waters with different physical and chemical characteristics. Rynearson and Armbrust have initiated studies to answer these questions, although you could make some predictions based on the work presented here.

MATH MINUTE 3.2 • HOW ARE POPULATIONS MODELED?

To build the model of diatom genomic diversity illustrated in Figure 3.5, the investigators have assumed exponential growth of each isolate. In other words, the population size at time t, denoted $P(t)$, is modeled by the equation

$$P(t) = P_0 e^{rt},$$

where P_0 is the initial population size and r is the population growth rate. The growth rate for each isolate under each light condition is assumed to be constant at the mean rate shown in Figure 3.4. Since the growth rate is given in units of 1/days, the time t in the equation is in days. For example, the population size after six days of isolate 1 grown with 166 μmol photons m^{-2} s^{-1} of light is given by $P_0 e^{1.37 \times 6} = 3714.5 P_0$.

Since P_0 is assumed to be equal for all isolates, its value does not affect the relative abundance of each isolate. This is because when the population size of a particular isolate is divided by the total population size (in order to calculate the relative abundance), P_0 is a factor in both the numerator and denominator and is canceled out. Therefore, the graphs in Figure 3.5 can be produced by arbitrarily assigning P_0 to be 1.

A spreadsheet is ideal for modeling the relative abundance of each isolate over time, using the exponential equation given above. The resulting population sizes (rounded to the nearest integer) during the first 14 days of growth under medium light intensity are shown in Table M3.1. Dividing each isolate population size on each day by the total

population size on that day (i.e., the sum of the row) produces the data in Table M3.2.
Figure 3.5b is a graph of the data in Table M3.2, extended through day 100.

TABLE M3.1 • Population sizes of the eight isolates grown at 66 μmol photons m^{-2} s^{-1} of light for 14 days.

Day	Isolate 1	Isolate 2	Isolate 3	Isolate 4	Isolate 5	Isolate 6	Isolate 7	Isolate 8
0	1	1	1	1	1	1	1	1
1	3	2	2	2	2	3	3	2
2	8	4	5	6	6	7	7	5
3	22	9	12	13	13	18	17	12
4	62	18	27	31	31	47	44	27
5	172	37	61	72	73	124	112	62
6	483	76	138	170	171	325	288	143
7	1,353	156	313	400	403	852	741	327
8	3,790	320	712	942	950	2,235	1,905	747
9	10,615	658	1,618	2,217	2,237	5,861	4,895	1,708
10	29,733	1,353	3,678	5,219	5,271	15,367	12,582	3,905
11	83,283	2,782	8,358	12,283	12,419	40,296	32,338	8,928
12	233,281	5,722	18,996	28,912	29,261	105,662	83,117	20,414
13	653,436	11,766	43,174	68,050	68,941	277,063	213,630	46,677
14	1,830,317	24,197	98,125	160,171	162,430	726,505	549,080	106,724

Whenever all growth rates are held constant and one isolate grows faster than the others, that isolate will eventually comprise 50% or more of the population. The relative magnitudes of the growth rates determine the number of days until this happens. In this example, isolate 1 reaches 50% on day 14, the day marked with an arrow in Figure 3.5b. If you would like to explore population models further, you can download the modeling spreadsheet and modify the isolate growth rates to produce the curves in Figure 3.5a or 3.5c.

TABLE M3.2 • Relative abundance of each isolate in the population, calculated from Table M3.1.

Day	Isolate 1	Isolate 2	Isolate 3	Isolate 4	Isolate 5	Isolate 6	Isolate 7	Isolate 8
0	0.125	0.125	0.125	0.125	0.125	0.125	0.125	0.125
1	0.145	0.106	0.118	0.122	0.122	0.136	0.133	0.118
2	0.167	0.090	0.110	0.118	0.118	0.146	0.140	0.111
3	0.190	0.075	0.102	0.113	0.113	0.156	0.147	0.103
4	0.215	0.063	0.093	0.107	0.108	0.165	0.153	0.096
5	0.242	0.052	0.085	0.101	0.102	0.174	0.157	0.088
6	0.269	0.042	0.077	0.095	0.095	0.181	0.161	0.080
7	0.298	0.034	0.069	0.088	0.089	0.188	0.163	0.072
8	0.327	0.028	0.061	0.081	0.082	0.193	0.164	0.064
9	0.356	0.022	0.054	0.074	0.075	0.197	0.164	0.057
10	0.386	0.018	0.048	0.068	0.068	0.199	0.163	0.051
11	0.415	0.014	0.042	0.061	0.062	0.201	0.161	0.044
12	0.444	0.011	0.036	0.055	0.056	0.201	0.158	0.039
13	0.473	0.009	0.031	0.049	0.050	0.200	0.154	0.034
14	0.500	0.007	0.027	0.044	0.044	0.199	0.150	0.029

DISCOVERY QUESTIONS

1. Would you expect the diatom genomic diversity to be equally high at different locations throughout Puget Sound? Would you expect to find the exact same genotypes, or different ones? Explain your reasoning.

2. The only genomic data we have for diatoms are the nine microsatellite loci. What would you need to know in order to identify individual alleles that might be responsible for variations in physiological responses to changing environments?

3. If you could sequence the entire genome of one diatom species, would this reference genome contain all the information you would need to understand diatom blooms? Explain your answer.

The completion of a reference genome sequence is an important starting place, but understanding what might happen when the ocean is seeded with iron would require a greater knowledge of the genomic variation in isolated populations. A population may exhibit an average phenotype, but this average underestimates the diversity of genomes that remain hidden until changes in the environment favor a different genotype. The interaction of genotype and phenotype diversity is complex and may not respond to human intervention the way we expect it to.

Summary 3.1

The case study in this section illustrates how genomic methods will affect our understanding of environmental and ecological issues. As often happens in science, in answering one question, we are surprised by an unexpected discovery. The genomic variation in photosynthetic diatoms is greater than we expected, indicating that the oceans are extremely dynamic environments. Diatoms have evolved diverse genomes to thrive as a species, by mitotically reproducing those genomes that temporarily best suit local conditions. Given this evolutionary response to a complex ecosystem, we should be cautious when devising simple solutions to complex problems such as global warming. By using genomic methods to answer ecological questions, we will become more aware of the environmental impact of genomic variations.

3.2 Human Genomic Variation

How much variation is there in the human genome, and are these variations informative? At the heart of the Human Genome Project is this question of variation in the human genome. You can tell by looking around that humans exhibit genome-encoded differences. The biomedical field is interested in disease-causing variations. What we often consider as "simple" diseases have complex genomic underpinnings (see Chapters 10 and 11). In this section, we examine how genomic variations are used to determine the causes for complex phenotypes and how they influence effective medical interventions.

How Much Variation Is in the Human Genome?

Up to this point, we have been examining microsatellite variations in genomes. They are very common and are useful for ecological studies. However, another type of polymorphism has gained more attention lately—**single nucleotide polymorphisms, or (SNPs).** SNPs (pronounced "snips") are single bases at a particular locus where individual people have differences in their sequences. SNPs are another form of genomic variation in a population that may occur anywhere in the genome, and each person will have many SNPs. Let's imagine that 90% of all humans have the following sequence at a particular location on one chromosome (i.e., at a unique locus):

```
GCATGCATGCATGCAT
||||||||||||||||
CGTACGTACGTACGTA
```

but 10% of all alleles have this slightly different sequence:

```
GCATGCAaGCATGCAT
||||||||||||||||
CGTACGTtCGTACGTA
```

This locus has a single nucleotide polymorphism and thus is one example of a SNP. SNPs differ from microsatellites because microsatellites involve multiple nucleotides and the patterns vary significantly (see Figure 3.2a). Furthermore, microsatellites are prone to continued mutations since they are caused by "slippage" of the DNA polymerase replicating tandem repeats. In contrast, SNPs are more stable since they differ by single nucleotides and do not induce additional errors by the DNA polymerase.

With the draft human genome reference sequence, we are in a position to enumerate all the SNPs. The International SNP Map Working Group is a consortium of the Cold Spring Harbor Laboratory in New York; the **National Center for Biotechnology Information (NCBI)** in Washington, DC; The Sanger Center in Britain; Washington University in St. Louis; and the Whitehead Center for Genome Research in Cambridge. The SNP mapping group has identified 1,433,393 SNPs, an average of 1 SNP every 2 kb of genomic sequence (Table 3.2). Since it had been estimated that SNPs occur about once every 1–2 kb, we are nearing the estimated density. Based on preliminary data, each ethnic group has its own collection of SNPs. Human SNPs have been classified in one of two ways—major or minor alleles. SNP minor alleles are less common in all humans than other SNPs at the same location. However, within a particular ethnic group, 20% of its allelic variation is due to SNP minor alleles. This means that SNPs classified as minor alleles for all humans may be SNP major alleles for an ethnically defined population of humans. Therefore, generalizations for the entire species based on genome-wide definitions of SNP minor alleles should be avoided since these generalizations lead to assumptions for all ethnic groups that may not be correct.

Exploring SNP Data Let's take a look at some SNP data that are compiled on a database at Cold Spring Harbor Laboratories. Go to **The SNP Consortium (TSC)** web site and follow these directions.

1. Under "Important Links," "Browse" the data and click on chromosome X to see how many SNPs have been found here. Mouse over the tallest red bar graph and a small balloon will appear that tells you the number of SNPs (count = 583 in June 2002) between bases 81,044,887 and 82,097,418. Go back, select chromosome 7, mouse over the tallest red bar, and note the number of SNPs. Although chromosomes 7 and X are both 163 Mb long, their frequencies of SNPs are not equivalent. (Sizes for chromosomes presented in Table 3.2 represent available **finished sequence** lengths and not the final, full-length sequence.)

2. You need to use a browser that works with the Java applet; on a Macintosh, this applet is supported only by Internet Explorer. Click on the "Search" link at the top of the page and enter SNP ID# "19265." Click on the TSC code and use the help link to determine which bases have been found at this location.

3. Click on "View screening details" and you will see the chromatograms for the three alleles that have been sequenced for this SNP. Be patient and allow the applet to fully load.

4. Go back to the SNP Report of TSC0019265 and click on "2669205" located to the right of "dbSNP ID." This will take you to the NCBI page with information about this particular SNP. In the left column, under the heading of "search," click on the population link that says "Detail."

5. Search the "submitter population id" from the menu and enter "MDECODE-3" in the blank box. The result will give you a description of the population from which this SNP was isolated. Based on what you learned about human evolution in Chapter 2, would you predict there will be many or few SNPs in this population compared to the reference genome? Click on the hyperlink "MDECODE-3," then click on "View associated SNP" to see if your prediction was correct.

TABLE 3.2 • Human SNP distribution by chromosome.

Chromosome	Length (bp)	SNPs	kb per SNP
1	214,066,000	129,931	1.65
2	222,889,000	103,664	2.15
3	186,938,000	93,140	2.01
4	169,035,000	84,426	2.00
5	170,954,000	117,882	1.45
6	165,022,000	96,317	1.71
7	149,414,000	71,752	2.08
8	125,148,000	57,834	2.16
9	107,440,000	62,013	1.73
10	127,894,000	61,298	2.09
11	129,193,000	84,663	1.53
12	125,198,000	59,245	2.11
13	93,711,000	53,093	1.77
14	89,344,000	44,112	2.03
15	73,467,000	37,814	1.94
16	74,037,000	38,735	1.91
17	73,367,000	34,621	2.12
18	73,078,000	45,135	1.62
19	56,044,000	25,676	2.18
20	63,317,000	29,478	2.15
21	33,824,000	20,916	1.62
22	33,786,000	28,410	1.19
X	131,245,000	34,842	3.77
Y	21,753,000	4,193	5.19
RefSeq	15,696,000	14,534	1.08
Totals	2,710,164,000	1,419,190	1.91

Source: The International SNP Map Working Group. 2001. *Nature.* 409: 929.

6. From the MDECODE-3 list of SNPs, click on SNP "ae1163." SNP ae1163 is in the coding region of *Apolipoprotein E*, a gene associated with early-onset Alzheimer's disease. Under the heading of "Allele Frequency Information:", notice the frequency of the different bases for each of the four populations (e.g., 21% of MDECODE-3 have a C at this position). Click on the links for MDECODE 1, 2, and 4 and correlate the four populations with the frequency of this SNP. What did you learn about populations and frequency of SNPs?

LINKS
The SNP Consortium (TSC)

7. Go back to the list of SNPs for the MDECODE-3 population. Compare the SNP frequencies for ae1163 with another SNP (e.g., ae560) in the same gene (any that start with "ae" on this page). Does ae1163 exhibit more or less heterogeneity among the four MDECODE 1–4 populations than other SNPs in this gene? Do you think this is due to chance or selection pressures?

We have begun an initial exploration of the SNP Consortium database. As more SNPs are cataloged, investigators will comb through this collection of human genome

METHODS
Chromatograms
LINKS
Warren Gish

diversity to gain new insights into why we are all different from each other despite having the same size genome with the same number of genes.

DISCOVERY QUESTIONS

4. Which chromosomes have the lowest density of SNPs (see Table 3.2)? any striking patterns? Can you formulate hypotheses to explain this observation?

5. Predict some research uses for a public-domain SNP database.

6. Go back to TSC and Browse the data, then click on chromosome 21 and select the region from 24250023–24549406 by mousing over the bar graphs and using the scale below to help you get started. What gene have you found? Are there any SNPs in the exons of this gene? Zoom in until you see the individual SNPs associated with this infamous gene. Based on what you know from your daily life, do you think genomic variation might be significant for this disease?

MATH MINUTE 3.3 • ARE ALL SNPs REALLY SNPs?

A SNP is a position in a genome at which two or more different bases occur in the population, each with a frequency greater than 1%. In general, a SNP can be found by first aligning a set of overlapping DNA sequences and then identifying positions in the alignment at which the same base does not occur in every sequence. For example, the following five sequences appear to have a SNP at position 8. However, is this enough evidence to conclude position 8 is a SNP?

```
GCATGCAaGCATGCAT
GCATGCAcGCATGCAT
GCATGCAaGCATGCAT
GCATGCAaGCATGCAT
GCATGCAaGCATGCAT
```

An important aspect of the search for SNPs is the assessment of false positive and false negative rates, and probability theory is a powerful tool for making these assessments. There are two ways in which a base may be falsely classified as a SNP: (1) inclusion of paralogs in sequence alignments, and (2) errors in sequencing. Let's walk through an approach to SNP finding proposed by investigators in Warren Gish's lab at Washington University in St. Louis, and see how they minimized false positive SNPs.

The investigators began with approximately 1.3 Mb of finished human genome reference sequence. They found 1,954 hits to the reference sequence in the EST database when they restricted the search to ESTs for which chromatograms were available. By aligning the ESTs and identifying sequence overlaps, they clustered the 1,954 ESTs into 147 contigs that aligned with 80,469 positions in the reference sequence.

Some of the ESTs that aligned with a particular segment of the reference sequence may have come from paralogs. To screen out paralogs, the investigators used the fact that a paralog would have less similarity to the reference sequence than would the same gene containing SNPs. Specifically, the frequency of variation in paralogs is thought to be approximately 1 base in every 50, while the frequency of SNPs is thought to be approximately 1 base in every 1,000. For example, in the following five sequences, EST 2 appears to be a paralog because of the relatively large number of differences between it and the reference sequence. Note that if EST 2 were removed from consideration, there would no longer appear to be a SNP at position 4 or 8.

```
EST 1       GCAtGCAaGCATGCAT
EST 2       GCAgGCAcGCATGCAT
EST 3       GCAtGCAaGCATGCAT
EST 4       GCAtGCAaGCATGCAT
Reference:  GCAtGCAaGCATGCAT
```

ESTs with significantly more than the expected number of mutations from the reference sequence were classified as paralogs and removed from further consideration. After paralogs were removed, 69,756 positions of the original 80,469 remained to be searched for SNPs.

The second step of their process was designed to screen out positions that appeared to be SNPs due to sequencing errors. To do this, the investigators considered the chromatograms of the 1,954 ESTs and the reference sequence. They determined the probability that a particular position was truly a SNP, considering (1) the bases that were observed at that position, (2) the depth of coverage at that position (i.e., how many ESTs were aligned to that position), and (3) the base quality value at that position in each sequence. For example, in position 8 of the five sequences shown above, the bases are {a,c,a,a,a} and the depth of coverage is 4. The base quality values come from processing each chromatogram with the PHRED base-calling program (see page 18). The expected rate of polymorphism (0.001) also entered into their calculations. All positions for which the probability of being a true SNP was estimated to be greater than 0.4 were designated candidate SNPs. There were 59 candidates, with an average probability of 0.78 of being a true SNP.

The candidate SNPs were examined at the corresponding positions in an independent collection of DNA sequences (the validation set). Twenty-three of the 54 candidate SNPs were excluded from validation for technical reasons. Twenty of the remaining 36 candidate SNPs (56%) were found to be polymorphic in the validation set. The fact that the other 16 candidates were not polymorphic in the validation set does not prove that they are not SNPs, but there is no supporting evidence that they are.

SNPs are being added to databases at a rapid pace, but are they all real? The SNP finding process followed by these investigators illustrates that the quality of the data must be taken into account before concluding a SNP is really a SNP.

Why Should We Care About SNPs?

As we saw in Chapter 1, only 2% of our genome encodes protein, so you might think that 98% of these SNPs are not interesting. If you define a gene as the coding region plus 10 kb upstream, then 98% of all genes are within 5 kb of a SNP, 93% of all annotated genes contain at least one SNP, 59% of human genes have five or more SNPs, and 39% have ten or more SNPs. It is interesting to note that the sex chromosomes have lower densities of SNPs than other chromosomes, although no one is sure why. Perhaps there is stronger selection on them not to change, but this is speculation.

There are four popular ideas of how SNP data will be useful. The first is in the study of evolution. By comparing variations in different subpopulations, we may be able to better understand when mutations arose and why they have been maintained in certain populations but not others. For example, some alleles may prove advantageous in one environment and maladaptive in another. A second use of SNP data will be in DNA fingerprinting for criminal or parental verification. Third, SNPs will be used as markers to map polygenic traits, as illustrated in the examples below. The fourth and most popular possibility is genotype-specific medication. Most genes contain at least one SNP, some of which may have functional consequences. The key for all four uses of SNP information will be identifying large numbers of SNPs in the human genome.

SNPs are reliable markers that may allow us to determine which combination of coding alleles are associated with particular diseases. This discussion requires that we talk about three terms that are often confusing—linkage, linkage disequilibrium, and haplotype. **Linkage** refers to how close two loci are to each other on a chromosome. If they are near each other, we say the two loci are linked. **Linkage disequilibrium** describes alleles rather than loci. If two alleles (or two SNPs) tend to be inherited together more often than would be predicted, we say the alleles are in linkage disequilibrium (i.e., inherited together more often than other possible allele pairs). **Haplotype** refers to the set of alleles on one particular chromosome. Each person has two haplotypes in a given region, and each haplotype will be passed on as a complete unit unless recombination occurs to separate this particular set of alleles to form two new haplotypes.

These three terms are used extensively when describing how we deduce which genes are involved in polygenic diseases. For example, let's imagine that a particular population has a higher-than-average incidence of Alzheimer's disease. If we could track SNPs that are in linkage disequilibrium and correlate with the phenotype, we would have SNP markers that form a particular haplotype at several loci involved in causing Alzheimer's. To begin, let's look at a hypothetical example where two SNPs and one gene are associated with a monogenic disease (Figure 3.6). Notice how the two different populations have different ratios of the three possible

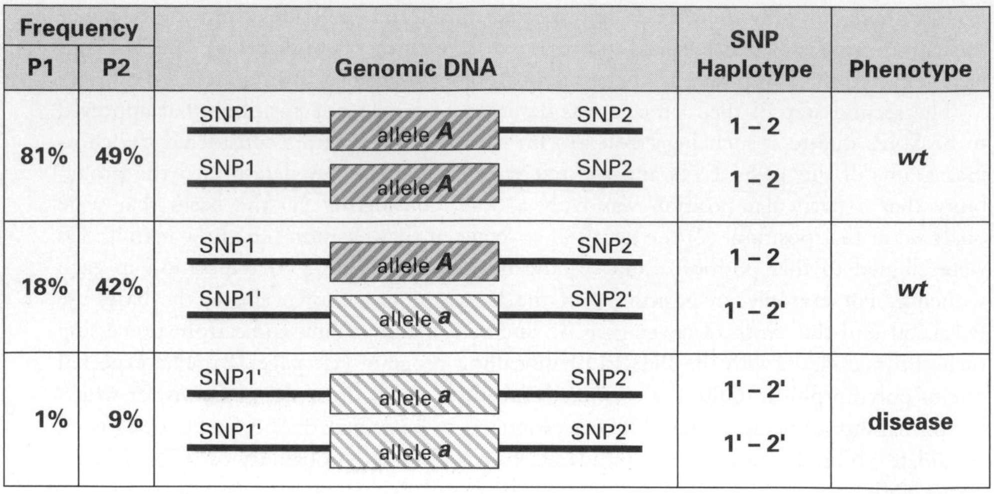

FIGURE 3.6 • Comparison of SNP data for two populations. Two populations (P1 and P2) have different frequencies of a single-locus disease. Included in this figure are diagrams of the genomic DNA showing two SNPs (1' and 2') and a recessive allele *(a)* that are in linkage disequilibrium. In this figure and Figure 3.7, an allele and its flanking SNPs define one locus.

Frequency		Genomic DNA	SNP Haplotype	Phenotype
P1	**P2**			
81%	49%	SNP1 — allele *A* — SNP2	1 – 2	*wt*
		SNP1 — allele *A* — SNP2	1 – 2	
18%	42%	SNP1 — allele *A* — SNP2	1 – 2	*wt*
		SNP1' — allele *a* — SNP2'	1' – 2'	
1%	9%	SNP1' — allele *a* — SNP2'	1' – 2'	**disease**
		SNP1' — allele *a* — SNP2'	1' – 2'	

LINKS

HGP Glossary

genotypes, which is what you would expect to see in a population that experiences a higher-than-average incidence of the disease. Even if you did not know allele *a* was the causative agent, you could evaluate the SNP haplotypes and deduce that SNPs 1' and 2' are linked to the disease allele, and that this is a recessive disease since an individual has to be homozygous to contract it.

Now let's look at a more complex trait where three different genes influence the phenotype (Figure 3.7). In this figure you can see the SNP data, but you cannot see the alleles located near these SNPs that contribute to the trait. In this

case, the phenotype could be something similar to cancer or Alzheimer's where the severity of the disease is similar in all cases but the age of onset varies. (In reality, cancer and Alzheimer's involve more than three genes.) Traits that are polygenic but can be measured in some quantitative manner can be genetically mapped to **quantitative trait loci (QTL).** By studying the inheritance of certain haplotypes that contain identifiable SNPs, we could begin to dissect the genomic causes of QTL for traits such as schizophrenia, high blood pressure, diabetes, as well as some diseases with genetic components such as cancer and Alzheimer's. Many investigators want to use similar SNP mapping techniques for

---

BOX 3.1 • *What's the Difference Between a Mutation and an Allele?*

Since nine different people donated their DNA to the Human Genome Project (HGP) for sequencing, there are bound to be some differences in their DNA. This leads us to a philosophical question: Is there such a thing as *the* human genome? No two humans have the exact same DNA sequence. Although identical twins start with the same DNA, by the time they reach sexual maturity, their gametes will not be identical. So it would be impossible to say there is such a thing as *the* human genome. Instead, we use the term reference genome to describe a composite genome, but we have recorded all the polymorphisms discovered along the way (about 1.4 million so far).

Comparing individual genomic variation has led to an interesting problem. What is a polymorphism, vs. an allele, vs. a mutation? The HGP glossary defines the terms as follows:

Allele: Alternative form of a genetic locus; a single allele for each locus is inherited separately from each parent (e.g., at a locus for eye color, the allele might result in blue or brown eyes).

Polymorphism: Difference in DNA sequence among individuals. Genetic variations occurring in more than 1% of a population would be considered useful poly-

morphisms for genetic linkage analysis.

Mutation: Any heritable change in DNA sequence.

First, you can see that these definitions are fuzzy. An allele is a heritable change in DNA sequence, and it may occur at a frequency above 1% of the population. Alleles and polymorphisms are heritable changes in DNA sequence. What's missing from the list is a definition for "population" and a clarification of "change"—compared to what? So if you think these terms are confusing, you're right. Completely sequencing *the* human genome only made these three terms seem less precise.

	Locus 1	Locus 2	Locus 3	% P1	% P2	Phenotype
1	1 – 2 1 – 2	a – b a – b	X – Y X – Y	20	3	0
2	1 – 2 1 – 2	a – b a – b	X' – Y' X – Y	12	3	2
3	1 – 2 1 – 2	a – b a – b	X' – Y' X' – Y'	12	3	2
4	1 – 2 1 – 2	a' – b' a – b	X – Y X – Y	20	5	0
5	1 – 2 1 – 2	a' – b' a – b	X' – Y' X – Y	12	5	1
6	1 – 2 1 – 2	a' – b' a – b	X' – Y' X' – Y'	10	5	1
7	1 – 2 1 – 2	a' – b' a' – b'	X – Y X – Y	2	4	1
8	1 – 2 1 – 2	a' – b' a' – b'	X' – Y' X – Y	0.01	3	5
9	1 – 2 1 – 2	a' – b' a' – b'	X' – Y' X' – Y'	0.001	3	5
10	1' – 2' 1 – 2	a – b a – b	X – Y X – Y	2	3	3
11	1' – 2' 1 – 2	a – b a – b	X' – Y' X – Y	1	3	4
12	1' – 2' 1 – 2	a – b a – b	X' – Y' X' – Y'	1	3	4
13	1' – 2' 1 – 2	a' – b' a – b	X – Y X – Y	0.01	5	3
14	1' – 2' 1 – 2	a' – b' a – b	X' – Y' X – Y	0.01	5	4
15	1' – 2' 1 – 2	a' – b' a – b	X' – Y' X' – Y'	0.01	5	4
16	1' – 2' 1 – 2	a' – b' a' – b'	X – Y X – Y	0.005	3	6
17	1' – 2' 1 – 2	a' – b' a' – b'	X' – Y' X – Y	0.001	3	7
18	1' – 2' 1 – 2	a' – b' a' – b'	X' – Y' X' – Y'	0.0001	3	7
19	1' – 2' 1' – 2'	a – b a – b	X – Y X – Y	2	3	3
20	1' – 2' 1' – 2'	a – b a – b	X' – Y' X – Y	1	3	4
21	1' – 2' 1' – 2'	a – b a – b	X' – Y' X' – Y'	1	3	4
22	1' – 2' 1' – 2'	a' – b' a – b	X – Y X – Y	2	5	3
23	1' – 2' 1' – 2'	a' – b' a – b	X' – Y' X – Y	1	5	4
24	1' – 2' 1' – 2'	a' – b' a – b	X' – Y' X' – Y'	1	5	4
25	1' – 2' 1' – 2'	a' – b' a' – b'	X – Y X – Y	0.001	3	6
26	1' – 2' 1' – 2'	a' – b' a' – b'	X' – Y' X – Y	0.0001	3	7
27	1' – 2' 1' – 2'	a' – b' a' – b'	X' – Y' X' – Y'	0.00001	3	7

FIGURE 3.7 • **Prevalence of a trait in two populations (P1 and P2).** The phenotype is rated on a 0–7 scale with 0 indicating no adverse phenotype (i.e., *wt*); 1 a mild phenotype; and 7 a severe phenotype. Three genetic loci are illustrated with a total of 6 different SNPs, 2 per locus, for a total of 27 genotypes. Each combination of SNPs is color coded to permit easier visualization of the distribution of the haplotypes.

genes associated with behaviors such as learning, homosexuality, and daily cycles that determine who needs five hours of sleep each day compared to someone who needs nine hours.

The hypothetical example in Figure 3.7 highlights several relevant points. First, this is a simplistic case with only three genes and two alleles for each gene for a total of 27 different genotype combinations possible. Therefore, mapping polygenic QTL will require large numbers of people in order to detect linkage disequilibrium. Another point is that some populations will be more informative than others when trying to deduce the causative genes in a genome. For example, individuals in P1 have the disease infrequently compared to P2. Therefore, P2 would be the better one to study since you would be more likely to discover informative linkage disequilibrium. Finally, in this simplistic example, there are only two haplotypes per locus, while real loci might have many more haplotypes, making it very difficult to determine which SNPs were associated with the trait of interest.

DISCOVERY QUESTIONS

7. How many loci contribute to the phenotype in Figure 3.7? Can you tell from these data?

8. If the phenotype were cancer, which SNPs are linked to disease-causing alleles? Which disease alleles are dominant and which are recessive?

9. How many more genotypes would be added to these data if one more locus were included that had only two SNPs at this locus? How much more complexity would be added to these data if the three loci each had a third SNP?

Are There Any Known Examples of SNPs That Cause Diseases?

Increasingly, we can document point mutations that lead to diseases. You probably have already studied at least one case, sickle cell disease, which was the first genetic disease to be understood at the genomic level. Sickle cell disease is caused by a SNP that results in one amino acid difference in the β subunit of hemoglobin. This SNP leads to a different conformation of the protein, which does not function as well as the **wild-type** protein. Examples such as this are not new any more, so we do not need to spend time enumerating multiple cases. Instead, we will study three types of SNPs with less direct consequences.

Protein Targeting People with primary hypomagnesemia are unable to maintain sufficient levels of magnesium in their serum. Magnesium is necessary for normal cellular func-

tions, so its concentration must be carefully controlled. Because hypomagnesemia is a polygenic trait, understanding the cause has been difficult, but one gene has been identified that contributes to the problem. You are probably familiar with the sodium/potassium pump, often called the **Na/K ATPase,** since it establishes the membrane potential in all cells. What you probably did not know is that the Na/K ATPase is composed of three subunits called α, β, and γ. The α subunit performs the pumping action, β ensures that the α subunit goes to the correct location within a cell, and γ modulates the activity of the α subunit. The Na/K ATPase internalizes K^+ and extrudes Na^+ across the plasma membrane. In your kidneys, you must excrete excess salt that you consume, but retain ions in your blood. The Na/K ATPase must be located in the plasma membrane of your kidney cells in order to function properly.

A large team of Dutch investigators identified a family with a dominantly inherited form of hypomagnesemia that had been linked to a few markers near 11q23. When they looked around this portion of chromosome 11, they realized the Na/K ATPase γ subunit was in this region. They sequenced several affected members of the family and determined that they all had the same SNP, 123G→A, which means nucleotide 123 in the coding DNA was changed from G to A. This SNP resulted in an amino acid change from glycine to arginine, and this amino acid was located in the only transmembrane domain of the γ subunit. When they compared the protein sequences of γ subunits from 14 different vertebrate species, they found that all transmembrane domains had the conserved glycine, which indicated that selection had maintained this particular amino acid from frogs to humans. When the investigators put a mutant form of the γ subunit gene into tissue-cultured cells, they found that the mutant γ subunit did not reach the plasma membrane, nor did the α or β subunits. All three subunits were retained in the cytoplasm. Previously, other investigators had shown that a kidney Mg^{2+} channel was responsible for an influx of Mg^{2+} ions, and this channel functioned properly only within a narrow range of K^+ ion concentrations. The Dutch team concluded that the mislocalization of the γ subunit prevented the Na/K ATPase from regulating the appropriate level of K^+ in kidney cells to permit the Mg^{2+} channels to retain sufficient levels of Mg^{2+} ions in the cytoplasm. Therefore, one molecular cause for hypomagnesemia is a SNP that resulted in inappropriate subcellular localization.

DISCOVERY QUESTIONS

10. Look at the structural difference between the amino acids glycine and arginine. Is this change in protein primary structure a substantial change or a minor one?

11. In their paper, the investigators reported that individuals with deletions of the γ subunit from one of their chromosomes did not suffer from hypomagnesemia. What can you deduce from this finding?

12. Let's assume their conclusion is correct. Can you now predict other genes that might be responsible for the polygenic nature of hypomagnesemia?

14. Go to OMIM and search for "mitochondrial disease." Choose one example from the list.
    a. Is the protein you found involved in ATP production?
    b. Which tissue or tissues are affected the most? Does this make sense, given what you know about the tissue's relative need for ATP?

LINKS
OMIM
Adrian R. Krainer

Mitochondrial SNPs Mitochondria produce most of our cells' ATP, and thus their function is critical for our well-being. Over 100 proteins are involved in oxidative phosphorylation. Most of these proteins are encoded by nuclear genes, but 13 are encoded by mitochondrial genes. Each mitochondrial gene requires the proper function of 22 tRNA genes and 2 rRNA genes that are also encoded in the mitochondrial genome. The phenotypes associated with defective oxidative phosphorylation tend to affect tissues that require the greatest amount of energy. Therefore, cardiac muscle, skeletal muscles, and tissues that have to pump a lot of ions are most often affected. More than 50 different disease-causing mitochondrial SNPs have been identified, and this number is expected to increase as we become more proficient at detecting SNPs.

A panel of 2,000 patients suspected of having diseases due to mitochondrial SNPs was screened for 44 known SNPs. Of these patients, 108 were determined to have a known disease-causing SNP. This research led to several conclusions. First, these 108 patients could be treated more effectively since the precise nature of their illness was apparent. Second, the few remaining SNPs that were not included in the original screen would need to be tested. Third, maybe some of these 2,000 patients did not have mitochondrial mutations. Fourth, there are non-SNP mutations that would have gone undetected in this screen, making it necessary to perform different tests. Finally, there are probably additional SNPs that have not been discovered yet that could be responsible for some of the undiagnosed cases. This study illustrates the need for high throughput methods for identifying as many SNPs as possible through basic research. Eventually, SNP discoveries could be transferred into diagnostic tests for detecting SNPs in clinical settings.

DISCOVERY QUESTIONS

13. If you were to consider the pedigree of a mitochondrial gene mutation (rather than a nuclear gene) that led to a disease, what feature would be diagnostic?

Incorrect mRNA Splicing Since high school, you have been aware that mRNA production requires splicing to join the exons and exclude the introns. In Chapter 1, we learned that most human genes have alternative splicing that results in two slightly different proteins. In 2001, a team of investigators, led by Adrian R. Krainer from the Cold Spring Harbor Laboratory in New York, determined the mechanism responsible for some unsuspected mutations in the breast cancer gene called *BRCA1*. When *BRCA1* encodes a normal protein, breast and ovarian cancers are less likely to occur. However, if the BRCA1 protein is absent or nonfunctional, then many women in the family (and occasionally men) are very likely to develop breast cancer and women may develop ovarian cancer as well. We have already considered SNPs that lead to the incorporation of inappropriate amino acids, but there is another form of SNPs that can also lead to diseases.

When we find a SNP in an exon, we immediately determine whether it alters the amino acid composition of the encoded protein. If the amino acid sequence is unaffected, we tend to think of this SNP as "silent" and call this type of point mutation a "silent mutation." Krainer and his colleagues identified a series of silent mutations in exons of *BRCA1* that lead to alternative splicing. Splicing of RNA to produce a mature mRNA involves the 5' and 3' ends of each exon, but there are internal sequences that are required as well. Although the consensus sequences are uncertain, exonic splicing enhancers (ESEs) and exonic splicing silencers (ESSs) are located within exons and are distinct from the terminal splicing junctions. When working properly, ESEs are bound by proteins involved in splicing and the bound exon will be included in the mRNA. If a SNP occurs in an ESE, then an exon that would normally be included in the mRNA will be skipped, which means the encoded protein will not retain its appropriate structure or function. Conversely, ESSs are DNA sequences bound by proteins that prevent the inclusion of the exon that contains a functional ESE. If a SNP disrupts the binding site of an ESS, then the exon will be inappropriately included in the mRNA and disrupt normal protein function. Krainer found examples of ESE and ESS mutations in *BRCA1,* which probably explained why women with silent mutations developed breast and ovarian cancer. This case illustrates that even silent SNPs can have profound influences on phenotypes, including polygenic traits such as cancer.

Are There Any Known Changes in Nondisease QTL Due to SNPs?

The three SNP cases on pages 82–83 illustrate the diversity of SNPs that can produce diseases. You now realize that incorporating the wrong amino acid in a nuclear-encoded gene is only one category of SNP that must be characterized if we are to fully appreciate the consequences of more than 1.4 million SNPs in the human genome. Let's consider some SNPs that do not lead to diseases but produce altered phenotypes nonetheless.

Fava Beans and a SNP May Kill You About 2,000 years ago, a Roman by the name of Lucretius Caro described the consequences of this SNP: "What is food to some men may be fierce poison to others." It turns out that some people experience a lysis of their red blood cells when consuming fava beans (a large white legume). Approximately 10% of people in the world are susceptible to this malady because they fail to make glucose-6-phosphate dehydrogenase (G6PD), a metabolic enzyme found in the cytoplasm of every cell in 90% of the population. G6PD produces NADPH, which is used in part to regenerate the enzymes used to neutralize the cellular toxin hydrogen peroxide. The fava bean contains a variety of substances that increase the red cells' sensitivity to oxidants such as hydrogen peroxide. Your red blood cells have only one way to produce NADPH—via G6PD, which is encoded on your X chromosome. Females would need a pair of SNPs, but males only need to inherit one SNP from their mothers to block the production of G6PD in their red blood cells and place them in 10% of the world's human population lacking G6PD. One SNP (376A→G), which produces G6PD with normal activity, is found in 20% of African males. A second SNP (202G→A), which reduces G6PD activity by about 10%, is found in about 20% of African alleles. Another SNP (563C→T) called G6PD

Mediterranean produces an enzyme with nearly undetectable activity, and it is found in about 20% of the alleles in Caucasians living around the Mediterranean Sea. Many other G6PD SNPs have been identified in populations around the world.

Variations in Medication Responsiveness Many of the drugs used as medications in humans are not in their final and active form when first administered to a patient. The drugs are metabolized in a predictable way, and the enzymatic product is the therapeutic compound. One famous example is AZT, which needs to be modified by human enzymes before it can block HIV replication. One enzyme that metabolizes a large number of these "pre-drugs" is called cytochrome P450. People fall into one of three classifications: typical metabolizers, poor metabolizers, and ultra-rapid metabolizers. Drug studies are performed on large panels of people to determine the optimum dosage for the "average" person. However, any one person may not have the average metabolism, so the ideal dosage for him or her may not be the average dosage. For example, poor metabolizers fail to convert the pre-drug into its active form and wind up excreting the inactive version before it can be converted. Conversely, ultra-rapid metabolizers quickly convert nearly all the pre-drug into its active form, which may result in an overdose if the "average" person normally converts a lower percentage of the pre-drug.

Cytochrome P450 is encoded by two separate genes called *2C19* and *2D6*. The gene *2D6* is on chromosome 22, has 9 exons and 8 introns, and 12 SNPs have been identified that lead to altered enzyme activity. The most common mutation is a G→A substitution within exon 4 that alters splicing in mRNA formation and results in no protein being produced. More than 40 different pre-drugs require 2D6p to become active, including antiarrhythmics (heart medication), antidepressants, antipsychotics, and pain killers in the opioid family.

The gene *2C19* is known to metabolize mephenytoin, which is used to treat epilepsy. Poor metabolizers do not activate this pre-drug and thus do not receive the same benefit as the "average" person. Only 2–3% of Caucasians are poor 2C19p metabolizers but about 23% of Asian populations are poor metabolizers. *2C19* is located on chromosome 10, and the most common mutation is a SNP in exon 5 (681G→A) that creates an aberrant splice site. Therefore, when drugs are administered to different populations, it is important to determine a population-specific recommended dosage.

DISCOVERY QUESTIONS

20. How could you determine if you were a poor or an ultra-rapid metabolizer?

21. Is there any practical way to respond to the information presented in this case study? Who is a poor metabolizer and who is not? What could a physician do if a patient's genotype were known?

BOX 3.2 • *Patent Law and Genomics*

No doubt you have heard that it is possible to patent genes, proteins, and even whole organisms. Since these objects are produced by normal biological processes, many people become enraged when they think about this aspect of biotechnology. "It is not right for anyone to patent proteins in my body!"

In order to understand patents granted for genetic material, we need to understand the principle of patent law. The purpose of a patent is to encourage development of new products and methods by granting "sole right to exclude others to make, use, sell, or import an invention for a limited period of time," currently 20 years. Though this sounds like an unfair monopoly, patent law is designed to reward the investment of time and money needed for research and development required to bring an invention to market. Exclusive rights come with a cost, though; to obtain a patent, you must disclose what the invention is and how it can be used. Since patents are public records, you have just revealed all your secrets to everyone. Human clotting factor made by bacteria would be patentable, but after 20 years, anyone could produce it without any legal restrictions. To see an example, go to the U.S. Patent and Trademark Office and search for patent number 6,180,371. Click on the link and you will see what has been patented and why we want to

encourage this type of research and development.

Biological patents have been granted for over one hundred years, before Mendel was famous or Watson and Crick were born. Aspirin is a natural product of willow trees, but in its purified form, a form not found in nature, it was a patentable product. Likewise, antibiotics are natural products of fungi, but for medicinal purposes they must be purified, which requires human intervention. Human intervention is the critical distinction between the genes and proteins found inside your body and the DNA and protein sequences that have been patented. Likewise, genetically modified organisms can be patented because they would not exist in nature without human intervention, even though they continue to reproduce naturally.

Patents are granted only to inventions that are novel, nonobvious, and useful. A particular DNA sequence produced by PCR is not a natural product (i.e., it is novel), therefore it could be patented as long as the inventor could explain its utility. If you were to randomly create a sequence of DNA, it could not be patented unless you could demonstrate a use for it. However, if you wanted to file a patent on a product that has already been published or used by others, this is considered "prior art," which means you cannot have exclusive rights to this "invention." Likewise, if someone

patented a cDNA sequence, you could not patent the mRNA since that would be considered "obvious."

Therefore, it is possible for a multinational corporation or an undergraduate student to patent a human gene sequence as long as its utility can be demonstrated and it is not obvious/prior art. It's too late, for instance, to file a patent on hemoglobin. However, if you discovered that the hemoglobin gene can be alternatively spliced to produce a different protein with a novel function, you might be able to obtain a patent even if you do not understand the mechanism of this novel function. If you wanted to conduct some research using a patented invention, there is one loophole in the law. If you are using the invention "merely for philosophical gratification," you can do so freely. This archaic phrase is open to interpretation and has been the source of more than one lawsuit, so be careful. If you were working on your thesis and presented your results at a local scientific meeting, you probably would be safe. If you published your results, no one would want to sue you, though they could. If you were paid for this work or obtained a patent based on work using someone else's invention without their permission . . . well, you might want to have a lawyer handy, because this does not sound like "philosophical enjoyment."

Written with advice from Cathryn Campbell, Ph.D., J.D.

LINKS
William Evans
Mary Relling

Food and Drug Interactions So far, we have learned that some foods can kill and some drugs can be ineffective if you have certain SNPs. Now we are going to examine the interaction of foods with medications—for example, grapefruit juice can alter your ability to absorb drugs.

Typically, pills pass through your stomach and dissolve in your small intestine, where the medication is absorbed. One protein involved in this is P-glycoprotein, which pumps drugs into intestinal cells. A second protein is cytochrome P450 3A, which metabolizes drugs (the body's way of detoxifying potentially harmful compounds) and converts the potentially beneficial drug into a form that is more readily excreted. One glass of grapefruit juice can block P-glycoprotein and cytochrome P450 3A for as long as 24 hours. Cytochrome P450 3A is inactivated by an unknown component of grapefruit juice, causing the enzyme to be destroyed. When volunteers were given a medication along with a typical "breakfast" amount of juice, the "average" person had only slightly elevated levels of the cholesterol-lowering drug in their blood. When a different panel of volunteers was given double-strength grapefruit juice, three times a day for two days, on average the same drug was elevated 12-fold in the blood. The immunosuppressant cyclosporin, which is given to organ transplant patients, was shown to be internalized better in the presence of grapefruit juice. Finally, in a small trial of 13 individuals, St. John's Wort, which is popular in health food stores, was found to enhance the activity of cytochrome P450 3A, which would reduce the effectiveness of drugs used to control cholesterol and seizures and transplanted organ rejection. There is very little research in the area of genomic interactions with drugs and food, even though they may affect human health.

DISCOVERY QUESTIONS

22. Can you imagine any SNPs that occur outside the coding region of cytochrome P450 3A genes that might also lead to differences in drug responses?

23. If grapefruit juice causes a drug to be more concentrated in the blood, which enzyme is affected, P-glycoprotein or cytochrome P450 3A? Explain your answer.

24. What genes are likely to be activated by grapefruit juice, given its effect on cytochrome P450 3A?

Why the SNP Frenzy? Pharmacogenomics!

At this point, you should be keenly aware that genomic variations, ranging from large changes such as aneuploidy to small SNPs, can have a profound impact on your personal physiology. Your collection of genomic variations make you a unique human being and contribute to your potential to learn, your predisposition to disease and drug addiction, and your response to pharmaceutical interventions. You should also be aware that there are variations within as well as between populations. With all this variability, learning anything significant seems problematic if we study only the reference human genome sequence. In fact, the variation between individual genomes has sparked a biotech boon in the area of SNP discovery.

When the relationship between genetic variation and drug efficacy was first recognized, the term **pharmacogenetics** was coined. Now, the term **pharmacogenomics** has been applied to genome-wide variations, though conceptually there is no difference between the 1950s and the 21st-century terminology. Pharmacogenomics refers to the complete list of genes that determine overall efficacy, while pharmacogenetics originally focused on single genes that metabolized drugs. Pharmacogenomics is trying to take into account all the genes that include drug metabolism, transporters, receptors, signaling pathways, etc. You can imagine the high degree of complexity in humans if each of these components can be polymorphic. Let's look at a simple example that appeared in a review written by William Evans and Mary Relling from St. Jude's Children's Hospital in Memphis, Tennessee (Figure 3.8). In this example, we will consider the efficacy and toxicity of a drug that requires only two genes, each with only two alleles. In this scenario, there are nine possible combinations of genotypes, though there are less than nine efficacy phenotypes. Therapeutic effects depend on the genotype of drug receptors in combination with the amount of active drug in circulation. The graphs overlay the efficacy with toxicity and it becomes apparent that with a homozygous mutation for drug metabolism, all three drug receptor genotypes suffer from toxicity. This example highlights the complex web of protein interactions that pharmacogenomics hopes to decipher.

Figure 3.8 illustrates how drug response is not monogenic but polygenic, and so we need new technologies to help us understand the connections between relevant proteins involved in drug responses. However, it is difficult to discern genomic causes for the range of human responses seen in clinical trials and routine patient care. There are two main categories of drug-metabolizing enzymes—those that oxidize and those that acetylate the drug (Figure 3.9, page 88). The examples shown here are ones known to have multiple alleles that can produce clinical variations. Different combinations of these polymorphic enzymes can produce a wide range of phenotypes in patients taking medication. As described above, cytochrome P450 2D6 is involved in metabolizing pain-killing medication such as codeine. There is ample clinical data to support the evidence that 2–10% of the population are homozygous for null alleles and thus cannot use codeine for pain relief. For most drug-metabolizing enzymes,

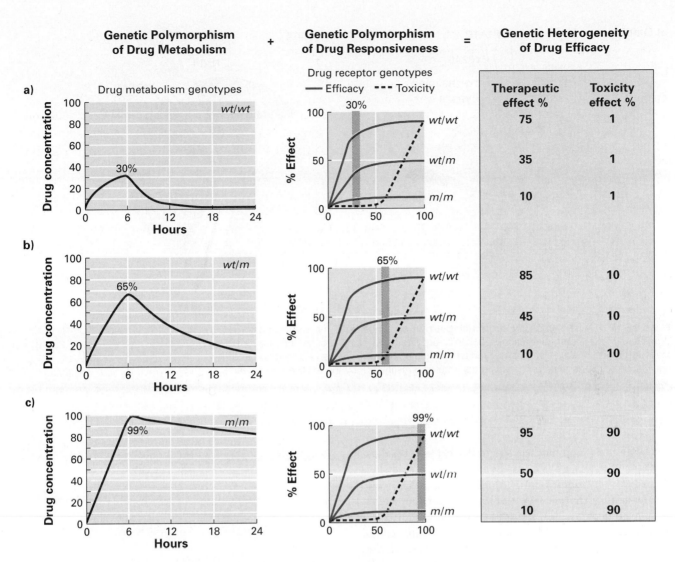

Genetic Polymorphism of Drug Metabolism	+	Genetic Polymorphism of Drug Responsiveness	=	Genetic Heterogeneity of Drug Efficacy	

Drug receptor genotypes
— Efficacy - - - Toxicity

		Therapeutic effect %	Toxicity effect %
a) Drug metabolism genotypes	wt/wt 30% → wt/wt, wt/m, m/m	75	1
		35	1
		10	1
b) wt/m 65%	65% → wt/wt, wt/m, m/m	85	10
		45	10
		10	10
c) m/m 99%	99% → wt/wt, wt/m, m/m	95	90
		50	90
		10	90

FIGURE 3.8 • Genotype and phenotype interaction for drug efficacy. Drug metabolism will be one of three genotypes: **a)** homozygous *wt*, **b)** heterozygous, or **c)** homozygous mutant (*m*). For each metabolic genotype, efficacy and toxicity are graphed in the middle column with three lines representing the genotypes of drug receptor locus. The third column summarizes the net effect for each of the nine genotypes.

we do not know any other phenotypes that alert us when a patient will not respond appropriately to a particular medication. It has been hypothesized that cytochrome P450 2D6 poor metabolizers are less tolerant of pain because they cannot produce endogenous morphine, but this is not a particularly helpful phenotype for screening patients.

Many enzyme polymorphisms lead to clinically significant consequences. Enumerating these polymorphisms is the pharmacological equivalent to listing all the genes in a genome—we are only beginning. Many pharmaceutical companies are spending a lot of money to discover clinically relevant SNPs in order to produce better medications. Traditionally, drug development was aimed at delivering medications that were effective and safe for everyone. Pharmacogenomics opens a new

paradigm where individuals with particular combinations of SNPs will be given a genotype-specific medication. Industry is still debating this approach and whether it makes financial sense to develop genotype-specific drugs since the combination of alleles is so large. If genotype-specific medication becomes a reality, when you are diagnosed, the physician will need to know your genotype to determine the appropriate medication and dosage for optimal therapy.

Once you begin a mass screening of genomes, a number of important ethical issues arise. Who should store patient-SNP data? Can a patient refuse genotyping and still receive adequate treatment? Should actuary tables for insurance companies be based on SNP data? To compound the problem, most physicians do not have a sufficient understanding of genetics

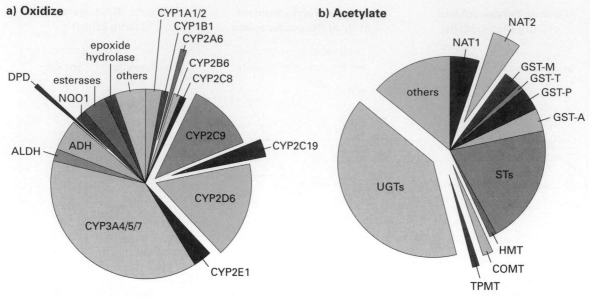

a) Oxidize

CYP1A1/2
CYP1B1
CYP2A6
CYP2B6
CYP2C8
epoxide
hydrolase
others
DPD
esterases
NQO1
ALDH
ADH
CYP2C9
CYP2C19
CYP2D6
CYP3A4/5/7
CYP2E1

b) Acetylate

NAT2
NAT1
GST-M
GST-T
GST-P
GST-A
others
STs
UGTs
HMT
COMT
TPMT

FIGURE 3.9 • **Polymorphic drug metabolizing enzymes. a)** Some enzymes oxidize drugs and make them more reactive. **b)** Other enzymes add acetyl groups onto the most reactive portion of drugs and typically inactivate them. The percentage of oxidization and acetylation of drugs that each enzyme contributes is estimated by the relative size of each section of the corresponding chart.

LINKS

Incyte Genomics

and biochemistry to be able to counsel patients adequately to make informed decisions. Genetic counselors will be needed to supplement patient care, which will cost more money and require many more trained counselors than are currently available. In short, genomic variation leads to many ethical questions that we address later in this chapter. Nevertheless, pharmaceutical companies are pursuing SNP-based drug development. For example, Incyte Genomics has identified 95 SNPs that affect the binding of ligands to the peroxisome proliferator-activated receptor γ (PPAR γ; a protein discussed in Chapter 11's obesity case study). PPAR γ is a key regulator of adipocyte differentiation and energy metabolism, which is central to the control of obesity. PPAR γ also plays a role in inflammation and is the drug target of a new class of compounds used to treat type-2 diabetes. By utilizing their database of SNPs that affect ligand binding, investigators at Incyte may be able to develop genotype-specific drugs to treat a more comprehensive range of diabetes patients. Ultimately, physicians would need to match the genotype of PPAR γ with the range of drugs to optimize patient care. Genomic medicine is coming, with all its inherent complexity.

DISCOVERY QUESTIONS

25. Using the simplistic case from Figure 3.8, outline a protocol to test the efficacy and toxicity of a new drug.

26. What impact will pharmacogenomics have on clinical trials for new drugs?

27. Do you think it is a realistic goal to decipher all the genomic variations involved in drug treatment? Do you think it will happen in your lifetime?

28. Imagine this scenario: You and a sibling have been diagnosed with the same disease. You are both genotyped, and the physician tells you there is an effective drug for your sibling but not for you. How would you respond? What if the situation were reversed?

Summary 3.2

The human genome might be described more accurately as "the many different human genomes." Each of us carries many SNPs, and most of our genes are near at least one SNP. A lot of effort has been focused on the discovery of SNPs to identify important QTLs. With this information, it is hoped that we can diagnose and treat more diseases more effectively. However, the genome variation that may identify QTLs may also prove problematic, since drug effectiveness is affected by genomic variations.

3.3 The Ultimate Genomic Phenotype—Death?

The only certainty in life is that eventually all living things die. Each organism has a genetically programmed life expectancy range, which can be affected by individual behavior. However, behavior alone cannot alter what evolution has selected—a predetermined maximum life span. Aging and death are universal phenomena and we know a lot about individual genes involved in aging. However, aging is a genome-wide process and it appears there are unseen consequences for extended life span. In Section 3.3, we briefly survey genes involved in aging in a number of model organisms and examine in detail longevity in the worm *C. elegans*. Finally, we examine an evolutionary theory that explains why each species has a genomically determined life span.

Why Do We Age?

Bessie and Sadie Delany were sisters who lived for more than 100 years. They were born in Raleigh, North Carolina, in 1891 and 1890, respectively. Their father was an emancipated slave and their mother a mixture of Native American and African American. The Delany sisters never married or had children, they were highly educated, and each pursued distinguished professional lives. Bessie died 25 September 1995 at age 104, and Sadie died 25 January 1999 at the age of 109. What was the secret of their longevity?

Why do some people live longer than others? Did the Delany sisters' genomes permit them to survive while all their age-mates died? How did they live cancer free for so long? Research in the field of aging and life span has revealed that many genes are involved in the aging process. The model organisms of yeast, *C. elegans*, *Drosophila*, and mice have revealed that the rate at which an individual ages is controlled by genomic variations. The list of known genes in each species is relatively short, although it grows each year. There are some common themes, though. First, caloric restriction slows aging. This appears to be a universally conserved evolutionary adaptation to permit individuals to delay reproduction until food is more plentiful. The mechanism is not fully understood and may vary between species. For example, in yeast, starvation appears to control gene silencing (a form of transcriptional repression controlled by the gene *Sir2*). Why gene silencing prolongs life is still unclear.

In multicellular animals, the worm *C. elegans* has been especially informative. In addition to caloric restriction, worms may have as many as four different signaling pathways that result in longer life spans. One pathway is regulated by the gene *Daf2* and involves hormonal signaling via a protein similar to insulin. Individuals with certain mutations in this pathway live longer and are more resistant to various forms of stress including oxidative damage, UV light, and heat. Reproductive cells responsible for gamete formation also regulate aging through a second pathway. In the third pathway, mitochondria signal the nucleus to slow the "rate of living" in mutants that live longer. A fourth pathway involves a tyrosine kinase that when mutated can prolong life by about 60%, but it does not affect overall development or reproduction the way the other pathways do.

LINKS
Bessie and
Sadie Delany

The famous *Drosophila* mutant allele *methusela* can prolong the life of individual flies by about 35%. The encoded protein is predicted to be an integral membrane protein that is coupled to G proteins. However, this particular gene has no known orthologs, and we do not know what ligand normally binds to this receptor. Therefore, this pathway may be similar to ones found in worms, or may be a fifth pathway.

"Ames dwarf mice" live longer than *wt* mice, but the molecular mechanism is still a mystery. These mice have altered hormonal levels, which may indicate a similar mechanism to the insulin-like pathway seen in long-lived worms. In *wt* mice, a protein called p66^{shc} helps relay stress signals that initiate cell death. In a different mutant mouse strain, individuals lack a functional version of p66^{shc} and live longer, though no one knows why.

Two general conclusions have emerged from all the research on aging in model systems. First, specific genes selected to promote *aging* are unlikely to exist. Second, aging is not the function of genes but results from accumulation of tissue damage, owing to limited energy used in tissue maintenance and repair. Longevity is regulated by genes controlling the degree of cellular damage and repair.

As you can see, there are many remaining questions to be answered. It is worth noting that the genomes of the model organisms of yeast, worms, and flies have been sequenced fully, and yet we know very little about aging, the most universal of phenotypes. Nevertheless, what has been learned so far has changed the way people think of aging and the diseases associated with it such as cancer. If the aging process can be slowed, might cancer also be postponed? Did the Delany sisters have the right combination of SNPs in their genome that permitted them to live longer than average? Can medicines be developed to prolong life and reduce the incidence of diseases? If humans could experience similar changes as seen in model organisms, it might be possible to feel 45 at the ripe age of 90. Is this scenario really possible? Does it contain hidden costs that we cannot predict? What ethical issues would be forced upon us in such a long-lived society?

It seems counterproductive for individuals to get old and die. If the ultimate selection pressure is to produce as many offspring as possible, then why do we age? What selection pressures have resulted in the progressive loss of functions and fertility, and eventual death? There are two main

LINKS

Gordon Lithgow

hypotheses that attempt to explain this apparent evolutionary oxymoron. The **mutation accumulation** hypothesis states that random mutations accumulate gradually and can become fixed by genetic drift in genes that are not subject to selection pressure. Under this model, adaptation to one environment and loss of adaptation to another are caused by distinct and unrelated variations in the genome. **Antagonistic pleiotropy** (antagonistic denotes opposition and pleiotropy indicates more than one function for a single gene) proposes there are trade-offs in evolution. Under this model, a SNP that produces a benefit in one environment will produce a detrimental consequence in another. Science progresses through experimental testing of hypotheses such as these.

From the discussion of aging in model organisms, it seems clear that many genes are involved and that aging is a consequence of other processes, rather than the activation of an "aging gene." The only area of contention, then, is the mechanism as postulated by these two hypotheses. The existence of a cost for longer life has been documented in many species. Trade-offs seen in model organisms with longer life spans include lowered fecundity, lower viability, delayed reproductive maturity, and prolonged development. These are well documented, but the genomic mechanisms are still unknown. Let's look at two experiments that indicate which hypothesis is more appealing.

Are There Hidden Costs for a Prolonged Life?

A team of investigators led by Gordon Lithgow from Manchester, England, wanted to test the trade-off for longevity in the *C. elegans age-1* mutant, which has an altered kinase involved in the insulin signaling pathway. These mutants can

live as much as 80% longer than *wt* worms. When grown at 20°C, they have similar appearances, developmental rates, activity levels, and total fertility. When *wt* and *age-1* worms were grown together, at three differential ratios (90% *age-1* and 10% *wt*; 50% *age-1* and 50% *wt*; 10% *age-1* and 90% *wt*), the proportion of *age-1* mutants remained essentially unchanged over ten generations (Figure 3.10). These experiments, performed twice with similar results, support the mutation accumulation hypothesis since the change in aging had no apparent negative consequence.

However, Lithgow recognized that standard laboratory conditions with a constant environment and unlimited food did not resemble life in the natural world where *C. elegans* normally lives. Therefore, his team repeated the experiment, but this time they allowed the worms to consume all their food to mimic temporary starvation for four days. Each starvation cycle was continued for six cycles of feeding and starving. At the conclusion of the experiment, the investigators determined the proportion of the surviving worms that were *age-1* or *wt* (Figure 3.11). They initiated this experiment with a 50:50 mix of *wt* and *age-1* mutant worms, and the experiment was conducted five times. Unlike the previous result, this time the ratio of mutant worms in a mixed population decreased dramatically, from 0.5 to 0.06. Upon examination, it was determined that only young adults were able to lay eggs during the starvation cycles, which indicated the benefit of prolonged life for *age-1* mutants was lost in reproductive capacity during starvation.

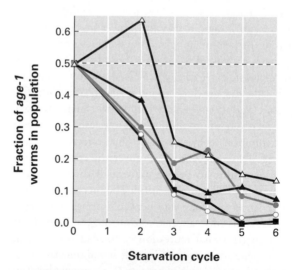

FIGURE 3.11 • **Comparison of fitness between food-restricted long-lived *age-1* mutants and *wt C. elegans.*** The worms were maintained at 20°C and were permitted to consume all the available food and then go without food for four days to complete one starvation cycle. The initial ratio of *age-1* worms to *wt* was set at 0.5 and repeated five times (each line represents a separate experiment). A reduction of 0.44 allele frequency over 12 generations is estimated to indicate a relative fitness of 0.77 compared to *wt*. The dashed line marks the initial ratio.

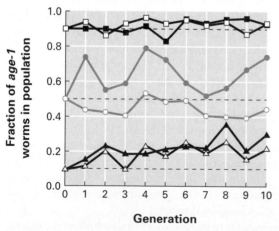

FIGURE 3.10 • **Relative fitness of long-lived worms.** Long-lived *age-1* mutants and *wt C. elegans* were maintained at 20°C and given unlimited food. The initial ratios of *age-1* to *wt* were set at 0.9 (black), 0.5 (gray), or 0.1 (blue), and each experiment was duplicated (filled and unfilled shapes). Horizontal dashed lines mark the three initial ratios.

These experimental conditions appear more realistic than standard laboratory conditions, and the data support the antagonistic pleiotropy hypothesis. Therefore, Lithgow and his colleagues concluded that the genomically determined longevity is regulated by antagonistic pleiotropy and not by mutation accumulation.

DISCOVERY QUESTIONS

29. Do you know of any drug that does not have a side effect in some individuals?

30. Do you think it would be possible to extend human life span by medication and not experience negative side effects?

Do Bacteria Experience Genomic Trade-offs Too?

When trying to explain mechanisms for the evolution of aging, you would like them to apply to all organisms. The longevity experiments with *C. elegans* supported the antagonistic pleiotropy hypothesis, and now we will look at evolution in the model bacterium *E. coli*. Richard Lenski at Michigan State University has been studying real-time evolution in cultures of bacteria that have been growing and evolving continuously since 1988 (over 3,000 days), which covered 20,000 generations. Lenski and his graduate student Vaughn Cooper wanted to test the two competing hypotheses to explain evolution.

They grew 12 different populations of *E. coli* for 20,000 generations in the presence of glucose as the only carbon source. During this time, they removed aliquots and froze them for later analysis. Once they had obtained aliquots from different time points over the 20,000 generations, they measured the ability of each aliquot of bacteria to utilize nonglucose energy sources. They wanted to know if there had been a concomitant trade-off for evolving optimum metabolism of glucose, as predicted by antagonistic pleiotropy. Or perhaps the cells experienced genetically distinct changes when glucose was the only energy source, so that optimization of glucose metabolism and the loss of other catabolic functions were unrelated, as predicted by mutation accumulation. For each of the aliquots of evolved bacteria, the investigators measured fitness as defined by improved metabolism and growth in the glucose-only environment (Figure 3.12a). The rate of adaptation to glucose was rapid in the first 5,000 generations and then slowed.

In 64 separate experiments, they also measured the ability of each aliquot to grow under new conditions where different sources of energy were available but glucose was not. The

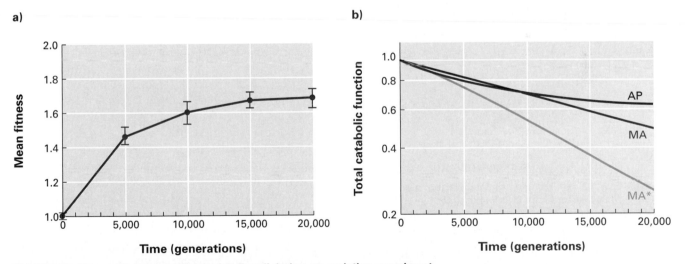

a)

b)

FIGURE 3.12 • **Changes in fitness for *E. coli* during an evolution experiment.**
a) Measurement of mean fitness during 20,000 generations in minimal glucose medium. Each point is the mean of all 12 populations, with each population tested five times. Error bars are the 95% confidence intervals, and Y-axis is a linear scale. **b)** Computer-simulated trajectories of evolving metabolic specialization. AP (antagonistic pleiotropy) in which functional decay is inversely proportional to gains in fitness. MA (mutation accumulation with original mutation rate) in which loss of catabolic potential occurs at a constant rate and is independent of the pace of adaptation. MA* (mutation accumulation at accelerated mutation frequency) occurs when a population becomes an accelerated mutator strain. Total catabolic function is shown on a logarithmic scale, with mutation accumulation graphed linearly.

ability of each aliquot of cells to catabolize a nonglucose source of food was compared to the original parental strain of bacteria that had not been subjected to glucose-only metabolism. If the loss of nonglucose catabolic potential occurred at the same rate as the increase in fitness, then antagonistic pleiotropy would be supported. If alternative catabolic pathways were lost at a linear rate, then mutation accumulation would be supported (Figure 3.12b).

Over the course of the 20,000-generation experiment, the loss of catabolic potential was rapid initially, and then slowed (Figure 3.13). During the course of the experiment, 3 of the 12 populations of bacteria developed mutations, so that they accumulated new mutations at a faster rate than the parental strain; these were called "accelerated mutators." The nine normal mutators and the three accelerated mutators lost their ability to catabolize other energy sources at the same rate that they adapted to growth on glucose. These data support the antagonistic pleiotropy hypothesis to explain the mechanism of evolution.

We began this section on aging wondering if genomic variations could explain differences in life spans. The growing body of evidence supports a genomic role for aging. We also have abundant data indicating that genomic variations

produce variable phenotypes. In the two experiments presented above, we begin to see why we age and why death is a biological certainty. There are trade-offs to all adaptations, and reduced fecundity and fitness appear to be the trade-off for increased longevity. Similar explanations can be proposed for disease alleles that persist in certain populations, such as a reduction in glucose-6-phosphate dehydrogenase and resistance to malaria (see page 84). In February 2001, a genetically engineered mouse that was smarter than normal mice (in learning and memory tests) was shown to be more sensitive to chronic pain. Evolution is a balance between gains and losses, and aging is no exception.

DISCOVERY QUESTIONS

31. In Figure 3.12b, antagonistic pleiotropy (AP) was a curve while mutation accumulation (MA) was a straight line. Do you think it would be easy to distinguish between the lines drawn for AP and MA?

32. Look at the data in Figure 3.13. Given the variation for the replicated experiments, do you think the two curves drawn for the two types of cells accurately summarize the consequences of adapting to glucose-only metabolism?

33. Do we have enough information to determine whether antagonistic pleiotropy or mutation accumulation hypotheses have been supported by these experiments? Explain your answer.

When a mutation produces a selective advantage early in life, we may also accumulate a cost that is exhibited late in life. When model organisms are mutated to prolong life, they lose fitness, meaning the mutation would not spread in a natural population under normal selection pressures. One remaining question about aging is whether we humans have controlled our environment to a similar degree as found in constant laboratory conditions. The Delany sisters did not marry, so we do not know if their longevity came with a loss of reproductive potential. Could medical techniques be used to produce long-lived humans who could thrive in a well-controlled environment? Would a pharmaceutical "fountain-of-youth" pill provide us with a capacity to live longer, just as *age-1* worms did under stable conditions? What would the trade-offs be if our environment were suddenly changed either locally (e.g., restricted diet) or globally (e.g., increased temperatures and CO_2 levels)?

All these futuristic scenarios have a caveat. Genetically identical animals maintained in identical environments do not die at the same time. As will be discussed in Chapter 6, all proteins exhibit stochastic behaviors such that two identical enzymes do not have identical kinetics. Even if a fountain-of-youth pill were available and we cloned humans,

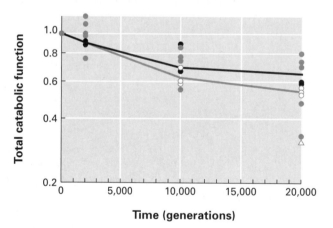

Time (generations)

FIGURE 3.13 • **Evolution of total catabolic potential during 20,000 generations in minimum medium supplemented with glucose.** Total relative catabolic function is calculated as a weighted average for growth on 64 different substrates, with 1.0 representing the starting point of catabolic potential. Values are shown on a logarithmic scale so a mutation accumulation outcome would have appeared linear. Each point is the mean of three clones from each population. The black line indicates the mean value for the eight populations with original mutation rate, while the gray line indicates the mean for the three accelerated mutators. Closed circles are populations with original mutation rate; open circles are populations with accelerated mutation rate; open triangle is a late accelerated mutator population. An accelerated mutator appeared late in the experiment and was excluded from the mean calculations for both lines.

there would still be a range of ages when people die. So variation is inherent in the **proteome** as well as the genome.

Summary 3.3

Genomic variation is integral to the survival of any species in a changing environment. From the largest mammals to the smallest phytoplankton, genomes vary. These variations can be used to study population dynamics, including human evolution and diseases. However, a few lessons must be kept in mind when thinking about genomic variation: (1) not all variations will have an impact on phenotypes; (2) environmental factors influence the "interpretation" of genomes; (3) even genetically identical individuals under identical environments will exhibit phenotypic differences; (4) evolutionary advances come with trade-offs; and (5) legitimate ethical concerns need to be addressed before technical progress presents us with unanticipated dilemmas.

The final portion of this chapter focuses on ethical concerns. With the exception of murder, incest, and other extreme situations, there are few universally accepted ethical conclusions. We use case studies where unanimous opinions are impossible but consensus may be possible. Then we examine some cases where ethical objections will be clear to most people, but individuals may proceed nonetheless.

3.4 Ethical Consequences of Genomic Variations

In this section, we examine a few of the ethical issues that surround genomics. The chapter on genomic variations is the best choice for this topic because if all genomes were identical, there would be no ethical dilemmas to consider. The topics discussed in this section highlight subjects that can ignite angry debates, but are not intended to include every ethical issue. The topics are divided into three general categories: genetically modified organisms, genetic testing, and cloning humans.

Are Genetically Modified Organisms Bad?

Genetically modified organisms (GMOs) are plants, animals, and prokaryotes that had their genomes modified through the addition of new genes and/or removal of endogenous genes. This topic is seen as completely evil by some people, and enormously beneficial by others. Let's examine six cases and try to see both sides of the arguments. As with any new knowledge, there is the potential for great improvements and disastrous consequences. The balance of these two potentials will be determined on a case-by-case basis, and outcomes may depend on your point of view. Humans have been genetically modifying other species for thousands of years through

selective breeding. Rice, corn, potatoes, meats, and fruits have all been modified from their original forms to enhance qualities deemed valuable by humans. Now we have the ability to accelerate this process and design traits that may never have been possible through selective breeding.

LINKS

Syngenta

Greenpeace

It is worth mentioning one common argument against genetically modified foods—consumers will be forced to eat DNA and proteins from species other than the intended food. This has been true since humans evolved, because bacteria and other microbes live everywhere: in your kitchen, at restaurants, and on your skin. Our digestive and immune systems have evolved to take care of dietary freeloaders. DNA is readily destroyed by the low pH of our stomachs and the DNases we produce. So consuming more DNA should not be considered a realistic hazard. But there are other potential hazards that are worth further discussion.

Golden Rice Rice is the staple food for 124 million people worldwide, many of whom suffer from vitamin A deficiency. In Southeast Asia, it is estimated that 250,000 children go blind each year due to vitamin A deficiency. Oral vitamin pills are not feasible due to cost and limitations in infrastructure such as transportation. In January 2000, a team from Switzerland and Germany genetically engineered a strain of rice that can produce β-carotene (normally absent in rice), which humans can convert into vitamin A. To construct this strain, they transformed rice with three genes: two encoded enzymes from daffodil and one encoded a bacterial enzyme. Rice that makes β-carotene appear golden in color, and the name "Golden Rice" is associated with this strain.

In February 2001, the first strains were shipped to the International Rice Research Station in Los Baños, Philippines. The research station will conduct safety studies and then use traditional breeding techniques to introduce the new phenotype into varieties that perform well in local growth conditions. Similar shipments are planned for research institutes in Africa, China, India, and Latin America. The inventors of Golden Rice transferred all commercial right to the world's largest agribusiness called Syngenta, which in turn will give the rice free of charge to all subsistence farmers (earning less than US$10,000 annually). Reaching this agreement was not easy since Golden Rice utilized technology covered in 70 patents held by 32 different companies and universities.

It sounds too good to be true—at least that's what groups such as Greenpeace think. They are concerned that this project will enable agribusinesses like Syngenta to introduce more GMOs into developing countries, where organized resistance is limited. The inventors of Golden Rice counter that only strains consumed in developing countries will be bred, so that it does not become a backdoor way to grow GMOs in poor countries for consumption in rich countries. There are other institutions using biotechnology to help

STRUCTURES

Antibiotic
Resistance Genes

LINKS

CAMBIA

Metabolix

*Scientific
American* Article

Charles Arntzen

Nextran

farmers in developing countries. The Center for the Application of Molecular Biology to International Agriculture (CAMBIA) in Melbourne, Australia, is one example.

Biological Plastics A large number of labs around the world are busy engineering plants and bacteria that can produce polyhydroxyalkanoate (PHA), a naturally occurring form of polyester. The first PHA was identified by the French microbiologist Maurice Lemoigne in 1925, and currently over one hundred different PHAs have been described in the scientific literature. Rather than having to use petroleum as the raw material, PHA uses renewable resources such as sugars (for bacterial production) or CO_2 and sunlight (for plant production). In addition to consuming less petroleum, these plastics are also biodegradable. So by using GMOs to produce a less toxic and more degradable plastic, we may be able to keep using plastic products and do less harm to the planet at the same time. One company working on this is Metabolix, and you can learn more from the company's perspective at their web site. You can read more in the *Scientific American* article to see if the hype has real potential.

Pharmaceutical Produce Beginning in the first half of the 1990s, potatoes and other edible plants were engineered as delivery mechanisms for vaccinations. By engineering plants to produce viral or bacterial peptides, Charles Arntzen and his research group were able to produce the world's first prescription potato. Initially, they used raw potatoes since heat is known to denature most proteins. However, most people (especially children) prefer cooked potatoes. Arntzen's team tested fried potatoes and found that they retained 50% of their potency. You cannot force most children to eat more than one bite of raw potato, but they will eat a bag full of chips, so the loss of potency could be compensated for by increased consumption. New trials are underway to develop bananas instead of potatoes, since they are grown in more places that need access to vaccines and because everyone likes raw bananas.

Sterile Fruit In 1997, a patent was granted for the technology that became known as "Terminator." Terminator biotechnology is the production of plants that produce sterile seeds. Monsanto announced a merger with one of the two patent holders (the other being the U.S. Department of Agriculture) with the intention of using Terminator technology to "protect the investment companies make in developing genetically improved crops, as well as possibly providing other agronomic benefits." This may not sound too bad since you don't think about eating seeds, but to farmers, the consequences are significant. Since humans began farming

thousands of years ago, we have saved some seeds to plant next year. If seed companies produced only products that were sterile, farmers (and thus everyone) would become dependent on seed companies to supply next year's crops. This would put the world's food supply in the hands of a few multinational seed suppliers, which does not appeal to most farmers.

In October 1999, Robert Shapiro, CEO of Monsanto, declared its intentions "not to commercialize sterile seed technologies, such as the one dubbed 'Terminator.'" However, Monsanto does hold patents on technologies for gene protection that only inactivate the transgene introduced by Monsanto. Shapiro indicated that Monsanto continues to develop these technologies for possible future use. You can read the full patent by going to the U.S. Patent and Trade Office and searching for patent number 5,723,765.

Pest-Resistant Plants Many companies have produced plants that contain toxins normally made in bacteria that are capable of killing insect pests. The best-known example is *Bacillus subtilis* toxin (BT). The attraction of this design is twofold. First, since the toxin evolved through natural selection, it is believed that resistance to BT will be slower to appear in insects. Second, if the plants produce the pesticide, then farmers will not spray their crops as much, thereby saving money and protecting the environment. There were some concerns that BT corn was killing monarch butterflies, but this has been shown to be untrue.

We read in Chapter 1 that some human genes may have been acquired by lateral transfer from bacteria. Similarly, many other species have exchanged genes through nonsexual mechanisms, making pest-resistant crops potentially hazardous. Since plants are pollinated by animals or wind, we have no control over their gene flow. Furthermore, we barely understand how genes are transmitted by horizontal transfer. Therefore, if a pest-resistant gene were to find itself in a weed, we might be unable to control it in our agricultural fields. Bioengineered genes are often transferred along with antibiotic resistance genes in the process of genetic engineering. What if antibiotic resistance was transmitted to a pathogen that we currently control by antibiotics? Finally, insects are becoming resistant to BT, negating the original assumption that evolution had produced a compound that would never be ineffective on insects.

Xenotransplants Each day in the United States, 13 people die while waiting for an organ transplant. Xenotransplants are organs taken from one species and put into another. For example, in 1984, Dr. Leonard Bailey transplanted an infant baboon's heart into a newborn human identified as Baby Fae, who lived 20 days with the animal heart in her chest while she waited in vain for a human heart donor. Several companies are genetically engineering pigs to be better donors for organ transplantation to humans. One company, Nextran, has modified pigs so their plasma membranes contain human proteins and thus would not be rejected as quickly. How is this research progressing? Here is what Nextran said:

Nextran recently concluded a Phase I clinical trial that used transgenic pig livers as an *ex vivo* (outside the body) support system for patients with acute liver failure. The pig liver was used to bridge the gap between organ failure and obtaining an appropriate human liver for transplantation in these patients. The results were encouraging, and our researchers have learned a great deal about the ability of the transgenic organs to withstand the human immune response and about the development of clinical protocol and various assays for the monitoring of patients in this groundbreaking area. With the data gathered from this *ex vivo* trial, together with information from our ongoing preclinical research, we plan to develop protocol recommendations for a Phase I *in vivo* (inside the body) clinical trial and to submit those recommendations to the FDA.

Many scientists think xenotransplants may unleash known and unknown pathogens. All pigs contain several copies of porcine endogenous retroviruses (PERV). In September 2000, pig tissue was transplanted into immunodeficient mice, which became infected with PERV. Furthermore, PERV was able to infect human cells in culture. This demonstrated that PERV was able to infect a new species in vivo and human cells in vitro. Since it would be unethical to perform in vivo experiments in humans, this is as close as we will get under controlled conditions to finding out if pig viruses could infect human patients.

It is very unlikely that we have identified every *human* pathogen—new ones are discovered each year. We know even less about pig pathogens, so it seems reasonable to think there may be porcine pathogens that could jump to humans due to increased exposure. If you were the recipient of a pig organ, you would be given very strong immunosuppressants (e.g., cyclosporin) to prevent your immune system from rejecting the new organ. Under these conditions, a pathogen may be able to mutate into a more human-compatible variation, producing a new human pathogen that could spread rapidly to other humans. The people living around you would most likely become infected first. Although the need for organ donors is compelling, the potential for species-wide harm is equally so.

DISCOVERY QUESTIONS

34. Do you think there are any positive uses for GMOs? Do you think there are any inappropriate uses for GMOs? Cite examples to support your answers.

35. Are there more potential uses for GMOs in developing countries, or is this an immoral way to sneak GMOs into the economies of developed countries?

36. If you were a CEO of a biotech company, what criteria would you use to evaluate which products to develop and which not to develop? Remember, you would have a legal obligation to maximize profits for your shareholders.

LINKS
ELSI

Is Genetic Testing Good?

From the very beginning of the HGP, it was recognized that ethical issues would need to be studied and discussed. In 1990, a portion of the HGP budget was dedicated to ethical issues under the heading of **Ethical Legal and Social Issues (ELSI)**. ELSI focuses its efforts in four areas:

- *Privacy and Fairness in the Use and Interpretation of Genetic Information.* Activities in this area examine the meaning of genetic information and how to prevent its misinterpretation or misuse.

- *Clinical Integration of New Genetic Technologies.* These activities examine the impact of genetic testing on individuals, families, and society and inform clinical policies related to genetic testing and counseling.

- *Issues Surrounding Genetics Research.* Activities in this area focus on informed consent and other research ethics related to the design, conduct, participation in, and reporting of genetics research.

- *Public and Professional Education.* This area includes activities that provide education on genetics and related ELSI issues to health professionals, policy makers, and the general public.

In this section, we examine some real-world examples of how genome sequence information can lead to ethical dilemmas. Again, this is not meant to be an exhaustive list but one that illustrates issues we all will face.

Every week a new disease-associated gene is identified, which means a new genetic test could be available soon. If you decide to have children, you may be confronted with a long list of possible genetic tests that could be performed on your developing fetus. Even before conception, you and your partner could be tested for the potential of producing a child with a genetic condition that most people consider unappealing. Since genomics has revealed to us that most medical conditions are not as simple as one gene–one phenotype, there is a growing interest in collecting genome-wide SNPs from diverse populations. Depending on the individual being tested (i.e., family history, ethnicity) and the genetic test being considered, the value of a particular test will vary (Figure 3.14). Genetic testing is expensive and may subject individuals to societal pressures to conform. Is it a new version of eugenics? Is this the best way to spend limited medical budgets? When

FIGURE 3.14 • **Continuum for the utility of a particular genetic test.** Tests for diseases that can be treated easily have a high utility, while tests for incurable diseases offer less utility or might even be harmful.

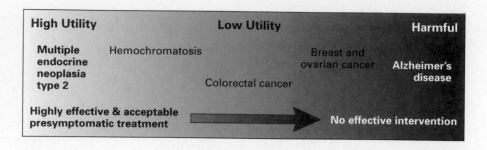

LINKS

Steve Hyman

Dorothy Nelkin

Peter Neufeld

Innocence Project

are genetic tests reasonable and cost-effective? Who has a right to know the results of your tests? Who has a right to obtain some of your DNA for genetic testing? Is it acceptable for someone to grab a handful of hair from your brush since you were going to dispose of those dead cells anyway? The answers to these questions are less obvious than the ones raised in the section on GMOs.

Some experts in scientific, ethical, and legal implications of genomics were asked about ethical concerns in a New York forum in October 2000. There was a consensus that most people, including journalists, behavioral scientists, and physicians, either were unaware of the new understanding of complexity in the genome or were confused and believe in a new form of **genetic determinism.** Steve Hyman, director of the National Institutes of Mental Health, remembered a newspaper article that stated the "worry gene" had been cloned, but reminded the audience that complex personality traits involve many genes as well as developmental and environmental influences. Dorothy Nelkin from New York University reported her dismay at the large proportion of physicians who "know nothing about genetics" and often believe we humans are determined by our genes alone. (Her concerns were supported by a paper that appeared in the April 2001 edition of the *British Medical Journal* that emphasized the need for a multifaceted approach to assist physicians in coping with the new genetic tests available.) Peter Neufeld, director of the Innocence Project in New York, soberly stated that advances in genetics will not overcome racism and classism. He predicted that insurance companies and the judicial system will increasingly use genomic data. "There is no doubt that this body of work will be abused."

Let's look at three examples to illustrate some of their concerns.

British Life Insurance As of 13 October 2000, life insurance companies in Britain may use data from a new Huntington's disease (HD) genetic test. This becomes the eighth test for alleles associated with diseases including Alzheimer's, breast cancer, and colon cancer. The Genetics and Insurance Committee established by the UK Department of Health made the recommendation that HD be added to the list of acceptable tests. If the test result is negative, they must use rates based on whole-population actuary tables. The Committee

recommended that the test not be used for life insurance policies of less than £100,000 since this is often required before buying a house. However, the Committee managed to sidestep what happens to individuals who test positive for HD. Chris Smith, an actuary with the insurance company Swiss Re, stated that these people are usually denied insurance. Insurance companies believe that genetic tests are no different from other risk factors, such as smoking and family pedigrees, that are used universally to calculate insurance rates. Yet the degree of certainty with HD alleles seems to run counter to the very concept that everyone should be able to obtain insurance based on whole-population statistics.

This HD case was published in November 2000 as an editorial in *Nature Genetics.* The specifics may be unique to Britain, but the fundamental principles and the desire of insurance companies to test applicants are global. In the United States, applicants for life insurance submit blood to be tested for HIV, but other tests could be performed with the same sample. What tests should be permitted, and how should the results be used? Doesn't it seem fair that those who test negative for a disease trait should pay less than those who test positive? What if you were the one who tested positive and were denied insurance? These issues affect everyone who wants life and medical insurance.

DISCOVERY QUESTIONS

37. If you were asked to write a policy statement for the government, what would you conclude about the use of genetic testing and insurance applications?

38. If you outlawed such tests, would you also outlaw blood tests for HIV? What about family background information? Where would you draw the line?

39. If you permitted such tests, would you include less certain tests such as *BRCA1* and *BRCA2* for breast and ovarian cancer? *ApoE* screening for Alzheimer's? Where would you draw the line?

40. Since tests are already in use, would you prohibit denial of insurance but allow the companies to set the prices based on the genetic knowledge?

Universal Screening for Cystic Fibrosis Soon, every pregnant woman in America will be informed about the availability of a genetic test for the recessive disease cystic fibrosis (CF). In October 2001, the American College of Obstetricians and Gynecologists and the American College of Medical Genetics recommended that CF screening information be provided to "couples who are in lower-risk racial and ethnic groups and have no known ancestors from high risk groups." This recommendation is made even though the detection rate is 69% for African Americans, 57% for Hispanics, and 30% for Asian Americans due to population allele frequencies.

More than 400 genetic tests are available, but none have been implemented nationally until now. CF is the most common genetic disease for Caucasians but not other populations. It is worth noting that the U.S. Census Bureau estimates that by 2100, 60% of the American population will be composed of non-Caucasians, so universal screening may not be appropriate as a national policy. Many studies have been conducted on the cost/benefit ratio of national genetic tests, and most everyone agrees they should not be performed. However, a physician who does not inform his or her patients may be vulnerable to legal action should a child be born with CF, which happens in the United States about once in every 2,500 Caucasian live births and once in every 17,000 African American live births.

For most people, the first time they would learn about CF would be shortly after being told the woman was pregnant. Although CF sounds like a straightforward genetic disease, it is not. Over 900 different CF alleles have been identified, and the consequences are known for about 100. The most common mutation, called ΔF508, accounts for about 70% of all mutant alleles in Caucasians and another 20 alleles account for an additional 15% of all mutant alleles. Therefore, the accuracy of the test would be approximately 85% for Caucasians if all 21 alleles were tested. Physicians would need to explain the test, the relative risk for each partner's ethnic background, the potential outcomes, and the choices facing the couple depending on the outcomes. Furthermore, the long-term prognosis for CF patients has improved in recent years, and childhood death is preventable with aggressive intervention.

The California-based HMO Kaiser Permanente has already conducted a small-scale pilot test under well-controlled conditions. They offered the test to all Caucasian women and their partners. If both tested positive, the expecting parents could choose to test the fetus as well. If the fetus proved to be homozygous, the couple had to decide what action to take, if any. Approximately 90% of these cases ended in abortion of the fetus. Kaiser has screened over 18,000 women to date.

The Kaiser pilot test was well controlled, but there is concern that a national program will not be as uniform or as reliable. Furthermore, a national screening program requires that every physician become fully informed about the possible consequences, and they need to be taught how to communicate this information effectively to a diverse audience. The testing was free to patients in the Kaiser pilot program, but a national program would probably cost patients at least part of the expense, which ranges from $20 to $300. Sixty U.S. laboratories test for different numbers of alleles, from 1 to 87. Different labs will have different error rates, so a single national standard for accuracy seems unlikely.

Although most experts agree that a universal test for CF does not make sense (see National Institute of Health [NIH] recommendation), it is coming anyway. Physicians will be pressured to inform their patients, who may feel pressured to submit to the test. No one knows what will happen as the testing becomes widespread. Physicians will probably rely on well-written booklets to inform patients about CF and genetic testing, but will that be adequate education for everyone? You will become participants in this experiment, and the results will be tabulated as couples face new choices and uncertainties. No doubt you will know people who get tested, or you might be tested. What used to be an abstract hypothetical case will soon be very real and personal. What will you do?

LINKS

American College of Medical Genetics

U.S. Census Bureau

NIH Recommendation

DISCOVERY QUESTIONS

41. Should every woman be informed about the CF test? only high-risk Caucasian women? only those who ask?

42. Currently, U.S. funding is denied to any institution in a developing country that informs women about abortion options. In the United States, Title 10 money is denied to facilities that perform abortions. What impact might national CF screening have on the supply and demand of abortions in America? Would a similar genetic screen be advisable in developing countries?

43. CF may be the first of many national screening programs. What standard should be used to determine which genetic tests are appropriate and which should be limited to high-risk populations?

44. If physicians are unable to adequately educate their patients, due to lack of training or limited time, what steps are needed to assist them? Is there a national plan to address any anticipated shortcomings?

45. If a woman's insurance company will not pay for a CF test, who should pay for it? If a woman's insurance company does pay for a CF test, who has the

LINKS

deCODE

right to know the results of her test? the insurance company? the woman's blood relatives? her partner?

46. Given the guidelines for the utility of genetic testing (see Figure 3.14), what would your advice be for CF screening a Caucasian couple with no family history of CF? what about a couple of a Caucasian man and an Asian woman when the man had a cousin with CF?

Genomic Diversity Banks and Small Populations In 1996, Kari Stefansson started a company called deCODE with 20 employees. Their mission was to utilize population genetics to discover new disease-associated genes. This does not sound unique until you learn that their target population is the entire country of Iceland. There are 275,000 living Icelanders; the vast majority are descendants of a few European explorers who arrived about 1,000 years ago. The population is very homogenous, so finding significant differences that lead to medical conditions will be much easier than in a heterogeneous population. Icelanders have a rich tradition of maintaining family trees, so the whole country can be compiled into one large pedigree. Iceland has a single healthcare provider, and all the medical records are stored in one database. deCODE has purchased the medical records from the government and has correlated family relationships with medical records. Finally, every citizen will give blood to be used to create "genomic fingerprints," unless an individual opts out of the program.

In addition to finding new disease-associated genes, deCODE will be able to track the progress of patients with similar genotypes who are taking newly developed drugs. Roche Pharmaceuticals has already begun collaborating with deCODE to develop data mining software for this purpose. The data collected by deCODE can be sold to other companies that also want to use this island nation to develop new drugs. As of June 2002, deCODE employs over 800 people, making it the largest Icelandic corporation. Furthermore, by employing high-tech workers, deCODE reduces the "brain drain" experienced in Iceland as many of its highly trained citizens have sought employment in other countries.

Although deCODE quickly obtained endorsement from the national government, it does have its critics. Some physicians fear their patients will be less forthcoming with information now that they know it will be stored in commercial databases. Some feel the patient-physician trust has been broken by selling all the records as the default with informed *dissent* being the only option out, which 5% of the nation has chosen. Finally, some think physicians will refuse to comply with a law that requires them to deliver new clinical information to the database, and thus undermine the entire plan.

Stefansson does not agree with the critics, as he explained during an interview with CNN. "Recognize that knowledge is never evil in and of itself. If you run the world by forbidding new discoveries, you are controlling the world in an unpredictable manner. You are putting yourself in the position of God."

Internationally, the reaction by other investigators has been mixed. In Estonia, formerly a part of the Soviet Union and located in the Northwest region of Europe, a group of faculty at the country's only medical school want to establish a similar program. There are 1 million citizens and the health records are also centrally controlled. Estonian critics of the plan think too much money will be diverted from basic health-care needs. They also worry that people are not educated enough to make informed decisions and are likely to conform to national policy as they had done under communist rule.

DISCOVERY QUESTIONS

47. Did deCODE violate a basic rule of human research, which requires informed *consent* of the individual rather than informed *dissent*?

48. Should the people of Iceland share in the profits since their information and DNA is being used to generate profits?

49. deCODE says that DNA from nondissenting donors will be handled under the strictest conditions of anonymity. Do you think this is practical given the small population size and the very complete national pedigree?

50. Everyone has the same health insurance in Iceland, so the motivation to decipher the anonymity might seem minimal. However, there are competing life insurance companies in Iceland. Now can you imagine a motivation to discover the identity of donors?

51. Let's assume you think the deCODE approach is worthwhile, despite the protests of a few fringe physicians and citizens. Does that mean it is also worthwhile for Estonia to launch a similar program, given its differences in economy, health care, and education?

52. Is this a situation where it is okay for a developed country but not for a developing country to attempt similar projects?

Who Will Benefit from Genomic Medicine? We have all heard of cases where a health insurance company has refused to pay for "experimental treatment" that *may* save a patient's life. Genomic tests are expensive and not 100% accurate. Genomics has revealed many secrets; one is that biology is a complex web of interconnecting proteins, carbohydrates, lipids, etc. If you and your physician feel a new genomic-based test or treatment might help you, who should pay for it? If gene therapy for a rare disease costs $1,000,000 for each patient, should we all pay higher insurance premiums to pay for the treatment of a few patients? If there are limited dollars in the economy that can be used for health care, how should this money be spent and who should decide? Are there some medical cases that warrant no treatment even if there is a high probability of a cure? What if there is only a 1% chance of a cure? or a 0.001% chance? Is it appropriate to allow those who can afford expensive new treatments to be the only recipients? Is it appropriate for only developed countries to benefit from genomic medicine, since developing drugs for poor people will not result in a profit for the pharmaceutical companies?

There are many questions and no easy answers. This section was intended to provoke thought and discussion. It is important that those of you who understand the science also think about the consequences. It is important to share your knowledge and facilitate discussions. If you are reading this book, you are in a very small and elite minority. What responsibility do you have, given your place in society?

LINKS

Google.com

Are There Simple Applications for Complex Genomes?

In Aldous Huxley's *Brave New World,* human embryos were manipulated to occupy particular niches in society. Today, we are confronted with situations that are eerily similar to Huxley's science fiction scenario. Here are three cases that are neither fictional nor futuristic. They are all cases taken from the recent past and deserve our consideration.

New _____ (fill in the blank) Gene Discovered! We have all seen or heard headlines touting the discovery of a gene that encodes a complex trait. If you go to Google.com, you can

---

BOX 3.3 • *Should I Get a Genetic Test?*

The answer to this simple question is anything but simple. The utility of a genetic test hinges on two main factors: who is asking the question and which test is under consideration. James Evans (an oncologist), Cécile Skrzynia (a genetic counselor), and Wylie Burke (a bioethicist) published a short list to consider when debating the utility of genetic tests.

Increased Utility of Genetic Tests

- High morbidity and mortality of disease

- Effective but imperfect treatment for the disease

- High predictive power of the test (i.e., high **penetrance** for the disease)

- High cost or burdensome nature of nongenetic screening and surveillance methods

- Preventive measures are expensive or associated with adverse effects.

Decreased Utility of Genetic Tests

- Low morbidity and mortality of disease

- Highly effective and acceptable treatment

- Poor predictive power of genetic test (i.e., low penetrance)

- Screening and surveillance methods that are effective, inexpensive, and acceptable

- Treatment that is inexpensive, effective, and acceptable (e.g., vaccinations)

Most physicians are unprepared to cope with the hundreds of genetic tests that are commercially available. Here are the basic skills needed to integrate successfully genetic testing with primary care:

- Recognition of common inheritance patterns

- Appreciation for the role ethnicity plays in the assessment of risk for particular diseases

- Communication and counseling with patients in a nondirective manner

- Awareness of the inherent limitations in genetic tests and the consequences of testing for insurance coverage

Unfortunately, most physicians are not prepared to face the challenges of genetic testing in part because their training was inadequate. It is very difficult to teach medical students the latest information in a field that changes rapidly, and the number of genetic tests available continues to increase. Since keeping track of all this information is daunting, a new model is being evaluated in the UK—the use of community genetic counselors. These counselors act as a liaison between patients and physicians, perform the initial screening to determine when testing is appropriate, and conduct counseling sessions before and after testing. A suggested alternative model, that one physician in a group practice be chosen as a genetics expert, might lead to problems with scheduling and the ability to move from one job to another. More study needs to be conducted to determine the best way to integrate genetic testing with primary care.

LINKS

Leland M. Heller, MD

Panos Zavos and Severino Antinori

NBC News

search and find reputable web sites that will inform you about a "gay gene," "smart gene," "fat gene," "worry gene," "Alzheimer's gene," "cancer gene," even a "fountain-of-youth gene." By this point in the book, you must realize that our genomes are more complex than this. Personality traits and most genetic diseases do not fit neatly into tidy packages such as *the* _____ gene. As discussed in Chapter 10, even diseases and conditions that are relatively simple such as muscular dystrophy are not monogenic. So next time you hear a stop-the-presses headline about a new gene, be skeptical, very skeptical.

Maybe you would expect such claims of simple genomic perceptions from tabloids, but they are not alone. For example, in 1999, ABC News.com ran a story entitled "Where did the gay gene go?" In this story, ABC traces the history of the "gay gene," first popularized in 1993. There is good evidence of a genetic component to homosexuality, but probably only 50% of this phenotype. This story has fueled the left and right sides of the political and ethical spectra and led to confusion, misunderstandings, and misconceptions. In 2000, the Associated Press published a story entitled "Scientists Find Fat Gene," which is different from the other fat genes previously reported (see Chapter 11). CNN's Health Story Page announced "Scientists discover cellular 'fountain-of-youth.'" CNN was careful to say in the second paragraph that the finding won't make people any younger or allow them to live forever, but the researchers say it could keep them healthier longer.

Okay, we can't blame these news organizations; their job is to sell a product, and who wants to buy a paper that claims "Investigators characterized a SNP that is associated with obesity in mice"? But we tend to think physicians would not fall into a similar trap. If you log onto the web site of Leland M. Heller, MD, you can search for the term "BuSpar" to find the following amid some medical advice: "This [serotonin receptor #1] is a genetic worry gene." The sentence does not say *the* worry gene, but to the average reader the difference is unnoticeable. Is oversimplifying biological information just part of a Western tendency to hype science to sell products? The *Indian Express* ran a headline "Patients of Alzheimer's take heart, 'smart gene' is on the way." It appears that in India, at least, oversimplification is as common as it is in the United States.

We know that news organizations are in business to sell products. But at least academics are careful in the way their research is publicized locally, right? In the March/April 1997 issue of *Stanford Today,* you would find the following: "From a vibrating wing's serenade to copulation, the survival of the species depends on one gene" and "Breast cancer gene isolated"—which was neither of the other two "breast cancer genes" called *BRCA1* and *BRCA2.*

The problem seems to arise from time constraints. "Who has enough time to listen to the entire story anymore? Just give me the main point." Genomics is a study of complexity, dynamic interactions, and changing patterns. What we learn does not fit neatly into a 30-second news summary aimed at mass markets. If you take the time to talk with most scientists, they are careful not to hype their findings. On occasion, however, you may hear scientists make a claim that sounds a bit too bold, too amazing, too . . . simple. The point of these case studies was to caution you to listen skeptically to news stories, find out more information, and then be prepared to explain any oversimplifications to people you know who do not understand the technical jargon.

DISCOVERY QUESTIONS

53. Go to any news web site that you prefer (newspapers, TV, magazines, etc.). Search for the word "gene" and discover the genetic hype. Does the story appear too simplistic?

54. Do scientists share any of the responsibility if a news story distorts their findings?

55. Should we rely on news organizations to be the interpreters of genomic discoveries?

56. Can we blame news organizations for using hyperbole in order to boost sales if we, the consumers, continue to reward such behavior by buying their products?

57. Can you imagine a scenario where a scientist might intentionally hype his or her findings?

Cloning a Human Being In February 2001, two physicians announced they intended to **clone** a human being by 2003. Panos Zavos from Kentucky and Severino Antinori from Italy discussed a case where a man who had been castrated in an accident still wanted to have "his own biological children." In an interview with NBC News, Zavos explained his point of view on cloning humans.

Zavos was asked if there would be failed attempts to clone humans just as there have been for all other mammals. Zavos replied, "Mishaps. And we're ready to face those mishaps. We're going to try to limit those mishaps. We are not perfect, only God is." When asked how he could minimize "mishaps" Zavos said, "Quality control is something we scientists apply all the time. It's Q.C. It's a way of measuring your success. . . . We can find out if that particular embryo carries any hereditary diseases, any genetic deficiencies, and then we decide that embryo should not be implanted." If developmental problems develop after the first trimester, Zavos indicated that parents would be allowed to choose the outcome of these embryos even though it would not reflect well on his team's cloning efforts. "Those are the realities and this is how you learn from any procedure."

Zavos acknowledged there will likely be some miscarriages; when asked how many discarded embryos, miscarriages, or abortions would be allowed for each couple wanting a cloned offspring, his response was, "We hope none. We will be expecting, but obviously we're going to take all precautionary measures to see that this doesn't happen." Zavos sees any potential problems as the price that has to be paid in order to achieve progress. "It's part of any price that we pay when we develop new technology."

In summation, Zavos said, "[Human cloning] is here and it is for the world to harness. We hope to educate the world and tell the world that this technology can be used safely and securely enough without being abused. And that's really what people are afraid of. They don't want monsters. They don't want armies of cloned individuals out there. That's not going to happen."

It is worth mentioning that another group, the religious group called Raelians, has announced it too will clone a human. The Raelians believe that all plants and animals, including humans, are bioengineered products of extraterrestrial beings. Zavos said that cloning a human should not be a race with the Raelians and that his group will follow strict scientific guidelines. His efforts will be carried out in an undisclosed country and by a team of individuals who are not identified, though recently he suggested it might be conducted on a boat floating in international waters since so many countries are outlawing human cloning.

Ian Wilmut, the Scottish principal investigator who cloned Dolly, and Rudolf Jaenisch (an expert in **epigenetic regulation** of the genome) wrote a strong condemnation of Zavos and Antinori in the 30 March 2001 issue of *Science*. They outlined the number of failed attempts in work done with farm animals and neonatal deaths caused by respiratory and circulatory problems. The few cloned animals that have survived are often oversized, a condition referred to as "large offspring syndrome." Those apparently normal cloned animals may suffer from immune, kidney, or brain malfunctions and may develop abnormally.

Wilmut and Jaenisch proposed that the most likely cause for failed attempts in animals is inadequate genomic reprogramming. This is a series of epigenetic changes (changes that do not alter the DNA sequence) that alter the expression of certain genes (see imprinting in Chapter 1). In fact, expression of normally imprinted genes was significantly altered when mouse or sheep embryos were cultured in vitro prior to implantation. The are no methods available to screen embryos for epigenetic alterations, which could lead to severe consequences. Therefore it is impossible to establish substantial "quality control" on cloned human embryos.

Finally, Wilmut and Jaenisch make the point that attempts to clone a human will so repulse the public that legitimate efforts to use stem cells for growing new neurons, hearts, and other tissues will suffer. They conclude by saying that attempts to clone humans while there is so little known

about animal cloning would be dangerous and irresponsible.

Many governments, scientific societies, and bioethics communities have uniformly denounced human cloning. And yet the technology is available and inexpensive. The concerns of investigators who first cloned animals is reminiscent of physicists in the 1930s who helped develop atomic energy technology and later urged that atomic weapons not be developed. As Zavos said, the cloning genie is out of the bottle. In addition to technical experts, there must be women who are willing to carry a cloned fetus to term. What thoughts, feelings, and concerns will they have? What impact will the woman's emotional condition have on the hormonal state in which the fetus will grow? If Zavos is successful, will he keep it a secret? Will he publicly disclose his failures too, as one would expect if "scientific guidelines" are followed? Are there only two groups trying to clone humans, or only two that have gone public? Outside of those immediately involved, it is hard to imagine anyone who supports this enterprise. What childless couple would be willing to attempt this method to produce another human? Yet Zavos's team does have a web site where they claim to receive outside support.

LINKS
Raelians
Ian Wilmut
Rudolf Jaenisch

DISCOVERY QUESTIONS

58. Do those who developed animal cloning bear any responsibility for human cloning?

59. Is there any situation where cloning an animal would be a good thing? What if this were the only way to raise enough individuals to prevent the extinction of a species?

60. Since cloned animals are not natural products, they can be patented. Could a human clone be patented?

61. Some think cloning headless humans would be an acceptable way to supply the world's need for organ donors. What do you think?

62. Zavos admits that a cloned human would not necessarily be identical to the parental donor of the nucleus. Explain why he is right.

63. Would it be possible to produce cloned twins that would truly be identical?

Summary 3.4

All of us will be affected tomorrow by genomic research being conducted today. We will benefit from or suffer the consequences of new genetic tests, GMOs, and those who want to utilize the human genome as objects for simplistic understanding and manipulations. Agronomic companies want to develop better foods to improve the human condition and maximize their profits. Pharmaceutical companies want to create drugs that are genome specific. Prolonging life is the objective for much of our medical intervention, and yet there may be unexpected consequences. The questions are important and the answers elusive. However, sometimes debating the questions is more important than arriving at the "right" answer.

Chapter 3 Conclusions

Variety is the spice of life, and genomic variation is the engine of evolution. As you saw in Chapters 1 and 2, reference genome sequences will reveal many answers that will take decades to decipher. However, many more discoveries lie hidden in the genomic variation found within a species. For example, breeds of dogs or cats exhibit wonderful variations in their phenotypes, not because they have different genes but because they contain allelic variants of the same genes. The variation you see in people is due to genomic variation since we all share the same genes. How can some people who never smoked get lung cancer, while others who smoke heavily stay cancer free? Why do some people exposed to HIV never develop AIDS? The data are all around us, yet we know very little about the effects of genomic variations.

References

Environmental Case Studies

Baker, C. S., and S. R. Palumbi. 1994. Which whales are hunted? A molecular genetic approach to monitoring whaling. *Science.* 265: 1538–1539.

Comstock, Kenine E., Samuel K. Wasser, and Elaine A. Ostrander. 2000. Polymorphic microsatellite DNA loci identified in the African elephant *(Loxodonta africana). Molecular Ecology.* 9: 1004–1006.

López-García, Purificación, Francisco Rodríguez-Valera, et al. 2001. Unexpected diversity of small eukaryotes in deep-sea Antarctic plankton. *Nature.* 409: 603–607.

Moon-van der Staay, Seung, Rupert De Wachter, and Daniel Vaulot. 2001. Oceanic 18S rDNA sequences from picoplankton reveal unsuspected eukaryotic diversity. *Nature.* 409: 607–610.

Roca, A. L., N. Georgiadis, et al. 2001. Genetic evidence for two species of elephant in Africa. *Science.* 293: 1473–1477.

Rynearson, Tatiana, and E. Virginia Armbrust. 2000. DNA fingerprinting reveals extensive genetic diversity in a field population of the centric diatom *Ditylum brightwellii. Limnology and Oceanography.* 45(6): 1329–1340.

Human Genomic Variation

Dequeker, Elisabeth, and Jean-Jacques Cassiman. 2000. Genetic testing and quality control in diagnostic laboratories. *Nature Genetics.* 25: 259–260.

Hagmann, Michael. 1999. A good SNP may be hard to find. *Science.* 285: 21–22.

Ho, P. C., K. Ghose, et al. 2000. Effect of grapefruit juice on pharmacokinetics and pharmacodynamics of verapamil enantiomers in healthy volunteers. *European Journal of Clinical Pharmacology.* 56: 693–698.

International SNP Map Working Group. 2001. A map of human genome sequence variation containing 1.42 million single nucleotide polymorphisms. *Nature.* 409: 928–933.

Kanazawa, S., T. Ohkubo, and K. Sugawara. 2001. The effects of grapefruit juice on the pharmacokinetics of erythromycin. *European Journal of Clinical Pharmacology.* 56: 799–803.

Kruglyak, Leonid. 1999. Genetic isolates: Separate but equal? *PNAS.* 96: 1170–1172.

LaForge, K. Steven, Valentin Shick, et al. 2000. Detection of single nucleotide polymorphisms of the human mu opioid receptor gene by hybridization or single nucleotide extension on custom oligonucleotide gelpad microchips: Potential studies of addiction. *American Journal of Medical Genetics.* 96: 604–615.

Lee, M., D. I. Min, et al. 2001. Effect of grapefruit juice on pharmacokinetics of microemulsion cyclosporine in African American subjects compared with Caucasian subjects: Does ethnic difference matter? *Journal of Clinical Pharmacology.* 41: 317–323.

Liang, Min-Hui, and Lee-Jun C. Wong. 1998. Yield of mtDNA mutation analysis in 2,000 patients. *American Journal of Medical Genetics.* 77: 395–400.

Lin, Yvonne. 2001. Double-whammy: Food-drug interactions. *Northwest Science and Technology.* Winter: 42–43.

Maquat, Lynne. 2001. The power of point mutations. *Nature Genetics.* 27: 5–6.

Marth, Gabor T., Ian Korf, et al. 1999. A general approach to single-nucleotide polymorphism discovery. *Nature Genetics.* 23: 452–456.

McIntosh, J. Michael, Ameurfina D. Santos, and Baldomero M. Olivera. 1999. Conus peptides targeted to specific nicotinic acetylcholine receptor subtypes. *Annual Review of Biochemistry.* 68: 59–88.

McLeod, H. L., and William E. Evans. 2001. Pharmacogenomics: unlocking the human genome for better drug therapy. *Annual Review of Pharmacology and Toxicology.* 41: 101–121.

Monnens, A. H., Lambert P. W. J. van den Heuvel, and Nine V. A. M. Knoers. 2000. Dominant isolated renal magnesium loss is caused by misrouting of the Na^+, K^+-ATPase γ-subunit. *Nature Genetics.* 26: 265–266.

Morrow, K. John. 2000. Improving patients' responses to drug therapies. *Genetic Engineering News.* 20(10): 18, 75, and 86.

Mullikin, J. C., Hunt, F. E., et al. 2000. A SNP map of human chromosome 22. *Nature.* 407: 516–520.

Olivera, Baldomero, Lourdes J. Cuz, and Doju Yoshikami. 1999. Effects of *Conus* peptides on the behavior of mice. *Current Opinion in Neurobiology.* 9: 772–777.

Roberts, Leslie. 2000. SNP mappers confront reality and find it daunting. *Science.* 287: 1898–1899.

Rubin, Edward M., and Alan Tall. 2000. Perspectives for vascular genomics. *Nature.* 407: 265–269.

Spahn-Langguth, H., and P. Langguth. 2001. Grapefruit juice enhances intestinal absorption of the P-glycoprotein substrate talinolol. *European Journal of Pharmacological Science.* 12: 361–367.

Weiss, Kenneth M., and Joseph D. Terwilliger. 2000. How many diseases does it take to map a gene with SNPs? *Nature Genetics.* 26: 151–157.

Pharmacogenomics
Evans, William E., and Julie A. Johnson. 2001. Pharmacogenomics: The inherited basis for interindividual differences in drug response. *Annual Review of Human Genetics*. 2: 9–39.

Evans, William E., and Mary V. Relling. 1999. Pharmacogenomics: Translating functional genomics into rational therapeutics. *Science*. 286: 487–491.

Aging
Chakravarti, Aravinda. 1999. Population genetics—making sense out of sequence. *Nature Genetics Supplement*. 21: 56–60.

Cooper, Vaughn S., and Richard E. Lenski. 2000. The population genetics of ecological specialization in evolving Escherichia coli populations. *Nature*. 407: 736–739.

Guarente, Leonard. 1997. What makes us tick? *Science*. 275: 943–944.

Guarente, Leonard, and Cynthia Kenyon. 2000. Genetic pathways that regulate aging in model organisms. *Nature*. 408: 255–262.

Kirkwood, Thomas B., and Steven N. Austad. 2000. Why do we age? *Nature*. 408: 233–238.

Kathleen McFadden. 2000. The week's famous and infamous women: Centenarian sisters Bessie (1891–1995) & Sadie (1890–1999) Delany. <http://writetools.com/women/stories/delany_bessie.html>. Accessed 11 June 2001.

Lanza, Robert P., Jose B. Cibelli, et al. 2000. Extension of cell lifespan and telomere length in animals cloned from senescent somatic cells. *Science*. 288: 665–668.

Ly, Danith H., David J. Lockhart, et al. 2000. Mitotic misregulation and human aging. *Science*. 287: 2486–2492.

Michikawa, Yuichi, Franca Mazzucchelli, et al. 1999. Aging-dependent large accumulation of point mutations in the human mtDNA control region for replication. *Science*. 286: 774–779.

Pennisi, Elizabeth. 1998. Single gene controls fruit fly life-span. *Science*. 282: 856.

Rogina, Blanka, Robert A. Reenan, et al. 2000. Extended life-span conferred by cotransporter gene mutations in *Drosophila*. *Science*. 290: 2137–2140.

Walker, David W., Gawain McColl, et al. 2000. Evolution of lifespan in *C. elegans*. *Nature*. 405: 296–297.

Ethical Consequences of Genomic Variations
Bowen, David, Thomas A. Rocheleau, et al. 1998. Insecticidal toxins from the bacterium *Photorhabdus luminescens*. *Science*. 280: 2129–2132.

Breast cancer gene isolated. 1997. *Stanford Today*. March/April. 32.

Celera. 2001. Test case for genetic testing. <http://www.celera.com/genomics/news/articles/03_01/Test_case_CF.cfm>. Accessed 11 June 2001.

Chamberlain, Claudine. 1999. Where did the gay gene go? <http://abcnews.go.com/sections/living/DailyNews/gaygene990422.html>. Accessed 11 June 2001.

Cibelli, Jose B., Steve L. Stice, et al. 1998. Cloned transgenic calves produced from nonquiescent fetal fibroblasts. Science. 280: 1256–1258.

CNN.com. 1998. Scientists discover cellular "fountain of youth." <http://www.cnn.com/HEALTH/9801/13/life.extention/>. Accessed 11 June 2001.

CNN.com 2000. Iceland sells its medical records, pitting privacy against greater good. <http://www.cnn.com/2000/WORLD/europe/03/03/iceland.genes/index.html> Accessed 11 June 2001.

deCODE Genetics. 2.1. Company web site. <http://www.decode.com/company/history/> Accessed June 11 2001.

Editorial. 2000. Testing times. *Nature Genetics*. 26: 251–252.

Emery, Jon, and Susan Hayflick. 2001. The challenge of integrating genetic medicine into primary care. *British Medical Journal*. 322: 1027–1030.

Environment News Service. 1999. Monsanto sacks sterile seeds. <http://www.wired.com/news/lycos/0,1306,31786,oo.html>. Accessed 11 June 2001.

Evans, James P., Cecile Skrzynia, and Wylie Burke. 2001. The complexities of predictive genetic testing. *British Medical Journal*. 322: 1052–1056.

Frank, Lone. 1999. Storm brews over gene bank of Estonian population. *Science*. 286: 1262–1263.

Finkel, Elizabeth. 1999. Australian center develops tools for developing world. *Science*. 285: 1481–1483.

Heller, Leland M. 2000. Is BuSpar really safe? <http://www.biologicalunhappiness.com/AskDoc/IsBuSpar.htm>. Accessed 11 June 2001.

Jaenisch, Rudolf, and Ian Wilmut. 2001. Don't clone humans! *Science*. 291: 2552.

Kahn, Patricia. 1996. Coming to grips with genes and risk. *Science*. 274: 496–498.

NBC Dateline. 2001. Brave new world? <http://www.msnbc.com/news/525661.asp?cp1=1>. Accessed 11 June 2001.

Ochman, Howard, Jeffrey G. Lawrence, and Eduardo A. Groisman. 2000. Lateral gene transfer and the nature of bacterial innovation. *Nature*. 405: 299–304.

Patients of Alzheimer's take heart, "smart gene" is on the way. 2000. *The Indian Express*. <http://www.expressindia.com/ie/daily/20000621/iin21008.html>. Accessed 11 June 2001.

Polejaeva, Irina A., Shu-Hung Chen, et al. 2000. Cloned pigs produced by nuclear transfer from adult somatic cells. *Nature*. 407: 86–90.

Ritter, Malcolm. 2000. Scientists find fat gene. <http://abcnews.go.com/sections/science/DailyNews/fatgene000327.html>. Accessed 11 June 2001.

Schiermeier, Quirin. 2001. Designer rice to combat diet deficiencies makes its debut. *Nature*. 409: 551.

Sex and the single gene. 1997. *Stanford Today*. March/April. 31–32.

Steinberg, Douglas. 2000. N.Y. panel explores genomics issues. *The Scientist*. 14(20): 9.

Tian, X. Cindy, Jie Xu, and Xiangzhong Yang. 2000. Normal telomere lengths found in cloned cattle. *Nature Genetics*. 26: 272–273.

van der Laan, Luc J. W., Christopher Lockey, et al. 2000. Infection by porcine endogenous retrovirus after islet xenotransplantation in SCID mice. 2000. Nature. 407: 90–94.

Wei F., G. D. Wang, et al. 2001. Genetic enhancement of inflammatory pain by forebrain NR2B overexpression. *Nature Neuroscience*. 4(2): 164–169.

Wu, C. 1997. Weight control for bacterial plastic. *Science News*. 151: 23.

Ye, Xudong, Salim Al-Babili, et al. 2000. Engineering the provitamin A (β-carotene) biosynthetic pathway into (carotenoid-free) rice endosperm. *Science*. 287: 303–305.

CHAPTER 4

Basic Research

with DNA

Microarrays

Goals for Chapter 4

4.1   Introduction to Microarrays

Learn DNA microarray methodology.

Understand how data are presented.

Utilize microarrays to learn basic biology.

4.2   Alternative Uses of DNA Microarrays

Discover genome instability with microarray data analysis.

Improve genome annotation with DNA microarrays.

Perhaps more than any other method, **DNA microarrays** (often referred to as **DNA chips**) have transformed genomics from a discipline restricted to large sequencing labs to a cottage industry practiced by labs of all sizes. DNA microarrays allow investigators to measure simultaneously the level of transcription for every gene in a genome. With DNA chips, we can collect data from an entire genome as it responds to its environment. Although DNA microarrays were invented elsewhere, their popularity grew in response to publications written by the graduate students and postdocs working with Pat Brown and David Botstein at Stanford University; therefore, many of the case studies in Chapter 4 are taken from their publications, which use genes from the first eukaryote to have its genome sequenced, the yeast *Saccharomyces cerevisiae*. The last case study uses DNA microarrays to improve the annotation of the human genome draft sequence.

4.1 Introduction to Microarrays

This section uses a fictitious story to guide you through one case study: A student trying to make home brew in the dorm has a problem with the biochemistry of fermentation. In this story, you will learn how DNA microarrays are created, experiments are performed, and data analyzed. You will see how DNA microarrays provide insights into the dynamics of a living genome. The Discovery Questions will lead you through public domain databases that contain raw data you can mine. Since the data sets are immense, much of the data remain unanalyzed. By posting them online, the original investigators have made it possible for you to make your own original discoveries.

What Happened to My Home Brew?

When you were in third grade, your grandfather gave you a microscope. The next year he gave you a junior chemistry set. Since then, you could not stop "playing" with science toys. You're the type of person who still likes to put an avocado pit into water so you can watch it sprout roots. Now you are in college, and you are taking real chemistry courses. You cannot believe they actually let you design experiments and give you academic credit for it! But you could not give up your hobby of dabbling in science, so you decided to brew your own beer. It was your understanding that it was illegal to *buy* alcohol, not make it (what you know about legal issues could fit in a shot glass). You are also taking cell biology, so it was just a short hop from cell cycle mutants to yeast biochemistry. You had intended to take biochemistry, but it did not fit into your schedule, and you thought of brewing beer as a form of "distance learning" in metabolic pathways. All of this made sense at the time, but now. . . .

It was about 3 A.M. on a Tuesday when it happened. You were sleeping soundly when the first one went—bam! It sounded like an explosion with an aftersound of suds running down your walls. Then another— Bam! Soon, you couldn't count them fast enough—BAM! BANG! CRASH! It sounded like the 4th of July, except no one said ooooh, ahhhhh. Your first batch of home brew beer was gone in a 21-gun salute, of sorts. You decided there was nothing left to do but roll over and clean up in the morning.

It was a terrible sight to see all your home brew pooled on the floor. You got a shop-vac from the custodian and cleaned it all up. You didn't feel like eating breakfast so you went straight to class—cell biology. As classmates asked you what happened, you confirmed some rumors and put others to rest. The instructor entered the class and wrote on the board:

Beer vs. Bread: yeast metabolism

You cannot believe the cruel joke fate has played on you. The topic for today is how yeast cells can switch from fermentation to oxidative phosphorylation. Inside your notebook sits the unread paper that would have saved your beer—Joseph DeRisi, Vishwanath R. Iyer, and Pat Brown's "Exploring the Metabolic and Genetic Control of Gene Expression on a Genomic Scale." If only you had read it before starting your brew. . . . Oh well, no use crying over spilled beer. But you can't help wondering, what went wrong?

The paper you should have read examined the genomic shift in metabolism, but you have no idea what is in this paper. Your look of confusion catches the instructor's eye, and she calls on you to start today's discussion.

"Uhh, errr . . ."—you start to think of something witty to say and she interrupts.

"Too busy cleaning the suds off your wall to read, huh?"

You can't believe she knows, too. She calls on a student next to you, who explains "DNA chips are tools used to measure simultaneously the transcription level of every gene in a cell (in this case yeast). By comparing an experimental **transcriptome** (a complete collection of all transcripts in a cell at any given time point) with a reference transcriptome, you can detect changes in transcription levels for every gene in the genome." You never would have come up with an answer like that. The student continues by saying that this past summer she conducted research using yeast DNA chips (microarrays) as a part of the **Genome Consortium for Active Teaching (GCAT),** a consortium of undergraduates and their faculty mentors. (Now you don't feel so badly that her answer was perfect.)

Your instructor begins to explain how microarrays are made. Since the yeast genome has been sequenced, most of the 6,200 or so genes have been identified. All 6,200 genes are amplified by **polymerase chain reaction (PCR),** using a set of primers that are commercially available. The PCR

LINKS

Joseph DeRisi

Vishwanath Iyer

Pat Brown

GCAT

Commercially
Available

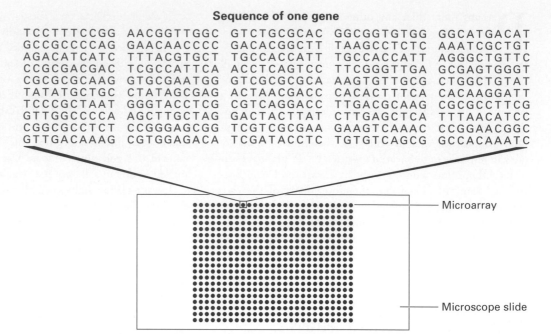

Sequence of one gene

```
TCCTTTCCGG   AACGGTTGGC   GTCTGCGCAC   GGCGGTGTGG   GGCATGACAT
GCCGCCCCAG   GAACAACCCC   GACACGGCTT   TAAGCCTCTC   AAATCGCTGT
AGACATCATC   TTTACGTGCT   TGCCACCATT   TGCCACCATT   AGGGCTGTTC
CCGCGACGAC   TCGCCATTCA   ACCTCAGTCC   TTCGGGTTGA   GCGAGTGGGT
CGCGCGCAAG   GTGCGAATGG   GTCGCGCGCA   AAGTGTTGCG   CTGGCTGTAT
TATATGCTGC   CTATAGCGAG   ACTAACGACC   CACACTTTCA   CACAAGGATT
TCCCGCTAAT   GGGTACCTCG   CGTCAGGACC   TTGACGCAAG   CGCGCCTTCG
GTTGGCCCCA   AGCTTGCTAG   GACTACTTAT   CTTGAGCTCA   TTTAACATCC
CGGCGCCTCT   CCGGGAGCGG   TCGTCGCGAA   GAAGTCAAAC   CCGGAACGGC
GTTGACAAAG   CGTGGAGACA   TCGATACCTC   TGTGTCAGCG   GCCACAAATC
```

Microarray

Microscope slide

FIGURE 4.1 • **DNA microarray.** Each blue spot indicates the location of a PCR product on the glass slide. One particular spot has been chosen to illustrate the presence of one gene's sequence. On a real microarray, each spot is about 100 μm in diameter.

METHODS

Microscope Slide

PCR

cDNA

LINKS

Make Your Own

products are verified, purified, and spotted onto an ordinary glass microscope slide by a robot. The robot spots all 6,200 PCR products onto 144 slides in about 12 hours. The spotted DNA is denatured and linked by covalent bonds to the glass slide. After being spotted, the microscope slide does not look any different, but it has an invisible gridlike pattern (an **array**) of spots, each containing many amplified copies of a single gene (Figure 4.1). If you fog a microarray with your breath, you can see tiny dots about the size of the period at the end of this sentence. The microarray is covered (all 6,200 spots will fit under a 22-mm slipcover), and stored dry at room temperature.

Another student asks why the DNA doesn't degrade at room temperature. This time you know the answer and say, "Because polymerized DNA is a very stable molecule; that's why it can be found intact in bacteria 250 million years old. So a few days sitting on the shelf should be fine." Answering this question makes you feel a little better about not reading the paper. You get another boost when your instructor says that the Brown lab has provided all the necessary directions to make your own one-armed microarray robot. Since you love to build gadgets, this sounds like a great idea for an independent study project.

DNA microarrays are the functional equivalents to 6,200 Southern blots, in that single-stranded DNA has been immobilized onto a solid support. However, on a Southern blot, the immobilized DNA is usually a mixture of DNAs,

such as total genomic DNA. A probe of a known sequence is labeled and allowed to hybridize to the immobilized DNA. For this reason, your instructor explains, there is some debate in the field whether to call the immobilized DNA on a DNA microarray "probes" (because the sequences are known) or "targets" (since the DNA is immobilized). Luckily, your instructor tells you that this nuance is insignificant, so you don't worry about it. (Sometimes she can be kind.)

Where's the Probe? Now that the yeast DNA chip has been made, you begin to wonder what will be used to probe the equivalent to 6,200 Southern blots. You cannot imagine making that many probes individually, but you're not about to ask since you didn't do the reading. She continues her lecture with a description of experiments that are performed with a yeast DNA chip. Cells are grown in two different conditions, such as in the presence and absence of oxygen. The two populations of mRNA are harvested from each population of cells and separately converted into **cDNA**s. The nucleotides used to make the cDNA include either a green dye called **Cy3** or a red dye called **Cy5.** Therefore, the two populations of cDNAs are colored either green or red, each color representing the transcriptome from one population of cells (Figure 4.2a on your CD-ROM).

FIGURE 4.2 • **Production of cDNA probes for a DNA chip.**

Go to the CD-ROM to view this figure.

The two populations of cDNAs (green and red) are mixed and incubated with the DNA chip. In theory, the colored cDNAs will bind to one strand of one PCR product—the one that represents its gene. During the incubation, there is a competition of sorts (like musical chairs), with all the different cDNAs trying to find places to bind, except in this competition, there are more than enough places to land (Figure 4.2b). After a long incubation (typically overnight), the cDNAs that did not bind to any spots are washed off and the chip is allowed to dry in the dark, since the fluorescent dyes are easily bleached by light.

When the microarray is dry, it is put into a scanner that uses light and sensors to detect the dyes and record their locations and intensities. Scanning takes about 20 minutes to obtain a high-resolution image. When completed, the two color images (one green and one red) are stored in a computer and the intensities of the two colors are determined. The computer also generates a merged image, with yellow spots indicating which open reading frames (ORFs) are transcribed in both transcriptomes (Figure 4.3 on your CD-ROM). Your instructor turns down the lights and shows a DNA microarray Flash animation that summarizes the procedure.

FIGURE 4.3 • Results from a single DNA chip.
Go to the CD-ROM to view this figure.

The yellow spots are a visual way of depicting a red to green ratio of 1:1. But it is unusual for a ratio to be exactly 1:1. More typically, the merged image will be a bit more green or a bit more red. Figure 4.4 on your CD-ROM illustrates examples of real data.

FIGURE 4.4 • Measuring fluorescence on a DNA chip.
Go to the CD-ROM to view this figure.

DISCOVERY QUESTIONS

1. Why is there a dark center in the middle of each spot? (See Figure 4.4.)

2. How is cDNA made? Why does the DNA on the array need to be denatured?

3. What differences and similarities does a DNA chip have with a Southern blot?

4. Do you foresee any problems spotting the entire ORF on the chip instead of only part of it?

From the colored spots, the data are converted to numbers that represent the light intensity of red dye, green dye, and the ratios of red to green (Table 4.1).

A long table of numbers is difficult to read and understand. Humans have evolved as visual creatures, and we process visual data much faster than numerical data. Therefore, the group at Stanford converted the numerical ratios of red to green into a visually comprehensible system. They created a color scale to represent the numerical ratios of red to green, but unfortunately they chose a red to green scale that can lead to some confusion (Figure 4.5 on your CD-ROM).

METHODS
DNA Microarray

FIGURE 4.5 • Red-green color scale for changes in transcription.
Go to the CD-ROM to view this figure.

If there was no change in a gene's transcription between control and experimental growth conditions (i.e., an mRNA ratio of 1:1), the gene is given a black color for that time point. If a gene was **induced** (stimulated to produce more mRNA) in the experimental condition, it is given a red color for that time point. The greater the induction, the brighter the red color. If a gene was **repressed** (transcribed less) in the experimental condition, it is given a green color for that time point. The greater the repression, the brighter the green color. This color scale makes sense if you follow two simple rules:

1. The cDNA produced from the control population of cells is green and cDNA from cells grown in the experimental condition (e.g., absence of oxygen) is red.

2. Always place the numerical value for the red dye (from the experimental cells) in the numerator and the green value (control cells) in the denominator when calculating the ratio.

```
red experimental mRNA ÷ green control mRNA
```

Following rule number one, mRNA isolated from cells grown in the presence of oxygen is converted into green cDNA, and mRNA isolated from cells grown in the absence of oxygen is converted into red cDNA (i.e., green = with O_2 and red = without O_2). Following rule number two, red number ÷ green number will result in a ratio that represents the change in transcription for one gene at one time point. Let's look at two particular cases:

1. Gene A is expressed in control cells (1,200 light units of green dye bound) but is repressed in the absence of oxygen (only 400 light units of red dye bound). Numerically, expression of gene A would look like this: 400 ÷ 1200 = 0.33, a 3X (X stands for fold) repression of gene A. Using the ratio conversion color scale (see Figure 4.5), we see 3X on the repression scale would be converted to a dark green color.

TABLE 4.1 • **Summary table of data for one yeast DNA microarray.** The fluorescence intensity for each color was determined, and the ratio of red to green signal was calculated. This table shows data for only 14 of the 6,200 genes on the full microarray. Also contained in the data is the location for each spot. When a microarray is fabricated, it is important to know where each gene is located on it.

Block	Column	Row	Gene Name	Red	Green	Red: Green Ratio
1	1	1	*tub1*	2,345	2,467	0.95
1	1	2	*tub2*	3,589	2,158	1.66
1	1	3	*sec1*	4,109	1,469	2.80
1	1	4	*sec2*	1,500	3,589	0.42
1	1	5	*sec3*	1,246	1,258	0.99
1	1	6	*act1*	1,937	2,104	0.92
1	1	7	*act2*	2,561	1,562	1.64
1	1	8	*fus1*	2,962	3,012	0.98
1	1	9	*idp2*	3,585	1,209	2.97
1	1	10	*idp1*	2,796	1,005	2.78
1	1	11	*idh1*	2,170	4,245	0.51
1	1	12	*idh2*	1,896	2,996	0.63
1	1	13	*erd1*	1,023	3,354	0.31
1	1	14	*erd2*	1,698	2,896	0.59

METHODS
Northern Blots

2. Gene B is expressed in control cells (180 light units of green dye bound) but is induced in the absence of oxygen (1,800 light units of red dye bound). Numerically, expression of gene B would look like this: 1,800 ÷ 180 = 10, a 10X induction of gene B. Using the ratio conversion color scale, we see 10X on the induction scale would be converted to a medium red color.

Therefore, we are interested in numerical ratios for each time point that is represented by a color. The color scale is meant to be easier to read when we have thousands of data points to examine, as you will see later.

Microarray Data Look Good, but Are They Real? One problem with inventing a new method is that it is difficult to be sure whether it produces meaningful data. Your colleagues will be skeptical and require evidence to convince them. Recognizing scientific skepticism, the Brown group compared their microarray data with more traditional data—Northern blots (Figure 4.6 on your CD-ROM). **Northern blots** measure the amount of mRNA produced one gene at a time, have been used for many years, and researchers are very comfortable with them.

FIGURE 4.6 • Comparison of Northern blots with DNA microarray data.
Go to the CD-ROM to view this figure.

DISCOVERY QUESTIONS

5. Choose two genes from Figure 4.6 on your CD-ROM and draw a graph to represent the change in transcription over time.

6. Look at Figure 4.7, which depicts the loss of oxygen over time and the transcriptional response of three genes. These data are the ratios of transcription for genes X, Y, and Z during the depletion of oxygen.
   a. What is the fold change for each gene at each time point?
   b. Using the color scale from Figure 4.6, determine the color for each ratio in Figure 4.7b.

7. Were any of the genes in Figure 4.7b transcribed similarly?

a)

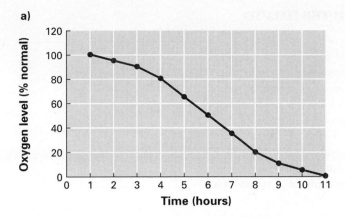

b)

	1 hour	3 hour	5 hour	9 hour
gene X	1.0	2.2	1.0	0.15
gene Y	1.0	4.5	0.95	0.05
gene Z	1.0	1.5	2.0	2.0

FIGURE 4.7 • Transcriptional response of three genes to the gradual loss of oxygen. a) Graph of oxygen consumption over time by yeast growing in a closed container. **b)** DNA microarray data given in the form of ratios. To calculate a ratio, one gene's activity in cells gradually consuming all the available oxygen was divided by its activity in control cells with unlimited oxygen.

Data Analysis As you can see, microarray data are not difficult to obtain. You grow cells in two different conditions, make fluorescent cDNA of two colors, mix them, and allow them to **hybridize** (form base pairs with complementary strands) to the appropriate spots, wash, dry, scan the chip, and have a computer produce ratios and the appropriate color scheme. When all this is done to one chip, you have 6,200 spots, each with its own numerical ratio that is converted to a color. However, biologists tend to be more interested in the *process* of adaptation and not just its end point. We'd like to measure the cellular response every 2 hours to a depletion of oxygen over a 10-hour period. At the end of the experiment, we would have produced 31,000 gene expression ratios. No one can comprehend a spreadsheet of this magnitude, nor 31,000 colored spots on microarrays, each one representing a gene expression ratio. Similarly, it would be difficult to evaluate 6,200 different graphs, each with five data points. Are there any genes that responded in similar ways to the depletion of oxygen? To answer this fundamental question, you would want to organize the data into meaningful patterns.

To facilitate the visualization and comprehension of huge data sets, and to look for genes that respond in similar ways, the Stanford microarray team devised a new computer program to help them. Their software looks at each gene's response over time and tries to find other genes with similar responses. Let's look at a simple example of how this can be done. Table 4.2 contains 12 genes organized by name, which makes it difficult to discern gene expression patterns.

TABLE 4.2 • Data showing fold change (experimental ÷ control) in mRNA production for 12 hypothetical genes (C – N).

Name	0 hours	2 hours	4 hours	6 hours	8 hours	10 hours
gene C	1	8	12	16	12	8
gene D	1	3	4	4	3	2
gene E	1	4	8	8	8	8
gene F	1	1	1	0.25	0.25	0.1
gene G	1	2	3	4	3	2
gene H	1	0.5	0.33	0.25	0.33	0.5
gene I	1	4	8	4	1	0.5
gene J	1	2	1	2	1	2
gene K	1	1	1	1	3	3
gene L	1	2	3	4	3	2
gene M	1	0.33	0.25	0.25	0.33	0.5
gene N	1	0.125	0.0833	0.0625	0.0833	0.125

MATH MINUTE 4.1 · HOW DO YOU TRANSFORM THE DATA TO AVOID FRACTIONS?

You have seen how red to green intensity ratios are computed from the intensity levels determined in the scanning process, and how they represent repression or induction of the genes. For example, a 4-fold repression in gene expression results in a ratio of approximately 0.25. Similarly, a 16-fold repression results in a ratio of 0.0625, a 4-fold induction 4.0, and a 16-fold induction 16.0. But there is a much greater numerical difference between 4.0 and 16.0 than there is between 0.25 and 0.0625. Graph the expression patterns for genes C and N from Table 4.2 on a single set of axes, and notice how the induction stands out much more clearly than the repression, even though the repression is of the same magnitude. In addition to biasing our visual interpretation of gene expression patterns, fractional ratios cause problems with mathematical techniques of analyzing and comparing gene expression patterns (Math Minute 4.2). Furthermore, to interpret a fractional ratio as a fold repression, you must take its reciprocal (e.g., $1/0.0625 = 16.0$), an operation that most people find difficult to do quickly and accurately in their heads.

To avoid these problems, investigators often use a log transformation of the ratio data. Log transformation is illustrated in the following example. Suppose the ratio is 0.0625. Take the base 2 logarithm of this number:

$$\log_2(0.0625) = \log_2(1/16) = \log_2(1) - \log_2(16) = -\log_2(16) = -4.$$

Since the $\log_2$ of 1/16 is the negative of the $\log_2$ of 16, a 16-fold induction and a 16-fold repression have the same magnitude (one positive and one negative) in the $\log_2$-transformed data. The magnitude of the number is the power of 2 needed to get the induction/repression number (e.g., $16 = 2^4$). The $\log_2$ transformations of the ratios in Table 4.2 are given in Table M4.1.

TABLE M4.1 · $\log_2$ transformation of gene expression data in Table 4.2.

Name	0 hours	2 hours	4 hours	6 hours	8 hours	10 hours
gene C	0	3	3.58	4	3.58	3
gene D	0	1.58	2	2	1.58	1
gene E	0	2	3	3	3	3
gene F	0	0	0	−2	−2	−3.32
gene G	0	1	1.58	2	1.58	1
gene H	0	−1	−1.60	−2	−1.60	−1
gene I	0	2	3	2	0	−1
gene J	0	1	0	1	0	1
gene K	0	0	0	0	1.58	1.58
gene L	0	1	1.58	2	1.58	1
gene M	0	−1.60	−2	−2	−1.60	−1
gene N	0	−3	−3.59	−4	−3.59	−3

Sometimes $\log_{10}$ is used instead of $\log_2$; in this case, the magnitude of the transformed data is the power of 10 needed to get the induction/repression number. Transforming the same ratios as above leads to the following:

$$\log_{10}(4) \approx 0.6 \qquad \log_{10}(.25) \approx -0.6$$
$$\log_{10}(16) \approx 1.2 \qquad \log_{10}(0.0625) \approx -1.2$$

In log$_2$-transformed data, a value of 2 corresponds to a ratio of 4; however, you would be surprised to see a value as large as 2 in log$_{10}$-transformed data, since 2 corresponds to a ratio of 100. In general, a log$_2$ transformation helps you easily identify doublings or halvings in ratios, while a log$_{10}$ transformation helps you see order of magnitude changes. The key attribute of log-transformed expression data is that equally sized induction and repression receive equal treatment visually and mathematically.

MATH MINUTE 4.2 • HOW DO YOU MEASURE SIMILARITY BETWEEN EXPRESSION PATTERNS?

In Table 4.2, the transcriptional responses of genes G and L are clearly similar, since they have the same numerical values at each time point. But what about gene D? How similar is its response to that of gene G and L? A common way to measure the similarity between gene expression patterns like those in Table 4.2 is with the Pearson **correlation coefficient,** abbreviated with the letter r.

Correlation quantifies the extent to which the expression patterns of two genes go up together and down together over several time points or experimental conditions, even if the numbers are not of the same magnitude. A correlation coefficient of 1.0 between two genes means that their expression patterns track each other perfectly. A correlation coefficient of -1.0 between two genes means that their expression patterns track perfectly, but in opposition to one another (i.e., one goes up while the other goes down). A correlation coefficient near zero means that the expression patterns of the two genes do not track each other at all.

In the following example, we will work with the expression values given in Table 4.2 to further illustrate the disadvantages of analyzing raw (untransformed) expression ratios, as discussed in Math Minute 4.1. To find the correlation between genes D and L, denoted by $r(D,L)$, first compute the mean and standard deviation of the expression values for each gene (i.e., each row):

$$\overline{X}_D \approx 2.83 \qquad \overline{X}_L = 2.5 \qquad s_D \approx 1.067 \qquad s_L \approx 0.957$$

Subtract $\overline{X}_D$ from each value in the D row, and divide each result by s_D. The result is a row of *normalized* values in the D row:

$$D_{norm} = -1.715, 0.1593, 1.097, 1.097, 0.1593, -0.7779.$$

Do the same in the L row, this time subtracting $\overline{X}_L$ and dividing by s_L, to produce the following normalized row:

$$L_{norm} = -1.567, -0.5225, 0.5225, 1.567, 0.5225, -0.5225.$$

Now multiply the first number in D_{norm} by the first number in L_{norm}, the second number in D_{norm} by the second number in L_{norm}, and so on, keeping a running sum of these products. (You might recognize this operation as the dot product of the two vectors D_{norm} and L_{norm}.) Finally, divide this sum (5.386) by the number of elements in each row (6) to get the correlation coefficient $r(D,L) = 0.897$.

Notice that the expression ratios for genes H and M are the reciprocals of the ratios for genes L and D, respectively. In other words, gene H is repressed to exactly the same extent that gene L is induced, and gene M is repressed to the same extent as gene D is induced. We would thus expect $r(D,L)$, which compares the patterns of induction, to be the same as $r(H,M)$, which compares the patterns of repression. However, $r(H,M) = 0.97$, which is quite a bit larger than $r(D,L) = 0.897$. This strange behavior is because correlation is sensitive to the relative magnitudes of the patterns, and can be prevented by first log-transforming the data (see Math Minute 4.1). The correlation coefficients of each pair of genes in Table M4.1, computed with the log-transformed data, are shown in Table M4.2, page 114. Note that $r(H,M)$ is now the same as $r(D,L)$.

cluster [LG] is gene C, with r(C,G) = r(C,L) = 0.95. However, gene C and cluster [LG] are not the two most similar objects; rather genes C and E are, with r(C,E) = 0.96. Thus we join genes E and C to form cluster [EC].

At the next iteration, we need to know the correlation of each object with the average log-transformed expression patterns of genes E and C: 0, 2.5, 3.29, 3.5, 3.29, 3. The correlations of all available objects with this pattern are as follows:

D	F	H	I	J	K	M	N	[LG]
0.90	−0.48	−0.93	0.32	0.33	0.32	−0.90	−0.99	0.93

The most similar object to [EC] is cluster [LG]. Gene D is even more similar to [LG], since r(D,G) = 0.94. But the two most similar objects now are genes N and H, with r(N,H) = 0.95. Therefore, we join genes N and H to form cluster [NH].

The hierarchical clustering process for these 12 genes is summarized in Table M4.3. Note that the final object created is the clustering of all 12 genes shown in Figure 4.8.

TABLE M4.3 • Summary of the hierarchical clustering algorithm applied to the 12 genes in Table 4.2.

Iteration	Two Most Similar Objects		Correlation	New Object
	Object 1	Object 2		
1	L	G	1.00	[LG]
2	E	C	0.96	[EC]
3	N	H	0.95	[NH]
4	M	[NH]	0.95	[MNH]
5	[LG]	D	0.94	[LGD]
6	[EC]	[LGD]	0.94	[ECLGD]
7	I	F	0.60	[IF]
8	J	[ECLGD]	0.29	[JECLGD]
9	K	[JECLGD]	0.19	[KJECLGD]
10	[KJECLGD]	[IF]	−0.12	[KJECLGDIF]
11	[MNH]	[KJECLGDIF]	−0.96	[MNHKJECLGDIF]

The hierarchical clustering process can also be summarized in a dendrogram similar to those discussed on pages 41–46. Figure M4.1 shows the dendrogram for the hierarchical clustering detailed in Table M4.3 and Figure 4.8. Notice that genes L and G are consolidated into a single node in the tree. The depth (from right to left) at which a node connects two objects represents the similarity between them. At any node that joins two branches, the top and bottom branches can be exchanged without changing the interpretation of the tree. Therefore, many different orderings of the leaves are consistent with the branching structure of a particular dendrogram. Dendrograms will be used extensively in Chapter 5 to represent clusters.

Hierarchical clustering is the most popular method for finding trends in gene expression data, but there are several others. Another common method is the k-means cluster algorithm, which tries to find the best partition of the entire set of genes into precisely

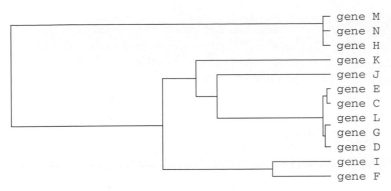

FIGURE M4.1 • Dendrogram of clustered genes from Table M4.3 and Figure 4.8.

k groups. Software for clustering gene expression data with hierarchical, k-means, and other methods is freely available for academic use. Each cluster algorithm may result in a different overall clustering of the data. Although there are mathematical methods for evaluating the extent to which clusters agree with the input similarity measurements, the last word in cluster evaluation belongs to the investigators who form and test hypotheses based on the clusters.

Now you have a pretty good understanding of DNA chips, but you have not figured out why your beer bottles exploded. You had hoped this paper would offer you some practical advice. But class is over, and no one has told you the bottom line. You decide you'll have to read it yourself.

You begin reading DeRisi's paper and are impressed that over 43,000 expression ratios are summarized. Surely you can find the answer you're looking for in this much data. In this experiment, they grew two populations of yeast. The experimental cells gradually depleted the glucose, while the control cells had ample glucose in their flask (Figure 4.10). As indicated in the figure, over the course of the experiment they took **aliquots,** or

small volumes, from the glucose-limited population of cells. The resulting cDNA was labeled red and compared to green cDNA produced from control cells with unlimited glucose.

LINKS

Software

You recognize the first figure as a "get acquainted" figure that explains what microarrays are and how to read them (Figure 4.11 on your CD-ROM).

FIGURE 4.11 • Example of real DNA microarray data.

Go to the CD-ROM to view this figure.

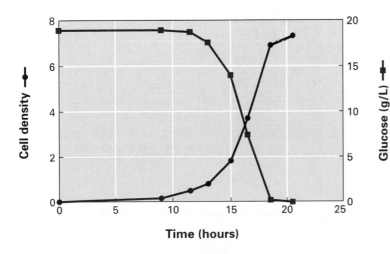

FIGURE 4.10 • Growth of cells as glucose is consumed. The graph shows the loss of glucose and the concurrent rise in cell density (OD_{600}) during the time course of the experiment by DeRisi et al.

METHODS
Metabolism of
Glucose

TCA Cycle

LINKS

Function Junction

DISCOVERY QUESTION

8. Go to Function Junction and choose two genes labeled in Figure 4.11, one repressed and one **constitutive,** or always active, as represented by a ratio of 1:1. Find out what each gene does. To do this, leave all the boxes selected to search all the different databases, and enter the abbreviated name of the first gene you have chosen. Look at the metabolism of glucose and see if you can figure out why each gene would be repressed or constitutive.

Figure 4.12 on your CD-ROM is a close-up view of the white-boxed area in Figure 4.11 and illustrates gene regulation over time. Some genes are induced or repressed earlier than others.

FIGURE 4.12 • **Time course of gene regulation.**
Go to the CD-ROM to view this figure.

DISCOVERY QUESTIONS

9. Why would most spots be yellow at the first time point?

10. Go to Function Junction, search for *Tef4*, and you will see it is involved in translation. Look at the time point labeled OD 3.7 in Figure 4.12, and find the *Tef4* spot. Over the course of this experi-

ment, was *Tef4* induced or repressed? Hypothesize why *Tef4*'s gene regulation was a part of the cell's response to a reduction in available glucose (i.e., the only available food).

It is interesting to note that about half the genes that change their expression patterns have no known function yet. In the DeRisi paper, few genes are singled out for comment because it would be impossible to analyze and discuss all the data in one publication.

The first genes discussed in the paper are *Ald2* and *Acs1,* which function together to convert the products of alcohol dehydrogenase into acetyl-CoA, which enters the tricarboxylic acid (TCA) cycle. Simultaneously, the gene encoding pyruvate decarboxylase is repressed while pyruvate carboxylase is induced, which shifts pyruvate away from acetaldehyde and toward oxalacetate and into the TCA cycle. Further, phosphoenolpyruvate carboxykinase and fructose 1,6-biphosphatase are both induced. These two proteins control the two irreversible steps in glycolysis, which in turn reverses the overall flow of carbon toward glucose-6-phosphate. Trehalose synthase and glycogen synthase are also induced; these enzymes convert glucose-6-phosphate into storage sugars.

Over the course of the experiment, the yeast cells responded as if they sensed the glucose concentration was being reduced and thus they might soon starve. As a result, genes in entire pathways were induced or repressed in concert (Figure 4.13). Genes involved in protein synthesis were coordinately repressed.

Therefore, microarray data have revealed their potential for new insights. First, they can uncover entire pathways

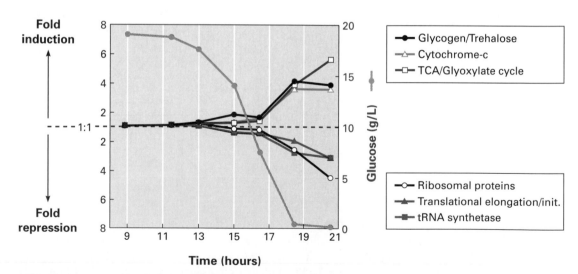

FIGURE 4.13 • **Coordinated regulation of genes encoding enzymes that work together.**
This graph shows that as the glucose was consumed, some genes changed their expression levels as a group. Energy metabolism genes were induced while protein production genes were repressed.

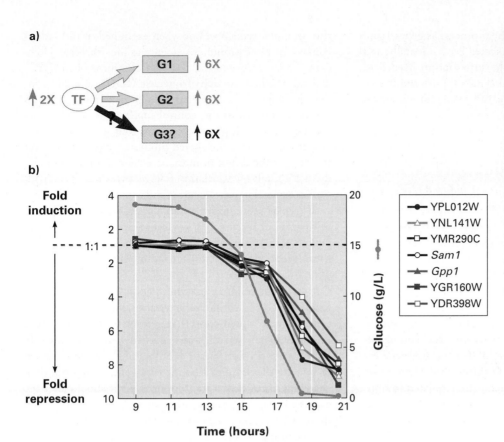

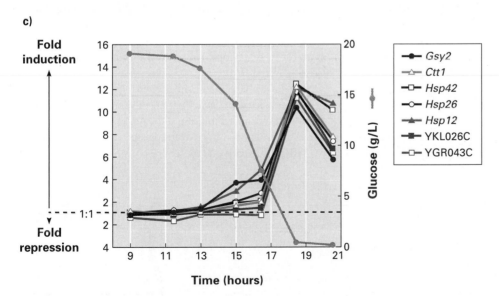

FIGURE 4.14 • Guilt by association. a) The transcription factor (TF) is induced twofold and is known to activate genes 1 (G1) and 2 (G2), which are both induced sixfold. Gene 3 (G3) is also induced sixfold. Is it regulated by TF? Does it participate in the same circuit as G1 and G2? **b)** and **c)** Clustering of genes suggests functions for unknown genes. In b) known genes cluster with unknown genes, all of which are co-repressed in response to glucose consumption. In c) known and unknown genes are induced before the glucose is completely consumed, but the level of induction drops when the glucose is gone.

that respond in concert to changing conditions. Second, genes that currently have no known function can cluster with genes of known function. When an unknown gene and a known gene are regulated the same way, you can make a good guess that the unknown gene has a function related to the known gene's. This **guilt by association** method can lead to predictions of possible gene functions

and eventually experimental testing of these predictions (Figure 4.14). In Figure 4.14, a transcription factor is induced twofold and genes 1 and 2 are induced sixfold. Genes 1 and 2 are known to be induced by the transcription factor. A third gene with unknown function is also induced sixfold, which reveals two possibilities. First, gene 3 may be regulated by the transcription factor; second, it may be involved in the same cellular process as genes 1 and 2. These two possibilities can be tested experimentally, and the microarray data provided the information needed to make testable predictions. Real data support the use of guilt by association to propose functions for unknown genes (Figure 4.14b and c).

> ## DISCOVERY QUESTIONS
>
> 11. Why would TCA cycle genes be induced if the glucose supply is running out?
>
> 12. What mechanism could the genome use to ensure genes for enzymes in a common pathway are induced or repressed simultaneously?

Chips Reveal Regulatory Sequences The genome contains a lot more information than just the coding sequences, such as regulatory regions that control gene expression. By clustering genes that exhibit similar expression patterns, we can predict which genes are regulated in similar ways. Because the entire yeast genome has been sequenced, we can examine the promoter regions of clustered genes and look for conserved sequences. By searching for these sequences, it is possible to identify the transcription factor binding sites responsible for regulating these clustered genes.

A common regulatory sequence was identified in one particular cluster of genes with a striking expression profile (Figure 4.15). This cluster of seven genes is induced at least tenfold between the 18- and 21-hour time points. All seven genes were known to be repressed in the presence of glucose, and five of them were known to share a common promoter sequence. The gene *Acr1* had not been reported to contain

this promoter sequence, but when examined, it did indeed contain the same promoter sequence as the other five. However, *Idp2*, which encodes the cytosolic form of NADP⁺-dependent **isocitrate dehydrogenase (IDH)**, does not contain the common promoter motif. Thus, there must be an additional mechanism to control *Idp2*'s transcription. Furthermore, the yeast genome contains four genes not in this cluster that contain the shared promoter sequence, and yet their expression differs from the seven genes clustered in Figure 4.15. Guilt by association follows two rules: genes with similar expression profiles have similar promoters, and genes with similar promoters have similar expression profiles. In this cluster of seven genes, we found two exceptions to the guilt by association rules. Do we keep the rules and live with these exceptions? ignore these data? modify the rules?

Can We Formulate Testable Predictions with These Data? When you think you have a system figured out, it is time to make a prediction that is testable. The data above indicated that, based on expression patterns, we can identify shared transcription factor binding sites. If transcription factors are the proteins that regulate gene expression, then changing the level of a transcription factor should alter the expression pattern of associated genes. What would happen if we delete a transcription factor gene, or overexpress a transcription factor? These questions are not hypothesis driven but rather discovery based. Cell and molecular biology have been powered by hypothesis-driven research for many years, but with the advent of genomic methods such as microarrays, people are asking different types of questions—"What if we . . . ?"

The Brown lab asked similar questions: What would happen if we deleted the gene that encodes the repressor Tup1p? (p indicates we are talking about the protein and not the gene.) What would happen if we overexpressed the transcription factor Yap1p? **Tup1p** is one protein responsible for the repression

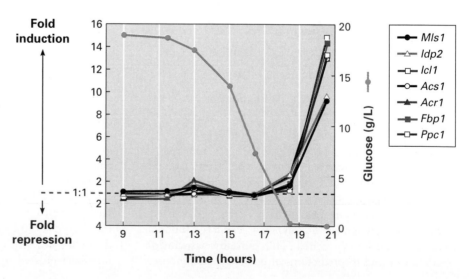

FIGURE 4.15 • **Cluster of seven genes with similar expression patterns.** The light blue line with filled circles indicates the level of glucose present, while the level of induction for the seven genes is indicated by the remaining lines as noted in the legend.

of glucose-repressed genes. Of course, the sugar glucose cannot block the transcription of a gene directly, so several proteins are needed, including Tup1p. **Yap1p** is a transcription factor known to confer resistance to environmental stresses such as hydrogen peroxide, heavy metals, and osmotic shock.

DISCOVERY QUESTIONS

13. Given rule one on page 109, what color would you see on a DNA chip when cells had their repressor gene *Tup1* deleted?

14. What color spots would you expect to see on the chip when the transcription factor Yap1p is overexpressed?

15. Could the loss of a repressor or the overexpression of a transcription factor result in the repression of a particular gene?

16. What types of control spots would you like to see in this type of experiment? How could you verify that you had truly deleted or overexpressed a particular gene?

When the analysis was performed on the two mutant strains, as predicted, there were many changes in gene expression. A large number of red spots are seen in the *Tup1* deletion (**Δ Tup1**) strain. Likewise, overexpression of the transcription factor Yap1p (*yap1+++*) resulted in many red spots (Figure 4.16 on your CD-ROM)

FIGURE 4.16 • Transcriptional effects in mutant strains.
Go to the CD-ROM to view this figure.

DISCOVERY QUESTIONS

The following series of directions and questions will show you how to mine the expression databases for the *Tup1* deletion and *Yap1* overexpression experiments. You will be directed to find particular genes and consider the data obtained for each one.

17. Go to the Tup1 database and determine the change in *Tup1* expression in the *Δ Tup1* mutant strain. You must enter "Tup1" in the gene name box. Then hit "start search." Describe what you find. Is it what you expected? Notice the amount of variation in the replicated experiments (this will be discussed later). *Technical note:* The *form* you will see is delivered by a web server at Stanford. However, the *database* is delivered from a nonstandard server at the University of California, San Francisco. The

LINKS
Tup1
Yap1

IP address for the database server is 128.218.121.63, but it connects via a nonstandard port called 591. Most web servers connect via port 80, and many institutions have default firewalls that block any server using nonstandard ports. Thus, if you cannot connect to the Tup1 database, contact your IT department to see if your firewall is blocking access to it.

18. Go back to the Tup1 database to find genes in the *Δ Tup1* strains that were repressed rather than induced. To perform this type of search, delete the Tup1 from the gene box. Go down to the "Average fold induction in Tup1 deletion" box, choose "less than" from the menu and enter the number "1." Don't sort the results for now. There will be over 3,000 genes in this list, but only 25 on the first screen. From those first 25, find the gene with the greatest average repression. Is the variation in the two trials high or low?

19. From the list you have of 3,000+ hits, go to record number 52 (click on "next page" a couple more times). Look at its average repression. Then look at the variation in the two trials. Is its variation high or low? What is the function of gene number 52 from this list?

20. Go back to search Tup1 database and find the gene with the greatest repression. To do this, have it sort by "average fold induction." What gene was repressed the most by deleting a repressor? What gene was second?

21. Search Tup1 database again to find the gene that was *induced* the most. This time change the menu to "greater than," sort by "average fold induction," and choose "descending" for the order of display. You will see a common gene family in the most induced genes. What is the family? Which gene was induced the most? (*Hint:* You might want to have two browser windows open for the next series of questions, one for Function Junction and one to explore the Tup1 database.)

22. Go to the Function Junction and enter the name of the gene family that describes the trend for Discovery Question 21 (but don't add the "p" which stands for *protein*). Describe the role this family plays in the cell.

23. Based on your answer from Discovery Question 22, is there a significant impact on the cell when one repressor is deleted?

24. Go through a similar process for the *Yap1* overexpression set in the Yap1 database.

a. Which gene was induced the most? Was there variation in the two trials?

b. Go to Function Junction and find out what this most induced protein does. Does it affect any other genes, or is it more of a dead end?

c. Which gene was repressed the most? Was there variation in the two trials?

d. Go to Function Junction and find out what this most repressed protein does. Does it affect any other genes, or is it more of a dead end?

e. How much was *Tup1* induced or repressed?

f. How much was *Yap1* induced?

g. How many genes were induced when *Yap1* was overexpressed? How many genes were repressed when *Yap1* was overexpressed?

In Discovery Question 24, you should have discovered that the most induced gene is YNL331C, which by sequence similarity appears to be an aryl-alcohol reductase, although there are no experimental data on its real function. However, if you followed the Function Junction links long enough, you would have stumbled on a 1998 reference that had not been published when the DeRisi experiment was performed. Go to PubMed and enter the phrase "AAD gene set yeast." Click on the December 1999 reference that comes up, and read the abstract by Delneri et al.

Based on the Delneri data, the aryl-alcohol reductase story has not been completely told. What is so amazing and exciting is that you have followed up only one of the 6,200+ stories buried in data that contain many gems still waiting to be mined. The cell is a complex web with many interconnections that are unknown, even in one species of the "simple" yeast. When first analyzing huge data sets, the best stories may be missed by the original investigators. For example, what are the odds that a randomly chosen book in your library would be interesting to you? The clustered data tell you which genes share expression patterns, but they do not select the most interesting stories. That's up to you.

Microarrays Seem Too Good to Be True—Are They? The DeRisi paper had a huge impact on the scientific community, and launched a wave of labs learning how to perform microarray experiments. But the world of microarrays has been in such a hurry to produce results and make new discoveries that some shortcuts have been taken that need to be addressed.

Remember what you found when answering Discovery Questions 18–24 about the variation in the data for the *yap1+++* and *Δ Tup1* mutant strains? Some genes produced very "tight" data, while others did not. DeRisi repeated the experiments only once. How many times have you repeated

experiments once and gotten different answers? Does this mean your entire methodology is wrong? Does it affect your level of confidence? Or does it mean you simply need to repeat the experiment a few more times? Although microarray experiments have been presented to make them look simple, they are actually tricky, and expensive. Repeating experiments once when microarrays were being developed must have seemed sufficient. As chips, protocols, and reagents have become more available, people are rethinking the reproducibility of their data.

The DeRisi paper contained a short discussion about the variation of their data. The correlation coefficient for the duplicated experiment was 0.87. Was this variation due to variation in experimental procedure? or "cellular indecision" for a subset of genes (i.e., stochastic genetic circuits that can choose more than one outcome)? DeRisi and his colleagues also stated that their data did not always agree with previous publications. Do contradictory data indicate the method was flawed, or something more biologically significant? We don't know, and won't know, until someone does the experiments necessary to find out. That's why many call this "discovery science."

Let's look at Table 4.4, which appeared in the DeRisi paper. It lists genes that are induced when Yap1p is overexpressed. Topping the list, as you know, is the putative aryl-alcohol reductase. But notice that while the fold increase is 12.9, the standard error is not reported. Interestingly, footnote number 50 associated with this table says:

> In addition to the 17 genes shown in [Table 4.4,] three additional genes were induced by an average of more than threefold in the duplicate experiments, but in one of the two experiments, the induction was less than twofold (range 1.6- to 1.9-fold).

This footnote tells us 3 out of 20 genes (about 15%) were induced an *average* of threefold but showed induction of less than twofold in one of the two trials. These three genes were all induced, but the magnitude varied. Why? Does this indicate a need for more replications and/or standard errors for future publications?

DISCOVERY QUESTIONS

25. Now that you have a more complete understanding of how microarray data are obtained and the potential problems in reporting them, here is a challenging question.

   If full-length PCR products are used for spotting, does this create a problem? If one gene is 300 bp and another is 3,000 bp long, you would expect the shorter cDNA to have only 0.1 as much dye as the longer cDNA. Does the length of the PCR product present a problem when looking at ratios? clustering? or comparing the expression levels of two genes? Explain your answers.

TABLE 4.4 • Genes induced by Yap1p overproduction. This list includes all the genes that were induced over twofold in both trials of the same experiment, and the average is over threefold.

ORF	Gene	Description	Fold Increase
YNL331C		Putative aryl-alcohol reductase	12.9
YKLO71W		Similarity to bacterial csgA protein	10.4
YML007W	*Yap1*	Transcriptional activator involved in oxidative stress response	9.8
YFLO56C		Homology to aryl-alcohol dehydrogenases	9.0
YLL060C		Putative glutathione transferase	7.4
YOL165C		Putative aryl-alcohol dehydrogenase ($NADP^+$)	7.0
YCR107W		Putative aryl-alcohol reductase	6.5
YML116W	*Atr1*	Aminotriazole and 4-nitroquinoline resistance protein	6.5
YBR008C		Homology to benomyl/methotrexate resistance protein	6.1
YCLX08C		Hypothetical protein	6.1
YJR155W		Putative aryl-alcohol reductase	6.0
YPL171C	*Oye3*	NAPDH dehydrogenase (old yellow enzyme), isoform 3	5.8
YLR460C		Homology to hypothetical proteins YCR102c and YNL134c	4.7
YKR076W		Homology to hypothetical protein YMR251	4.5
YHR179W	*Oye2*	NAP(DH) oxidoreductase (old yellow enzyme), isoform 1	4.1
YML131W		Similarity to *A. thaliana* zeta-crystallin homolog	3.7
YOL126C	*Mdh2*	Malate dehydrogenase	3.3

Source: DeRisi et al. 1997. *Science.* 278: 684.

26. Let's imagine the green dye labeling of the cDNA did not work as well as it normally does. To compensate, you adjust the scanner so that it brightens the signal from the green channel. Does this present any problems with data interpretation? Explain your answer.

OK, Why Did the Beer Blow? It is approaching dinnertime, and you are ready to head back to your room. You only have one figure left to read (Figure 4.17 on your CD-ROM), and you are hoping the answer will be clear to you so you won't waste all that good beer next time.

FIGURE 4.17 • Yeast cells utilize reversible pathways to metabolize sugar quickly.
Go to the CD-ROM to view this figure.

Now you have finally gotten to the heart of the matter. You know your beer shattered the bottles because the pressure was too great. If you had used thicker glass bottles, it might have worked. The downside is that the beer would have spewed the way champagne does when opened. But the question is, what increased the pressure more than normal? After studying Figure 4.17, it hits you . . .

DISCOVERY QUESTIONS

27. How did the metabolism of glucose generate the increased pressure inside the bottle?

28. There are two reasons why the bottles may have blown up. The more likely answer is that there was too much sugar left when the beer was bottled. But Figure 4.17 offers another possible explanation. What is it?

LINKS
David Botstein
December 2000

You know that the increased bottle pressure was produced by metabolic waste product of CO_2. You also know that CO_2 is a by-product of fermentation and of the TCA cycle. Fermentation cannot continue in the absence of glucose, but could the TCA continue even if all the glucose were gone? The microarray data revealed that the genome sensed the loss of glucose and the presence of alcohol, which the yeast produced by fermentation. Later, the cells switched from fermentation to utilization of the ethanol. There must have been oxygen present in your bottles so the yeast was able to convert the alcohol into glycogen and trehalose. The genes needed to metabolize ethanol were induced as glucose was depleted. You suspect your bottles were airtight and trapped the CO_2 as it was produced by metabolizing ethanol to produce acetyl-CoA, which entered the TCA cycle. The continuation of metabolism caused your bottles to explode.

Why would yeast waste its time to make alcohol first only to consume it later? Yeast cells are not multicellular, so each cell is doing its best to survive by outcompeting its genetically identical siblings. Cells that accumulate the most energy (in the temporary storage form of alcohol) have the resources necessary to produce more offspring. Once the glucose is gone, the cells convert the somewhat toxic alcohol to safer storage molecules such as glycogen.

You put the paper down and wonder, "Wow, how did yeast get to be so smart?" Of course they have had a few hundred million years to optimize their genomes to adjust their metabolism as needed. Hopefully, it will not take you as long to learn how to brew beer.

What Can We Learn from Stressed-out Yeast?

Yeast as Supermodel By now it must be clear that yeast has a well-deserved reputation as a model organism. The Human Genome Project sequenced several model organisms as well as the human genome. Knowing the nonhuman genome sequences has already proven useful in the case of yeast. *Drosophila* and *C. elegans* genomes have begun to influence medical research, too. With the full human and mouse genomes, we will see many new uses of microarrays in biomedical research and diagnosis. Which brings up another story worth telling. . . .

It has been said that all yeast can do is reproduce and excrete, but this is an oversimplification. Yeast can do so much more than you might expect. Think about what a yeast cell is. It is a complete eukaryotic organism with about 6,200 genes. It can live as either a diploid or a haploid organism and can switch from aerobic to anaerobic metabolism when it needs to. Yeast can reproduce through sexual and asexual methods, and its metabolism has given us cheese, yogurt, and ethanol. That is an impressive

assortment of behaviors for a simple fungus with about half the number of genes as a fly.

The most amazing thing about unicellular organisms is how they sense their surroundings and respond appropriately. They contain most of the same cell-signaling systems that humans possess, without all that excess baggage of cellular specialization and tissue formation. Yeast has become one of the darlings of cancer research because mitosis is almost identical in all species. Our understanding of the human cell cycle is due in large part to yeast biologists, because yeast cells and human cells have many genes in common. That is why another paper by Pat Brown's lab is of such interest to us at this point.

The Brown lab has teamed up with many others, and some members of his lab have moved on to start their own labs. But of all Brown's collaborations, perhaps none have proven more successful than his close interactions with David Botstein. Botstein and Brown have become a "dynamic duo" by rapidly publishing many groundbreaking microarray papers. In December 2000, they published a rather weighty paper that examined the genomic response of 6,200 genes under 142 different experimental conditions, which required analyzing over 880,000 gene expression ratios. This tour de force contained more information than they described, but they did highlight a few interesting trends. Botstein and Brown have led the way in keeping scientific publications and raw data sets in the public domain. Due to their concern for free access to scientific information, you are able to analyze their work and continue to mine some of the richest microarray data sets available. The Yap1 and Tup1 databases were two examples; we will explore others in upcoming Discovery Questions.

Stressed-out Fungus In the Gasch et al. (2000) paper, the combined labs of Botstein and Brown performed three types of experiments. First, they measured the time course of gene expression in cells exposed to changes in their environment. Second, they looked at the genome responses in cells subjected to a gradient of heat shocks. Third, they examined gene expression profiles in cells that had already adapted to growth in these stressful conditions. Here is a list of the environmental challenges used in their study: heat shock, hydrogen peroxide, exposure to a superoxide-generating drug called menadione, a sulfhydryl oxidizing agent called diamide, a disulfide reducing agent called dithiothreitol (DTT), hyperosmotic shock, amino acid starvation, and nitrogen starvation. The severity of the treatments was designed so that about 80% of the cells would survive. To live, survivors had to change their protein makeup by altering gene expression. The final stress condition was growth into **stationary phase,** which means that the cells had exhausted the nutrients and were neither dead nor dividing.

From this collection of data, the investigators observed four classes of phenomena:

1. It became possible to hypothesize the functions of genes with previously unknown functions when they clustered repeatedly with genes of known function.

2. Genes that clustered together often contained conserved promoter sequence motifs.

3. Clustered gene expressions revealed the components of genomic circuits that work in concert to perform a single "task."

4. Individual genomic circuits could be observed to work in conjunction with other individual circuits to reveal integrated circuits working as larger units to accomplish complex "tasks."

The type of hierarchy of understanding from single genes to integrated circuitry described above is an excellent example of why genomics is more than a new set of tools. It permits us to envision more complete and complex models of how cells function as dynamic and evolving organisms.

By now you are beginning to wonder what all the data look like (Figure 4.18 on your CD-ROM).

FIGURE 4.18 • Expression of genes in response to 142 environmental perturbations.
Go to the CD-ROM to view this figure.

Even at low magnification, you can observe one trend. There are two large clusters of genes that mirror one another: the repressed cluster F is bright green when the induced cluster P is bright red. Environmental stress has already killed 20% of the survivors' genetically identical siblings, so it's time to adapt or die. The integrated genomic circuits are working collectively to save the remaining cells from death. The collective effort of the 900 genes in clusters F and P was dubbed the **environmental stress response** (**ESR**). What is so surprising is that these 900 genes altered their expression patterns in a coordinated fashion and in response to many, though not all, of the environmental perturbations (Figure 4.19 on your CD-ROM). How do they do this? Is yeast smarter than we are? Could you consciously initiate 900 different responses to changes in your environment? Suddenly the tiny fungus doesn't look so dumb.

FIGURE 4.19 • Stress-induced expression of 900 genes clustered into the ESR.
Go to the CD-ROM to view this figure.

What Goes Up . . . Although it is difficult to notice, there is an interesting phenomenon buried in these data. With the exception of starvation conditions, all of the ESR gene expression profiles were transient (Figure 4.20 on your CD-ROM). You can see from the heat shock experiments that both mild and harsh temperature shifts initiated the ESR (gene repression and induction), but for only a short time before the level of transcription returns toward previous levels.

FIGURE 4.20 • Transient ESR response.
Go to the CD-ROM to view this figure.

DISCOVERY QUESTIONS

29. What is the selective advantage for a transient ESR, even though the "shocking" stimulus is continuous? Why is the ESR substantially reduced once the cells have reached steady-state growth at the elevated temperatures?

30. What differences do you see between mild and harsh heat shock ESR profiles (Figure 4.20)? Propose a mechanism to explain these differences.

Once again we see how smart the yeast genome is. It would be wasteful (and thus not adaptive) to constantly respond to an environmental shock when it is no longer shocking. For example, in winter you get used to colder temperatures. If you flew to the other side of the world where it was summer, you would be shocked by how hot the day felt. However, after a few days that same temperature would not be shocking to you because you would have adapted to the hotter temperatures. Your body would not respond to a hot day the same way if it occurred in the winter vs. the summer. Once the appropriate physiological adjustments are made, there is no need for cells to transcribe the genomic equivalent of 911.

Why Are There So Many Copies of Some Genes but Not Others?

Like all organisms, the yeast genome contains genes that encode for **isozymes,** or enzymes that share a lot of sequence identity and appear to perform redundant functions. In the Delneri abstract (page 122), we saw that aryl-alcohol reductase is encoded by seven genes that produce seven isozymes. Why do some genes appear multiple times in a genome with only slight differences in their coding regions? Yeast seems to understand isozymes better than we do (Figure 4.21 on your CD-ROM). Although isozyme coding sequences are very similar and isozymes catalyze the same biochemical reaction, the yeast genome treats isozyme-encoding genes differently. Some are repressed while their functional equivalents are induced. Why?

FIGURE 4.21 • Differentially regulated isozyme genes in the ESR.
Go to the CD-ROM to view this figure.

DISCOVERY QUESTIONS

31. Why would an organism want to regulate differentially genes that encode isozymes? What does the yeast genome know that we don't?

32. Look at the IDH isozyme web page. Examine the expression profiles of these five apparent isozymes that all encode isocitrate dehydrogenase.
    a. Are any of them differentially regulated?
    b. *Should* they be differentially regulated? How would your interpretations change if you knew the following:
       • Idp1 uses NADP$^+$ as a coenzyme and is located in mitochondria.
       • Idp2 uses NADP$^+$ as a coenzyme and is located in the cytosol.
       • Idp3 uses NADP$^+$ as a coenzyme and is located in peroxisomes.
       • Idh1 uses NAD$^+$ as a coenzyme, is located in mitochondria, and is one subunit from a heterodimer with Idh2.
       • Idh2 uses NAD$^+$ as a coenzyme, is located in mitochondria, and is one subunit from a heterodimer with Idh1.
    c. The ESR database has all five isozymes listed as participating in the TCA cycle. Is this the complete story? Go to GenomeNet, select "PATHWAY" from the search menu, and one at a time find these three pathways: sce00020, sce00720, and sce00480. Click on the pathway link and you will see a metabolic pathway map. IDH has the E.C. number 1.1.1.42, and it will be highlighted. Is IDH restricted to the TCA cycle?

33. Go to Expression Connection, enter a gene name from Figure 4.21 (e.g., *Hxk1*) in the blank *below* the list of available experiments, and begin the search. The resulting page will contain lists of genes that had expression profiles similar to your chosen gene. Can you find any isozymes that were clustered with your gene?

34. If you mouse over the colored boxes and click on any gene, you will retrieve a new set of 20 genes that clustered with the one you clicked on. For further exploration of any particular dataset, click on the link that says "Visit the Website."

There are three reasons why a smart genome like yeast's might want to differentially regulate isozyme genes. Perhaps what appears to us as enzymes that do the same job may do different jobs, given their subcellular localization. Or isozymes catalyze the same reactions *in vitro*, but their *in vivo* activity might be quite different. A third reason might be that they catalyze the same biochemical reaction, but the cell needs complex regulation of gene expression. For example, when the cell is stressed, it may need isozyme #1 for a global response, but for a more subtle change in the environment, a more specific, non-ESR, response by isozyme #2 might be needed. If IDH performs only one reaction, but there are four different situations that require IDH, evolution may have selected four separate promoters (with four coding regions) rather than one promoter with four sets of transcription factors.

DISCOVERY QUESTIONS

35. Look at Yeast Stress Figure 4 and you will see six subsets of genes that are part of the ESR expression profile and have related functions.
    a. Design one common promoter for one of the six subsets of genes (just draw a box with labels; don't worry about the DNA sequence) that would allow all these genes to be up-regulated at one time.
    b. Add to your design from Discovery Question 35a so that when needed, each of the six genes could be regulated separately from the ESR.
    c. How many different transcription factor genes do you imagine are needed to provide collective and individual control over gene expression?

How Well Do Promoters Control Gene Expression?

In Discovery Questions 35a–c, you designed different combinations of transcription factors and promoters to elicit the adaptive responses present in the yeast genome. Some transcription factors have been the subject of intensive investigation, so we know a lot about how they work. Three transcription factors that have been characterized in detail are Msn2p, Msn4p, and Yap1p. DNA microarray data for 15 genes that contain binding sites for Msn2p/Msn4p and Yap1p were analyzed for their responses to heat and hydrogen peroxide when *Msn2/Msn4* and *Yap1* were deleted (Figure 4.22 on your CD-ROM). Three strains of yeast were used for this study—*wt*, an *Msn2 Msn4* double deletion strain, and a *Yap1* deletion strain.

FIGURE 4.22 • Dependence of ESR gene regulation on transcription factors.
Go to the CD-ROM to view this figure.

DISCOVERY QUESTIONS

36. What role does the Yap1p transcription factor play in hydrogen peroxide and heat shock responses? What role do the Msn2p and Msn4p transcription factors play in hydrogen peroxide and heat shock responses?

37. How can this level of gene regulation be used to provide the flexibility a genome needs when responding to environmental stress vs. less drastic needs of a cell? It appears that all three transcription factors can activate these 15 genes. Why do we need two types of transcription factors to do the same job?

38. When you think of cells, you probably imagine a functional unit that is very efficient and does not waste its resources. That's certainly what you would expect after billions of years of evolutionary refinement, but the ESR appears to be an exception to the rule of efficiency. Genes that do reciprocal jobs are induced en masse as a part of the ESR. For example, transcription was stimulated for the genes encoding the positive effectors of protein kinase A and the negative regulators of protein kinase A. Hypothesize the evolutionary advantage to this apparent extravagance.

39. An old question revisited: Are there any problems associated with spotting PCR products from the entire gene rather than smaller portions of a gene?

The yeast genome has been stressed and shocked for a billion years. It is so used to shocks that it no longer panics every time the red alert warning is sounded. Genomes sense a change in the environment that could prove lethal if not responded to quickly and appropriately. So the 900 ESR genes are induced or repressed according to standard emergency procedure. This includes genes that govern activation and repression of a common critical controlling factor such as protein kinase A. Producing an activator and a repressor at the same time seems wasteful until you realize that the environment can change quickly. Let's imagine that two yeast cells are riding on the back of a deer that has stopped to feed in the sunlight. It gets very hot in the summer sun, so both yeast cells initiate the heat-induced ESR set of 900 genes with a few exceptions. Yeast number one senses the heat and, rather than producing both an activator and a repressor, it

"knows" that *heat* shock only requires the repressor in order for the cell to survive. Therefore, yeast number one begins the specialized "I'm too hot" response and makes no allowances for alternative responses. Yeast number two responds in the traditional way by producing both an activator and a repressor.

These two yeast cells represent a classic example of genetic variation where two cells express slightly different proteins and thus phenotypes. If the deer stays in the sunlight longer, yeast number one has a slight head start, but yeast number two will eventually respond in the same way. However, if the deer decides to lick its back and a very basic saliva coats our number two yeast, there may be a different outcome. Yeast number one decided too soon that this was an "I'm too hot" emergency and committed full speed in one direction. Yeast number two decided to wait a little longer and collect more information before committing. In this instance, yeast number two has both of its regulatory proteins ready to respond either way and is capable of initiating very quickly a response to the new "It's too basic!" emergency. In other words, after a billion or so years, it appears that the yeast genome has been selected to respond in stages. First is the red alert and the full 900-gene ESR, followed by the more finely tuned and specific response. The yeast genome appears to live by the motto: "Always be prepared." In the lab, when the investigator is applying only one environmental stress at a time, the yeast ESR seems wasteful. But perhaps in the real world where yeast evolved, it is advantageous to respond by simultaneously producing activators and repressors of a few critical genomic "circuit switches" so that the cell can respond appropriately as it collects more information.

Are Promoters Able to Work in Reverse?

Most of what you have learned has been under the heading of "What happens if . . . ?" But as we learn more about how cells respond, we are able to build models that are testable. Because these data sets are so rich, the data to test the models are already available. For example, if evolution has produced promoters that work well in response to heat shock, they should be able to respond in the opposite direction when the cell is subjected to cold shock. It would be a disaster for a cell to respond the same way for both heat and cold shock. Imagine being able to sense the weather (hot and cold), but being hardwired so that your only response was to add more layers of clothing, even in the summer. Luckily, you and the yeast genome can sense these two different stimuli and respond in opposite ways (Figure 4.23 on your CD-ROM). When the cells are shifted from cold to hot (Figure 4.23a; solid ramp), about 600 genes are repressed and 300 induced. Conversely, when cells are shifted from hot to cold (open ramp), the

"cloud" of points near the origin in Figure 4.24b signifies a low correlation between the expression ratios of the two mutants relative to *wt*.

In some experimental designs, you might expect two distinct columns of an expression matrix to track one another. For example, the two columns may represent the same phase of the cell cycle, one from the first cycle and another from the second. Or, as we will see in Chapter 5, two columns may represent two different samples of the same tissue type. However, in the experiment shown in Figure 4.24, the strong similarity between the two mutants is surprising. Ratios that were significantly elevated in both mutants (almost exclusively genes on chromosome 7) were the primary cause of the high correlation. Thus, identifying an unusually high correlation between columns led immediately to the hypothesis of aneuploidy.

With these results, we now have one more type of information we can extract using microarrays—**aneuploidy.** Aneuploidy is the presence of an abnormal amount of chromosomal material in a particular cell. In this case, both mutants had duplicated some of chromosome 7. It is worth pointing out that these duplication events occurred separately in separate strains that were isolated in different labs. It points to a new possibility, that whenever a mutant strain of yeast is made, there is a chance for aneuploidy to occur.

Once you observe a new phenomenon, you want to know how widespread it is. Hughes and his colleagues examined 290 mutant strains and found 22 (about 8%) exhibited aneuploidy! That is a much higher number than anyone expected. How can aneuploidy be so common in yeast that duplicating a chromosome happens in 8% of all deletion mutant strains? Through a careful analysis of their strains, the team has developed a reasonable hypothesis (Figure 4.25). The growth rates for deletion mutants compared to their parental strains show that the deletion mutants often grow slower.

DISCOVERY QUESTIONS

52. On which chromosome(s) would you *guess Rnr1* and *Rps24a* are located?

53. Go to the MIPS database and search for the two deleted genes: *Rnr1* and *Rps24a*. What do they have in common?

54. Hypothesize why chromosome 9 would be duplicated in *Rnr1* and *Rps24a* mutant strains.

55. Looking at Figure 4.25a, why do you think there is a high percentage of aneuploidy in deletion strains maintained in labs? Explain your answer in evolutionary terms.

Once again *Rnr1* and *Rps24a* have nothing in common except their chromosomal locations. Surprisingly, they are both located on chromosome 5, not chromosome 9 as you might have guessed. If they had been on chromosome 9,

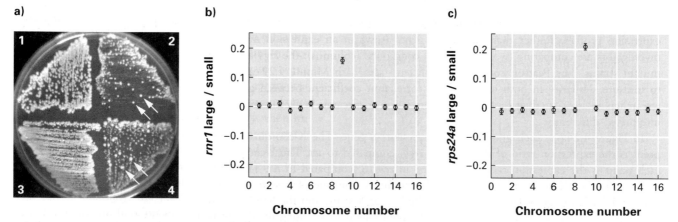

FIGURE 4.25 • **Development of aneuploidy in deletion strains. a)** Two different mutant strains (2 is *rnr1* and 4 is *rps24a*) were compared to their parental strains (quadrants 1 and 3) for their ability to grow in standard lab conditions. Two different-sized colonies (small and large) are denoted by white arrows. **b)** and **c)** Relative chromosomal content of deletion strain was calculated and the $\log_{10}$ of the ratio was plotted on the Y-axis; chromosome number is plotted on the X-axis. Error bars represent error of the mean $\log_{10}$ (ratio).

though, the duplication of a deleted gene would not have helped the cells at all. Why would both mutants have duplicated chromosome 9? Both of these genes have isozymes encoded on chromosome 9. *Rps24a* encodes a protein 97% identical to Rps24b. Rnr1p has an isozyme (Rnr3p) with 80% amino acid identity. So the only reason both deletion strains duplicated chromosome 9 is that each had functionally redundant genes on chromosome 9. The fact that both strains duplicated the same chromosome does raise the possibility that certain chromosomes may duplicate more readily, or be less toxic than others when duplicated. The possibility of chromosome bias for duplication will require further analysis of many more deletion/aneuploidy mutants. By increasing the amount of mRNA produced from isozymes of the deleted genes, these strains had evolved under lab conditions to grow faster, which is what every researcher subconsciously wants since faster-growing strains produce faster results. However, we now know that deletion strains have different agendas from investigators.

It would be interesting to determine whether aneuploidy is common or not. Luckily for Hughes, there were public data sets available from Brown and Botstein that could be mined further to determine if other labs had similar strains with aneuploidy. If you refer back to Figure 4.16, you will see that one of the strains DeRisi used had deleted the repressor gene called *Tup1* (*Δ Tup1*), which is located on chromosome 7. When Hughes and his team analyzed DeRisi's expression ratios, they found a surprising result (Figure 4.26).

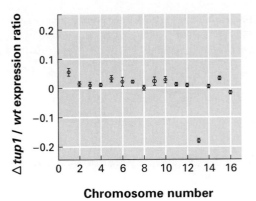

FIGURE 4.26 • **Re-analysis of data produced by DeRisi et al.** (pages 120–122) Spots indicate the average gene expression ratio ($\log_{10}$ on the Y-axis) in *Δ Tup1* compared to the parental *wt* strain for each chromosome (X-axis). Error bars indicate the error of the mean $\log_{10}$ of the published expression ratios.

DISCOVERY QUESTIONS

56. What do you conclude about the *Δ Tup1* strain used by DeRisi et al.?

57. What additional information would you want to have before saying you are certain that aneuploidy exists in *Δ Tup1*'s genome?

58. Propose a mechanism to explain why *Δ Tup1* resulted in a chromosomal deletion rather than a duplication.

59. What analysis does this work suggest should become standard in all microarray experiments?

From the data, it appears that DeRisi was working with a deletion strain that had lost more than he realized. *Tup1* is on chromosome 7, and in response, chromosome 13 lost some of its DNA. The chromosome 13 aneuploidy may have accelerated the growth of the *Δ Tup1* strain, and DeRisi never knew his strain was less than he had expected. Based on this analysis by Hughes, aneuploidy is more common than we expected and needs to be investigated more rigorously.

Can Cells Verify Their Own Genes?

Although the exact definition of a gene is debated, most people agree that an RNA molecule should be produced by a real gene. If the gene produces mRNA, DNA microarrays should be able to detect it. Rosetta Inpharmatics near Seattle specializes in performing DNA chip experiments. They have helped develop an "inkjet" printer that allows them to synthesize DNA **oligonucleotides** (short polymers of single-stranded DNA) for each spot. Instead of ink, the printer sprays **dNTPs:** dGTP, dCTP, dATP, and dTTP. Rosetta can produce DNA microarrays composed of 25,000 different oligonucleotides, each one 60 nucleotides long (called **60-mers** which means poly*mer* of 60 bases; Figure 4.27).

For this work, Rosetta investigators focused on chromosome 22, the first human chromosome to be fully sequenced and annotated (published in 1999). A total of 8,183 exons had been annotated, and the Rosetta team decided to verify these predictions by using labeled cDNA pairs from 69 different experimental conditions. This level of analysis is very rigorous since each exon was tested by two different 60-mers. For each experiment, cDNAs were produced from two different sources; sometimes they were biopsied tissue samples and other times human cell lines.

The investigators replicated their experiments (Figure 4.28a on your CD-ROM) by labeling the cDNAs with the opposite color dye to ensure probe production did not introduce a bias in their data. If everything was working properly, you would expect the colors to be reversed in the two replicated experiments, as shown in Figure 4.28b and 4.28c. These duplications were performed 69 times on a total of 138 DNA chips.

FIGURE 4.27 • Designing DNA microarrays for determining exon boundaries. From left to right, increased magnification of chromosome 22, one BAC clone, 10 kb of one BAC, a 150 bp exon, and two 60-mer oligonucleotides. Two 60-mers for each annotated exon were spotted on the chip to create a comprehensive chromosome 22 DNA microarray.

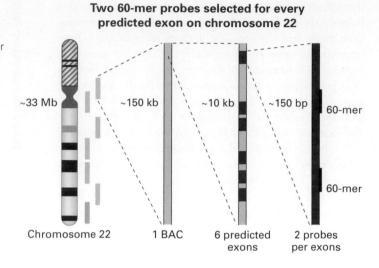

Two 60-mer probes selected for every predicted exon on chromosome 22

~33 Mb   ~150 kb   ~10 kb   ~150 bp   60-mer / 60-mer

Chromosome 22   1 BAC   6 predicted exons   2 probes per exons

FIGURE 4.28 • Chromosome 22 DNA microarray data.
Go to the CD-ROM to view this figure.

DISCOVERY QUESTIONS

60. How many different data points would this experiment generate if you count each color for a given spot as a different data point?

61. What would you expect to see in Figure 4.28 if a gene contained nine exons and the spots were designed so that each 60-mer pair is side by side and each 60-mer pair is placed in the same order in which they are arranged on the chromosome?

62. What would you expect to see if a gene were turned on in only a subset of the 69 experimental conditions?

63. What would you expect to see if a predicted gene were not a real gene and thus was never transcribed?

64. What would you expect to see if a predicted gene were real but produced very little mRNA, or were transcribed in a condition not included in the 69 used for this experiment?

65. What would you expect to see if a gene contained only one exon?

66. What would you expect to see if a gene were alternatively spliced?

67. How could you distinguish a real gene (Discovery Question 64) from a pseudogene (Discovery Question 63)?

Once all these data points were collected, they needed to be displayed in a comprehensible manner (Figure 4.29a on your CD-ROM). The average expression ratio (control vs. experimental) was represented as a color-coded line represented with the 69 experiments listed on the Y-axis and 16,366 60-mers listed on the X-axis. Each exon was represented by a pair of 60-mers, so we should look for pairs of data points with identical color patterns. For example, when a gene with five exons was transcribed, we should see ten spots with similar color patterns.

FIGURE 4.29 • DNA microarray validation of predicted exons on chromosome 22.
Go to the CD-ROM to view this figure.

Which Predicted Genes Are Real and Which Ones Aren't

If a particular gene were transcribed only under certain experimental conditions, we would be hard-pressed to summarize that information because Figure 4.29a is overwhelming. For single-gene analysis, investigators can reorder the data to place next to each other experiments that affect particular genes in the same way (Figure 4.29b–e). In Figure 4.29b, we see a known gene induced in 39 experimental conditions (red boxes) and repressed in 8 experimental conditions (green boxes). In c), we see the results for another known gene, but one predicted exon (denoted by arrow) does not appear to be included in the mRNA for this gene. In d), we see two different EST sequences transcribed under similar conditions, which indicates they might be encoded by the same gene. Finally, e) shows one set of exons that is transcribed at high levels only in experiments 21 and 29. Experiment 21 compared mRNA from a collection of cell lines to mRNA from testes, while experiment 68 compared testes to uterus mRNAs. These are the only two experimental conditions that examined testes mRNA, the only two that have clear color patterns for these 60-mers. These predicted exons are expressed only in the testes and appear to constitute a real gene.

LINKS
Sanger Center

DISCOVERY QUESTIONS

Here are a series of questions related to Figure 4.29. For these questions, let's assume there are no errors due to experimental technique.

68. In panel b), why do some individual spots appear black even when they occur in the middle of a red or green patch? Why might *pairs* of spots appear black even when they occur in the middle of a red or green patch?

69. One interpretation for c) is that this exon is not a true one. Propose an alternative interpretation.

70. In panel d), one pair of spots appears black most of the time. What does this indicate?

71. In panel e), the exons had not been annotated as a gene, but this experiment indicates that they form a real gene. How many exons would be in this gene?

DISCOVERY QUESTIONS

72. How many different 10 bp overlapping 60-mers did Rosetta have to produce in order to cover both strands of the 113,000 bp fragment of chromosome 22?

73. Why did they produce 60-mers for *both* strands of the DNA?

74. Would the DNA chip data alone be sufficient to determine exactly which specific nucleotide began or ended an exon? Explain your answer.

From Proof of Concept to Full-Scale Analysis After the Rosetta group had produced a list of microarray-confirmed exons for chromosome 22, they decided to survey the complete human genome. Using the **frozen genome** from 15 June 2000, Rosetta produced 50 DNA microarrays that contained 1,090,408 60-mers to cover all 442,785 predicted exons (Figure 4.31a on your CD-ROM). They probed these DNA chips using colored cDNA produced from two cell lines, a human colon cancer and a human lymphoma. With these two cDNA populations, they were able to detect expression from 58% of the confirmed exons and about 123,000 additional exons out of 364,299 predicted but previously unconfirmed exons. They estimated a 5% false positive rate, which may reduce the number of confirmed exons, but their pioneering work is very impressive (Figure 4.31b–d). By devising these experiments, they have introduced a method for large-scale and experimental annotation of the human genome.

Refining Exon Boundaries with DNA Chips Computer programs are very good at rapidly performing tasks we tell them to do, but the programs have no more insight into the truth than the people who program them. Unfortunately, no one fully understands what defines the beginning or ending of an exon. We know that most exons are preceded by the dinucleotide AG and followed by the dinucleotide GT, but these dinucleotides also occur by chance away from exons. We must rely on other traits with vague rules to help us locate exons, but we make mistakes. Since the Rosetta team had identified a new gene based on predicted exons, they wanted to determine experimentally the exact size of the six exons. The Sanger Center (one of the biggest genome labs in the world) had defined the length of each potential exon with their software. The Rosetta team made a new set of DNA chips, but this time they produced a new 60-mer every 10 bp over the entire length of a 113,000 bp piece of DNA that contained the newly discovered testes-specific gene (Figure 4.30, page 134). After creating this microarray of overlapping 60-mers, they probed the 60-mers with cDNAs that contained the transcript of interest and determined which spots bound the labeled cDNAs. By designing a systematic approach, the team was able to confirm some exon predictions (exons 1, 2, 4, and 5) and correct the mistakenly short predictions for exons 3 and 6.

FIGURE 4.31 • Whole-genome DNA microarray to validate every predicted exon.
Go to the CD-ROM to view this figure.

DISCOVERY QUESTIONS

75. If only 60% of the "confirmed" exons were detected, does that mean the method is not very accurate or the "confirmed exons" are not really exons? Explain your answer.

76. The investigators detected mRNA from some predicted exons but not all of them. Interpret these data.

Summary 4.2

One yeast DNA chip produces as much data as about 12,000 Northern or Southern blots. With this much information, you can appreciate why data analysis and visualization are fundamental to our ability to understand the data. Since we

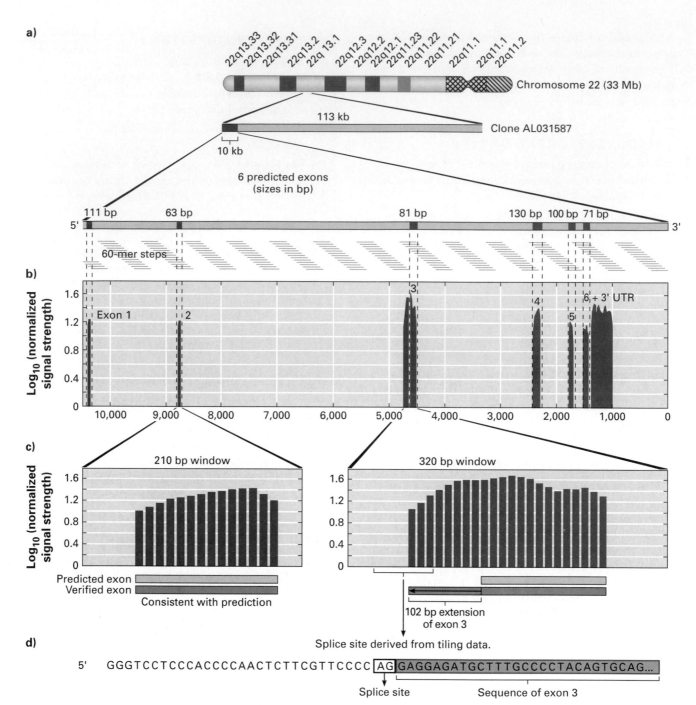

FIGURE 4.30 • **Defining exons of newly discovered gene. a)** Exons for the testes-specific gene (Figure 4.29e) were defined by 60-mers every 10 bp over a 113,000 bp piece of chromosome 22. **b)** Six exons were identified over a 10 kb region as shown. **c)** Two exons are shown at higher magnification with each 60-mer represented as a separate bar graph. Below the graph are comparisons of exon predictions (light blue) using current software technology and the microarray exon detection (dark blue). **d)** Sequence from the 5' end of exon 3 showing the consensus splice site that is consistent with the microarray prediction for the 5' boundary of exon 3.

often use guilt by association to propose functions for nonannotated genes, variations in clustering software can produce different associations between genes. The way we display this information may limit our perceptions or open new views of the data that lead to new interpretations. With carefully designed microarray experiments, we saw how genomes can lose or gain parts of their chromosomes through aneuploidy. As the resolution and production of DNA microarrays improve, we can use these high throughput methods to better understand genomic sequences, as was demonstrated by the first use of chips on the draft human genome sequence.

Chapter 4 Conclusions

DNA microarrays are powerful tools to analyze genomes in vivo. We can see how a genome inside a nucleus responds to an ever-changing environment—the cell. When one component of the extracellular environment changes, the genome adapts by altering its cellular environment. Genomic research is repainting the picture we see inside cells. Traditionally, we have altered single genes and then measured the cellular phenotype. DNA microarrays have shown us that genomes are capable of altering themselves through aneuploidy, so we unknowingly study substantially different cells controlled by dynamic genomes—ones that can alter their expression and composition. When applied to the human genome, we can discover new genes, or refine gene annotation. The success of DNA microarrays depends in large part on how we analyze and visualize the data. Mathematicians, computer scientists, and biologists are collaborating to create bioinformatics tools that will improve our understanding of genome-wide behaviors. In Chapter 5, several examples illustrate how basic research methods can be applied to better understand cancer, how medicines work, which medications should be used for individual patients, and how leptin modulates obesity and many other genome circuits.

References

Yeast Chips
Chu, S., J. DeRisi, et al. 1998. The transcriptional program of sporulation in budding yeast. *Science.* 282: 699–705.

DeRisi, Joseph L., Vishwanath R. Iyer, and Patrick O. Brown. 1997. Exploring metabolic and genetic control of gene expression on a genomic scale. *Science.* 278: 680–686.

Gasch, Audrey P., Paul T. Spellman, et al. 2000. Genomic expression programs in the response of yeast cells to environmental changes. *Molecular Biology of the Cell.* 11: 4241–4257.

Sherlock, Gavin, Tina Hernandez-Boussard, et al. 2001. The Stanford Microarray Database. *Nucleic Acids Research.* 29(1): 152–155.

Walker, Graeme M. 1998. *Yeast physiology and biotechnology.* New York: John Wiley and Sons.

Winzler, Elizabeth, Daniel Shoemaker, et al. 1999. Functional characterization of the *S. cerevisiae* genome by gene deletion and parallel analysis. *Science.* 285: 901–906.

Clustering and Visualization of Microarray Data
Conklin, Bruce. 2001. GenMapp: Gene MicroArray Pathway Profiler. <http://gladstone-genome.ucsf.edu/introduction.asp>. Accessed 8 February 2002.

D'haeseleer, Patrik, Shoudan Liang, and Roland Somogyi. 2000. Genetic network inference: From co-expression clustering to reverse engineering. *Bioinformatics.* 16(8): 707–726.

Eisen, Michael, B., Paul T. Spellman, et al. 1998. Cluster analysis and display of genome-wide expression patterns. *PNAS.* 95: 14863–14868.

Fuhrman, Stefanie, M. J. Cunningham, et al. 2000. The application of Shannon entropy in the identification of putative drug targets. *Biosystems.* 55(1–3): 5–14.

Getz, Gad, Erel Levine, and Eytan Domany. 2000. Coupled two-way clustering analysis of gene microarray data. *PNAS.* 97: 12079–12084.

Gilbert, David R., Michael Schroeder, and Jacques van Helden. 2000. Interactive visualization and exploration of relationships between biological objects. *Trends in Biotechnology.* 18: 487–494.

Heyer, Laurie J., Semyon Kruglyak, and Shibu Yooseph. 1999. Exploring expression data: Identification and analysis of coexpressed genes. *Genome Research.* 9: 1106–1115.

Kyoto Encyclopedia of Genes and Genomes (KEGG). 2001. <http://www.genome.ad.jp/kegg/>. Accessed 8 February 2002.

Lee, Mei-Ling Ting, Frank C. Kuo, et al. 2000. Importance of replication in microarray gene expression studies: Statistical methods and evidence from repetitive cDNA hybridizations. *PNAS.* 97(18): 9834–9839.

Lemkin, Peter F., Gregory C. Thornwall, et al. 2000. The microarray explorer tool for data mining of cDNA microarrays: Application for the mammary gland. *Nucleic Acids Research.* 28(22): 4452–4459. <http://www.lecb.ncifcrf.gov/MAExplorer/>.

Liao, Birong, Walker Hale, et al. 2000. MAD: A suite of tools for microarray data management and processing. *Bioinformatics.* 16(10): 946–947. <http://pompous.swmed.edu/>.

Manduchi, Elisabetta, Gregory R. Grant, et al. 2000. Generation of patterns from gene expression data by assigning confidence to differentially expressed genes. *Bioinformatics.* 16(8): 685–698.

National Center for Genome Resources. 2001. GeneX: A collaborative Internet database and toolset for gene expression data. <http://www.ncgr.org/research/genex/>. Accessed 8 February 2002.

Aneuploidy Chips
Hughes, Timothy R., Christopher J. Roberts, et al. 2000. Widespread aneuploidy revealed by DNA microarray expression profiling. *Nature Genetics.* 25: 333–337.

Human Genome Analysis
Shoemaker, D. D., E. E. Schadt, et al. 2001. Experimental annotation of the human genome using microarray technology. *Nature.* 409: 922–927.

Chip Reviews and Commentaries
Barry, Clifton E. III, and Benjamin G. Schroeder. 2000. DNA microarrays: Translational tools for understanding the biology of

Mycobacterium tuberculosis. *Trends in Microbiology.* 8(5): 209–210.

Berns, Anton. 2000. Gene expression in diagnosis. *Nature.* 403: 491–492.

Brazma, Alvis, and Jaak Vilo. 2000. Gene expression data analysis. *FEBS Letters.* 480: 17–24.

Brent, Roger. 1999. Functional genomics: Learning to think about gene expression data. *Current Biology.* 9: R338–R341.

Dalton, Rex. 2000. DIY (Do it yourself) microarrays promise DNA chips with everything. *Nature.* 403: 234.

GCAT: the Genome Consortium for Active Teaching. 2001. <http://www.bio.davidson.edu/GCAT>. Accessed 8 February 2002.

Geschwind, Daniel H. 2000. Mice, microarrays, and the genetic diversity of the brain. *PNAS.* 97: 10676–10678.

Harris, Thomas M., Aldo Massimi, and Geoffrey Childs. 2000. Injecting new ideas into microarray printing. *Nature Biotechnology.* 18: 384–385.

Lockhart, David J. 1998. Mutant yeast on drugs. *Nature Medicine.* 4: 1235–1236.

Lockhart, David J., and Carrolee Barlow. 2001. Expressing what's on your mind: DNA arrays and the brain. *Nature Reviews.* 2: 63–68.

Madden, Stephen L., Clarence J. Wang, and Greg Landes. 2000. Serial analysis of gene expression: From gene discovery to target identification. *Drug Discovery Today.* 5(9): 415–425.

Marx, Jean. 2000. DNA arrays reveal cancer in its many forms. *Science.* 289: 1670–1672.

Perou, Charles M., Patrick O. Brown, and David Botstein. 2000. Tumor classification using gene expression patterns from DNA microarrays. *New Technologies for the Life Sciences: A Trends Guide.* 67–76.

Pinkel, Daniel. 2000. Cancer cell, chemotherapy, and gene clusters. *Nature Genetics.* 24: 208–209.

Schaffer, Jeff Landgraf, Miguel Pérez-Amador, and Ellen Wisman. 2000. Monitoring genome-wide expression in plants. *Current Opinion in Biotechnology.* 11: 162–167.

Velculescu, Victor E. 1999. Tantalizing transcriptomes: SAGE and its use in global gene expression analysis. *Science.* 286: 1491–1492. <http://www.sagenet.org/>.

Young, Douglas B., and Brian D. Robertson. 1999. TB vaccines: Global solutions for global problems. *Science.* 284: 1479–1480.

Additional Readings of Interest

New Microarray Fabrication Methods

Brown, Patrick O., et al. 2001. The Mguide. <http://cmgm.stanford.edu/pbrown/mguide/>. Accessed 8 February 2002.

Kane, Michael D., Timothy A. Jatkoe, et al. 2000. Assessment of the sensitivity and specificity of oligonucleotide (50mer) microarrays. *Nucleic Acids Research.* 28(22): 4552–4557.

Kumar, Anil, and Zicai Liang. 2001. Chemical nanoprinting: A novel method for fabricating DNA microchips. *Nucleic Acids Research.* 29(2): e2.

Okamoto, Tadashi, Tomohiro Suzuki, and Nobuko Yamamoto. 2000. Microarray fabrication with covalent attachment of DNA using bubble jet technology. *Nature Biotechnology.* 18: 438–441.

Bacterial and Viral Papers

Arimura, Gen-ichiro, Kosuke Tashiro, et al. 2000. Gene responsiveness in bean leaves induced by herbivory and by herbivore-induced volatiles. *Biochemical and Biophysical Research Communications.* 277: 305–310.

Cho, Raymond J., Micheline Fromont-Racine, et al. 1998. Parallel analysis of genetic selections using whole genome oligonucleotide arrays. *PNAS.* 95: 3752–3757.

Cummings, Craig A., and David Relman. 2000. Using DNA microarrays to study host-microbe interactions. *Genomics.* 6(5): 513–525.

Khodursky, Arkady B., Brian J. Peter, et al. 2000. DNA microarray analysis of gene expression in response to physiological and genetic changes that affect tryptophan metabolism in Escherichia coli. *PNAS.* 97(22): 12170–12175.

Khodursky, Arkady B., Brian J. Peter, et al. 2000. Analysis of topoisomerase function in bacterial replication fork movement: Use of DNA microarrays. *PNAS.* 97: 9419–9424.

Selinger, Douglas W., K. J. Cheung, et al. 2000. RNA expression analysis using a 30 base pair resolution *Escherichia coli* genome array. *Nature Biotechnology.* 18: 1262–1268.

Talaat, Adel M., Preston Hunter, and Stephen Albert Johnston. 2000. Genome-directed primers for selective labeling of bacterial transcripts for DNA microarray analysis. *Nature Biotechnology.* 18: 679–682.

Ye, Rick W., Wang Tao, et al. 2000. Global gene expression profiles of Bacillus subtilis grown under anaerobic conditions. *Journal of Bacteriology.* 182(16): 4458–4465.

CHAPTER 5

Applied

Research with

DNA Microarrays

Chapter 4 introduced **DNA microarrays** and how they can be used for basic research. This chapter focuses on biomedical applications for DNA chips. The first half of the chapter focuses on cancer. You will see how DNA microarrays were used to better understand two forms of cancer, a lymphoma and breast cancer. Ideally, microarrays would allow us to diagnose and treat patients more successfully. In addition, we would like to understand more fully the genomic circuitry underlying the diversity and complexity found in clinical settings. In the second half of the chapter, you will examine case studies where microarrays were used to better understand how medications work and how we can improve their effectiveness. The final case study explores the complexity of the fat-regulating hormone leptin (see Chapter 11). In each of the case studies, basic and applied research are interwoven as we try to understand the basic biology in order to develop more effective treatments.

METHODS

DNA Microarrays

cDNA

LINKS

Pat Brown

David Botstein

Louis Staudt

Ash Alizadeh

Mike Eisen

5.1 Cancer and Genomic Microarrays

As life expectancy increases, there appears to be a trade off—increased incidence of diseases. You probably know someone who has been diagnosed with cancer. Many times the patients are elderly, but not always. Since some cancers run in families, we know that an important component of cancer formation is determined by the genome. Can we use the dynamic activity of a person's genome to cure? Are some cancers more amenable to DNA microarray diagnosis than others? When a cell becomes cancerous, can we use microarrays to search for chromosomal instability? The three case studies in Section 5.1 illustrate the potential for DNA microarrays, and their limitations.

Are There Better Ways to Diagnose Cancer?

Microarrays are being touted as the next great tool for diagnosis of disease. Pat Brown's and David Botstein's laboratories teamed up with several other labs, including Louis Staudt at the National Cancer Institute in Frederick, Maryland, to determine if DNA microarrays could be used to diagnose diffuse large B-cell lymphoma (DLBCL). This work involved several investigators such as Ash Alizadeh, who did much of the microarray work, and Mike Eisen, a former postdoc at Stanford who now runs his own lab at the Lawrence Berkeley National Laboratory. Eisen is the person most responsible for converting the ratios into more comprehensible clustered gene sets. Each member of the team provided unique expertise.

DLBCL is a very aggressive malignancy of B cells. Each year in the United States, over 25,000 new cases are reported, accounting for about 40% of all non-Hodgkin's lymphoma cases. DLBCL is diagnosed by a loose combination of morphological and molecular characteristics. For example, the **immunoglobulin** (antibody) genes in non-Hodgkin's lymphomas show signs of **somatic hypermutation,** which is a process of increased changes in the DNA sequence of the variable regions of antibody genes. B cells undergo somatic hypermutation to improve the ability of antibodies to bind to their antigens. Somatic hypermutation occurs in the germinal centers of secondary lymphoid tissue such as lymph nodes. Sadly, only about 40% of all DLBCL patients are treated successfully with chemotherapy. No one is sure why some respond and others do not.

To conduct this work, the team had to construct human DNA microarrays consisting of 17,856 spots each. They produced 128 microarrays and converted mRNA samples into **cDNA** probes from 96 normal and malignant lymphocytes to collect about 1.8 million ratios (Figure 5.1 on your CD-ROM). Although it looks like a mess, Eisen's software has extracted some interesting trends. In these experiments, the 96 human samples were used individually to produce 96 different red cDNAs and compared to green control cDNA produced from 9 different lymphoma cell lines, which were pooled and used as a consistent source of cDNA.

FIGURE 5.1 • **DLBCL gene expression analysis.**
Go to the CD-ROM to view this figure.

The dendrogram shown in Figure 5.1 was produced by first clustering the genes according to expression pattern similarities and then clustering the 96 samples based on their similarity of gene expression profiles. The result produced "**signature genes,**" indicated by colored bars on the right of the figure. Signature genes can be used to describe a particular cell type. For example, activated B cells are known to induce and repress particular genes and these genes constitute the activated B-cell signature genes.

The first notable feature is that a subset of genes was able to classify the samples based on cell type (e.g., germinal center B cell) or cell process (e.g., proliferation). It is striking to see the DLBCL samples clustered into one of the two main branches of samples. When only the data from the six sets of signature genes are examined, some genes do not have a known function (Figure 5.2 on your CD-ROM) but their expression profiles were similar to the known ones included in the signature genes.

FIGURE 5.2 • **Biologically distinct DLBCL gene expression signatures.**
Go to the CD-ROM to view this figure.

DISCOVERY QUESTIONS

1. Which three types of samples have the highest overall expression of proliferating genes (see Figure 5.2)?

2. Are all the DLBCL samples clustered together? Hypothesize why.

3. Which of the signature genes might be the best to subdivide the DLBCL into subcategories?

4. Go to Explore Lymphoma Figure 1 web site. This is an interactive version of the complete dataset from the DLBCL paper. In the left frame is a "radar" view that is zoomed out. In the right frame is a "zoom" window that gives you a closer look at 100 genes at a time. The top frame allows you to change views. Perform the following steps:

   a. In the top frame, "Change zoom X factor to:," enter the number 3, and click on "submit." This will zoom out the view on the right.

   b. In the "radar" frame, scroll down and click on the orange bar to view all the GC-like genes.

   c. Search for a gene/protein that, based on its expression profile, might be a good target for killing DLBCL but not normal B or T cells. Explain why you have chosen your particular target.

All but three samples from the DLBCL collection were clustered into one large group. The three outliers were clustered apart from the others because a majority of the cells in these three biopsies may have contained larger numbers of T cells or less lymph tissue than the other DLBCL samples. DLBCL cells tended to express a lot of proliferation genes, which is consistent with DLBCL being an aggressive malignancy. The germinal center B-cell signature genes appear to

LINKS

Explore Lymphoma

be the most heterogeneous, and thus might prove useful for reclustering DLBCL into new subcategories. Furthermore, the germinal center signature genes might help reveal different stages of B-cell development that had occurred in the original cancerous cell.

The DLBCL samples and two germinal center B-cell lines were reclustered using only the germinal center B-cell signature genes (Figure 5.3a on your CD-ROM). The DLBCLs were clustered into two large groups. About half of the samples (colored orange) cluster so that they have induced germinal center (GC) genes, while the other half (colored blue) have repressed GC genes. The two normal GC B cells (colored black) clustered with the DLBCL half that had induced GC genes. Using this organization of samples, Eisen returned to the full collection of genes and reclustered the expression profiles of all 17,856 genes based on the orange/blue subclassification of DLBCL (Figure 5.3b). The 2,984 genes that were induced (marked by orange side bar) or repressed (marked by blue side bar) in the DLBCL samples were reclustered while maintaining the DLBCL dendrogram, as shown in Figure 5.3a. Based on the gene expression profiles of DLBCL samples, about half of them looked like GC B cells, while the other half looked like activated B cells (Figure 5.3c). Some heterogeneity still existed in these two halves of DLBCLs, which indicated that additional subdivision might be possible.

FIGURE 5.3 • Discovery of DLBCL subtypes.
Go to the CD-ROM to view this figure.

MATH MINUTE 5.1 • WHAT ARE SIGNATURE GENES, AND HOW DO YOU USE THEM?

In Math Minute 4.3, we saw how genes can be grouped using a hierarchical clustering algorithm. With the experimental design of the DLBCL study, genes expressed in particular types of cells or processes are grouped together; the investigators called them "signature genes." You can identify signature genes by searching for a single gene whose role and/or subcellular location matches certain criteria, and including genes clustered nearby. However, since all genes are eventually clustered into one dendrogram, you need a systematic way to identify nearby genes.

One definition of a gene's neighbors is the set of genes that are joined to the original gene at a similarity level above an arbitrary threshold. For example, in Figure M4.1 and Table M4.3, gene C has only gene E as a neighbor at the 0.96 threshold, but it has genes E, L, G, and D at the 0.94 threshold. Defining clusters in this way is sometimes referred to as "cutting the tree." A vertical line drawn in Figure M4.1 at a predefined similarity level would cut the tree into disjoint subtrees, each one a cluster formed at the predefined similarity level or above.

As shown in Figure 5.1, hierarchical clustering was also applied to the columns of the gene expression matrix, resulting in the clustering of the 96 samples. Samples that are clustered together have similar expression profiles across all 17,856 genes. The clusters labeled A and G in the dendrogram of Figure 5.1a are obtained by cutting this tree.

To find genes useful for classifying the clinical prognosis of DLBCL samples, you want to find gene clusters (signature genes) that have patterns of red in some DLBCLs and green in others. (Note that this cannot be done without first clustering the genes.) It appears that the investigators identified the signature genes by carefully studying the patterns in Figure 5.1b, but the same job could be done mathematically as follows. Create a fictitious gene with an expression pattern consisting of twofold repression in each of the first 6 columns (coinciding with cluster A), twofold induction in each of the next 10 columns (coinciding with cluster G), and no change in expression in the remaining 80 columns. The germinal center B signature genes have a large positive correlation with the fictitious gene. The activated B-cell signature genes have a large negative correlation with the fictitious gene. The correlations of the remaining signature genes with the fictitious gene are closer to zero.

Once signature genes have been identified, they can be used to recluster the samples. For example, Figure 5.3a shows the results of a new sample clustering, formed by using only the germinal center B signature genes to assess similarity between samples. Note that a strong linear relationship (i.e., high correlation) exists between the expression profiles of samples DLCL-0051 and DLCL-0033 when the comparison is restricted to germinal center B signature genes. Accordingly, DLCL-0051 and DLCL-0033 are clustered together in Figure 5.3d. But since most other genes did not exhibit a distinctive expression pattern in the DLBCL samples, including them in the comparison diluted the correlation. The dilution effect on correlation may explain why DLCL-0051 and DLCL-0033 were not clustered tightly together when all 17,856 genes were used to assess similarity between samples (see Figure 5.1a).

DISCOVERY QUESTIONS

5. Summarize the steps taken in the three parts of Figure 5.3.

6. Why did they bother doing steps b and c if they just wound up with the same organization of their DLBCL biopsy samples?

The microarray data supported clinical observations: Not all DLBCLs were alike. Although interesting, this was not clinically useful information. The team took their basic research results and analyzed the accompanying clinical data to see how the different patients fared after standard chemotherapy treatments. Forty patients had been treated over the course of many years, and 22 (55%) of them had died, which is close to the national mortality rate for DLBCL. When the patients were matched with their samples and charted, there was a startling outcome (Figure 5.4). Those whose cells appeared to be GC-like had a much higher survival rate than those who had activated B-cell-like DLBCL. However, classifying patients based on gene expression patterns is experimental, so it is important to compare microarray-based classification with a more familiar method. Therefore the patients were regrouped according to traditional clinical criteria (International Prognostic Index [IPI]) and charted (Figure 5.4b). The patients were divided into high- and low-risk categories. The IPI was also good at

distinguishing the types of patients, and it correlated well with the new experimental method.

DISCOVERY QUESTIONS

7. Were all 35,000 of the human genes included on the microarray? Might larger numbers of genes improve the resolution of this method?

8. Other than predicting survival rates, could microarray data be used in other ways that might be clinically useful?

9. Go to Lymphoma Images Figure 3 to see a list of genes that were determined to be the best at distinguishing the two subclasses of DLBCL. Scan the expression profiles and see if any gene/protein might be a good target for chemotherapy treatment of both classes of DLBCL.

10. One gene you may have noticed from Discovery Question 9 was *TTG-2*, which is highly induced in more than half of the DLBCL samples. Go to UniGene and search for *TTG-2*. Based on its tissue distribution, do you think this would be a DLBCL-specific target? Support your answer with the UniGene data.

At this point, you figure the study is over, but the team makes an intriguing suggestion. What would happen if the

a)

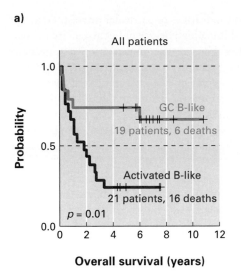

b)

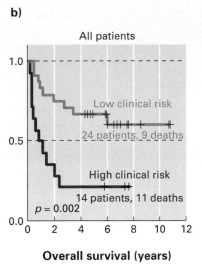

FIGURE 5.4 • Clinical distinctions of DLBCL. Kaplan-Meier plot of 40 patients being treated for DLBCL. **a)** When the patients were separated into GC-like or activated B-cell-like categories, the survival rates were noticeably different, as indicated. **b)** When more traditional IPI indices were used, the patients again segregated into high-risk and low-risk categories, as indicated. The X-axis is the number of years survival postdiagnosis, and the Y-axis is the fraction of surviving patients.

24 IPI-rated low-risk patients were reanalyzed using the GC B cell signature gene criteria? Would it be possible to characterize people with better accuracy than the IPI method? Using their clustering definition, it appears they were better able to predict who would benefit from standard chemotherapy and who might have benefited from bone marrow transplantation (Figure 5.5).

DISCOVERY QUESTIONS

11. Summarize the outcome illustrated in Figure 5.5.

12. What can you conclude about the potential for using DNA microarrays when diagnosing DLBCL?

13. What work remains to be done when you look at Figures 5.4 and 5.5?

We would like to use this information to predict which patients would benefit from chemotherapy and which would not. There is no point in subjecting a person to chemotherapy if he or she will not respond favorably. Furthermore, we might be able to utilize this information to develop more effective drugs for cancer therapy (see pages 152–154). Although this was only a pilot study, it indicates that DNA microarrays may be very helpful for improving diagnoses in the future.

Can Breast Cancer Be Categorized with Microarrays, too?

Botstein and Brown teamed up again to examine whether breast cancer treatment might benefit from microarray data as appears to be possible for DLBCL. With the discovery of genes that play a role in suppressing breast cancers (*BRCA1* and *BRCA2*), many people outside the research community think

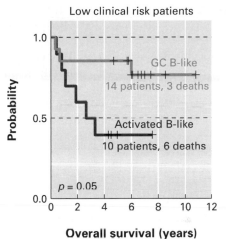

FIGURE 5.5 • Clinical distinction of IPI low-risk DLBCL patients. The 24 low-risk patients from Figure 5.4b were reanalyzed using the GC signature gene analysis and segregated into two categories as indicated.

the battle is nearly won. This impression is fostered by simplistic headlines and a lack of understanding of the complexity and diversity of breast cancer. The two lead authors on this publication were Charles Perou (a postdoc in Botstein's lab) and Therese Sørlie of Oslo, Norway. Unlike DLBCL, breast cancer does not exhibit two distinct clinical categories, so the challenges to categorize breast cancer were greater than for DLBCL.

LINKS

Charles Perou

Therese Sørlie

Molecular Portraits Histology

DISCOVERY QUESTIONS

14. Go to the Molecular Portraits Histology site and see the tissue biopsies used in this study. Click on at least three of the samples, and look at the diversity

of cell types. Be sure to view at least one of each of the grade levels of tumors, where 1 is the least and 3 is the most progressed. What problems do you foresee using tissue biopsies as the sources for sample mRNA?

15. Since experimental mRNAs need to be compared to a control source of mRNA, what would you choose as your control mRNA?

The international team collected 65 surgical specimens from 42 women with breast tumors. Twenty of the tumors were sampled twice, before and after chemotherapy with doxorubicin, a standard treatment. Two samples were obtained from lymph nodes where the cancer had metastasized. Human microarrays were printed that contained 8,102 human **open reading frames (ORFs)**. For control cDNA, the team produced green **Cy3**-labeled cDNAs from 11 different cell lines as a reliable source of reference mRNA. The cDNAs produced from biopsy samples were labeled red with **Cy5** dye. The investigators chose to focus on only 1,753 ORFs whose expression in at least three biopsies differed by at least fourfold compared to the control median expression (Figure 5.6 on your CD-ROM). In addition to the biopsies, 17 cell lines were analyzed that represented the types of cells found in typical breast cancer biopsies.

FIGURE 5.6 • **Variations in expression of 1,753 human ORFs.**
Go to the CD-ROM to view this figure.

DISCOVERY QUESTIONS

16. Are there any obvious trends in the breast cancer biopsies' gene profiles?

17. What pattern do you notice for most of the paired tissue samples? Are there any exceptions to it?

18. Find the normal breast tissue cluster. Where did they cluster relative to the tumors? What does this indicate?

19. Return to the Molecular Portraits Histology images. Examine the two available biopsies of the samples nearest to normal breast tissue (Norway 61 and Norway 101). Are they grade 1 as you might predict based on the clustering?

The data lack any obvious patterns in the hierarchical clustering of the biopsies. The normal tissues clustered right in the middle of the tumors and not by themselves, indicating that the tumors are very much like normal breast tissue.

There are a couple of features of particular interest. First, the tissue samples exhibit a high degree of heterogeneity in gene expression profiles. The variation in gene profiles appears to correlate with the variation in tumor growth rates and signaling pathways activated in each tumor (e.g., estrogen, interferon, etc.). Second, the before and after biopsies clustered together in all but three instances. The primary objective for this study was to determine which genes might prove useful in identifying subclasses of breast tumors and thus be clinically useful.

This study was especially well designed since it takes advantage of breast tumor biopsies before and after chemotherapy. This enabled the investigators to look for genes that were regulated in similar ways within a patient but not in other patients. The team identified 496 genes that meet their criteria for distinguishing one person's paired biopsies from biopsies taken from other patients. This "**intrinsic gene subset**" of 496 genes varied their expression more when comparing tumors from different patients than paired biopsies from one patient (Figure 5.7 on your CD-ROM).

FIGURE 5.7 • **Cluster analysis using the intrinsic gene subset.**
Go to the CD-ROM to view this figure.

DISCOVERY QUESTIONS

20. Go to the web version of GeneXplorer, and click on Figure 2 to explore the database. Set the radar for 400%, and then click on regions in the left frame where you would like to see the names of the genes. Do the genes you selected have anything in common?

21. Do you think that all the categories of breast cancer have been identified by the four subclasses in Figure 5.7?

22. In GeneXplorer Figure 2, type "ERB" in the blank box at the top and click on "GO." You will see a list of four *ERBB2* genes. *ERBB2* is on chromosome number 17. Click on one of the brightest red boxes in an *ERBB2* row. You will retrieve genes that were coexpressed with *ERBB2*. Notice the expression pattern for a gene called "FLOT2," and click on the blue *FLOT2* name to learn what this gene does. Does it have any functions that indicate it might lead to cancer if mutated?

23. Go to LocusLink. Enter "*ERBB2*" and notice the chromosomal location for the human gene. Perform a new search for "*FLOT2*," and note its chromosomal location.

a. What genomic event might explain several linked genes all with elevated expression profiles?

b. Other genes nearby include an apoptosis antagonist called MAP kinase kinase kinase, which is needed for inducing mitosis, and *BRCA1,* the "breast cancer gene." This region is often amplified in breast cancers. Do you think chromosomal amplification (i.e., aneuploidy) is a cause, an effect, or unrelated? How could you determine which?

It appears that the easiest way to classify the tumors was by their expression of the estrogen receptor (ER), with ER-positive cells comprising about half of the tumors. One class of tumors appears to be related to a genomic amplification of a portion of chromosome 17 that includes many important genes for breast cancer. Unfortunately, the sample size was too small to definitively identify most classes of breast cancer.

An interesting feature is that, for the most part, a particular tumor looks more like itself than like a member of a class. For example, almost all of the biopsy pairs look more like individual tumor signatures than class signatures. Of course, the intrinsic gene set was designed to maximize this effect, but even before this last round of clustering, biopsy pairs predominantly clustered near each other. The weeks of chemotherapy did not significantly alter the gene expression profile in 85% of the tumor pairs.

DISCOVERY QUESTIONS

24. Hypothesize why gene clustering may be more challenging to use for breast biopsies than for DLBCL.

25. What is the significance of tumor pairs (before and after chemotherapy) having gene profiles that are more similar to the person than the type of tumor? Do you believe each person's tumor is unique, or is there an alternative explanation?

26. Explain the rationale for using the intrinsic gene subset for clustering. Is this a valid approach? Support your opinion with data presented above.

What Genomic Changes Occur in Cancer Cells?

In the section about breast cancer and the yeast aneuploidy case study (Section 4.2), we learned that microarrays can detect chromosomal duplications and deletions if the investigators look for them. Robert Lucito and his colleagues working in the lab of Michael Wigler at Cold Spring Harbor

METHODS
PCR
LINKS
Michael Wigler

Laboratory wanted to examine directly the extent of aneuploidy in human cancers and if tumors have particular patterns of aneuploidy. To perform this work, they used a partial human genome DNA microarray. The human genome was not completely sequenced at the time the paper was written. But as Wigler's team points out, the method they developed does not require that the genome be fully sequenced. How can that be?

Wigler's method is a variation on the microarray methods we have seen so far. In the standard method, the target DNA on the glass slide is composed of genes. Since the human genome is too big and too complex to work with easily, they decided to print "**genomic representations**" of the genome instead. A genomic representation is designed to be a subset (about 2.5% of the total) of the whole genome, composed of pieces about 1 kb in length. To create a representation, they digest total genomic DNA with a "six cutter" restriction enzyme that recognizes a particular combination of six nucleotides. They chose Bgl II, which cuts the sequence AGATCT, as shown below, to create sticky ends.

```
5' . . . A-OH PO4-G A T C T . . . 3'
3' . . . T C T A G-PO4 OH-A . . . 5'
```

When complete, the entire human genome is converted into restriction fragments of varying lengths with the same sticky ends. Each genomic restriction fragment is ligated to a short stretch (a **linker**) of DNA that has the complementary sticky ends to pair with the genomic DNA. These fragments are amplified using **PCR** under conditions that favor the formation of short ($\leq$ 1 kb) fragments. The final outcome was 1,658 PCR products that ranged from 200 to 1,000 bp in size. These were spotted in duplicate on glass slides for a total of 3,316 spots, or **features,** on one microarray.

The production of colored probes also had to be rethought since tumors are small and heterogenous. And unlike yeast samples, the supply of cells was limited. Therefore, the investigators wanted to get away from working with RNA. They produced genomic representations for each tumor sample to create Cy3- and Cy5-labeled PCR products for probes on their microarrays. Genomic representations eliminated the need to know what was a real gene vs. what was not, as is required with typical DNA microarrays. The intensity of the green probe (X-axis) was plotted as a function of the red to green expression ratios (Y-axis) for each feature on the microarray (Figure 5.8, page 144).

Since this method is new, let's carefully describe how each experiment was performed. To do this, we will label different human genomes with different letters (i.e., genomes A, B, and B' are from two different people). In the first experiment (Figure 5.8a), the features were genomic representations from genome A, and the Cy3 and Cy5 probes were created from the same genomic representations from genome A. In the second experiment (Figure 5.8b), the features were genomic

METHODS

FACS

representations from genome A, but the Cy3 and Cy5 probes were genomic representations from genome B. The features on the third experiment (Figure 5.8c) were produced from the same genomic representations from genome A. The Cy3 probe was a genomic representation produced from genome B taken from wild-type cells. However, the Cy5 probe was a genomic representation from genome B' isolated from cancerous cells. Therefore, we are comparing the genome from a single person in cancerous vs. noncancerous tissue.

Anytime you invent a new method, you have to prove it works as advertised. First, Wigler compared the representative genome to itself, which resulted in a series of dots, one for each feature on the microarray (Figure 5.8a). For features that had low levels of dye (left side) or high levels of dye (right side), the ratio of the two colors was near 1:1, as you would expect. When a sample from a different person was used as a source for the representative probe and compared with itself, the resulting distribution is still closely clustered around the 1:1 ratio (Figure 5.8b), though not as uniformly as before. Finally, a breast cancer biopsy was taken from a patient, the tumor was separated into individual cells, and sorted by fluorescence activated cell sorting (FACS) into either cancerous or normal cells. These two cell types were used as the source for two different probes, and it is clear they do not have equivalent genomes, even though they were taken from a single person (Figure 5.8c). Keep in mind that the probes were derived from genomic DNA and not mRNA.

DISCOVERY QUESTIONS

27. Do you think genomic representation DNA microarrays is a valid method? Support your answer with the data from Figure 5.8.

28. Why is there more variation in the data when a second genome was compared to itself (Figure 5.8b) vs. the original genome compared to itself (Figure 5.8a)?

29. What kind of changes in genomic DNA do you see in the tumor (Figure 5.8c)?

30. How could you use this information to figure out which *genes* might be deleted or amplified?

Most of the DNA in Figure 5.8c that did not appear near the ratio of 1:1 appears to be duplicated, though there is one feature that is greatly reduced in this particular tumor. These features represent the entire genome and not just ORFs. To determine whether a gene is duplicated, the particular feature would need to be sequenced and mapped to the genome. You could use PCR and sequenced tagged site (STS) primers in the chromosomal region of the deleted feature to determine the size of the deletion. Resolving the size of the aneuploidy would require some tumor DNA as PCR

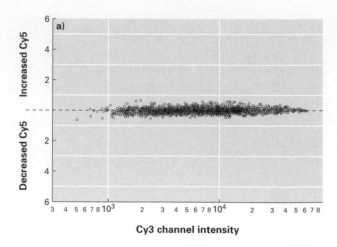

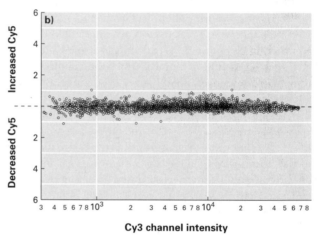

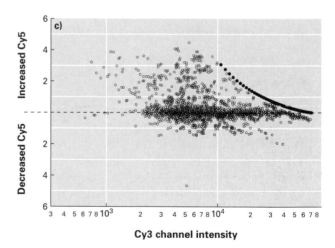

FIGURE 5.8 • **Representative genomic DNA microarrays to detect aneuploidy.** The X-axis (logarithmic scale) is the intensity of the Cy3 channel, and the Y-axis (linear scale) is the Cy5:Cy3 ratio. **a)** The same genomic representation was used to print the features and produce both Cy3 and Cy5 probes. **b)** The second experiment produced Cy3 and Cy5 probes from a single genome but different from the genome used to print the microarray. **c)** The third experiment compared a breast cancer tumor to normal cells from a single person with the features derived from a second person. The filled black circles that form a diagonal indicate that the scanner's light detectors were saturated and could not measure any more light.

template, so hopefully you did not use all of it in the earlier experiments.

If genomic representations worked once, they should work a second time. When two different breast cancer cell lines were compared in duplicate experiments, the results were reproducible (Figure 5.9). For a perfect correlation between the two experiments, we would have expected to see a diagonal line, and we do see a scatter of points that fall on a diagonal. However, we are not given the correlation coefficient, so it is difficult to compare this method with the precision of others.

Verifying the Data Talk is cheap, and so is a new method that does not directly compare its data to a more familiar form. If Wigler's method can truly detect aneuploidy in tumors, it should compare well to a traditional method such as a **Southern blot.** The investigators decided to compare some of the particular pieces of DNA on Southern blots using the same two breast cancer cell lines from Figure 5.9. Two types of Southern blots were performed; either genomic representations or total genomic DNA were blotted. These sources of DNA were probed with individual clones (identified by CHP numbers) and compared (Figure 5.10).

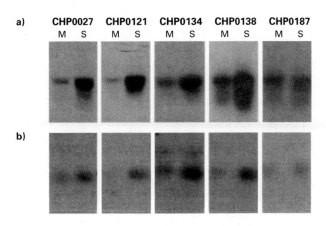

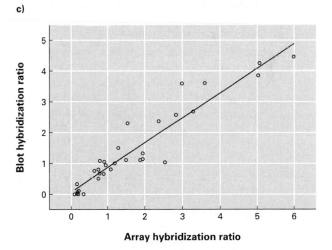

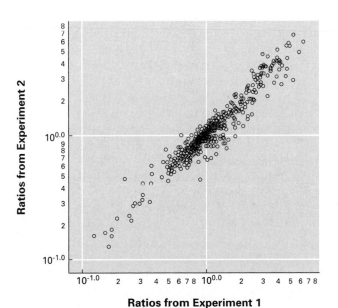

FIGURE 5.9 • Comparison of duplicate experiments. DNA from two breast cancer cell lines (MDA-MB-415 and SKBR-3) were used to generate representative probes and hybridized to an array of 938 genomic representation fragments, each spotted twice. The averaged ratios for each feature are plotted on the X-axis (experiment 1), and the averaged ratios for each feature are plotted on the Y-axis (experiment 2).

FIGURE 5.10 • Southern blot confirmation of genomic representation DNA microarrays. Five representatives (designated by CHP numbers) that proved to be present at different levels (see Figure 5.9) were used as probes for Southern blots using **a)** the 938 representative PCR products as target or **b)** total genomic DNA as target. CHP0187 was selected as an example of a feature that exhibited no differences between the two cell lines. The source of DNA in parallel experiments was either the MDA-MB-415 cell line (designated M) or the SKBR-3 cell line (designated S). **c)** A scatter plot comparing the ratios of DNA copy number obtained by microarray analysis (X-axis) vs. Southern blot hybridization (Y-axis).

Certainly by eye, the representation microarray method looks as though it gives comparable data to Southern blots, but the data are neither quantitative nor analyzed by statistical means. Nevertheless, genomic representations are a clever use of microarrays that will improve with time. The reason so many people want to know which genes are duplicated or deleted is to more accurately prescribe a medication that will work for a particular tumor. For example, if drug X works by binding to gene *Lmnop1,* then drug X will have no chance of helping a patient whose tumor has lost gene *Lmnop1.* So the incentive to devise the best way to detect genomic aneuploidy is much more than an academic exercise in technology. Hopefully the people you know who have been diagnosed with cancer may benefit from DNA microarrays as a part of their clinical assessment.

Summary 5.1

Like the mutated yeast genomes described in Section 4.2, human genomes that suffer from mutations are unstable. Uncontrolled cell growth is the consequence of an unregulated cell cycle, and genomes can suffer additional mutations, including aneuploidy. Breast cancer is a very complex compilation of cells and has proven difficult to categorize by microarrays. The degree of progression and the origin of the first cancerous cell makes breast cancer particularly difficult to understand and combat. However, DLBCL is a more homogeneous disease that is easier to diagnose and categorize. Although it was once thought of as a single type of cancer with two distinct clinical outcomes, you have seen how DNA microarrays have reclassified DLBCL into two distinct clusters. Recognizing two subtypes of DLBCL may help us better understand why patients with the same diagnosis might respond very differently to the same chemotherapy treatment. As we learn more about each type of cancer, we hope DNA microarrays will improve the prognosis for all cancer patients.

5.2 Improving Health Care with DNA Microarrays

Section 5.1 focused on disease; Section 5.2 focuses on treatments. The first case will utilize DNA microarrays to discover the evolution of the tuberculosis vaccine over the last 100 years. As the genome of an attenuated bacterium has evolved, the effectiveness of the vaccine has diminished. You will see how microarray data located the best source of a reinvigorated vaccine for tuberculosis. Later in the section, you will analyze DNA microarray data to determine how a drug works and why it can produce unintended side effects. From

analysis of a single medication to analysis of over 100 drugs, you will see an early attempt to combine several databases with DNA microarray data in order to predict which medication might be the most effective for a particular cancer. The dream of many investigators is to be able to tailor medication to an individual patient's genome. Finally, you will examine obesity and its possible treatment with leptin. Is it possible to treat obesity with a simple injection of leptin? Or does leptin regulate several integrated genomic circuits that are intimately connected?

Why Is the Tuberculosis Vaccine Less Effective Now?

DNA microarrays are versatile and powerful, but will they ever improve the development of medications? Whether the probe originates from RNA or genomic DNA will determine the type of information obtained. Using genomic DNA, we can compare entire genomes for ploidy differences in a single experiment. Cancer cells are not the only cells that add or delete large segments of DNA. Bacteria also experience duplications and/or deletions. If genome changes convey a selective advantage to the individual bacterium, the new mutation will become established in the population. Of course, this occurs every day in nature, but in the lab, we control the environment so much that prokaryotes rarely change their chromosomes and live to tell about it . . . at least, that is the commonly held assumption.

The bacterium *Mycobacterium tuberculosis* is the causative agent for the lethal and highly contagious respiratory disease **tuberculosis (TB).** TB kills 2 million people in the world each year. A vaccine was first produced around 1900 when two French researchers, Albert Calmette and Camille Guérin, began a collaboration in Lille, France. They used *M. bovis,* a strain of bovine (cattle) tuberculosis, as a starting place (similar to using cowpox to vaccinate for smallpox). Over 13 years, they maintained the bacterium through a series of 231 subcultures before producing an attenuated strain of the bovine tubercle bacillus. The culture medium they used to grow their strain consisted of potato, glycerine, and bovine bile salts, and the vaccine was called *bacille de Calmette et Guérin* (BCG). From 1908 to 1921 (and throughout World War I), the bacillus was maintained in culture until they demonstrated the vaccine was harmless to the tuberculosis-susceptible guinea pig. The first and successful prophylactic use of BCG was carried out in 1921, when a physician persuaded Calmette to allow BCG to be used to protect a newborn child delivered from a mother with TB. Nine years later, after many successful vaccinations, something went wrong.

In 1930, the public health director in Lübeck, Germany, asked Calmette to send a stock of BCG. The stock was grown for vaccine production and administered to 251 children. Within a year, 207 of the children had developed TB and 72 of them died. Skeptics of the BCG vaccine became outraged and very vocal in their criticism of Calmette and his vaccine. However, physicians who had successfully used different batches of BCG to protect people from TB rallied to support BCG and Calmette. An inquiry led by German health officials determined that the BCG stock had become contaminated in the Lübeck laboratory where the vaccine had been prepared. The BCG vaccine was determined to be safe and effective, while the German physicians responsible for the contamination were punished with imprisonment.

Since its earliest use, the BCG vaccine has been amazingly successful. It was not until the 1960s (about 1,000 vaccine generations later) that the strain called BCG, which had been maintained in the Pasteur Institute, could be stored in a frozen state to minimize future changes. By this time, it had been sent around the world, where different labs continued to grow their own stock supplies (only live bacteria produce robust immunity). From 1924 to when it was frozen, some BCG strains occasionally were reported to be ineffective as a vaccine. This inconsistency led to the hypothesis that the locally prepared vaccine had changed because BCG had been successfully used elsewhere to prevent TB.

Marcel Behr and his colleagues at McGill University decided to examine the genome of *M. bovis* to see if it had changed over the years. To perform this study, Behr took advantage of the complete genomic sequence for the closely related species *M. tuberculosis*. He microarrayed 4,896 features that represented 3,902 ORFs, all but 22 of the species' ORFs. Of these, 3,756 ORFs produced significant signal for analysis. Probes were made of total genomic DNA isolated from *M. bovis* and each of the 13 remaining vaccine-producing BCG strains he could collect from around the world. In each experiment, he used random priming DNA polymerization reactions to produce red (Cy5-labeled) copies of *M. bovis* genomic DNA as a reference and green (Cy3-labeled) copies of the BCG vaccine strains as the experimental probe. Therefore, the genome of each BCG strain was compared to the genome of a recently isolated wild-type strain of *M. bovis* (Figure 5.11). Every feature bound by both genomes was yellow, and anywhere the vaccine had lost some genomic DNA was red.

> **FIGURE 5.11 • *M. tuberculosis* DNA microarray probed with *M. bovis* (red) and BCG (green).**
> Go to the CD-ROM to view this figure.

For *M. bovis* vs. BCG genome comparisons, the investigators used software to highlight regions of the *M. bovis* genome that were missing from the BCG strains (Figure 5.12). Regions that contained two or more deleted ORFs that produced a different signal (red instead of yellow features) on the microarray were identified. Regions of the chromosome that contained these ORFs from the vaccine strains were sequenced to confirm the loss of genomic fragments.

METHODS
Random Priming
LINKS
Marcel Behr

DISCOVERY QUESTIONS

36. Using Behr's approach, would it be possible to detect genes that are not present in the *M. bovis* genome but are present in *M. tuberculosis*? What would the value of this be?

37. Would it be possible to detect genes that are not present in the *M. tuberculosis* genome but are present in BCG vaccine genomes? Could this information have any practical consequence?

38. What types of mutations in the BCG strains would not be possible to detect by this method?

Based on their DNA microarray results and the sequenced *M. tuberculosis* genome, Behr's group determined 16 regions had been deleted; they were designated RD1–RD16 (Regions Deleted). When examined at higher resolution, it was possible to estimate how many ORFs had been deleted in each region (Figure 5.13).

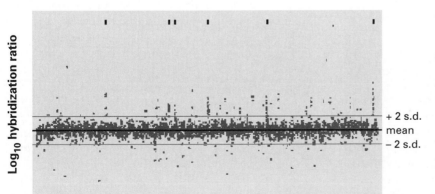

FIGURE 5.12 • Genomic differences between *M. bovis* and BCG. Computer plot of all features (X-axis) and the ratio of red to green signals. The mean ± two standard deviations (s.d.) is indicated. Blue bars at the top indicate regions where two or more features differed from the mean ratio by more than twice the standard deviation.

FIGURE 5.13 • Examples of BCG genomic deletions. Expanded view of RD15 and RD2 (blue bars), which are separated by 335 bp. Each deleted *M. bovis* ORF is indicated by a blue dash. The length of the blue dashes indicates the relative size of the ORF.

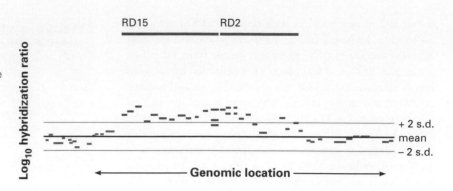

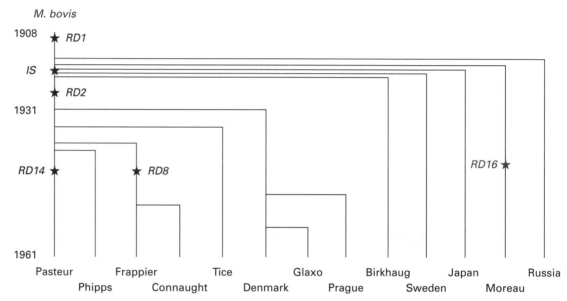

FIGURE 5.14 • BCG genealogy deduced from microarray-detected deletions in the 16 currently available strains. The vertical axis represents time, and the horizontal axis distributes different geographical locations of the strains used in this study. Six deletions are marked by the blue stars on branches of the dendrogram.

LINKS

BCG Genome

Deleted ORFs

DISCOVERY QUESTIONS

39. Go to the BCG Genome web site and you will find a complete list of RD1–RD16. What are the smallest and largest deletions in terms of base pairs? in terms of ORFs?

40. At the BCG web site, you will find a list of all the deleted ORFs. Use the "Find" function of your browser to locate these three ORFs: Rv985c, Rv1773c, and Rv3405c. (The "c" stands for Crick, one of the two strands of DNA.)
   a. What kind of molecules are these?
   b. Do you think their loss will have a small or large impact on the total protein content of the bacteria? Explain your answer.

In addition to finding particular deletions, it was possible to trace the evolution of the laboratory strains over the 40-odd years of growth in culture (Figure 5.14). The regions were lost at various times, with the earliest one (RD1) occurring before the vaccine had been distributed around the world.

DISCOVERY QUESTIONS

41. How do they know that RD1 was lost before any of the strains were distributed?

42. Which currently available strain(s) most closely resemble(s) the original strain of BCG? (The original strain was lost during World War I.)

43. Order the genomes from closest to most distant to *M. bovis*.

44. Why would evolution drive the bacteria to lose pieces of its genome?

Rarely do we get to see so clearly the consequences of genome instability. The bacterium was grown for over 1,000 generations in laboratory conditions outside its normal habitat. Under these artificial conditions, it was subjected to new selection pressures. We saw the same evolution in yeast when it altered its own genome in response to a single gene selectively deleted by an investigator (see pages 129–131). The bovine pathogen was grown in a cell-free environment with potatoes, sugar, and bile. Nature always produces genomic variation in a population, and in the lab environment, the old genome was no longer the most fit. As a result, laboratories around the world selected for strains that grew well in culture rather than working best as vaccines. Behr's group studied BCG's evolution and correlated it with efficacy of immunization. It should be possible to go back to the most "primitive" BCG strain and see if it produces the most robust immunity.

But the BCG story is not complete. These DNA microarrays could not detect any mutation less than 2 kb in length. Furthermore, they could not detect any chromosomal rearrangements, point mutations, deletions of duplicated genes, repetitive regions, or intergenic deletions. Would it be possible for the least deleted strain to produce the least protective vaccine? The genome of *M. bovis* is being sequenced, which will provide more detailed information to help us improve the TB vaccine.

How Does This Drug Work?

Yeast is an ideal model system for many research questions. In Chapter 4, you saw how quickly and successfully it adjusts to its environment. It has a sequenced genome, is easy to grow, and we know more about it than perhaps any other eukaryote. As stated earlier, it has helped us understand human cancer better than any other model system has. But there is much more it can do in a genomic world. Since we know how it responds to environmental stresses, would it be feasible to use yeast cells to help develop new drugs?

Using yeast as a starting place in drug development is not a new idea. However, a substantial collection of labs collaborated to use yeast DNA microarrays to determine how different drugs work. The paper was written by Matthew Marton of Rosetta Inpharmatics and some mentioned earlier in the text, such as Joseph DeRisi, Vishwanath Iyer, Pat Brown from Stanford, and Lee Hartwell from the Fred Hutchinson Cancer Research Center. Until recently, drugs were developed by randomly testing every known compound until you found one that decreased the severity of symptoms.

With this approach, you would have no idea how the drug worked or if there were possible side effects due to the drug disrupting more than one metabolic circuit. That was the pregenomic method; the genomic method is significantly different.

Let's imagine you know a particular pathway that causes a disease, for example, cancer. We know a lot about the cellular circuits cells use to progress through the cell cycle and mitosis. If we could find a drug that bound to a particular component of one circuit, we might be able to determine whether disabling this one circuit produced the desired consequence and no undesired side effects. Once again yeast was chosen as the model system, but this time the investigators wanted to study two immunosuppression drugs, cyclosporin A (CsA) and FK506. These two drugs are used extensively in organ transplant patients and sometimes in those suffering from autoimmune diseases. You may find it surprising to test immunotherapy drugs on an organism with no immune system, but the yeast genome contains genes that are highly conserved in humans. Both drugs inactivate molecules used in human cellular communication circuits that have evolved essentially unchanged over the last 1 billion years.

You need to understand a bit about the system before we begin. The signaling pathway under investigation is centered on a protein kinase called calcineurin (Figure 5.15). Calcineurin is comprised of three parts—the catalytic part, called calcineurin A (CnA), and two calcium binding parts called calcineurin B (CnB) and calmodulin (CMD). When CnB and CMD bind calcium, they bind to, and activate, CnA. CnA removes phosphates from many substrates, which eventually leads to the activation of several genes needed for a normal and vigorous immune response. Two proteins are capable of inhibiting the calcineurin complex if the inhibitors are complexed with the immunosuppressant drugs FK506 or CsA. When these drugs are present, the normal signaling pathway is blocked, resulting in suppression of the normal immune response.

Marton and his colleagues grew *wt* yeast in the presence and absence of the drug FK506 and computed their gene expression ratios (Figure 5.16a). They also computed gene expression ratios for untreated *wt* cells vs. mutant cells (*cna*) that had their calcineurin genes (*Cna*) deleted (Figure 5.16b). In theory, FK506 treatment should have the same effect on the genome as the *Cna* deletion. When the gene expression ratios from these two experiments were compared, most of the ratios were very similar (gray dots in Figure 5.16c). The degree of similarity was quantified with a correlation coefficient of 0.76 ± 0.03. Notice two things about this correlation. First, it is lower than the 0.87 correlation cited in the landmark DeRisi paper in Section 4.1. Second, the investigators included the variation of their data by providing the 95% confidence interval which is very helpful.

LINKS

M. bovis

Rosetta Inpharmatics

Joseph DeRisi

Vishwanath Iyer

Lee Hartwell

FIGURE 5.15 • **Calcineurin signaling circuit.** The phosphatase calcineurin is activated when the calcium-binding proteins calmodulin (CMD) and calcineurin B (CnB) bind to calcineurin. Calcineurin initiates a regulatory circuit that leads to the production of new mRNAs. When FKBP (encoded by the gene *Fpr1*) or CyP (encoded by the gene *Cph1*) has the appropriate drug bound, they inhibit the activation of calcineurin.

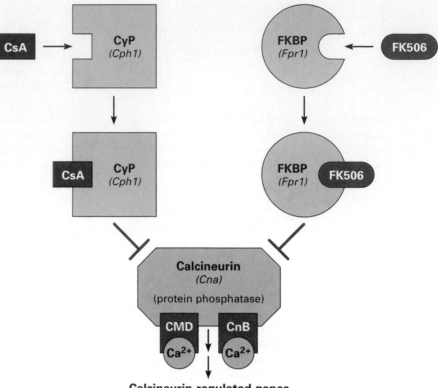

Calcineurin-regulated genes

FIGURE 5.16 • **Calcineurin signaling circuit DNA microarray data for drug-treated and mutant cells.**
Go to the CD-ROM to view this figure.

Gene expression ratios for *wt* cells ± drug treatment were compared to the ratios from a series of mutant strains ± drug treatment. By calculating the correlation coefficients for these comparisons, it is possible to detect the targets for FK506 as well as cyclosporin A (Figure 5.17). Each correlation shows the degree of similarity between the gene expression ratios of *wt* cells with and without drug exposure and the mutant strains with and without drug exposure. Notice that identical conditions of *wt* vs. *wt* produces correlation coefficients of 0.94 (CsA) and 0.93 (FK506).

When the drug target was deleted (e.g., *cna* strain), the correlation between the *wt* cells and the mutant cells was essentially zero, which meant that the way *wt* cells responded to the drugs was very different from the way *cna* mutants responded to them. This makes sense because the drug target was present in the *wt* cells but absent from the *cna* strain. Therefore, when the correlation between gene expression ratios of a mutant and *wt* cells is very low, the mutation has deleted the drug target. If the correlation is high, the mutation has not deleted the drug target.

DISCOVERY QUESTIONS

45. Explain what each number means in each table of Figure 5.17.

46. Why do you think that identical *wt* cells treated with the same drug did not yield a 100% correlation? Hypothesize where the variation may originate.

47. Do these experiments validate or invalidate the proposed binding sites of the drugs CsA and FK506 as shown in Figure 5.15? Explain your answer.

48. What would you expect to see if a drug also inhibited a secondary pathway by binding to a protein other than those shown in Figure 5.15?

The data from the two correlation signature tables show that when calcineurin or the drug targets are deleted, there is very low correlation, as you would expect. In these comparisons, we are looking at the transcriptomes of *wt* cells ± a drug compared to the transcriptomes of cells that lack the drug targets ± drug exposure. Mutants lacking the drug target would produce identical transcriptomes regardless of

a)

	Signature correlation of expression ratios as a result of FK506 treatment in various mutant strains				
	wt +/– FK506	*cna* +/– FK506	*fpr1* +/– FK506	*cna fpr1* +/– FK506	*cph1* +/– FK506
wt +/– FK506	0.93 +/– 0.04	–0.01 +/– 0.07	–0.23 +/– 0.07	0.12 +/– 0.07	0.79 +/– 0.03

b)

	Signature correlation of expression ratios as a result of CsA treatment in various mutant strains				
	wt +/– CsA	*cna* +/– CsA	*fpr1* +/– CsA	*cna cph1* +/– CsA	*cph1* +/– CsA
wt +/– CsA	0.94 +/– 0.04	–0.05 +/– 0.07	0.77 +/– 0.03	–0.11 +/– 0.07	0.18 +/– 0.07

FIGURE 5.17 • Drug effects on *wt* and mutant cells. Correlations comparing the effects of **a)** FK506 or **b)** CsA. The strain *cna* represents a calcineurin deletion, *fpr1* is an FKBP deletion, and *cph1* is a cyclophillin deletion. Each correlation represents *wt* cells with and without drug treatment compared to the indicated strains with and without drug treatment. The 95% confidence interval is shown for each correlation coefficient.

whether or not the drug was present. However *wt* cells would produce two significantly different transcriptomes—one normal and one in response to the drug-inhibiting calcineurin. When the deletion does not affect the drug's pathway, the correlation should be very high. By using this approach, we can quickly deduce that *Fpr1* encodes a binding site for FK506 and *Cph1* encodes a CsA binding site. These results validate the model depicted in Figure 5.15. You can see these effects more clearly in Figure 5.18 on your CD-ROM.

FIGURE 5.18 • Response of FK506 and CsA signature genes in deletion strains.
Go to the CD-ROM to view this figure.

In Figure 5.18, individual genes are shown to be induced (green) or repressed (red) in mutant strains compared to their expression in untreated *wt* cells. The first column shows which genes have altered expression when *cna* is deleted but no drug was administered. The second column shows the **drug signature genes** when *wt* cells were given the drug. Almost all of the FK506 signature genes have expression ratios near 1:1 (black) in deletion strains involved in pathways affected by FK506 (i.e., *cna, fpr1,* and *cna fpr1* mutant strains) but not in deletion strains in an unrelated pathway *(cph1)* or *cna* strain in the absence of FK506. Similarly, most of the CsA signature genes (*wt* + CsA) have expression levels comparable to untreated *wt* cells and appear black in mutants that lack *Cna* or *Cph1*. Deletions that do not disrupt the known signaling pathway (e.g., *cph1* for FK506 and *fpr1* for CsA) show expression patterns similar to drug-treated *wt* cells.

DISCOVERY QUESTIONS

49. If all the primary target genes are accounted for, why was the correlation between drug treatment and calcineurin deletion only about 0.76?

50. What can you say about the potential for secondary drug targets for these two drugs?

There appears to be an inconsistency that has not been explained yet. If we know the primary targets for these two drugs, why don't the expression profiles of strains lacking these targets match more precisely the profiles of *wt* cells treated with the drugs? Why isn't the correlation closer to 0.94? There are three possibilities: (1) the deletion strains still produce some target protein; (2) there are alternative targets; or (3) the variation is due to noise in the microarray method. The first possibility was easily ruled out by **Western blots,** so we need a way to test the second possibility. To facilitate this, the investigators applied higher doses of the drugs to see if changes previously undetected by the method could be detected. When the FK506 dosage was increased 50-fold, genes altered their expression that had not been detected by the lower dosage of FK506. When these high-dose-responsive genes were monitored for their response to FK506 in various mutant strains, the results were evident (Figure 5.19 on your CD-ROM).

METHODS
Western Blot

FIGURE 5.19 • Responses of FK506 secondary targets.
Go to the CD-ROM to view this figure.

LINKS

SGD

John Weinstein

NCI60

If a gene were a part of the primary pathway, it should appear black when the primary target was deleted and the drug administered (e.g., *cna* + FK506). If a gene were a part of a secondary pathway, its expression should be altered even when the primary target gene was deleted. Figure 5.19 is divided into four gene classes. The top class of three genes (*CNA*-dependent) is similar to what we have seen before where the drug does not affect gene expression in mutants of the primary pathway. These three genes are also present in Figure 5.18.

Thirty-two genes were induced even when *Cna* and *Fpr1* were deleted. However, in the presence of FK506, these genes were not induced when *Gcn4* was deleted. Look at the list of genes included in the *GCN4*-dependent class, and notice that many of the gene names are involved in amino acid synthesis (e.g., *HIS, ARG, MET*), though others are not (e.g., *IDP1* encodes an isocitrate dehydrogenase isozyme). The next class includes seven genes that are affected by FK506 even when *Gcn4* and FK506 primary target genes are deleted. The last class contains two genes that are not easily categorized.

DISCOVERY QUESTIONS

51. Are there any secondary targets for FK506? Explain your answer.

52. Go to *Saccharomyces* Genome Database (SGD) and search for the gene *Gcn4*. Click on the "1 gene name" link. What does the Gcn4 protein do?

53. Determine the function of *Pdr5* and *Snq2*, which are among the *CNA*- and *GCN4*-independent genes. Why does it make sense that they are induced when treated with FK506? Would you expect them to be induced by CsA?

54. How many different pathways are activated by FK506?

55. In general terms, how could you improve FK506 so it only binds to the primary target and not any secondary targets?

56. How could you use drug signature genes when developing new drugs?

Can We Predict Which Drugs Will Be Effective in Different Cancers?

The ultimate goal for many investigators using DNA microarrays analysis is to improve medical care of patients. Diagnosis is one avenue of research, and drug development is a second. Pat Brown and David Botstein teamed up with John Weinstein from the **National Cancer Institute (NCI)** to determine how microarrays might be used to improve drug development and selection for individual patients. Weinstein's group was led by Uwe Scherf, and Pat Brown's group by Douglas Ross. Their goal was to use microarray technology to determine if cancers could be better matched with the most effective medications.

Define the Cells First Ross and Scherf thought it would be prudent first to characterize the transcriptomes of the 60 cell lines used by NCI (called the **NCI60**) when testing new potential chemotherapy agents. They began by printing DNA microarrays of about 8,000 different human genes. The control cDNA, labeled green with Cy3, was produced from a mixture of 12 different human cell lines. The experimental cDNA, labeled red with Cy5, was produced from only 1 cell line at a time. They focused their attention on 1,611 ORFs that produced expression ratios at least sevenfold different from control in at least 4 of the NCI 60. Using these 1,611 ORFs, the 60 cell lines were clustered so that cell lines with similar expression profiles were placed next to each other (Figure 5.20 on your CD-ROM). Two cell lines were grown in triplicate to test the reproducibility of the method.

FIGURE 5.20 • **NCI60 gene expression dendrogram.** Go to the CD-ROM to view this figure.

DISCOVERY QUESTIONS

57. Do all cell lines cluster according to their tissue of origin?

58. Which tissues appear more difficult to cluster than others? Which are easiest?

59. With the data already collected, can you think of a way to identify genes that might be better at unifying difficult-to-cluster tissue types?

It is clear that some tissue types have more easily distinguished expression profiles than others, with breast tissue being especially hard to characterize, as we saw earlier (pages 141–143). If we wanted to cluster these hard-to-cluster cell lines, it might be useful to go back to the full set of ORFs. There we could select those genes that were expressed maximally in one of these three tissue types (breast, prostrate, and non–small lung) but not the other two tissue types, and add these new genes to the set of 1,611 genes for reclustering. In addition, many genes were not included on the original DNA microarrays, so there might be better ORFs that were not included in the data set. But for the cell lines with clearly identifiable tissue types, there are subsets of genes that allow them to be identified easily (Figure 5.21 on your CD-ROM).

FIGURE 5.21 • Gene cluster for melanoma cell lines.
Go to the CD-ROM to view this figure.

DISCOVERY QUESTIONS

60. Based on the results from the breast cancer study (pages 141–143), do you think it more likely that breast cancer and melanoma would have identical expression profiles or different profiles?

61. Hypothesize about possible roles for the 16 ORFs that coclustered in the melanoma set of genes.

62. Go to GeneCards to determine what tyrosinase does (tyrosinase is a member of the cluster in Figure 5.21). Type in "tyrosinase" and hit "Go." When the list of hits comes up, click on the "Display complete GeneCard" link to the left of the tyrosinase gene called "*TYR.*" Are there are any phenotypes associated with the loss of tyrosinase?

63. Why does it make sense that melanomas would express high levels of tyrosinase?

Drugs on Chips Having clustered the NCI60 and identified genes that distinguish these cell types, Scherf and Ross had a pretty good handle on the normal transcriptomes for the NCI60. But this was only half of their project. They wanted to find a way to utilize the NCI60 and microarrays to predict the potential utility of a particular drug for a given tumor type. Could microarrays be utilized to predict the success of one chemotherapy agent over another in a given patient?

To answer their question, Scherf and Ross integrated two data sets. The first was the NCI60 microarray data discussed above. The second was a drug activity profile that measured the effectiveness of 118 chemotherapy medications tested by NCI on each of the NCI60. Scherf and Ross used 1,376 genes from their microarray data that provided the most robust signals and produced the strongest differences among the NCI60. In the end, they produced a mammoth figure called a **clustered image map (CIM).** CIMs provide a visual way to interpret the intersection of two different data sets (Figure 5.22 on your CD-ROM). On the X-axis is the clustered data set of 1,376 genes based on their normal expression profiles in the NCI60. The Y-axis represents the clustering of 118 drugs based on their effectiveness in the NCI60.

FIGURE 5.22 • Clustered image map.
Go to the CD-ROM to view this figure.

LINKS
GeneCards
CIM

DISCOVERY QUESTIONS

64. Do the two drugs in the insets in Figure 5.22 work better or worse when their respective gene targets are highly expressed?

65. Find a large region on this CIM where the drugs work best when genes are highly expressed.

66. How could this CIM be used to determine in advance whether a new drug is likely to work or not?

67. Let's imagine that a new drug has been shown to be effective on some cancers. This drug is located at 15 on the Y-axis, and had a high negative correlation with genes around 780 on the X-axis. How could you use this information to screen cancer patients for the potential benefit of your new drug?

Scherf and Ross have analyzed parts of the data, though everyone acknowledges that there are more data than any one group could analyze completely. In their paper, they focused on two interesting cases. Certain malignant tumors, such as acute lymphoblastic leukemia (ALL), lack asparagine synthetase (encoded by the gene *ASNS*). If a drug could be given to ALL patients to deplete extracellular asparagine, the cancer would starve from a lack of the amino acid asparagine. One drug that has been on the market for years is L-asparaginase which, as the name implies, is an enzyme that cleaves asparagine. The two ALL lines (MOLT-4 and CCRF-CEM) expressed the lowest levels of *ASNS* and were the most sensitive to L-asparaginase. K-562 is a chronic myelogenous leukemia cell line that expressed the highest levels of *ASNS* and was the least sensitive to L-asparaginase (Figure 5.23, page 154).

The CIM revealed a negative correlation (−0.44) between expression of *ASNS* and the sensitivity of the NCI60 to L-asparaginase. There was a stronger correlation coefficient (−0.98) for the six leukemia cell lines.

DISCOVERY QUESTIONS

68. Why would a cell line (and by extrapolation a tumor) be more susceptible to L-asparaginase if it expresses less *ASNS*?

69. Does Figure 5.23 indicate that L-asparaginase is being used on the most responsive or the least responsive cancer types?

70. Ovarian cancer cell lines also had a negative correlation (−0.88) between *ASNS* expression and L-asparaginase sensitivity. What does this negative correlation suggest to you? Find a point not

LINKS

Jeffrey Friedman

labeled in Figure 5.23 that represents a cell line likely to be responsive to the drug L-asparaginase.

71. *DPYD* encodes an enzyme that degrades uracil and thymidine and also the chemotherapy medication 5-FU. Do you think 5-FU would work better in cells whose *DPYD* expression levels have a positive or negative correlation with the drug? Look back at CIM data in Figure 5.22b. What does the CIM indicate is the correct answer to this question?

72. What gene numbers (from the X-axis in Figure 5.22) have a positive correlation with the drug 5-FU? What does this indicate for potential use of 5-FU in cancer patients?

This CIM project has provided new insight into drug discovery and ways to maximize the utility of any given drug. Of course, you would not want to treat a patient based on microarray data alone, but these data do provide us with a model that leads to predictions that are testable in clinical drug trials.

What Happens When You Accumulate Fat?

Fat homeostasis is complex and difficult to regulate by pharmacological intervention (see Chapter 11). The biggest problems were all the unknowns since the protein **leptin** (encoded in mice by the gene ***ob***) and the **lipostat** are

connected to many "black boxes" including the brain, immune system, and sexual reproduction. Here are some reminders of the complexity involved:

- Untreated mice that are restricted in their food intake to match the food intake of leptin-treated mice lose significantly less weight than the leptin-treated mice. The food-restricted but non-leptin-treated mice are called **pair-fed** mice.

- Starvation causes a loss of both lean body mass and fat mass, while leptin treatment selectively reduces fat mass.

- Leptin treatment does not lead to the compensatory drop in energy expenditure seen in pair-fed mice.

- Therefore, leptin initiates a metabolic circuitry that is distinct from the one induced by food restriction.

How can we begin to understand the circuitry utilized by leptin signaling? Yeast won't work since there are no obese yeast. Neither flies nor worms have leptin genes. Grinding up people for research projects like this is frowned upon. What options are there?

How can we learn more about the hidden circuitry behind the complex genomic responses initiated by leptin? Jeffrey Friedman, who first cloned leptin, and his lab utilized a microarray approach to this problem, though they had some challenges to address before they could collect any data. First, the mouse genome had not been sequenced when they began their research. They started with what was known, some mouse genes and ESTs. Rather than using cDNAs to spot on the microarray, they chose to go with an

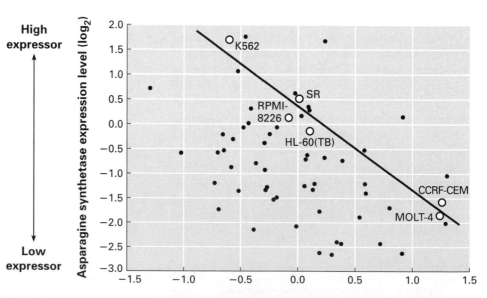

FIGURE 5.23 • Correlation between *ASNS* expression and chemosensitivity to L-asparaginase for each of the NCI60. The Y-axis represents the gene expression ratio ($\log_2$) of *ASNS* in each cell line, relative to control. The X-axis is a relative scale of sensitivity to L-asparaginase, as measured by GI_{50}, meaning that growth was inhibited by 50%. A large value on the X-axis indicates that a low dosage was required to achieve GI_{50}. Large blue circles highlight six different leukemia cell lines with their best fit line that had a correlation coefficient of −0.98 compared to a best fit line correlation of −0.44 for all NCI60.

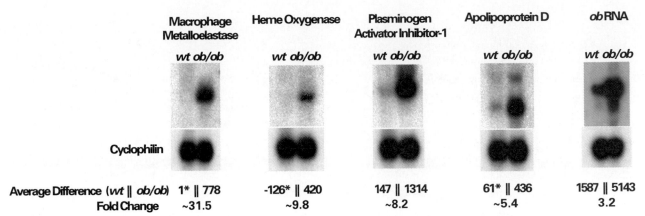

	Macrophage Metalloelastase	Heme Oxygenase	Plasminogen Activator Inhibitor-1	Apolipoprotein D	*ob* RNA
	wt ob/ob	*wt ob/ob*	*wt ob/ob*	*wt ob/ob*	*wt ob/ob*
Average Difference (*wt* ‖ *ob/ob*)	1* ‖ 778	-126* ‖ 420	147 ‖ 1314	61* ‖ 436	1587 ‖ 5143
Fold Change	~31.5	~9.8	~8.2	~5.4	3.2

FIGURE 5.24 • **Comparison of microarray and Northern blot data.** Northern blots of RNA isolated from *wt* and *ob* mice were probed for the indicated genes. The "fold change" at the bottom was determined from the intensity of signals on Northern blots for *ob* vs. *wt* mice. When the *wt* signal was essentially zero, the ~ symbol was used to indicate an approximation. The average difference indicates the averaged numerical signal from the microarray data in *wt* vs. *ob* mice that was used to determine the difference in gene expression for ratios. Cyclophilin was used as a control for the Northern blots.

oligonucleotide microarray developed by Affymetrix. The principle is the same as spotting cDNAs, except that single-stranded oligonucleotides are attached to the glass slide. There are some other minor differences, but essentially it is a standard microarray experiment.

DISCOVERY QUESTIONS

73. What advantages are there for using oligonucleotides instead of cDNAs?

74. One benefit of oligonucleotides could not be fully utilized when making mouse microarrays. Why?

The partial mouse genome microarray represented about 6,500 ORFs, of which about 2,000 were able to detect dye-labeled cDNAs at sufficient levels to be considered "above background." Using oligonucleotides instead of cDNAs has two advantages. First, you never have to denature the spotted DNA since it was never double stranded, and this saves time and one potential source of variation. Second, oligonucleotides can be chosen so that they do not bind to more than one gene's mRNA/cDNA, since areas of low similarity can be chosen. Choosing gene-specific regions is not possible with full-length cDNAs produced by PCR. However, since the mouse genome had not been sequenced, it was impossible to be sure the oligonucleotide sequences would bind to only one gene.

The dye-labeled cDNAs used in this obesity research were produced from fat tissue of **wild-type** and *ob/ob* (homozygous mice lacking a functional copy of *ob*) mice. Only 77 genes

produced ratios ≥ 3-fold when comparing *wt* to *ob* mice. As other groups have done previously, Friedman and his team wanted to make sure their microarray data could be confirmed by **Northern blots.** If anything, microarrays tend to underestimate the difference in expression ratios compared to the values obtained by Northern blots (Figure 5.24).

METHODS

Northern Blot

LINKS

Affymetrix

Comparing *wt* to *ob*

DISCOVERY QUESTIONS

75. How do you use the data provided by the cyclophilin blots? Is there a similar control for microarray data?

76. How are numerical values obtained for Northern blots?

77. Which fold change data are better, Northern blots or microarrays? Answer this question both at the whole-genome level and the individual gene level.

The first experiment was to examine the effect of leptin injections for *ob/ob* mice compared to *ob/ob* pair-fed mice (Figure 5.25, page 156). As previously reported, the leptin-treated *ob* mice lost weight faster than the *ob* pair-fed mice. The control *ob* mice continued to gain weight during the 12 days of this experiment. Food intake for the leptin-treated and pair-fed mice was identical, while the saline-injected *ob* mice got used to the injections and quickly regained their appetites.

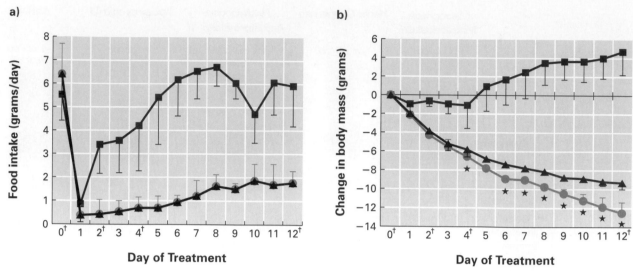

FIGURE 5.25 • **Effect of leptin on food intake and body mass for *ob/ob* mice injected with leptin, pair fed, and injected with saline. a)** The amount of food eaten in *ob/ob* mice injected with saline (black squares), *ob/ob* mice injected with leptin (gray circles), and *ob/ob* pair-fed mice (blue triangles). **b)** Body mass of the three groups of mice. Stars indicate significant differences ($p < 0.05$) comparing leptin-injected to pair-fed mice. The † indicates time points for which microarray data were collected.

DISCOVERY QUESTIONS

78. Summarize the data from both graphs in Figure 5.25.

79. Why were so many of the microarray time points taken early in the leptin treatment rather than evenly distributed over the 12 days?

The *ob/ob* mice in this study responded just as predicted and the control mice ate well and gained weight, which ruled out the influence of physical handling as a reason for any differences in the mice. These preliminary data allowed Friedman's group to proceed with the microarray analysis. Approximately 2,000 genes were clustered by their ex–pression profiles in white fat tissue from leptin-treated and pair-fed *ob* mice. The investigators identified 16 clusters of genes, each of which showed different responses to leptin (Figure 5.26 on your CD-ROM). The color scheme used by Friedman's lab is different from those we've seen before, but the principle is identical. In all cases, expression ratios were calculated by comparing gene expression in untreated *wt* mouse white fat to similar samples from the experimental mice. From the examples in Figure 5.26, you can see that some genes were induced by leptin (red-yellow), others were repressed (from dark blue to light blue), and some re-mained unchanged (black).

FIGURE 5.26 • Cluster overviews of leptin-treatment and pair-feeding on *ob/ob* white adipose tissue.
Go to the CD-ROM to view this figure.

When the identities of the genes were examined, there were some intriguing findings (Figure 5.27 on your CD-ROM). In this cluster, many of the genes were involved with fatty acid biosynthesis, which you might have expected to be induced but instead were repressed. Included in this cluster was the gene "homologous to *SREBP-1*," a transcription factor known to induce the expression of *Spot14*. Interestingly, cytochrome P-450 IIE1, known to metabolize drugs, was also repressed.

FIGURE 5.27 • Detailed cluster of genes repressed by leptin injections in *ob/ob* mice.
Go to the CD-ROM to view this figure.

DISCOVERY QUESTIONS

80. How do the data in Figure 5.27 compare with the summary graph in Figure 5.26c? Are there any differences between the whole-cluster trend and individual genes? Explain the differences, if any.

81. Look at the expression profile for cytochrome P-450 in Figure 5.27. Draw a quick graph to summarize its expression in leptin-injected and

pair-fed mice. Would you have placed it fourth from the bottom in the cluster if you were clustering by eye?

82. Why would *Srebp-1* be repressed very little but *Spot14* be repressed the most of any gene in this cluster?

The overall trend in Figure 5.26c was to be gradually repressed over time in the leptin-injected mice while only transiently repressed in the pair-fed mice. However, when we compare individual genes within the cluster (Figure 5.27), we see some genes were only transiently repressed (e.g., *cytochrome P-450*), or slightly repressed (*Srebp-1*) in these leptin-treated mice. Note how subtle changes in a transcription factor (e.g., Srebp-1) can have major impact on the transcription of genes that require the same factor (e.g., *Spot14*). All of these data help explain why the response to leptin is very complex and not a simple circuit composed of a few components. The leptin circuit complexity is made more apparent when we look at genes that were transiently *induced* by leptin treatment (Figure 5.28 on your CD-ROM).

FIGURE 5.28 • Detailed cluster of genes transiently induced by leptin injections in *ob/ob* mice.
Go to the CD-ROM to view this figure.

DISCOVERY QUESTIONS

83. Look at the wide range of genes represented in Figure 5.28. Have you seen *caveolin* and *calpain* before? (See Chapter 10.)

84. Go to OMIM and determine the roles for these genes: *integrin* (click on alpha 7); *Rab* (click on 5a); *synaptoporin*; *Lim* (click on actin-binding Lim protein 1); *nexin* (click on sorting nexin 1); *FK506 binding protein* (click on 5); *syntrophin* (click on alpha-1).

85. Does leptin affect the expression of genes with known roles in the CNS and immune systems? Are there any diseases associated with the short list of genes you found?

86. Pick one gene from Discovery Question 84 and graph its change in expression as revealed by leptin-treated and pair-fed mice.

What Effect Does Leptin Have on *wt* Adipose Tissue? We would expect leptin injections to alter gene expression in mice that lacked the ability to produce leptin. But what hap-

pens when *wt* mice are injected with leptin? Friedman knew that increasing either serum or cerebrospinal fluid concentrations of leptin in *wt* mice mimics a mouse that has accumulated excess fat. In response, a mouse will reduce its food intake and lose adipose mass but not muscle mass. How does leptin alter the transcriptome to accomplish these results in a *wt* mouse? Friedman's lab asked the same question and discovered that increasing leptin levels in a *wt* mouse also affects many genes, though not necessarily the same way as seen in *ob/ob* mice (Figure 5.29 on your CD-ROM).

LINKS
OMIM

FIGURE 5.29 • Cluster overviews of leptin treatment and pair-feeding in *wt* white adipose tissue.
Go to the CD-ROM to view this figure.

The first sign *wt* mice respond differently from *ob* mice is that the same 2,000 genes were clustered into 13 groups instead of 16. If you look at cluster A in Figure 5.29, you will see that 58 out of 2,000 genes responded similarly in leptin mice as pair-fed mice. In contrast, clusters B, K, D, G, and F reveal that some genes respond differently to reduced food intake vs. leptin treatment; cluster G contains the leptin gene. Cluster K contains *Srebp-1, Spot14,* and many of the fatty acid metabolizing genes that were repressed in *wt* mice just as they were in the *ob/ob* mice. So the answer to our original question is that genes in a *wt* mouse respond to leptin in many different ways. About 20 transcription factors were affected by leptin treatment, which means that approximately 1% of genes with altered expression regulate the transcription of many other genes. This genomic response reinforces the notion that simple solutions to obesity are not going to be possible since leptin interacts with many different circuits.

Although it is easy to measure the genomic response in the form of a transcriptome, measuring protein changes is more difficult. However, Friedman's group performed one last experiment to assess changes in the transcription factor *Srebp-1*. Since most cellular functions are carried out by proteins, people want to know what is happening to the proteome more than the transcriptome. (Chapter 6 focuses on **proteomics,** the study of all proteins present in a cell.)

Srebp-1 is translated into a protein that initially binds to the endoplasmic reticulum and thus cannot reach the nucleus. Upon stimulation by sterol molecules, a protease is activated and cleaves Srebp-1p so the cleaved protein is freed from the ER and can enter the nucleus. This smaller version of Srebp-1p is the active form. When the amount of active Srebp-1p was compared in *ob/ob* leptin-treated and pair-fed mice, the protein differences were substantial, even though the changes in transcription were minor (Figure 5.30 on your CD-ROM). Comparing protein to mRNA levels of one gene made it clear that genomic control of integrated circuits is not restricted to transcription.

FIGURE 5.30 • **Comparison of transcriptional and post-transcriptional regulation.**
Go to the CD-ROM to view this figure.

METHODS
Chuck Close
LINKS
Seurat's Classic
Up Close

DISCOVERY QUESTIONS

87. Graph the change in *Srebp-1* transcription for the *ob/ob* mice treated with leptin and *ob/ob* pair-fed mice. Then roughly estimate values for the Western blot and plot this on the transcription graph. Describe the differences you see.

88. The change in Srebp-1p seen in the Western blot does not reveal any correlation with transcription in *ob/ob* mice. Explain why.

89. Which form of data (protein or mRNA) gives you the best understanding of genomic responses to leptin? Support your position with data.

90. What impact does the Western blot have on your evaluation of microarray data?

This study of leptin genomic response has opened more doors to understanding how leptin works, but it leaves many unanswered questions. It also illustrates both the strengths and weaknesses of microarray studies. They provide clues to the functions of unknown genes, illustrating new circuits, and the connections between circuits. But the resolution is limited. An analogy for this limited resolution is the artistic expression perfected by Georges Seurat—**pointilism.** Go to the National Gallery of Art and look at one of Seurat's classic works. You can recognize patterns and trends in the overall picture, but when you look up close, you cannot recognize individual shapes as easily as you did from the global perspective. The same may be said about microarrays; they provide informative overviews, but gene-specific clarity is elusive. Nevertheless, as with the perspective provided by Seurat, microarrays are valuable and provide insights.

Summary 5.2

DNA microarrays are powerful tools for improving the quality of medicine. You saw how a well-established vaccine for tuberculosis might be reinvigorated now that we understand the genomic variations currently in use around the world. The mechanism of action for a particular drug can be more completely understood by studying how it affects a genome in vivo. Furthermore, the genomic response to a particular drug can reveal why some compounds produce unwanted side effects, which can lead to the development of more specific drugs. By compiling years of basic research, a clustered image map enabled you to predict which medication might be best for a particular person's cancer. However, the regulation of fat appears to be so deeply woven throughout our genome that the only effective treatment for obesity may be low-fat diets and exercise. Simple cures may produce unacceptable consequences because of the integrated circuitry inherent in a genome.

Chapter 5 Conclusions

The central focus of Chapter 5 is the application of DNA microarrays to better understand the human condition. Substantial progress has been made toward the use of DNA chips to improve the treatment of cancer. Likewise, the development and understanding of new medications may be enhanced when DNA microarrays are incorporated into the process. What are the limitations for DNA microarrays? The range of experiments is only as limited as the creativity and curiosity of the scientific community. Just as computers have become cheaper and faster, microarrays too will become increasingly accessible to more people, which will open the way for better experiments. Are DNA microarrays better at seeing the entire genome or individual genes? For another analogy from the artistic world, we might consider the work of Chuck Close and ask if we are seeing a comprehensive picture or just a bunch of dots.

If you choose a career path that leads you away from science, you will still encounter microarrays in your personal lives. Papers have been published about microarrays being used to study the genomic rivalry between a host plant and its parasitic pathogen. Pathogen-host interaction could be studied in any organism (including humans) that can be attacked by parasites, bacteria, viruses. Another research team examined the genomic diversity of the bacterium *H. pylori,* which is the causative agent for stomach ulcers. Cancer, acquired immunodeficiency syndrome (AIDS), TB, Ebola—the list of potential uses is as long as genomes are diverse.

References

Cancer Chips

Alzadeh, Ash A., Michael B. Eisen, et al. 2000. Distinct types of diffuse large B-cell lymphoma identified by gene expression profiling. *Nature.* 403: 503–5111.

Golub, T. R., D. K. Slonim, et al. 1999. Molecular classification of cancer: Class discovery and class prediction by gene expression monitoring. *Science.* 286: 531–537.

Perou, Charles M., Therese Sørlie, et al. 2000. Molecular portraits of human breast tumors. *Nature.* 406: 747–752.

Ross, Douglas T., Uwe Scherf, et al. 2000. Systematic variation in gene expression patterns in human cancer cell lines. *Nature Genetics.* 24: 227–238.

Scherf, Uwe, Douglas Ross, et al. 2000. A gene expression database for the molecular pharmacology of cancer. *Nature Genetics.* 24: 236–244.

Weinstein, John N., Timothy G. Myers, et al. 1997. An information-intensive approach to the molecular pharmacology of cancer. *Science.* 275: 343–349.

Aneuploidy Chips
Behr, M. A., M. A. Wilson, et al. 1999. Comparative genomics of BCG vaccines by whole-genome DNA microarray. *Science.* 284: 1520–1523.

Forozan, Farahnaz, Eija H. Mahlamaki, et al. 2000. Comparative genomic hybridization analysis of 38 breast cancer cell lines: A basis for interpreting complementary DNA microarray data. *Cancer Research.* 60: 4519–4528.

Hoslink. Pioneers in Medical Laboratory Sciences. 2002. <www.hoslink.com/Pioneers.htm>. 7 June 2002.

Hughes, Timothy R., Christopher J. Roberts, et al. 2000. Widespread aneuploidy revealed by DNA microarray expression profiling. *Nature Genetics.* 25: 333–337.

Lucito, Robert, Joseph West, et al. 2000. Detecting gene copy number fluctuations in tumor cells by microarray analysis of genomic representations. *Genome Research.* 10: 1726–1736.

Leptin Chips
Belkin, Lisa. 24 December 2000. The making of an 8-year-old woman. How do we understand puberty? Through the prism of our times. *New York Times Magazine.* 38–43.

Soukas, Alexander, Paul Cohen, et al. 2000. Leptin-specific patterns of gene expression in white adipose tissue. *Genes and Development.* 14: 963–980.

Bacterial and Viral Papers
Arimura, Gen-ichiro, Kosuke Tashiro, et al. 2000. Gene responsiveness in bean leaves induced by herbivory and by herbivore-induced volatiles. *Biochemical and Biophysical Research Communications.* 277: 305–310.

Cho, Raymond J., et al. 1998. Parallel analysis of genetic selections using whole genome oligonucleotide arrays. *PNAS.* 95: 3752–3757.

Cummings, Craig A., and David Relman. 2000. Using DNA microarrays to study host-microbe interactions. *Genomics.* 6(5): 513–528.

Khodursky, Arkady B., Brian J. Peter, et al. 2000. DNA microarray analysis of gene expression in response to physiological and genetic changes that affect tryptophan metabolism in Escherichia coli. *PNAS.* 97(22): 12170–12178.

Khodursky, Arkady B., Brian J. Peter, et al. 2000. Analysis of topoisomerase function in bacterial replication fork movement: Use of DNA microarrays. *PNAS.* 97: 9419–9424.

Selinger, Douglas W., K. J. Cheung, et al. 2000. RNA expression analysis using a 30 base pair resolution *Escherichia coli* genome array. *Nature Biotechnology.* 18: 1262–1268.

Talaat, Adel M., Preston Hunter, and Stephen Albert Johnston. 2000. Genome-directed primers for selective labeling of bacterial transcripts for DNA microarray analysis. *Nature Biotechnology.* 18: 679–682.

Ye, Rick W., Wang Tao, et al. 2000. Global gene expression profiles of Bacillus subtilis grown under anaerobic conditions. *Journal of Bacteriology.* 182(16): 4458–4468.

Clustering and Visualization of Microarray Data
Conklin, Bruce. 2001. GenMapp: Gene MicroArray Pathway Profiler. <http://gladstone-genome.ucsf.edu/introduction.asp>. Accessed 8 February 2002.

D'haeseleer, Patrik, Shoudan Liang, and Roland Somogyi. 2000. Genetic network inference: From co-expression clustering to reverse engineering. *Bioinformatics.* 16(8): 707–726.

Eisen, Michael B., Paul T. Spellman, et al. 1998. Cluster analysis and display of genome-wide expression patterns. *PNAS.* 95: 14863–14868.

Getz, Gad, Erel Levine, and Eytan Domany. 2000. Coupled two-way clustering analysis of gene microarray data. *PNAS.* 97: 12079–12084.

Gilbert, David R., Michael Schroeder, and Jacques van Helden. 2000. Interactive visualization and exploration of relationships between biological objects. *Trends in Biotechnology.* 18: 487–494.

Kyoto Encyclopedia of Genes and Genomes (KEGG). 2001. <http://www.genome.ad.jp/kegg/>. Accessed 8 February 2002.

Lee, Mei-Ling Ting, Frank C. Kuo, et al. 2000. Importance of replication in microarray gene expression studies: Statistical methods and evidence from repetitive cDNA hybridizations. *PNAS.* 97(18): 9834–9839.

Lemkin, Peter F., Gregory C. Thornwall, et al. 2000. The microarray explorer tool for data mining of cDNA microarrays: Application for the mammary gland. *Nucleic Acids Research.* 28(22): 4452–4459. <http://www.lecb.ncifcrf.gov/MAExplorer/>.

Liao, Birong, Walker Hale, et al. 2000. MAD: A suite of tools for microarray data management and processing. *Bioinformatics.* 16(10): 946–947. <http://pompous.swmed.edu/>.

Manduchi, Elisabetta, Gregory R. Grant, et al. 2000. Generation of patterns from gene expression data by assigning confidence to differentially expressed genes. *Bioinformatics.* 16(8): 685–698.

National Center for Genome Resources. 2001. GeneX: A collaborative Internet database and toolset for gene expression data. <http://www.ncgr.org/research/genex/>. Accessed 8 February 2002.

New Microarray Fabrication Methods
Brown, Patrick O., et al. 2001. The Mguide. <http://cmgm.stanford.edu/pbrown/mguide/>. Accessed 8 February 2002.

Kane, Michael D., Timothy A. Jatkoe, et al. 2000. Assessment of the sensitivity and specificity of oligonucleotide (50mer) microarrays. *Nucleic Acids Research.* 28(22): 4552–4557.

Kumar, Anil, and Zicai Liang. 2001. Chemical nanoprinting: A novel method for fabricating DNA microchips. *Nucleic Acids Research.* 29(2): e2.

Okamoto, Tadashi, Tomohiro Suzuki, and Nobuko Yamamoto. 2000. Microarray fabrication with covalent attachment of DNA using bubble jet technology. *Nature Biotechnology.* 18: 438–441.

Chip Reviews and Commentaries
Barry, Clifton E., III, and Benjamin G. Schroeder. 2000. DNA microarrays: Translational tools for understanding the biology of *Mycobacterium tuberculosis. Trends in Microbiology.* 8(5): 209–210.

Berns, Anton. 2000. Gene expression in diagnosis. *Nature.* 403: 491–492.

Brazma, Alvis, and Jaak Vilo. 2000. Gene expression data analysis. *FEBS Letters.* 480: 17–24.

Brent, Roger. 1999. Functional genomics: Learning to think about gene expression data. *Current Biology.* 9: R338–R341.

Dalton, Rex. 2000. DIY (Do it yourself) microarrays promise DNA chips with everything. *Nature.* 403: 234.

GCAT: the Genome Consortium for Active Teaching. 2002. <http://www.bio.davidson.edu/GCAT>. Accessed 7 June 2002.

Geschwind, Daniel H. 2000. Mice, microarrays, and the genetic diversity of the brain. *PNAS.* 97: 10676–10678.

Harris, Thomas M., Aldo Massimi, and Geoffrey Childs. 2000. Injecting new ideas into microarray printing. *Nature Biotechnology.* 18: 384–388.

Lockhart, David J. 1998. Mutant yeast on drugs. *Nature Medicine.* 4: 1235–1236.

Lockhart, David J., and Carrolee Barlow. 2001. Expressing what's on your mind: DNA arrays and the brain. *Nature Reviews.* 2: 63–68.

Madden, Stephen L., Clarence J. Wang, and Greg Landes. 2000. Serial analysis of gene expression: From gene discovery to target identification. *Drug Discovery Today.* 5(9): 415–425

Marx, Jean. 2000. DNA arrays reveal cancer in its many forms. *Science.* 289: 1670–1672.

Perou, Charles M., Patrick O. Brown, and David Botstein. 2000. Tumor classification using gene expression patterns from DNA microarrays. *New technologies for the life sciences: A Trends Guide.* pp. 67–76.

Pinkel, Daniel. 2000. Cancer cell, chemotherapy and gene clusters. *Nature Genetics.* 24: 208–209.

Schaffer, Jeff Landgraf, Miguel Pérez-Amador, and Ellen Wiseman. 2000. Monitoring genome-wide expression in plants. *Current Opinion in Biotechnology.* 11: 162–167.

Velculescu, Victor E., 1999. Tantalizing transcriptomes—SAGE and its use in global gene expression analysis. *Science.* 286: 1491–1492. <http://www.sagenet.org/>.

Young, Douglas B., and Brian D. Robertson. 1999. TB vaccines: Global solutions for global problems. *Science.* 284: 1479–1480.

CHAPTER 6

Proteomics

Goals for Chapter 6

6.1 Introduction

Apply multiple descriptors to individual proteins.

Use high-throughput methods to characterize cellular roles for proteins.

6.2 Protein 3D Structures

Think of proteins as solid objects that occupy space.

Utilize structural information to discern how proteins work.

Incorporate structural methods to develop better drugs.

Understand how subtle changes in shape can have profound consequences.

6.3 Protein Interaction Networks

Appreciate protein-protein interactions as central to a protein's function.

Adapt in silico and in vivo methods to comprehend whole proteome interactions.

6.4 Measuring Proteins

Define the parts of a proteome.

Evaluate methods for quantifying and comparing proteomes.

Convert microarray technology for proteomics.

Consider the role of single-protein molecules and single-cell proteomes.

Explore organismal consequences of subtle changes in proteomes.

Many researchers consider the 21st century the "**postgenomic era.**" By this, they are not implying we should stop sequencing genomes or analyzing existing sequences. Postgenomic era simply means that the technical barriers to obtaining genomic information have been resolved. The next big challenge is to understand **proteomes**—all the proteins of a cell at a given time.

Proteomics is a very difficult field because we lack methods for completely defining proteomes, partly because they present a moving target. Each of your cells has the same genome: the same DNA sequence you were born with and will have forever. However, each cell type in your body has a different proteome. In addition to cell-to-cell variation, any given cell will change its proteome over time. As you grow older, your proteomes change. When you have an infection, they are altered to meet the new challenge. Medications, exercise, diet—your proteomes change in response to fluctuations in the intracellular and extracellular environments. Chapter 6 uses a series of case studies to explore proteomic methods, research questions, and technical limitations.

METHODS

Gene Knockouts

Transposons

Homologous Recombination

LINKS

Gene Ontology

Mike Snyder

6.1 Introduction

Biologists want to know what each protein does (molecular function), to which cellular circuits each protein contributes (biological process), and where in the cell each protein is located (cellular component). One way to discover these attributes is to knock out, or inactivate, a gene and look for new phenotypes. Deleting a single gene to discern phenotype is classical genetics and may not sound "genomic" at all. The postgenomic approach is to knock out all the genes, one at a time, and perform massive screens for particular phenotypes. Section 6.1 examines three high-throughput approaches for deducing what each protein does in cells. Two cases focus on yeast, while the third presents a method optimized in the worm *Caenorhabditis elegans* that also works in vertebrate cells.

What Do All These Proteins Do?

Until recently, the term "function" was used to describe what a protein does, but a growing number of biologists feel that "function" is too limited. A consortium of genomic groups called **Gene Ontology** have collaborated to break the term into three more specific ones:

1. **Biological process** (why—why is this being done? e.g., movement of cell)

2. **Molecular function** (what—what kind of molecule is this? e.g., ATPase)

3. **Cellular component** (where—where is this located? e.g., Golgi, ribosome, etc.)

Using pregenomic methods, biologists have learned a lot about proteins and these methods will continue to play an important role in the postgenomic era. For example, they have learned what happens to a cell that lacks a particular gene—gene **knockouts**. Knocking out a gene allows the investigator to determine the consequences when a cell lacks a particular protein. Knockouts can be performed rather bluntly so that every cell lacks the protein at all times, or more subtly so that only certain cells in an organism at restricted times lack the protein. Either way, such research is not amenable to genome-wide analysis. What we need are methods that permit genome-wide analysis of biological processes, molecular functions, and cellular components.

How can we analyze every protein in three different ways? This daunting task was elegantly addressed by Petra Ross-Macdonald, Mike Snyder, and their collaborators at Yale University. Snyder's lab took advantage of **transposons** found in *Saccharomyces cerevisiae*. Transposons (also called jumping genes) are mobile pieces of DNA that can hop from one location in the genome to another. The location of the new insertion is random, as far as we know, which means a transposon can create random insertional mutations throughout the genome. But Snyder did not use an ordinary transposon. . . .

Snyder's team created a very useful transposon derived from the wild-type version called Tn3. This multipurpose minitransposon (called mTn) contained a number of genetic alterations to improve its utility in this genome-wide study of proteins (Figure 6.1). The inverted repeats (IR) on each of the ends are needed for mTn to insert into a chromosome. *lacZ* encodes the bacterial reporter enzyme β-galactosidase, which can turn a colorless substrate blue; when a cell turns blue, you know it has expressed β-galactosidase. However, this version of *lacZ* did not contain a promoter or the first codon, so β-galactosidase would not be expressed in most cells. You will notice there are two *lox* sites, which are binding sites for a viral protein that can excise all the DNA between two *lox* sites (more about this later). Finally there is

FIGURE 6.1 • **Diagram of mTn.** IR, inverted repeat; *lox*, a recombination site for excising intervening DNA; Xa, a restriction site; *lacZ*, a gene that encodes β-galactosidase; *URA3*, a gene required for uracil synthesis; *tet*, tetracycline resistance gene; *res*, required for transposition; 3xHA, three copies of the DNA encoding hemagglutinin epitope tag. To create the mutations, a yeast genomic plasmid library in *E. coli* was randomly mutagenized by insertion of mTn. Individual plasmid clones were isolated and used for homologous recombination to replace the *wt* gene with the mTn-mutated one in *URA3*-lacking diploid strains.

the site called 3xHA, which indicates the presence of three copies of the **hemagglutinin (HA) epitope tag** encoding DNA. Hemagglutinin is a viral protein recognized by a commercially available monoclonal antibody. Epitope tags are short stretches of amino acids that are recognized by an antibody for detection on blots or by microscopy. The nine-amino-acid stretch of HA epitope tag has no functional role other than as an epitope.

DISCOVERY QUESTIONS

1. Why would you want to include a *wt URA3* gene in mTn?
2. Why would you want to use a version of *lacZ* that lacks a promoter and start codon?
3. Why would you want to put in *lox* P sites that allow you to cut out all the intervening DNA?
4. When they examined a subset of the mutagenized strains to verify that homologous recombination had occurred, 47 out of 48 strains tested were the product of homologous recombination. What is the significance of this finding?

Using the mTn approach, Snyder's lab isolated 11,232 strains that were able to turn blue during vegetative growth. Since the *lacZ* gene lacked a promoter and start codon, the only way cells could turn blue would be if mTn landed downstream of a promoter in the yeast genome. DNA sequencing was used to isolate the insertion point for 6,358 strains, with most of these landing within 200 bp of an annotated open reading frame (ORF). Due to the random nature of the mutagenesis, some ORFs had zero insertions while others were mutated multiple times. In total, 1,917 different annotated ORFs were mutated at least once. Interestingly, 328 **nonannotated ORFs** (**NORFs**) were also mutated. When the yeast genome was sequenced and annotated, only ORFs of ≥ 100 codons were considered real genes. Furthermore, if two ORFs of at least 100 codons overlapped, usually only the larger ORF was annotated. Therefore, it is likely that some functional yeast genes were not annotated. Only by random mutagenesis would NORFs have been subjected to experimental analysis.

No one would want to analyze 11,232 strains individually, so Snyder's group needed a high-throughput method to screen them. They developed phenotype *macro*arrays, which meant they used robots to spot each strain individually onto agar plates containing growth media and determined which strains could grow under different growth conditions (Figure 6.2). YPD is normal growth medium for yeast, and you can see that there are some missing colonies in the macroarray of white colonies because some insertions were lethal (a total of 1,082 stains were inviable). Testing for

strains with lethal mutations must be performed in haploid cells, which is one of the benefits of working with yeast. Benomyl is a drug that disrupts microtubules, and you can see a few resistant colonies. High doses (Calc^R) and low doses (Calc^S) of another drug reveal strains that are resistant or hypersensitive, respectively, to calcofluor, a wall-binding dye. Hygromycin affects cell wall biosynthesis, and there are some strains that are sensitive to it. When glycerol was used as the only carbon source, most cells were viable, but some were not. Using these and 16 additional growth conditions, the investigators were able to rapidly screen thousands of strains, though there are obviously more than 20 growth conditions that could have been analyzed.

DISCOVERY QUESTIONS

5. How could Snyder's lab grow strains that had mutated viable genes? If the strains are inviable, wouldn't they have died before they could be spotted on the macroarrays?
6. Explain how it is possible to screen for calcofluor-sensitive and -resistant cells. How is it possible to test for these two extremes?
7. How can a strain be sensitive to one drug that blocks cell wall formation but not another?
8. The full list of phenotypes is on the TRIPLES web site; click on "Disruption Phenotypes" and then "Search Phenotype Data." Scroll through the list and design a new growth condition that was not tested by the group at Yale.

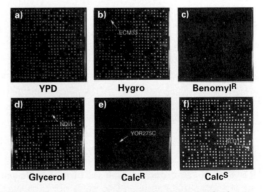

FIGURE 6.2 • Phenotype *macro*array analysis. Examples of 21 × 21 cm macroarrays testing growth on different media: **a)** YPD, **b)** YPD with 46 mg/ml hygromycin, **c)** supplemented with 20 mg/ml benomyl, **d)** YPGlycerol, **e)** YPD with 69.7 mg/ml calcofluor, and **f)** YPD supplemented with 12 mg/ml calcofluor. Arrows indicate strains mutated for genes functioning in cellular respiration (*Ndi1*) or cell-wall biogenesis (*Ecm33, Slg1,* YOR275C).

LINKS

Genomic View

Gene/Sequence
Resource

BLAST

A clever use of the yeast cell biology was to utilize the ability of yeast to turn red when metabolism by oxidative phosphorylation was functioning properly. All cells turned red unless they lacked one of the genes responsible for oxidative phosphorylation, in which case they would grow as white colonies (Figure 6.3). Another screen took advantage of the fact that all yeast cells contain the enzyme alkaline phosphatase in their vacuoles. Alkaline phosphatase modifies the colorless substrate BCIP into a blue product, in much the same way β-galactosidase can produce a blue product. When cells were placed on a weak detergent and BCIP, those with a weakened cell wall ruptured, spilling alkaline phosphatase, which created a blue halo around the cell wall mutants. Both of these phenotypes were easily scored, so that 11,232 strains could be screened in a couple of days.

After a few days of intense screening, Snyder's lab had collected a number of mutants that affected one or more of the 20 cellular processes they measured. A large number of mutants were identified, but there are some caveats worth noting. First, not all the 11,232 strains were tested for all conditions; only 5,760 were screened for cycloheximide hypersensitivity, for example. Furthermore, some of the mutants had multiple insertions in the same gene, since mTn's insertion site is random. Therefore, some of the mutant strains represent different alleles of the same mutated gene; those strains often, though not always, produced identical phenotypes. For example, there were 11 different strains with mTn inserted at different places in *Imp2*, a transcription factor involved in sugar utilization. Four of the insertions occurred in the 5' and middle portions of the gene (at

codons 46, 69, 230, and 270) and resulted in mutants unable to metabolize glycerol. However, one insertion, at codon 334 (11 codons prior to the step codon), created a mutant able to metabolize glycerol but unable to produce a cell wall. Insertions at codons 46, 230, 263, and 319 also produced the cell wall mutant phenotype. Therefore, *Imp2* needs its full length to stimulate the production of cell walls, but its carboxyl terminus is not necessary for glycerol metabolism.

Another useful component of the mTn method is the identification of NORFs. Snyder's team identified mTn insertions in 328 NORFs with a size range of 50–247 amino acids long. Some of the disrupted NORFs exhibited phenotypes in the 20 growth conditions screened. For example, an insertion at base 198,816 of chromosome 12 resulted in hypersensitivity to the microtubule-destabilizing drug benomyl.

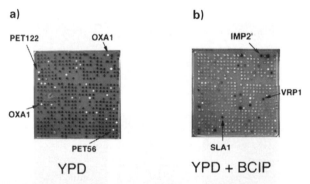

FIGURE 6.3 • **Macroarray analysis of metabolic pathways. a)** Metabolic mutants unable to carry out oxidative phosphorylation were identified as white (rather than red) colonies on YPD medium; all cells contained the *ade2* mutation, which leads to red pigment formation in *wt* metabolism. Respiratory genes identified by this method are labeled with arrows. **b)** Genes functioning in cell-wall maintenance were characterized through macroarray analysis of mutants grown on YPD overlaid with agar containing a mild detergent and BCIP. Dark colonies mark cells that lysed; arrows highlight genes known to be involved in cell-wall maintenance.

DISCOVERY QUESTIONS

9. Go to the *S. cerevisiae* Genomic View and mouse over yeast chromosome 12 until the selection window at the top indicates you are in the 100,000 to 200,000 range near the centromere, and click. Click again on the blue "*AAT2*" box and red "*SNR30*" box near base 198,816.
   a. What is the range in nucleotides of these two ORFs?
   b. What do these proteins do?
   Now go to the SGD Gene/Sequence Resource page. Enter nucleotides 198,416–199,219 and submit your search. This is an 800 bp window centered on the mTn insertion point at 198,816. The NORF described by Snyder's lab encodes 86 amino acids. Scroll down the results page a bit, and click "6-Frame Translation (with restriction map)." You will see the translation of all ORFs and NORFs in this region (Figure 6.4).

10. As you can see, there are two potential proteins disrupted by an insertion in the colored box region—one on the top strand (row a) and another on the bottom strand (row d).
   a. On this web page, locate the 86-amino-acid NORF described by Macdonald.
   b. What are the first five amino acids?
   Go back to SGD Gene/Sequence Resource page, enter the range of 198,728–198,988, and select chromosome 12, and click on "Submit Form." In the new window, click on the protein translation with no header. Copy that protein sequence.
   Go to BLAST to perform a *protein* search with BLAST. Paste in your protein sequence. BLASTp your NORF protein.

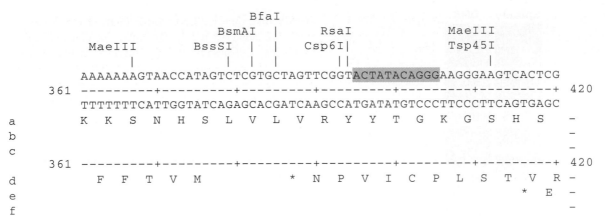

FIGURE 6.4 • **Region of interest in yeast genome, including putative NORF.** The site of mTn insertion is highlighted in blue. Above the double-stranded DNA sequence are restriction sites. Below the DNA are the six frames of translation (a, b, c top strand; d, e, f bottom strand) with a dashed line separating the three forward and reverse reading frames. Stars indicate stop codons.

11. What have you learned about the NORF protein? Are there any orthologs? any conserved domains?

12. How would you know if this NORF encodes a protein, or just looks like an ORF?

13. How would you know whether disrupting the NORF or the overlapping gene "*SNR30*" caused benomyl sensitivity?

The 20 different experimental conditions were clustered by Snyder's lab using software similar to what is used for microarrays (Chapters 4 and 5). In the phenotype macroarrays, each growth condition is a column and each mutant strain is a row. The color key is different from before, but the principle for clustering is identical—cluster the growth conditions based on the similarity of strains affected by each growth condition (Figure 6.5 on your CD-ROM). When the clustering was performed, we saw that some of the growth conditions affect more than one strain, as you would expect.

FIGURE 6.5 • **Clustered macroarray phenotype data.**
Go to the CD-ROM to view this figure.

DISCOVERY QUESTIONS

14. What advantage is there to clustering the phenotypes in this manner? Do you think it gives any new insight into gene function? Explain your answer.

15. Some of the genes identified in this analysis had no known function. How can clustering these data help us predict possible functions?

The predictive power of phenotype macroarrays goes beyond its initial phenotype identification. The investigators used the phenotypes to cluster strains and experimental conditions. Clustering allows us to visualize which proteins may be involved in more than one process, or identify a series of proteins that perform related functions in the same circuit. Proteins with previously unknown functions would also be clustered, allowing more detailed predictions about their possible functions. You may have detected an old habit that is going to be hard to break—the use of "function" as a catchall term. We have said nothing about biological process, molecular function, or cellular component. With careful analysis of these data, at least one biological process performed by each protein could be predicted with a high degree of certainty. Molecular functions of individual proteins might become evident with additional experiments to refine the precise phenotypes. The only aspect missing is the cellular component. Where are these proteins located?

The Snyder lab had anticipated wanting to know the cellular location of the proteins when mTn was constructed. Remember the *lox* sites in Figure 6.1? All the DNA between the two *lox* sites could be excised (*lox* recombination) placing the epitope tag-encoding DNA near the 5' end of the inserted DNA. If the mTn inserted within a coding region, then the *lox*-mediated excision could place the epitope tag onto the carboxyl end of the truncated protein. By tagging the protein with an epitope tag, the investigators could use the hemagglutinin monoclonal antibody to determine the cellular compartment that housed the tagged protein (Figure 6.6, page 166). As you can see, antibody-labeled proteins appear white in these immunofluorescence photomicrographs, and each pattern reveals its cellular compartment. Well, maybe you cannot see the differences, but yeast

METHODS
lox Recombination
Immunofluorescence

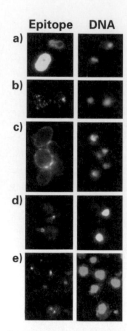

Epitope   DNA

a)

b)

c)

d)

e)

FIGURE 6.6 • **Localization of epitope-tagged proteins.**
(Left column) Examples of immunofluorescence patterns in
cells stained with monoclonal antibody against HA. **(Right
column)** The same cells stained with a DNA-binding dye
called DAPI. Numbers in parentheses indicate the number of
strains with similar labeling patterns: **a)** Diffuse cytoplasmic
labeling (189). **b)** A punctate pattern of cytoplasmic labeling
(10) **c)** Localization to the plasma membrane. **d)** A ring in the
mother-bud neck (2). **e)** The spindle body (5).

METHODS

Bar Codes

LINKS

NORF Database
Ron Davis

biologists can. Epitope tagging was performed on 1,340 strains (constituting about
10% of all mutant strains) and the cellular
localization for these proteins was recorded.
Interestingly, Snyder's team reported that a
few NORFs were also epitope tagged. They
cited one example located at base 1,236,754
on chromosome 4. When examined by immunofluorescence
microscopy, the tagged protein was localized to the nucleus.

DISCOVERY QUESTIONS

16. Go to *S. cerevisiae* Genomic View and find the location (base 1,236,754 of chromosome 4) of the
NORF identified by this research. Find the annotated gene that overlaps with the insertion site.

17. Go to SGD Gene/Sequence Resource page, retrieve the nucleotides on chromosome 4 ranging from 1,236,454–1,237,054, and click on
"6-Frame translation." The mTn insertion happened at base 300 of these 600 bases. How many
potential proteins are located at this site?

18. Go to the TRIPLES NORF database and search
for clone ID "V28B5," which is the name of the
NORF from Discovery Question 17 that is localized to the nucleus. What does it say under the
heading "Genome region"? What do you think of
this description? Click on the blue V28B5 text to
verify the phenotypes.

There is one more tool hidden in the mTn mutants. Because they were identified as blue colonies, they have landed
downstream of a promoter. As you know, protein production is regulated in time and produced only when needed.
Therefore, part of a protein's cellular role is determined by
the timing of its expression. By subjecting the mTn-mutated
cells to additional screens, it would be possible to determine
when each protein is produced—meiosis? mating? at certain
points during the cell cycle?

The Snyder lab produced the largest functional analysis of
proteins ever performed (as of summer 2001). It appears that in
screening so many different mutants, a few mistakes were made
in the analysis. By making their data publicly available, Snyder
has upheld the scientific standards that require reproducibility
and peer review, and enabled you to make a new discovery
about their research. The Snyder data are valuable tools that
contribute to a larger understanding of the yeast proteome.

Though this was the largest screen, not all the data have
been collected. Many of the mutants were not tested under
all 20 growth conditions, and new growth conditions could
be designed for all 11,232 strains. Complete analysis will require more time than any single lab could provide. In keeping with the spirit of scientific collaboration and free access
to reagents, the Snyder lab will share any strain of these mTn
mutants with interested scientists, including students. So
maybe there is a thesis project waiting for you in proteomics.
But before you hop on the internet to order your yeast, there
are other approaches you might find interesting. . . .

Which Proteins Are Needed in Different Conditions?

Everyone wants to know what all these proteins do, and there
are many different ways to find out. Elizabeth Winzeler,
working with the lab of Ron Davis at Stanford University,
was the first author on a paper with more than 50 coauthors
in a multinational collaboration, including Ross-Macdonald
and Snyder from Yale. In this mammoth project, 2,026 ORFs
were individually deleted by homologous recombination and
replaced with a selectable marker and two unique 20 bp sequences called UPTAGS and DOWNTAGS that functioned
like molecular **bar codes** (4^{20} is more than 10^{12} possible bar
codes). Flanking each unique UPTAG and DOWNTAG bar
code pair is a pair of common **polymerase chain reaction
(PCR)** primer sites so that the unique bar codes for all the

different mutants could be amplified with the same pair of PCR primers. In conjunction with these mutants, DNA microarrays of 4,052 UPTAG and DOWNTAG bar code spots were constructed so that each spot contained the DNA from one bar code. Therefore, each mutant strain had two, and only two, spots to which its genomic DNA could bind (Figure 6.7 on your CD-ROM). This bar code microarray method does not require the isolation of RNA for probe production. Instead, PCR products from genomic DNA template are labeled either red or green.

> **FIGURE 6.7 • Bar code microarray data.**
> Go to the CD-ROM to view this figure.

Ordinarily, it would have been a lot of painstaking work to analyze 2,026 different strains, but the team devised a high-throughput approach. They incubated 558 different strains in a single large flask. Immediately, an aliquot was taken and used as PCR template to produce a mixture of 558 red-colored pairs of bar codes. Six hours later, another aliquot was removed, amplified by PCR, and labeled green. These two mixtures of PCR products were combined, denatured, and added to the microarray (see Figure 6.7). Bar code DNA chips allowed investigators to see which strains were unable to compete under these growth conditions. This is seen in Figure 6.7 as red spots indicating which strains grew slower compared to their flaskmates.

Most gene deletions resulted in yellow spots, since their loss did not affect cell division under this growth condition. But a few spots showed shades of red, indicating that the number of cells containing this bar code decreased during

the six hours. Since the loss of cells can be monitored over time and under many different growth conditions, the investigators had created a system to study evolution—changes in the gene pool over time due to selective advantages of certain genotypes. In this experiment, the flask contained a population of cells with genomic diversity competing for limited resources. Those with the best proteomes continued to replicate, while those with inferior proteomes gradually became less abundant.

DISCOVERY QUESTIONS

19. In the top and bottom left corners of Figure 6.7 there is a checkerboard pattern. Why would these control spots be necessary?

20. There are some green spots present on the microarray as well. Explain what must be happening in the flask to account for this.

21. Are any NORFs represented on this microarray? Explain your answer.

The bar code method showed which proteins provided a selective advantage in a mixed population of cells. By having two bar codes for each strain, they had a built-in control, since each bar code within a particular strain should produce equal signals on the chip. This was verified by comparing the growth rates for each pair of bar codes at a given time point, and Figure 6.8a shows the correlation for cells grown in rich media. This is the first example you have seen where data from two spots of different sequences were used to validate

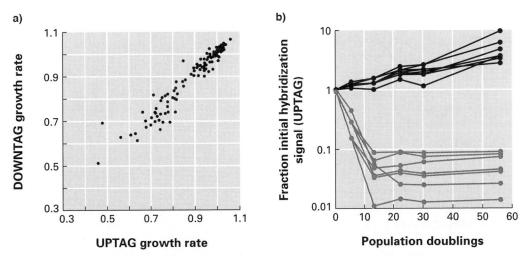

FIGURE 6.8 • Analysis of bar code DNA microarrays. a) Correlation of growth rate data obtained with both bar code (DOWNTAG and UPTAG) sequences for strains grown in rich medium. Data are shown for 331 strains that produced UPTAG and DOWNTAG hybridization signals that were both at least threefold over background at time zero. **b)** Normalized hybridization intensity data for the 10 slowest-growing (light blue) and 10 fastest-growing (dark blue) strains in rich medium. The data are presented as fraction of labeling intensity at time zero (log scale on Y-axis) over time in the form of population doubling (X-axis).

METHODS

Exponential Growth

the results for individual genes. This type of validation is very helpful in identifying potential problems with the data prior to analysis of the role for any particular protein. Most of the data points are clustered in the upper right corner, which indicates *wt* growth rates. The data points that are lower on the diagonal line indicate cells that grew slower due to their mutations.

The growth profiles for the 10 fastest and 10 slowest strains illustrate the types of outcomes observed when a strain lacked a particular protein (Figure 6.8b). In this graph, the Y-axis shows the change in signal for each time point compared to the original signal. Since the same number of cells

were sampled each time, strains growing at normal exponential growth rate would have produced a signal at 1 (i.e., yellow spots on the microarray) for each time point. If a strain grew faster than typical cells, its fraction of DNA microarray signal would be greater than 1, as is seen for the 10 fastest strains. If a strain grew slower than typical cells, it would be less abundant in the sample of cells removed for probe production and thus its fraction of signal would be below 1, as seen for the 10 slowest strains. If a particular bar code mutation caused a lethal phenotype, we would expect its representation in a sample from each time point to quickly become 0. Strains that failed to give any signal above background were excluded from consideration in Figure 6.8b.

MATH MINUTE 6.1 • HOW DO YOU KNOW IF YOU HAVE SAMPLED ENOUGH CELLS?

A hidden assumption in the bar code study is that the proportion of each strain in a sample is the same as the proportion of the strain in the entire population at the time the sample is taken. If this assumption does not hold, the growth profiles of the 558 strains cannot be accurately assessed. The validity of the assumption depends on how large the sample is, compared to the size of the population. For example, suppose a population of 10^{10} cells contains an equal number of cells from each of 558 strains. If 10^9 cells (10%) are sampled from the population, it is more likely that the sample contains an approximately equal number of cells from each strain than if only 10^8 cells (1%) are sampled. If one strain were to comprise only 1% of the population, a very small sample might miss the strain completely, or get too few cells from the strain to be detected on a DNA microarray.

You can calculate the probability of a particular sampling outcome (i.e., getting a particular number of cells from each strain in a sample) using the hypergeometric frequency function, a formula involving the sample size and number of cells from each strain in the population. Specifically, suppose a population contains N cells from M different strains, with n_1 cells from strain 1, n_2 cells from strain 2, . . . , and n_M cells from strain M. You can compute the probability that a random sample contains k_1 cells from strain 1, k_2 cells from strain 2, . . . , and k_M cells from strain M, with the hypergeometric formula:

$$\frac{\binom{n_1}{k_1}\binom{n_2}{k_2}\binom{n_3}{k_3}\cdots\binom{n_M}{k_M}}{\binom{N}{K}},$$

where K is the sample size (i.e., $K = k_1 + k_2 + \cdots + k_M$). For example, if a population contains 100 cells from each of three different strains (a total of 300 cells), the probability that a sample of 60 cells contains exactly 20 cells from each strain is

$$\frac{\binom{100}{20}\binom{100}{20}\binom{100}{20}}{\binom{300}{60}}.$$

In the hypergeometric formula, each pair of numbers in parentheses is a **binomial coefficient**

$$\binom{n}{k} = \frac{n!}{k!(n-k)!}$$

(read "n choose k"), which represents how many distinct sets of k cells of a particular strain can be chosen from the n cells of that strain in the population. For example, with $n_1 = 100$ and $k_1 = 20$,

$$\binom{n_1}{k_1} = \binom{100}{20} = \frac{100!}{20!(100-20)!} \approx 5.36 \times 10^{20}.$$

Many calculators, as well as various mathematical and statistical software programs, have a binomial coefficient function that can save you a lot of computation.

In a sample of 60 cells from a population of 300, the ideal sample would contain 20 cells from each strain, in perfect agreement with the population proportions of 1/3 for each strain. Evaluating the binomial coefficients in the above formula results in a probability of 0.017 of getting an ideal sample, i.e., one that contains exactly 20 cells from each of the three strains. However, for practical purposes, we are satisfied with getting *close* to 20 cells from each strain. Suppose we are willing to accept a deviation of up to 3 cells from the ideal number (20 ± 3) for each strain, corresponding to a deviation of $3/60 = 1/20 = 0.05$ from the ideal proportion ($1/3 \pm .05$). There are 37 different sampling outcomes that satisfy this criterion. (Can you identify all 37?)

You can find the probability of getting one of these 37 sampling outcomes by computing the probability of each outcome (using the hypergeometric formula) and adding the 37 probabilities. The result of these calculations—i.e., the probability that the sample proportions are all within 0.05 of 1/3—is approximately 0.47. In other words, we have a 53% chance of getting a sample that *differs* from the ideal by more than 3 cells in one or more strains.

To improve our chances of getting a good sample (one in which the sample proportions are close to 1/3), we must sample more cells. Alternatively, we could relax our maximum deviation criterion, accepting a deviation of up to 5 cells (20 ± 5), for example. The probability that all strains are represented within a given deviation from 1/3 is shown in Table M6.1 for three different sample sizes. The deviation is given as a proportion of the sample size; the deviation in number of cells is different for each sample size.

TABLE M6.1 • **Probability that sample proportions are within a specified deviation from 1/3 for sample sizes of 60, 120, and 180 cells.**

Deviation	Probability		
	Sample 60 Cells	Sample 120 Cells	Sample 180 Cells
0.025	0.113	0.345	0.502
0.05	0.470	0.766	0.954
0.075	0.649	0.954	0.998
0.1	0.887	0.995	0.999991
0.125	0.945	0.9997	≈ 1

This table of probabilities shows that we should sample at least 180 cells from our population of 300 cells to be 95.4% certain that the sample proportions will be $1/3 \pm 0.05$ for all three strains. However, if we were willing to accept errors as large as 0.125 (i.e., sample proportions ranging from 0.208 to 0.458), a sample of 60 cells would probably do.

Now suppose we have a population containing 200 cells from strain 1, 97 cells from strain 2, and 3 cells from strain 3. In other words, the population is the same size as in our previous example (300 cells), but strain 3 comprises only 1% of the population. What is the probability that a sample of 60 cells from this population contains at least one cell from strain 3? There are 61 sampling outcomes that have no cells from strain 3. By using the hypergeometric formula as above, you can sum the probabilities of these 61 outcomes to find that the probability of completely missing strain 3 is 0.511. Therefore, the probability that strain 3 is present in the sample of 60 cells is $1 - 0.511 = 0.489$. Once again, sampling more cells improves our chances of getting a good sample—in this case, one in which a cell from strain 3 is present. The probability that strain 3 is present is 0.785 when 120 cells are sampled, and 0.937 when 180 cells are sampled.

For very large samples, it is difficult to compute probabilities with the hypergeometric frequency function, and the probabilities are often approximated using the normal probability distribution (see Math Minute 8.1, page 234). Whether exact or approximate, these probabilities show investigators that growth trends such as those in Figure 6.8 are not merely artifacts of the sampling process.

LINKS

C. elegans

Andy Fire

RNAi

C. elegans Databases

When the bar code analysis was complete, the team classified by biological processes the role each protein played in vivo and which biological processes were controlled by more essential genes than others (Figure 6.9). When the investigators examined the chromosomal distribution of essential and nonessential genes, they found that some essential genes were clustered near each other but every chromosome contained essential genes. The bar code microarrays allowed the investigators to compare relative growth rates of mutant strains grown under any condition they chose. Therefore, they could measure the relative contribution of every protein in the proteome.

DISCOVERY QUESTIONS

22. Predict what might happen if only the slowest-growing strains were incubated together (see Figure 6.8b). Would their graphs be the same as when they were grown with all 558 strains?

23. Hypothesize why some of the strains leveled off lower than others and why they never reach zero on the Y-axis.

24. In Figure 6.9, you can see that DNA synthesis has some nonessential genes. How can DNA synthesis be considered "nonessential" in a growing population? Explain this apparent contradiction.

25. What are the pros and cons of the bar code method compared to the mTn method? Is one better than the other?

Can You Live Without Some Proteins?

Proteomics is a very new field with a lot of good methods being developed. There are several newer ones that can be used on a wide range of species to block the production of particular proteins. One worth noting before we move on is

RNA interference, which was first developed in worms. The first animal to have its complete genome sequenced was *C. elegans,* the nematode worm that is about the size of an eyelash. These worms, which are very easy to grow in the lab, feed on bacteria and are completely transparent. So we can see every cell in their bodies and know the cell fate for every single cell from fertilized egg until death (959 cells in the adult female and 1,031 in the adult male).

In 1998, Andy Fire at the Carnegie Institute of Embryology in Baltimore made a rather shocking discovery. If you inject a worm with **double-stranded RNA (dsRNA)** that is complementary to a particular gene, the worm will be unable to produce that protein. The process of dsRNA inhibition of protein synthesis is called **RNA interference (RNAi).** Even more surprising was that worms could *eat* dsRNA, and RNAi still worked. This allowed investigators to block the production whenever and for as long as they wanted. RNAi is more specific than either of the knockout approaches described above for yeast, because a *wt* worm could be tested for the loss of a single protein for short periods of time—for example, the role of a protein that is vital to development, and could be determined in adults using RNAi, but not knockouts.

To streamline the methodology, the clever community of worm researchers realized they could modify *Escherichia coli* to **overexpress** any dsRNA they wanted. Since *C. elegans* eats *E. coli,* and eating dsRNA works for RNAi, it would be possible to perform genome-wide studies. A British group and a German group used this streamlined method of RNAi to systematically block the production of each protein on chromosomes 1 and 3, respectively (about one-fourth of the entire proteome). They determined the phenotypes of worms that were unable to produce each protein at various stages of the worm's life cycle. Approximately 13.5% of the 4,619 proteins tested (out of about 19,000 total annotated genes) produced observable phenotypes at one or more stages in development. The list of RNAi-generated phenotypes is growing quickly, but there are web sites that allow you to examine the *C. elegans* databases. In 2001, several research groups applied RNAi methods to insects and vertebrates, including human cells grown in tissue culture. This method is also

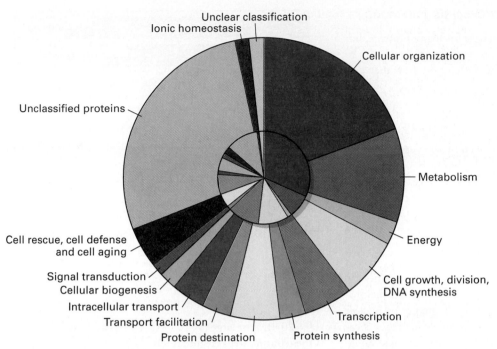

FIGURE 6.9 • Bar code analysis of biological processes. Distribution of functional classes of essential (inner circle) and nonessential (outer circle) genes using criteria from the Munich Information Center for Protein Sequences (MIPS).

referred to as short inhibitory RNA (siRNA), so you might want to search the web for nonworm siRNA/RNAi databases to learn the latest information about mammalian RNAi. In May 2002, *Science* highlighted recent advances in RNAi, which functions in all eukaryotes.

DISCOVERY QUESTIONS

26. Explain why it is desirable to block the production of a protein in an adult but not a larva.

27. Given that RNAi is sequence specific, could it be used to block the production of more than one protein at a time? multiple isoforms? Could it be used as a pharmaceutical tool?

Summary 6.1

Conceptually, the gene deletion approach is not "genomic," but the capacity to analyze entire genomes does provide us with more comprehensive perspectives. The two yeast whole-proteome analysis methods could be applied to any organism, but due to the small genome of yeast and its utility as a model eukaryote, yeast will continue to be a very popular system. Since the earliest days of *Drosophila* genetics, investigators have tried to learn what proteins do inside living cells. The first step is to characterize what every protein does (molecular function), why the cell requires this function (biological process), and where it takes place (cellular component). With the advent of RNAi, we will be able to understand dynamic proteomes by selectively blocking the production of any protein we choose at any time and for varying durations. The ability to control the timing of RNAi provides better resolution to understand the dynamic nature of proteomes, which change over time, even within single cells. However, to fully understand the proteome requires additional research tools and experimental designs, as you will see in the remaining sections of Chapter 6.

6.2 Protein 3D Structures

A major goal of biology is to describe how cells work by defining the rules by which they live. All cells come from preexisting cells. DNA is the heritable material. These rules have been validated repeatedly, and it is difficult to imagine a time when people doubted them. Another rule is that form meets function, which means that if you know the shape of a molecule, you can understand how it works. In this section, we examine several research efforts that focus on the 3D structure of proteins. Is there a high-throughput method for determining the 3D structure of entire proteomes? Can structure alone reveal how proteins work? Can we develop better medications once we know the 3D shape of a protein? And finally, can a change in 3D shape be the difference between life and death, even when the amino acid sequence remains unchanged?

Does a Protein's Shape Reveal Its Function?

We have sequences for many proteins, but amino acid sequence alone is insufficient to reveal what the proteins do, how they perform these roles, or with which proteins they interact. Many times, a protein's role is not fully understood until its 3D shape is known. Progress toward this goal has been sporadic and driven by the perseverance of many individual laboratories around the world. Funding agencies have concentrated their resources on high-throughput efforts in **structural proteomics.** (A good place to learn about viewing 3D structures is the Protein Explorer web page.)

In order to obtain detailed information about a protein's structure at the atomic level, we need to be able to purify the protein in large amounts (about 1 mg), crystallize the protein, and then use either X-ray crystallography or nuclear magnetic resonance (NMR) to determine the precise location of every atom in the protein. 3D structure determination is usually performed one protein at a time by individual labs. Structural proteomics demands faster methods, but progress has been slow. The first group to launch a proteome-wide 3D structure determination effort was the German Protein Structure Factory. About a year later, the National Institute of General Medical Sciences (NIGMS) initiated a $150 million program to support seven research centers to expand the global effort of structural proteomics.

The rate-limiting step for structural proteomics is the production of samples and crystals. Protein structural information is collected by the Protein Data Bank (PDB), an international repository for 3D structure files. The rate of 3D data collection has increased rapidly in recent years, but it will need to accelerate more if structural information is going to have a comprehensive impact on proteomics.

The importance of determining the 3D structure is recognized by most biologists, and there is no national monopoly on collecting data. For example, a Canadian team led by Dinesh Christendat, Adelinda Yee, Aled Edwards, and Cheryl Arrowsmith decided to determine the structure of all proteins in the archaeon *Methanobacterium thermoautotrophicum.* The genome of *M. thermoautotrophicum* contains 1,871 ORFs, which would present a formidable challenge to any group. As with any large project, the Canadian team used a "rational approach" to maximize their success. First, they eliminated membrane proteins (about 30% of the proteome) because they are notoriously difficult to crystallize. Next, they skipped any proteins that had obvious orthologs in the PDB (about 27% of the proteome) since they would not provide as much new information. A total of 424 proteins of the remaining 900 proteins were selected for structural analysis. The proteins were divided into small

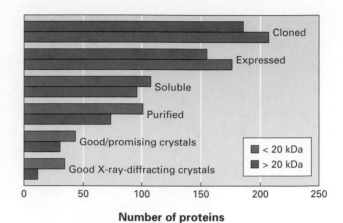

FIGURE 6.10 • **Structural proteomics of an Archaea.** Histograms of the number of *M. thermoautotrophicum* proteins at the end of each step in cloning, expression, and sample preparation. Proteins were divided into two classes based on molecular weight.

(< 20 kDa) or large (> 20 kDa) groups and carried through the requisite steps (Figure 6.10). As you can see, at each step, some proteins were excluded from the next phase, which illustrates the difficulty of the task.

Structures for the first ten proteins were determined, leading to several new discoveries. Some folding patterns had never been seen before, including binding sites for coenzymes such as flavin mononucleotide (FMN) and nicotinamide adenine dinucleotide (NAD^+), as well as cation cofactors such as of calcium, nickel, and magnesium. Binding sites for these ligands could not have been predicted by amino acid sequences.

We will look at one protein as an example of what can be learned from structural proteomics. MTH129 is a known ortholog of *Orotidine* 5' monophosphate *DeCarboxylase* (ODCase), which catalyzes the removal of one carbon dioxide in exchange for one hydrogen ion at carbon number six of the other product, uridine 5' monophosphate (UMP). This exchange of CO_2 for H^+ occurs 10^{17} times faster than it would in the absence of ODCase. MTH129 is the first member of this enzyme family to have its structure determined. The key to its catalysis is the unusual localization of four amino acids of opposite charges (KDKD), which line the binding site. A quick series of electrostatic interactions permits the removal of the CO_2.

To further illustrate how function can be deduced from structural proteomics, let's look at an indispensable but often neglected protein **aquaporin**-1. In any standard introductory textbook, the definition of a cell membrane includes the term "semipermeable." When you learned that membranes are made of phospholipids, you also learned that only hydrophobic molecules can pass through cell membranes. No charged or polar molecule can pass through, not even a proton. Yet you probably also learned that water, a polar molecule, can pass through a cell's membrane. This sets up a logical contradiction that is rarely discussed. In reality, water does not pass through our cell membranes; it travels through aquaporin and not the phospholipid bilayer. As the name implies, aquaporin is a protein channel that allows only water to pass. Aquaporin-1 was the first water channel gene to be cloned, and thus the number one was added to its name. You can interact with aquaporin-1 online to see how its structure reveals its function.

The characterization of aquaporin-1 answered many questions that had baffled biologists for years, but a big question remained: How can water (H_2O) pass through a channel when protons (H_3O^+) cannot? It would be a disaster if protons could sneak through with water, since membrane potentials and hydrogen gradients are vital for many cellular functions, including the production of ATP. Protons can move along a column of water by hydrogen bond exchange, which means that any stream of water rushing through aquaporin (at 3×10^9 water molecules per second) has to be able to prevent this movement of H^+. A multinational group led by Kazuyoshi Murata took on the task of determining the structure of perhaps the most important membrane protein.

Aquaporin **monomers** are composed of 269 amino acids that form six membrane-spanning domains. The functional unit of aquaporin is a **homotetramer** (composed of four identical subunits) that allows water to pass through each monomer. The key to its function is the protein's hourglass-shaped pore with the narrowest constriction only 3.0 Å wide (Figure 6.11). The size of this constriction is critical since a water molecule is about 2.8 Å in diameter. In addition, there are several hydrophobic amino acids that line the pore and help exclude other charged small molecules. Although Murata's team could not be certain, the structure allowed them to predict that water does not form a continuous stream as it passes through the pore. Instead, as one water molecule reaches the constricted portion, the hydrogen bonds that had connected one water molecule to the next were temporarily transferred to two asparagine amino acids that help form the constricted region (Figure 6.11b and c). Once the continuous column of hydrogen bonds is broken, protons cannot move along with water, and aquaporin has permitted water but not protons to pass through your membranes.

DISCOVERY QUESTIONS

28. How did the structure of ODCase lead to the discovery of a new functional motif (KDKD)?

29. Based on the 3D structure of aquaporin and its hypothesized mechanism of preventing protons from passing through the channel, predict two amino acids that should be conserved in every species.

Can We Use Structures to Develop Better Drugs?

By this time, you might be thinking that all this attention to structure is only for crystallographers and has no real-world applications. Let's look at one example of how the use of structures can lead to improved medical treatment.

a)

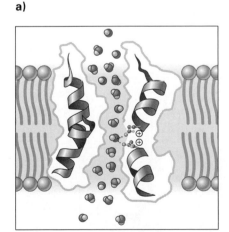

b)

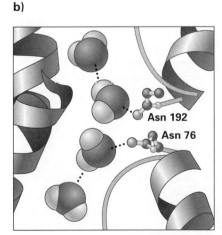

c)

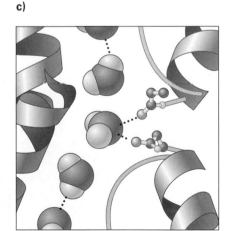

FIGURE 6.11 • Structural basis for aquaporin function. a) The charges from the helix control the orientation of the water molecules passing through the narrowest part of the channel. **b)** and **c)** The hydrogen bonding of a water molecule to asparagines 192 and 76, which extend their R groups to form the narrowest part of the channel.

STRUCTURES
Urokinase

LINKS
Vicki Nienaber

Vicki Nienaber and colleagues at Abbott Laboratories in Illinois had a very simple but clever idea. If you soak a crystallized protein in a solution of many different molecules, you might be able to cocrystallize compounds that bind in the active site (Figure 6.12). Using this approach, Nienaber's group wanted to identify a new compound that could inhibit the enzyme urokinase. Slowing urokinase activity has been shown to slow the growth and metastasis of tumors without the unpleasant side effects of traditional chemotherapy. If the team could find a compound that could be taken orally, this would reduce discomfort and expense for patients. Although the group did not find the ultimate drug they wanted, they did identify a compound that was able to enter the body after oral administration, and inhibited urokinase when present in the range of 2.5 μM. That new compound will be a good starting point for further drug development.

DISCOVERY QUESTIONS

30. Why is Nienaber's method better than current methods of drug discovery in which many compounds are given to cell lines to see which ones produce the desired effect?

31. Why is the cocrystallization method less appealing than the current method of drug discovery?

32. Go to the urokinase 3D tutorial. What similarities are there between the inhibitor in this 3D structure and the new one Nienaber discovered? What are the differences?

Can One Protein Kill You?

There is another interesting story about shapes, but this one does not have a happy ending. **Prions** are proteins that can change their shapes, which is nothing new for proteins. But

FIGURE 6.12 • Structural method to discover new medications. a) In an ideal situation, the active site is open to the solvent when the protein is crystallized. **b)** Different small molecules are mixed with the crystallized protein. **c)** Those with appropriate shapes will be incorporated into the crystal. In the structure web page of urokinase, you can see an inhibitor similar to aminopyrimidyl 2-aminoquinoline, which inhibits the enzyme at 0.37 μM.

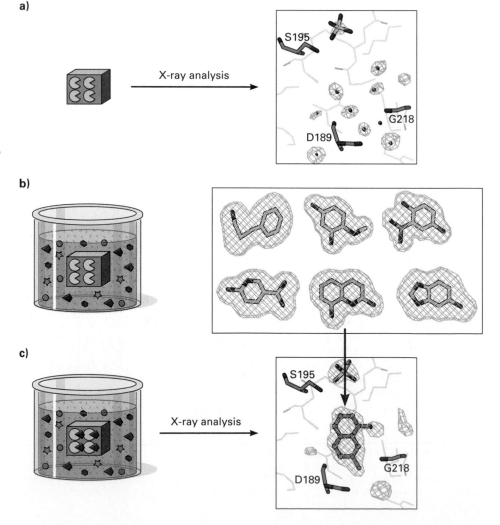

once a prion protein has changed its shape, it can make others do so. As the number of converted proteins grows, a new phenotype becomes evident. Converted prions are the cause of scrapie in sheep, mad cow disease in cattle, and Creutzfeldt-Jakob disease (CJD) in humans. All three of these illnesses lead to a rapid, progressive dementia and eventually death. If you eat prion-"infected" meat, you can develop CJD. That humans become infected with the bovine (cattle) form of prions indicates a similarity in 3D structure between human and bovine prion proteins, which is evident when you compare their structures.

The existence of a contagious disease that does not require any nucleic acid was unprecedented and considered impossible. Demonstrating this protein-only form of disease took many years, and eventually Stanley Prusiner's persistent efforts were rewarded with a Nobel Prize. The existence of highly conserved proteins whose only known function is to cause a lethal disease does not fit with our perception of evolution. Even more surprising has been the discovery of a prion-like protein in yeast, **Sup35.** Why does yeast have a prion protein that cannot play any neurological role?

Sup35 is a translation termination factor with a twist. Its carboxyl end binds to the ribosomal complex to terminate translation. The other two parts (amino end and middle section) can be deleted, and the translation termination function is left intact. But if Sup35 is converted to its infectious prion conformation, then the shape change spreads throughout the cell and is passed on to all daughter cells. When in the prion conformation, Sup35 causes ribosomes to read

through stop codons, extending the lengths of proteins and thus altering their shapes and functions as well. The loss of fidelity in translation does not sound evolutionarily adaptive, which makes us think prions should not have evolved. But they did evolve and are conserved in many eukaryotes. Could natural selection produce Sup35's ability to change shape and function?

Heather True and Susan Lindquist of the University of Chicago (now director of The Whitehead Institute at MIT) wanted to know why yeast would retain a protein with apparent detrimental translation capabilities. In their paper, they documented that Sup35 in the prion conformation conveys new phenotypes to cells (Figure 6.13a). As the translation fidelity is reduced, extended proteins are produced, some of which must provide the means to withstand antibiotics. Interestingly, the conversion of Sup35 to the prion conformation can occur spontaneously at a rate ranging from 10^{-5} to 10^{-7} (Figure 6.13b–d). The ability of converted Sup35 to produce new and heritable physiologies enables a single genotype to have multiple phenotypes. As cells grow in a changing environment, those with the prion form of Sup35 may be more fit and thus thrive while their genetically identical but phenotypically distinct sisters cannot survive. If this new phenotype were advantageous, genetic mutations might occur that lead to the stabilization of the new phenotype, and Sup35 might subsequently revert to its nonprion version. Whether this prion-induced advantage

STRUCTURES
Prions

LINKS
Stanley Prusiner
Nobel Prize
Susan Lindquist

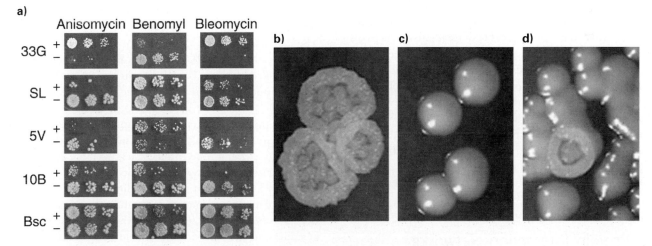

FIGURE 6.13 • Functional consequences of protein conformation. a) Sup35 prion conformation affects growth in different genetic backgrounds and under different conditions. Media contain fungicides anisomycin, benomyl, or bleomycin. Five different strains are grown in the presence of a prion-shaped Sup35 (+) or the absence of prion shape (−). **b)** Colony morphology is altered by prion conformation in Sup35. Cells were grown, diluted, and spotted onto plates containing potassium acetate and grown at 30°C. **c)** Cells lacking a prion form of Sup35 and **d)** the spontaneous appearance of a colony containing prion-shaped Sup35 amidst many colonies lacking prion-shaped Sup35.

exists in mammals is unknown, but it does lead to interesting speculation on why mammals express potential prion proteins that are poised to create neurological diseases. Nevertheless, it illustrates rather bluntly why structural proteomics is needed to augment sequence information.

DISCOVERY QUESTIONS

33. In what way did the 3D shape change of Sup35 alter its function? Is the functional change related to the change in colony morphology? Explain your answer.

34. Is the change in shape of Sup35 due to any changes in DNA sequence? How could new phenotypes (see Figure 6.13a) become established in the gene pool if Sup35 reverted back to its original shape and function?

6.2 Summary

The rule "form meets function" appears to be even more solid with the discoveries from structural proteomics. We can develop better medications once we know where an existing drug binds and the structure of its binding site. The ability to see the shape of aquaporin revealed how water molecules but not protons pass through the channel. Although we still do not know why eukaryotes have prion proteins, the prion-like protein Sup35 in yeast has offered us the first hint that a change in shape can lead to a beneficial change in function under certain circumstances. As we learned in Section 3.3, every adaptation comes with trade-offs. Perhaps on an evolutionary time scale, prions have proven to be advantageous for some individuals and deleterious to others as selection pressures change. As investigators continue to study the cellular role of prions, their structure has shown us how prion-based diseases can spread to different species, which may lead to effective treatments or vaccines. However, no protein acts in isolation, so to fully understand proteomes, we will need to define all the protein-protein interactions within cells.

6.3 Protein Interaction Networks

In order to understand what a person does for a living, you need to know more than a simple job description. For example, every college and university has a development officer, but what does this person really do? If you knew with whom the development officer interacted, you would have a better understanding of this role in the life of your institution. Similarly, to fully understand what each protein does, we need to determine with which other proteins each individual protein interacts. In the pregenomic era, **immunoprecipitation** was the primary means of determining protein-protein interactions (see Chapter 10). In the postgenomic era, we need high-throughput methods to catalog protein interactions. As you will see, two methods have proven especially helpful for this purpose. A more vexing problem is how to display interaction networks so we can comprehend them, gain new insights, and make testable predictions.

Which Proteins Interact with Each Other?

At the risk of sounding like a conspiracy theorist, proteins rarely act alone. Most proteins need a partner to either allosterically or covalently modulate them to change their conformations, but not in a contagious way as prions do. If a protein needs to be activated, often it must bind to another protein. When the cellular task is done, proteins frequently need to interact with another protein to become inactivated. Proteomics requires more than a list of the protein functions but also each protein's interactions within a cell. Two proteomic methods currently amenable to high-throughput screening are DNA sequence analysis and yeast two-hybrid assays. We'll look at DNA sequence analysis first and then turn our attention to the more prevalent two-hybrid method. Regardless of the method, the goal is to create a protein interaction map that will enable us to visualize cellular functions.

Good Sequences Make Good Neighbors A series of papers published in 1999 described how protein-protein interactions could be deduced by comparative genome analysis. The rationale is simple. If species A has a protein with two domains (called 1 and 2), and species B has orthologous domains (called proteins 1' and 2'), then proteins 1' and 2' probably interact to generate the same function as found in species A.

Using this approach, various groups have predicted over 100,000 protein-protein interactions in yeast, and thousands more in *E. coli* and other prokaryotes. As more eukaryotes have their complete genomes sequenced, this type of computer-based research and model building will continue.

A lot of useful proteomics information remains hidden in the published genome sequences. An in silico analysis of genomes provides us with reasonable models that can be tested in the lab, but is limited to proteins encoded by a single gene in one species and by two or more genes in another. Surely there must be protein interactions that do not fit these criteria, and we need a way to determine them as well. Sequence analysis is insufficient, so a more direct high-throughput method is needed.

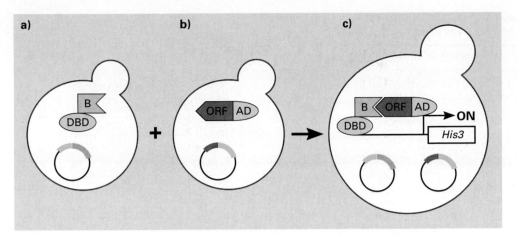

FIGURE 6.14 • **Yeast two-hybrid method. a)** The DNA binding domain (DBD) is fused onto the protein of interest "B." This construct, the "bait," is encoded by the plasmid shown in the same cell. **b)** The "prey" is encoded by its own plasmid and is composed of an ORF fused onto the activation domain (AD), which is capable of activating RNA polymerase. **c)** Both the bait and prey plasmids are inside the same cell, and if the B and ORF protein physically interact, the RNA polymerase will be able to transcribe a reporter gene, in this case *His3*.

How Can We Measure Protein Interactions?

In 1991, Stan Fields (currently at the University of Washington, Seattle, but previously at the State University of New York, Stony Brook) developed a new proteome-wide method to detect protein-protein interactions. It is worth noting that the method, called **yeast two-hybrid (Y2H),** was developed before the term proteomics was coined and thus it was way ahead of its time. In short, the Y2H method is designed to use a protein of interest as *bait* in order to discover proteins that physically interact with the bait protein; these proteins are called *prey* (Figure 6.14). Y2H experiments have been conducted in many labs to discover proteins that interact with each lab's favorite bait protein. In fact, Y2H has become a mini-industry as illustrated by Fields's web pages dedicated to Y2H.

Here's how the Y2H method works. A single transcription factor is cut into two pieces called the *DNA Binding Domain* (DBD) and the *Activation Domain* (AD), which stimulates the RNA polymerase to begin transcription. Fused to the DBD is the bait protein of interest (B), which cannot initiate transcription on its own. Fused to the AD is the prey ORF, which can be any known or unknown protein. The prey protein of AD + ORF fused together cannot initiate transcription either. When the bait and prey proteins are produced in the same cell, they might interact; if they do, transcription of the *His3* gene is initiated. As shown in Figure 6.14, the reporter gene is *His3*, which leads to the production of the amino acid histidine. Without *His3*, cells cannot grow unless histidine is added to the growth medium. Any ORF can be tested with Y2H, which means a

proteome-wide survey can be performed rapidly by transforming a genomic library into cells that contain bait plasmids. In this way, every protein in a proteome can be tested individually for its potential to interact with the bait. You can watch a QT movie that illustrates a variation of Y2H in which interaction is signaled by the production of a blue color rather than histidine synthesis. The blue color is produced by the enzyme β-galactosidase, which is encoded in the *lacZ* reporter gene.

METHODS
Y2H
lacZ Reporter
LINKS
Stan Fields
Mini-industry
His3

DISCOVERY QUESTIONS

35. The protein interactions detected by the Y2H method must take place inside the nucleus of the yeast cell. Why might this prove problematic?

36. What controls would you want to run to make sure the only positive cells are those with interacting proteins? In other words, how could you rule out that you had simply recloned the *His3* gene?

37. Since Y2H takes place inside yeast cells, is it restricted to yeast proteins only?

The Y2H method has been very popular and successful but not perfect. For example, what would the phenotype be of a yeast cell if the prey ORF were the *His3* gene? Of course, it would grow in the absence of interaction with the bait. As

TABLE 6.1 • The set of prey where Rpc19 was used as the bait protein.
Interactions are ranked according to plausibility.

Gene	ORF	Description (from the YPD Database)
Rpc40	YPR110C	Shares subunit of RNA polymerases I and III (AC40)
	YLR266C	Protein with similarity to transcription factors, has Zn[2]-Cys[6] fungal-type binuclear cluster domain in the N-terminal region
Mtr2	YKL186C	Involved in mRNA transport, has similarity to *E. coli mbeA*
	YOL070C	Protein of unknown function
	YFR011C	Protein of unknown function
	YHL018W	Protein with weak similarity to human pterin-4-alpha-carbinolamine dehydratase
Gtr1	YML121W	GTP-binding protein involved in the function of the Pho84p phosphate transporter
Mas1	YLR163C	Beta (enhancing) subunit of mitochondrial processing peptidase

Color Codes
Previously known interacting protein (positive control)
Possible interaction partner
Significance unknown
Unlikely interaction partner

Source: http://depts.Washington.edu/%7Eyeastrc/th_8.htm (Stan Fields's web site)

LINKS

Y2H Results

Interaction
Sequence Tags

Additional Y2H
Results

a control experiment, cells containing only prey must be tested on media lacking histidine. If a prey-only cell grows, interaction is not necessary for growth, though protein interaction between bait and prey cannot be completely ruled out. Also, not all proteins work well inside the nucleus, so there may be some false positive or false negative results due to improper protein folding (i.e., 3D structure; Table 6.1). However, the greatest benefit of Y2H is that yeast cells can express genes from almost any species, which means this is a powerful proteomics method for *Drosophila, C. elegans, Arabidopsis,* zebra fish, mice, humans, and of course yeast, too.

DISCOVERY QUESTIONS

38. Go to the web site that lists many of the Y2H results from the Fields lab. Do a find function on your browser for RPC19. You will see that RPC19 bait protein bound to four prey proteins. Look at the next bait entry called RPC40. What do you see as one of its prey?

    Compare RPC40 prey to the list of prey proteins when RPC19 was used as bait. Do you perceive any problems? What would you have expected to see? Hypothesize how this result could happen.

39. Go to Interaction Sequence Tags and you will see a small table of *C. elegans* proteins that have been tested by Y2H. Click on these entries (list below) and compare the results, but be aware that this database is a work in progress:

    β-Catenin
    presenilin (associated with early onset of Alzheimer's)
    MAP Kinase
    MAP Kinase Kinase

    a. Why might some proteins make bad bait?
    b. Are positive results always reproducible? What impact do the *C. elegans* data have on the Y2H method?
    c. Based on the results, do you think MAP Kinase or MAP Kinase Kinase is at the center of protein-protein interactions?

40. Fields has continued to push the envelope by systematically exploring the protein interactions in the yeast proteome. Go to the Y2H page that lists all the results, about 1,000 protein-protein interactions.

    a. Search for *His3* and see what you find. Any surprises?
    b. Now try the Additional Y2H Results web page. Search for Swi. The bait "SWI5" is in the first column, the prey gene is in the middle

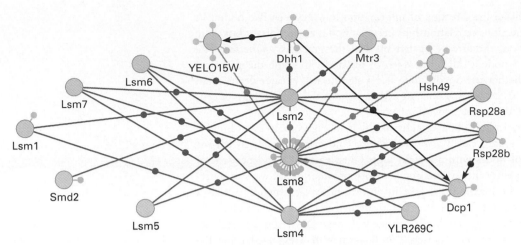

FIGURE 6.15 • **Proteomics circuit showing the interactions of RNA splicing proteins.** The proteins are indicated by large blue nodes and their interactions with lines. The dots on the lines help you follow each line. Blue lines show interactions that were detected by traditional Y2H array screens, black from multiple high-throughput screens, and gray from literature and array screens. The black arrows point away from the protein used as bait in the screens. Small gray nodes indicate other protein-protein interactions not highlighted here.

column, and the ORF name of this gene is in the third column.

Go to MIPS search page and find out the function for *Swi5* (search with YDR146c) and its prey with a known function (Mei5; search with YPL121c). Does this look like a probable interaction or artifact? Support your answer with information from the MIPS database.

In a rather large collaboration, Stan Fields and Peter Uetz at the University of Washington, Seattle, and CuraGen Corporation of New Haven, Connecticut, compiled a database of protein interactions. Many new interactions were detected which may lead to new insights about integrated circuits within the cell (Figure 6.15). When the analysis was completed, they compiled the putative interactions into the cellular roles. Since many proteins have no assigned role and the number of protein-protein interactions is very small, we have a very limited understanding of how the proteome functions.

A group from UCLA led by David Eisenberg has created a very rich database they call Database of Interacting Proteins (DIP). Using this database, they made some interesting predictions about the yeast prion-like protein Sup35 (Figure 6.16, page 181). Although the figure is too crowded to see all the subtleties, it is clear that Sup35 is in the center of an integrated circuit of protein-protein interactions. What is especially intriguing is that in addition to translation, as you would expect for a translation termination factor such as Sup35, the circuitry includes proteins involved in many other processes. This predicted interaction network can be tested experimentally, which is one reason for making such complex models. It also furthers the speculation that Sup35, and the mammalian orthologs, may have additional functions that are much more integrated into cellular circuitry than we realized. As research continues, our understanding of protein interactions will become increasingly more complex and interconnected.

MATH MINUTE 6.2 • IS Sup35 A CENTRAL PROTEIN IN THE NETWORK?

In Figure 6.16, Sup35 appears to be central to the network of interacting proteins. However, looks can be deceiving, particularly in such a complex network. We would like to quantify whether Sup35 interacts with an unusually large number of proteins, compared to other proteins in the network.

A mathematician would call the network of interacting proteins in Figure 6.16 a *graph*. In graph theory, the lines between nodes are called *arcs* or *edges*. A *directed* graph has arrows on the edges to indicate the direction of information flow between two nodes. In Figure 6.16 where the connections or relationships between nodes are important but there is no directional information, the arrows are left off and the graph is *undirected*. The number of edges touching a node is called the *degree* of the node. In graph theory terms, the question at hand is whether the Sup35 node has a significantly greater degree than is "typical" for this graph. If so, investigators are led to believe that Sup35 plays a central role in the function of the entire network of proteins.

One way to approach questions about graphs like that in Figure 6.16 is to build a probabilistic model of the graph, called a *random graph*. Under this model, an edge is drawn between each pair of nodes with a certain probability, each edge independent of the others. Advanced probability theory and graph theory provide precise ways to determine the probability that the maximum degree in an arbitrarily large graph exceeds a given number. However, under the simplifying assumption that the degree numbers are approximately normally distributed (see Math Minute 8.1, page 234), you can evaluate whether the degree of the Sup35 node is significantly greater than expected using the mean and standard deviation of the degree numbers. Since Figure 6.16 is too complex to illustrate this process, consider the following smaller network:

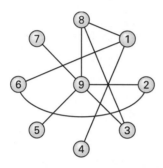

In this graph, nodes 4, 5, and 7 have degree 1; nodes 2, 3, and 6 have degree 2; nodes 1 and 8 have degree 3; and node 9 has degree 5. The mean degree is the average of the nine degree numbers (2.22). The standard deviation of the nine degree numbers (1.23) is used to determine if the degree of node 9 is unusual. Specifically, if node 9's degree is more than two standard deviations larger than the mean degree, it can be considered unusually large. Since $5 > 4.68 = 2.22 + 2 \times 1.23$, we conclude that node 9 has an unusually large degree. This same procedure could be applied to the graph in Figure 6.16 to help quantify whether Sup35 plays a central role in this network of proteins.

DISCOVERY QUESTIONS

41. There are several proteins with the name Cct and a number that appear in Figure 6.16. Go to search page, Genome Database, enter "Cct5," and find out what it does.

42. Now go to Eisenberg's DIP database (Javascript, use Internet Explorer) of protein interactions (requires registration, free to academic users). Enter "Cct5" in the name field and click "Query DIP" button (don't hit the return key). On the results page, click on the link under the heading "DIP Node." Then click on the word "graph" in the top right corner and you will see Cct5 as the red node. Can you find Sup35 interacting with Cct5? (Click on a few nodes and read the new window that appears.)

43. Enter "Sup35" in the search field and what do you find? Now try "Sup45" and see what role this protein plays. Click on the few nodes that directly interact with Sup45. Can you find Sup35 now? The thicker lines indicate interactions with more confidence. What problem do you see with the DIP database at this stage of its development?

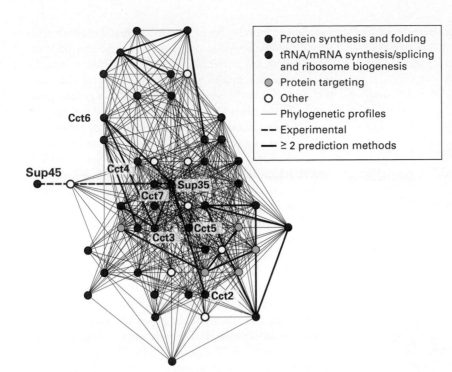

- ● Protein synthesis and folding
- ● tRNA/mRNA synthesis/splicing and ribosome biogenesis
- ◉ Protein targeting
- ○ Other
- — Phylogenetic profiles
- -- Experimental
- ━ ≥ 2 prediction methods

FIGURE 6.16 • Network of protein interactions with higher (bold lines) and lower (thin lines) confidence. This network is centered on the yeast prion protein Sup35. Many of the proteins in this network are involved in protein synthesis, including ribosomes, protein folding, sorting, modification, and targeting. To construct this model, proteins were treated as nodes and the edges as springs in order to position functionally related proteins close to each other.

Is It Possible to Understand Proteome-wide Interactions?

By now, you must realize that proteomes are very complex networks of more proteins than there are genes, and there are many useful methods to help us understand functional roles for each interaction. The goal is to synthesize all this information into an integrated protein circuit diagram that shows which proteins interact with each other.

Not many people would care to spend countless hours sifting through all the different databases to synthesize an interaction map for yeast. But Fields and Uetz have led the way in collaboration with Benno Schwikowski, a computer scientist at the Institute for Systems Biology. The three investigators collated 2,709 interactions among 2,039 different proteins. When they integrated these protein connections, they synthesized integrated protein circuit diagrams containing 2,358 interactions among 1,548 proteins in the yeast proteome. The figures are too large to reproduce here, so we will look at the PDF files instead.

Open the PDF file called "Benno Figure 1," which is from their paper. Here is the color key for the *lines:*

- red: cellular role and subcellular localization of interacting proteins are identical;
- blue: localizations are identical but functions differ
- green: cellular roles are identical but localizations differ
- black: cellular roles and localizations differ
- gray: localizations or cellular roles are unknown for one or both proteins

Here is the color code for the *boxes:*

- blue: membrane fusion
- gray: chromatin structure
- green: cell structure
- yellow: lipid metabolism
- red: cytokinesis
- light yellow: unknown role

SEQUENCES

Benno Figure 1

LINKS

Benno
Schwikowski

SGD

When the PDF file opens, zoom in to 800% magnification so you can read the names of the proteins. Use the find tool and enter Pex7. This protein is involved in lipid metabolism, as indicated by its yellow box. It is connected to several other proteins known to support lipid metabolism. Pex7 is also directly connected to two other Pex proteins. Go to SGD search page and determine the function of Pex7 and the other two Pex proteins of your choice. When you get the first results page, be sure to click on the link under the heading of "yeast hits." This will take you to a table that lists cellular roles. Does SGD agree with the categorization of these proteins as lipid synthesis proteins? Does this interaction circuit diagram portray at least part of the circuitry involved in the process described in SGD?

The purpose of a good model is not to confirm what you already know but to provide new insights. In the same figure, just above Pex7 is a cluster of proteins with no known function (e.g., YLR095c). The model predicts that YLR095c is a central player in a circuit that includes many proteins with known functions. Go to MIPS and determine the roles of Ami3 (YOL060c), Pup3 (YER094c), and Ilv6 (YCL009c). With this information, can you hypothesize the function of YLR095C and the other proteins in this circuit? What experiment could you design to test your hypothesis?

DISCOVERY QUESTIONS

44. In Benno Figure 1, locate Akr2. This protein is known to play a role in endocytosis. Can you tell which proteins are connected to Akr2 in this figure? You may need to zoom in further to see its connections.

45. The interaction network is so comprehensive that it is difficult to see individual connections. For example, Akr2 is connected to YHR105W, but you may not have been able to determine this. The investigators produced a separate figure to show Akr2's connections to a subset of the full proteome circuitry:

```
   Ypt1            Akr2           Yip1
    |               |              |
YPL246C ——— YHR105W ——— YGL161C
    |                           |
  Vam7                        Pep12
```

This separate figure is readable and makes clear one weakness of the full interaction diagram.

Describe at least one improvement you would like to see in their full protein interaction circuit diagram.

46. Open the following PDF files via their links on the Web site and use the find function to locate the prion protein "SUP35." For each file, see if Sup35 was given a particular function. Be sure to notice the color key in the upper left-hand corner of each figure before you zoom in to 800 or 1,600%.

   aging.pdf

   membrane.pdf

   degradation.pdf

   Go back to the aging file. Notice that Sup35 is hidden behind another protein called Kcs1. What is Kcs1's cellular role according to the aging.pdf network? Compare that with the role given in SGD. Its location on the network hints that it may interact with Sup35, but from this map we cannot be sure. Might Sup35 interact directly with Kcs1, based on their cellular roles?

As with microarrays, the power of a complex circuit diagram may be more in its overview than in individual nodes. For example, Schwikowski tested the predicted function of proteins based on their interactions with other previously characterized proteins and found that their circuit diagram was correct 72% of the time. Of course, this means that the model was incorrect 28% of the time, which reinforces the need to use this map as a general guide and a starting place for testing predictions. When they examined interactions between functional roles rather than individual proteins, they found a few surprises (see Benno Figure 2). On this map, locate the RNA polymerase II (pol II, which transcribes most genes) cluster. Prior to this study, you probably never would have predicted that transcription of genes interacts with so many other cellular roles. It is worth noting that Sup35 was included in protein synthesis, which interacts (directly or indirectly) with the pol II group. From Discovery Question 46, you learned that Kcs1 was involved with pol II transcription and was located very near Sup35 in the network. Perhaps protein synthesis and pol II groups interact directly. The Sup35 and Kcs1 example illustrate the power of a good model—it allows you to examine data in new ways and make predictions that can be tested experimentally. Data from these experiments can be used to refine your model and further your insights and understanding.

When Schwikowski redefined each protein by its cellular localization rather than its function (Figure 6.17), there were a couple of surprises. For example, there are 8 interacting pairs between the nucleus and the Golgi body and 27 interacting pairs between the nucleus and mitochondria.

DISCOVERY QUESTIONS

47. Does the cytoskeleton interact only with the plasma membrane, the nucleus, and the cytoplasm? Why does this look too simplistic given what we know about vesicle and organelle movement?

48. There are three obvious examples of missing connections between two closely associated organelles. Identify as many missing interactions as you can.

49. Why are obvious interactions missing from this network?

Schwikowski's team undertook a gigantic task and produced a circuit diagram of many interactions that will help us better understand proteomes. Of course, their work is not complete since the yeast proteome contains more than 6,000 proteins and this version of the interaction circuit contains only 1,548 proteins. Nevertheless, the first phase of their work provides us with new perspectives. It is a huge step forward in understanding the yeast proteome and by extension the proteomes of many other eukaryotes.

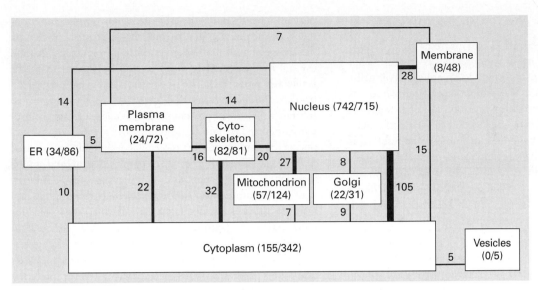

FIGURE 6.17 • Protein interactions grouped by cellular compartments. Numbers in parentheses indicate the number of interactions in the circuit diagram among proteins of this compartment/the total number of proteins in the same compartment. Lines connecting compartments indicate the number of protein interactions by the thickness of each line (numbers of connections are near each line). For example, there are 7 interactions between the 48 membrane proteins and the 72 plasma membrane proteins in "Benno Figure 1."

At this point, you should be impressed with what we can learn about protein interactions, though they may never occur in a living cell. For example, although protein A and B can interact in a Y2H experiment, they may never be present at the same time inside a *wt* cell. This leaves us with a problem that needs to be addressed: What proteins are present at any given time?

Summary 6.3

As noted in the introduction to this chapter, the term "postgenomic era" does not imply that whole genome sequence analysis is completed. By reanalyzing genome sequences, we can predict which proteins are likely to interact with each other, and these predictions can be verified experimentally. Yeast two-hybrid experiments allow us to screen large numbers of proteins or protein fragments to determine which proteins bind to each other. However, the Y2H has its limitations, which, as you will see on pages 195–198, **protein microarrays** might be able to offset. As we continue to accumulate new types of data in large quantities, it becomes more difficult to display them in a meaningful and comprehensible way. Protein interaction circuits are beneficial but currently limited by the 2D constraints of a printed page. New visualization methods are being developed to address this limitation. The last piece of the proteomics puzzle, addressed in Section 6.4, is listing and quantifying all the members of a given proteome.

6.4 Measuring Proteins

It is important to understand what proteins can do, with whom they interact, and their 3D shape. However, this information is meaningless taken out of context of a dynamic proteome. For example, let's imagine you knew the person who pumped up the basketballs used by Michael Jordan. You would know what both people do, where they perform their jobs, and that they interact via the basketball. However, they cannot play basketball together since they are never on the court at the same time. This simplistic example illustrates how important it is to distinguish between all possible interactions in all proteomes versus interactions that can actually occur because both proteins are expressed in a cell at the same time. Furthermore, it would be very instructive to know how many copies of each protein were present in a given proteome, just as it would to know if any basketball team had more than one Michael Jordan.

This section examines the identification and quantification of proteins. DNA microarrays have proven so successful that many investigators are developing protein microarrays. We will read about three different approaches to creating them. Finally, we consider single-cell proteomics and experiments that study one molecule at a time, which can affect the metabolome, the entire metabolism of an organism.

How Do We Know Which Proteins Are Present?

An accurate list of proteins present in a proteome at various times is critical. What proteins do brain cells contain? liver cells? muscles? What about a yeast cell while it is switching from one food source to another? completing each step of the cell cycle? mating? These questions need to be resolved if proteomics is going to have a significant impact. A primary objective of those working in this field is to identify and quantify every protein in a cell at a given time (i.e., define a proteome). Until recently, the best method for defining a proteome has been 2D electrophoresis.

2D Gels The proteomics workhorse since the mid-1970s has been **two-dimensional (2D) gel electrophoresis.** As shown in Figure 6.18, 2D gels separate proteins based on their isoelectric points (first dimension) and **molecular weights** (second dimension). Like any molecule, the net charge of each protein depends upon the pH of the surrounding environment. As the pH changes, the net charge changes. When the net charge of a protein is zero and the protein stops moving, the pH of the local environment will be equivalent to the **isoelectric point (pI)**. The isoelectric point of a protein is dependent upon its amino acid sequence, which allows us to separate a mixture of proteins based on charge. Once the proteins are sorted by isoelectric point, they are placed on a second gel and separated by molecular weight using standard SDS-PAGE. The end product is a pattern of spots and smears that defines each protein in a complex proteome. Unfortunately, some proteins do not resolve well by this method and thus they often escape detection. In particular, very basic proteins, very small peptides (Figure 6.18f), and proteins that are difficult to solubilize often escape detection by 2D gels.

Although 2D gels separate many proteins, this method's limitations frustrate investigators trying to define proteomes. Sample preparation is not a trivial component of the method, but this aspect depends in large part upon which proteome you want to study. The most troublesome issues

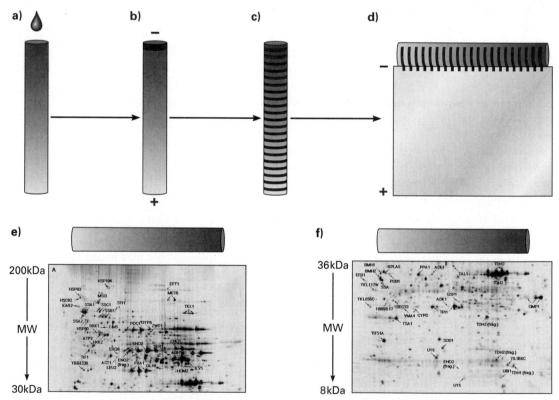

FIGURE 6.18 • **Two-dimensional (2D) gel electrophoresis.** Each column (a–c) with a gradation of gray shading represents the isoelectric focusing gel with a pH gradient. **a)** A mixture of proteins (blue drop) is applied to the isoelectric focusing gel and **b)** exposed to an electrical current. **c)** Proteins migrate to their isoelectric points (pI) and stop moving. **d)** This tubular gel is placed on top of a slab polyacrylamide gel that contains SDS and is subjected to electrophoresis (SDS-PAGE). Proteins migrate into the slab gel according to their molecular weights. Yeast cells were grown in rich media and subjected to 2D gel analysis. Using duplicate isoelectric focusing gels, large **e)** and small **f)** proteins were analyzed on separate gels. The same spots appear at the bottom of e) and the top of f). Molecular weights are resolved on the Y-axis and pIs on the X-axis. Panels e) and f) are from the Swiss 2D database at ExPASy.

are (1) detection of spots, (2) quantification of each spot, and (3) identification of each spot.

Spot Detection As you can imagine, no single method of labeling proteins is going to work for every possible protein. Silver and Coomassie staining are not sensitive enough to detect every protein, though they work well for the most abundant ones. Radioactivity can be used for some samples but is less successful with tissue biopsies taken from patients. Ideally, fluorescent dyes will be invented that bind to all proteins without altering protein mobility in 2D gels. As yet, no such dye is available. If you have an antibody to your favorite protein, then **Western blotting** of 2D gels can detect as few as 100 molecules per cell. However, Western blotting is labor intensive, limited by the availability of antibodies, and not suited for high-throughput proteomic analysis.

Quantification Figure 6.18 illustrates why quantification of each spot is difficult since some proteins smear, overlap other proteins, and are present in vastly different amounts than others. Fluorescent dyes hold the best hope for solving this problem as well.

Identification At the end of 2D gel analysis, there are several ways to identify each spot. Initially, investigators tried to create 2D "fingerprints" to compare one pattern against another, but this proved nearly impossible since 2D patterns were difficult to reproduce. The next approach was to cut out each spot and subject them to amino acid sequencing. This technique works better with the advent of improved methods for protein sequencing. **Mass spectrometry (MS)** is the preferred tool for determining the identity of the proteins. We will discuss MS in detail since it will be important for 2D gels as well as newer methods described later in this section.

Mass Spectrometric Identification of Proteins There are two key components to MS. The first is that all proteins can be sorted based on a **mass to charge ratio** (i.e., the molecular weight divided by the charge on this protein). The second is that proteins can be broken into peptide fragments, facilitating the identification of each protein. We will examine the four steps needed to identify each protein. Steps 2 and 4 utilize separate mass spectrometers and so the entire process is often referred to as tandem MS, or MS/MS.

Step 1: Ionization A protein sample is injected into a device that must ionize the proteins. Two methods are popular: matrix-assisted laser desorption/ionization (**MALDI**), and electrospray ionization (**ESI**). Both methods produce mixtures of proteins that have become charged, or *ionized*. In MALDI, a laser illuminates the protein mixture and the absorption of energy ionizes the proteins, which are sent on to a mass spectrometer (MS). In ESI, a strong electrical charge is imparted to the solvent containing the protein, after which

the solvent evaporates, leaving the proteins ionized (Figure 6.19a). In both MALDI and ESI, ionized proteins enter an MS.

METHODS

Western Blotting

Mass Spectrometry

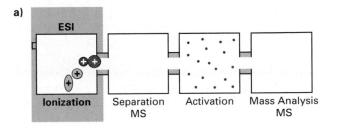
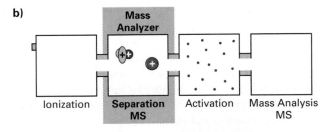
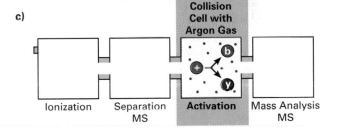
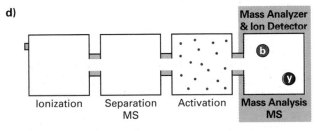

FIGURE 6.19 • Tandem mass spectrometry for protein identification. a) ESI creates ionized proteins, represented by the colored shapes with positive charges. Each shape represents many copies of identical proteins. **b)** The ionized proteins are separated based on their mass to charge ratio (m/z) and sent one at a time into the activation chamber. Separation and selection takes place in the first of the two MS devices. The solid blue protein has been selected for analysis while the other three are temporarily stored for later analysis. **c)** The group of m/z selected ionized proteins enters the collision cell that is filled with inert argon gas. The gas molecules collide with the proteins, which causes them to break into two peptide pieces (labeled b and y). **d)** Ionized peptide pieces are sent into the second MS device which again measures the m/z ratio. It is connected to a computer that compares this spectrum of peptide pieces to a database of ideal spectra to identify the original group of identical proteins.

LINKS

PROWL

Fat 2D Gel

ExPASy 2D

Step 2: Separation Once the proteins are ionized, their mass and charge must be determined (Figure 6.19b). This is a two-step process; first the ions are collected, and then the time it takes individual groups of identical proteins to move to the detector is measured. This timing determines mass (m) to charge (z) ratio (abbreviated as **m/z**). Detection of m/z is developing rapidly, but we do not need to focus on the details. Once the m/z ratios are determined for each protein, a computer allows identical proteins (defined by m/z ratios) from a single sample to be selected and sent to the activation chamber. This process is automated so proteins of different m/z ratios are sent to the activation chamber in quick succession.

Step 3: Activation After being selected by m/z ratio, proteins must be broken into smaller fragments (Figure 6.19c). This is accomplished by sending each protein into an argon-filled chamber. This inert gas collides with the ionized protein, and the vibrational energy of the gas causes the protein to break into two pieces. Each piece of the original protein, one containing the amino portion (labeled b) of the original protein and the other containing the carboxyl portion (labeled y), will carry a charge. Since each group of identical proteins can break in many places, more than one combination of peptide pieces will be produced. Bigger proteins can produce more combinations of peptide pieces since they can be broken anywhere along the protein's length.

Step 4: Mass Determination This is the second MS of the series, and from it the term **tandem MS** was derived since there are two adjacent MSs (Figure 6.19d). The final step is to determine the m/z ratios for the mixture of ionized protein pieces. Because there will be more than one fragment pair for each population of proteins (Figure 6.20a), the output of the second MS can be confusing (Figure 6.20b). However, a computer compares the spectrum of peptide pairs with ideal spectra for all known proteins in the database. The ideal spectrum with the best match identifies the original protein. This process is repeated rapidly (within seconds) so that in one complex sample, hundreds of proteins can be identified.

Now it may be a bit more obvious why sample preparation is critical. If you took whole cell extracts and put them into an MS/MS system, there would be too many proteins to sort and identify. One solution is to separate proteins into functional domains (e.g., nucleus, plasma membrane, etc.). In addition, you could run a 2D gel and then cut out spots to identify. In the end, 2D gels are labor intensive and cannot quantify how much of each identified protein was

present in a particular sample. We will return to this quantification problem later.

DISCOVERY QUESTIONS

50. Go to the PROWL database, where you can search a large range of species.
    a. Look under Metazoan (multicellular organisms), Chordata (animals with backbones), Mammalia (mammals), Rodentia (rodents), and select *Mus musculus.* In the "Enter Key Words" box, enter "ob" and hit the "Search Keywords" button.
    b. Click on the link to the left of "LEPTIN [MUS MUSCULUS]."
    c. Click on the link called "sequence," which will take you to a page with the mouse leptin protein sequence. At the bottom, you can choose a number of calculations. You want to execute the "Calculate Mass" function. What are the molecular mass and pI of mouse leptin? Write down this information.

51. Go to ExPASy Mouse White Fat 2D Gel and you will see a 2D gel of white adipose tissue from mouse. Look in the area where you expect to see leptin. Click on some of the red plus signs in this area to see if leptin has been identified in this gel.

52. You have just gone through the steps investigators perform in order to make an educated guess about which spot is their favorite protein. Based on your findings for Discovery Question 51, what needs to happen next in order to improve the PROWL database?

53. Search the ExPASy 2D database to find your favorite protein by entering its name. As a starting place, type in "IDH" and hit "submit." You will see that IDH has been identified in mouse liver and *E. coli.* Click on either one and scroll to the bottom of the results. There are three links you should visit, in this order:
    a. "Compute the theoretical pI/Mw"—You will get a long list. At the bottom, click on "submit" and you will have the information you need.
    b. "How to interpret a protein map"—You will see the kind of information you will get in the next step. This is a "help" page.
    c. Under "MapLocations:", click on any of the SPOT links to see where IDH is located. Compare this with the predicted pI and molecular weights. How did the prediction compare with the real gel?

a)

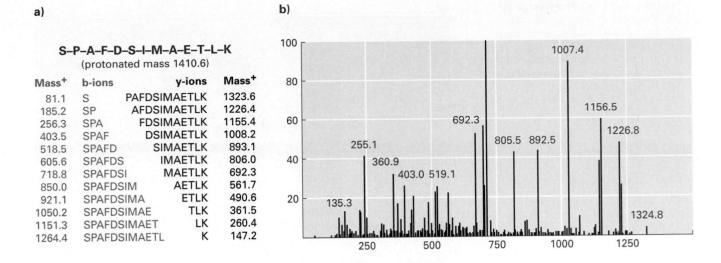

Mass⁺	b-ions	y-ions	Mass⁺

S–P–A–F–D–S–I–M–A–E–T–L–K
(protonated mass 1410.6)

Mass⁺	b-ions	y-ions	Mass⁺
81.1	S	PAFDSIMAETLK	1323.6
185.2	SP	AFDSIMAETLK	1226.4
256.3	SPA	FDSIMAETLK	1155.4
403.5	SPAF	DSIMAETLK	1008.2
518.5	SPAFD	SIMAETLK	893.1
605.6	SPAFDS	IMAETLK	806.0
718.8	SPAFDSI	MAETLK	692.3
850.0	SPAFDSIM	AETLK	561.7
921.1	SPAFDSIMA	ETLK	490.6
1050.2	SPAFDSIMAE	TLK	361.5
1151.3	SPAFDSIMAET	LK	260.4
1264.4	SPAFDSIMAETL	K	147.2

FIGURE 6.20 • Protein identification through peptide fragment formation and separation. When a group of identical proteins is broken into its peptide pieces, more than one pair of b and y peptides will be formed. **a)** One protein sequence and its calculated mass on top, with the b peptides/masses (gray) and the y peptides/masses (blue) below. **b)** An experimentally determined mass/charge spectrum from the peptide in a). Notice that some peaks are higher than others, which means that some b/y peptide pieces were more abundant than others. The spectrum is used to determine the peptide's amino acid sequence and protein identity.

What Proteins Do Our White Blood Cells Need to Kill a Pathogen?

Our immune system has many ways to kill pathogens. One way is to engulf the invading pathogen by phagocytosis and destroy it. Phagocytosis is a part of our innate immune system that we share with other warm-blooded animals and even nonvertebrates such as insects and sea urchins. Therefore, the process of phagocytosis evolved millions of years ago, and the proteins involved must be well conserved. The challenge has been to identify all the proteins involved. A French and Canadian collaborative effort was able to define the proteome of the phagocytotic vesicles called **phagosomes.**

Michel Desjardins led the group at the University of Montreal to define the proteome of the phagosome. The first challenge was sample preparation, since they did not want to define the complete proteome of the phagocyte. To do this, they "fed" phagocytes latex beads that could be engulfed by the cells. The cells were lysed, and the density of the latex beads allowed phagosomes to be separated from all the other cellular compartments. From this starting point, Desjardins's group used 2D gels to identify over 160 proteins associated with phagosomes (Figure 6.21, page 188). They identified most of the major and minor spots, though a few proteins were not visible on the 2D gel. They cut each spot out of the gel and identified the proteins using MALDI-MS/MS.

LINKS

Michel Desjardins

As we have discussed, some proteins are not easily resolved by this method, especially membrane-associated proteins. Therefore, Desjardins performed additional analysis that used a detergent (Triton X-114) and SDS-PAGE. Using this combination, 31 additional proteins were identified that had not been resolved by the previous 2D gel. Altogether, the team amassed a long list of proteins associated with phagosomes; some of the proteins had not been associated with phagosomes prior to their research.

Their accomplishment was substantial, but the group also wanted to determine which proteins were inside the phagosome (i.e., luminal side) and which were outside (i.e., cytoplasmic side). To determine which side, they incubated purified phagosomes with a **protease** that digests proteins. If the proteins were inside the phagosome, the protease would not digest them and they would still be visible on a 2D gel. If they were on the cytoplasmic side, then protease treatment would digest them and their spots would be missing or greatly reduced. As you can see from Figure 6.22, page 188, this method worked very well and helped refine the cellular location of these proteins.

Having established which proteins are associated with phagosomes and which are on the inside or outside, Desjardins and his collaborators studied one more component of phagosomes—development. We have discussed that proteomes can change over time. In this case, the proteome changes as the

FIGURE 6.21 • Phagosome 2D gel. Phagosomes were isolated from macrophages and subjected to isoelectric focusing (IEF) and SDS-PAGE. The pH gradient is presented above the gel and the molecular weight standards to the left. This gel was silver stained to locate the spots, which were excised and identified by tandem MS.

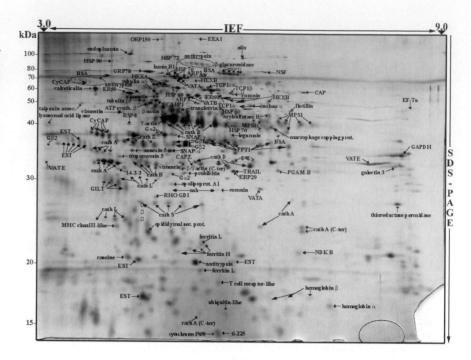

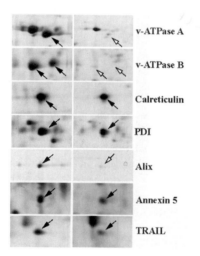

FIGURE 6.22 • Localization of phagosome proteins. Phagosomes were subjected to protease treatment to digest any proteins on the cytoplasmic side of the phagosome. Untreated (left) and protease-treated (right) phagosomes were compared on 2D gels. Proteins with substantial cytosolic portions are indicated with open arrows, while those resistant to protease are indicated with closed arrows.

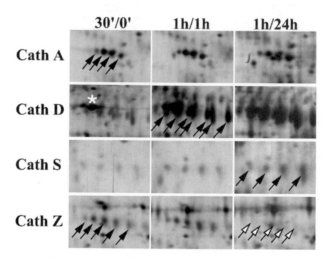

FIGURE 6.23 • Cathepsin composition during phagosome maturation. Phagosomes from macrophages were isolated after 30 minutes of exposure to latex beads (30'/0'), after 1 hour of incubation with latex beads followed by 1 hour of further development (1h/1h), or 1 hour of latex beads followed by 24 hours of further development (1h/24h). The cathepsins are labeled on the left, with loss of signal indicated by the open arrows. Multiple arrows point to the multiple spots for each cathepsin. The star denotes a precursor form of cathepsin D.

phagosome develops into a mature pathogen-killing organelle. The investigators were particularly interested in the accumulation of a family of digestive enzymes called cathepsins. From their data in Figure 6.23, you can see some proteases accumulate over time (e.g., cathepsins D and S), others are reduced (cathepsin Z), and some are relatively constant (cathepsin A).

Among their findings were a few surprises. For example, proteins normally thought of as endoplasmic reticulum

proteins were associated with phagosomes. Phagosome proteins normally associated with **apoptosis** and signal transduction also surprised the investigators. In their list of phagosome proteins are 17 with unknown roles and 7 **expressed sequence tags (ESTs),** providing us a clue to their cellular roles. The group created a "virtual phagosome" to see all the parts of its proteome, and which proteins might interact with each other. Many other labs will use the virtual phagosome information as a starting place to understand how some pathogens are killed by phagocytes and others escape destruction.

At this point, we have studied several ways to define a proteome. What proteins are present? What does each protein do? With which partners does each protein interact? But two problems still exist that are critical to understanding a complete proteome. How much of each protein is present? Can we measure protein modifications such as phosphorylation?

How Much of Each Protein is Present?

Defining the components of a proteome is a doable, though difficult, task. We can divide and conquer cells and tissues and gradually assemble the full contents of complex proteomes. But quantifying changes in abundance for each protein within a proteome has been substantially more difficult. Furthermore, we would like to know when a protein has been modified, which may be more significant than how many copies of a protein a cell contains. For example, some proteins must be phosphorylated in order to become activated; measuring such posttranslational modifications is a critical aspect to understanding a dynamic proteome. Though the task is difficult, a few groups are trying to devise new technologies to address the quantity and modification issues.

A Quantification Solution for Cultured Cells In Brian Chait's lab at Rockefeller University in New York, a team of biologists, chemists, and engineers has devised a new approach to protein quantification and modification problems. At the heart of their method is the use of stable isotopes to label proteins as they are formed. **Stable isotopes** are nonradioactive versions of atomic elements that vary in their atomic mass. A given molecule made with a heavy isotope will weigh more than the same molecule made with a lighter isotope. Using this principle, Chait's group developed a protocol (Figure 6.24) that allowed them to compare the quantities of heavy and light proteins in each spot on a 2D gel, followed by MALDI-MS/MS to identify each protein.

Two cell populations were grown under different conditions, similar to the way DNA microarray experiments are performed. One population of cells was grown in the presence of normal nitrogen (99.6% ^{14}N and 0.4% ^{15}N), while the other was grown in a medium enriched with heavy nitrogen (> 96% ^{15}N). When it was time to harvest the cells, they were combined into one mixture to ensure that all subsequent steps were performed equally to all proteins, assuring that changes in relative amounts of each protein were not due to differences in handling of the samples. The mixed samples were separated by 2D gels, from which spots were excised and subjected to sequence identification by MALDI-

MS/MS. Heavy and light versions of a particular protein would comigrate to the same spot on a 2D gel. However, unlike

LINKS

Brian Chait

traditional 2D gel samples, each spot separated into two m/z peaks, because the ionized peptide fragments were separated into heavy (^{15}N) and light (^{14}N) versions of the same molecule. In Figure 6.24, proteins A and B were present in equal amounts in the two cell populations, but protein C was more abundant in cell population 2 than in cell population 1. Relative protein abundance is indicated by the height of the pairs of peaks (gray and blue) at different m/z ratios. If someone accidentally spilled half of the sample along the way, the relative amounts for all the proteins in the mixed sample would remain the same. Therefore, we cannot say exactly how many molecules of protein C were in any given cell, but we can say that protein C was five times more abundant in population 2 than population 1. Using stable isotopes was a clever way to eliminate the problem of quantifying the intensity of spots in a 2D gel.

A slight variation of this method, illustrated in Figure 6.25, page 190, allows us to detect proteins that have been

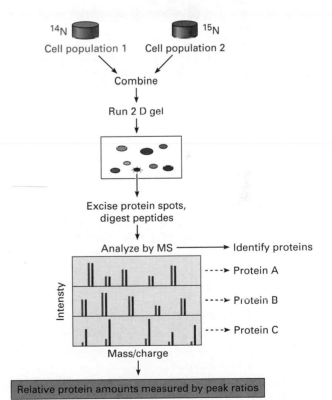

FIGURE 6.24 • Quantifying differences in proteomes. Two cell pools are grown in the presence (blue = cell population 2) or absence (gray = cell population 1) of heavy nitrogen ^{15}N. The proteins are extracted, pooled, and subjected to 2D gel analysis; spots excised; and proteins identified and quantified by MS/MS. The relative areas under the pairs of heavy and light peptide peaks indicate relative abundance of each protein pair.

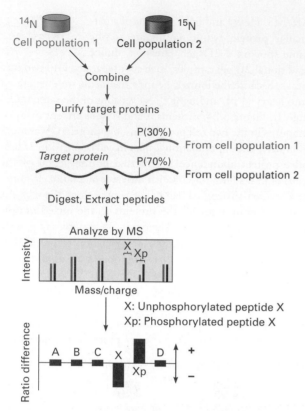

FIGURE 6.25 • **Quantifying relative levels of phosphorylation.** Bar graphs A, B, C, and D indicate no difference for proteins A–D in the two populations. Compared to population 1, unphosphorylated protein X was reduced in population 2. Phosphorylated protein X (Xp) was increased in population 2.

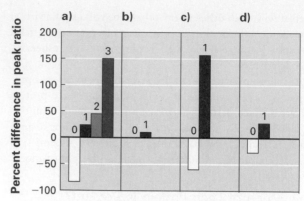

FIGURE 6.26 • **Phosphorylation of Ste20 requires Cln2.** Four Ste20 peptides are phosphorylated in *wt* cells but not *cln2* cells. The bar graphs show the loss of unphosphorylated Ste20 peptides (negative change) and the appearance of phosphorylated Ste20 peptides (positive change). The numbers above each bar indicate the number of phosphates added to each peptide.

phosphorylated. Spots of interest were excised from 2D gels, and the ratios of unphosphorylated and phosphorylated proteins were measured. As shown in the bottom of Figure 6.25, most peptides did not vary in their phosphorylation status. The bar graph was of equal height above and below the horizontal line, and peptides A–D were represented by only one mass on the horizontal line. Peptide X, however, was differentially phosphorylated (X and Xp), and there are two bar graphs to measure the relative amounts of unphosphorylated (X) and phosphorylated (Xp) peptide. The total amount of protein X was constant in both cell populations; however, 30% of protein X in population 1 was phosphorylated compared to 70% in population 2.

With Chait's method, we can determine the variation each protein experienced in relative quantity and degree of phosphorylation. As a proof of concept experiment, they tested whether a particular yeast protein (Ste20) can be phosphorylated. In this experiment, they compared the ability of two strains of yeast to phosphorylate Ste20. One strain lacked the protein Cln2, a cyclin that regulates a kinase during the G1 to S transition in the cell cycle, while the other

expressed Cln2. Mutants lacking Cln2 were labeled with ^{14}N, and *wt* cells were labeled with ^{15}N. In the analysis of the Ste20 spot on the 2D gel, they found that different peptides from Ste20 received different levels of phosphorylation (Figure 6.26). Panels a, b, c, and d represent different peptide fragments, phosphorylated very little (peptide b was near 0%), moderate amounts at one site (peptides c and d), and multiple times (peptide a could be phosphorylated on three different amino acids). These data showed that Cln2 was required for phosphorylation since cells that expressed Cln2 contained more phosphorylated Ste20 and less unphosphorylated Ste20.

DISCOVERY QUESTIONS

54. Using Chait's method, would it be possible to quantify the complete proteome of yeast? Explain your answer.

55. There is one extremely important limitation to this method. Can you figure out what it is? (Hint: What would you have to do to compare human biopsies using this method?)

ICAT Method Fits All Cells Chait's method was a huge improvement in quantification compared to traditional methods of measuring spot intensities on 2D gels. Using the stable isotope, we can compare relative amounts of each protein identified in two proteomes. However, there is a significant limitation—cells have to be grown in heavy isotope. Although this is not a problem for yeast, bacteria, or eukaryotic

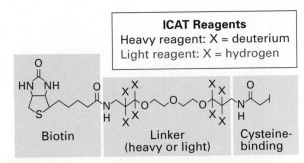

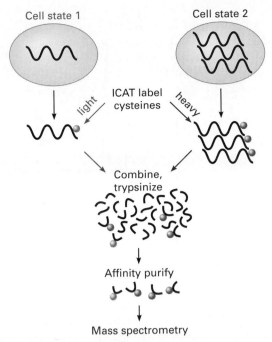

FIGURE 6.27 • **ICAT labeling reagent.** The reagent consists of three parts: an affinity tag in the form of biotin that binds irreversibly to avidin; a linker that contains eight stable isotopes; and a reactive group that will bind cysteines. The reagent exists in two forms: heavy contains eight deuterium atoms (blue), and light contains eight hydrogen atoms (gray).

FIGURE 6.28 • **Schematic of ICAT method.** Proteins from two cell populations are labeled with either the heavy (blue) or light (gray) ICAT reagent. The proteins are isolated, mixed, and digested with trypsin. ICAT-labeled peptide fragments are affinity purified and analyzed by tandem MS.

cells grown in tissue culture, it is much more difficult to accomplish with plants and large animals, and not feasible to perform on humans. The use of isotopes was a good idea, but the requirement for protein synthesis to incorporate them could be problematic.

Ruedi Aebersold from the Institute for Systems Biology in Seattle has developed a protocol that will work on any sample, even human biopsies: **isotope-coded affinity tags, (ICAT)**. At the heart of the ICAT method is the labeling reagent Aebersold's group created (Figure 6.27). ICAT labeling reagents come in two forms, heavy and light. The heavy version has eight deuterium atoms, which increases its mass by eight atomic units over the light reagent, which contains eight hydrogen atoms. Other than this difference in mass, the two reagents are identical. On one end of the labeling reagents is a chemically reactive group that can bind to the amino acid cysteine. This permits any protein that contains a cysteine to be labeled with either the heavy or the light ICAT reagent. The other end of the reagent contains **biotin,** which binds very strongly to a protein called **avidin**. Avidin and biotin are like two halves of Velcro except once they bind to each other, they almost never let go.

Anyone can take two protein populations and incubate one with the heavy ICAT reagent and one with the light ICAT reagent (Figure 6.28). All cysteine-containing proteins get labeled either heavy or light, and then the two populations of proteins are mixed, just as in Chait's method. But Aebersold's team wanted to simplify the complexity of proteins to be analyzed, so they included the biotin component of the ICAT reagent. First, all the proteins are cut with a protease called trypsin. Then all these trypsin-treated fragments are purified by avidin affinity chromatography, which retains only the ICAT-labeled trypsin fragments. This process reduced the complexity of the sample by about 90% without losing any protein identification or quantity information.

The biotin/avidin-purified, ICAT-labeled peptides are sent into a tandem MS where the peptides are ionized, sorted, and identified.

In addition to protein identification, the relative amount of each peptide (heavy ICAT–labeled vs. light ICAT–labeled peptides) can be determined (Figure 6.29, page 192). In this example, cells in population 1 only had one copy of the protein while cells in population 2 had three copies. Of course, this cartoon oversimplifies the situation because tandem MS cannot detect proteins in such small numbers, but it represents the relative quantification ICAT can determine. By eliminating the 2D gel, there is an increase in speed, and all proteins can be detected.

METHODS
ICAT
Avidin
Affinity Chromatography

LINKS
Ruedi Aebersold

Nice Idea but Does It Work? Aebersold's group had invented a completely new approach to a vexing problem—relative quantification of any two proteomes. To test their new method, they created two protein mixtures composed of the same six proteins, but the amount of each protein was different in the two mixtures. These two mixtures of six proteins were labeled with either heavy or light ICAT reagents, pooled, and subjected to MS/MS analysis (Table 6.2, page 192). As you can see, the method successfully identified the proteins and quantified the relative

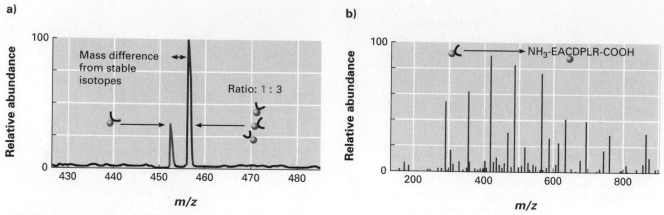

FIGURE 6.29 • Quantification and identification of proteins by ICAT. a) The heavy (blue) and light (gray) peptides are separated and quantified to produce a ratio for each protein from the two different cell populations. **b)** Each peptide bound to an ICAT reagent is subjected to MS analysis. Protein identification is performed by determining the amino acid sequence of the peptide and comparing it to a database of all possible trypsin fragments from the yeast proteome.

TABLE 6.2 • Sequence identification and quantification of six proteins in two mixtures. Two or three peptide fragments were identified for each protein and gene names are according to Swiss Prot nomenclature. ICAT-labeled cysteine residues are denoted by # signs. Ratios were calculated for each peptide as shown in Figure 6.29a. Expected ratios were calculated from the known amounts of proteins present in each mixture.

Gene Name	Peptide Sequence Identified	Observed Ratio (light/heavy)	Mean ± s.d.	Expected Ratio (light/heavy)	% Error
LCA_BOVIN	ALC#SEK C#EVFR FLDDLTDDIMC#VK	0.94 1.03 0.92	0.96 ± 0.06	1.00	4.2
OVAL_CHICK	ADHPFLFC#IK YPILPEYLQC#VK	1.88 1.96	1.92 ± 0.06	2.00	4.0
BGAL_ECOLI	LTAAC#FDR IGLNC#QLAQVAER #FDGVNSAFHLWC#NGR	1.00 0.91 1.04	0.98 ± 0.07	1.00	2.0
LACB_BOVIN	WENGEC#AQK LSFNPTQLEEQC#HI	3.64 3.45	3.55 ± 0.13	4.00	11.3
G3P_RABIT	VPTPNVSVVDLTC#R IVSNASC#TTNC#LAPLAK	0.54 0.57	0.56 ± 0.02	0.50	12.0
PHS2_RABIT	IC#GGWQMEEADDWLR TC#AYTNHTVLPEALER WLVLC#NPGLAEIIAER	0.32 0.35 0.30	0.32 ± 0.03	0.33	3.1

Source: Gygi, et al. 1999. *Nature Biotechnology.* 17: 996, Table 1.

abundance of all six proteins in the heavy and light protein mixtures. Since the investigators reported the standard deviation, we can evaluate the scatter of the data, giving us more confidence in their methodology.

Now that ICAT was working under controlled conditions, how would it perform on something as complex as two complete proteomes? To test this, Aebersold's team compared the proteomes of yeast cells grown in the presence of ethanol vs. galactose. In a one-hour time period, over 1,200 MS/MS scans were automatically recorded, and more than 800 different peptides were identified and quantified (Table 6.3).

TABLE 6.3 • **Protein ratios from yeast grown on 2% galactose (Gal) or 2% ethanol (Eth).** Gene names are according to the Yeast Proteome Database (YPD). Peptide sequences were identified by SEQUEST program. Protein expression ratios were calculated from the areas under the traces for each peptide as shown in Figure 6.29a. Data for only nine proteins are shown in this table.

Gene Name	Peptide Sequence identified	Observed Ratio (Eth : Gal)
Ach1	KHNC#LHEPHMLK	> 100 : 1
Adh1	YSGVC#HTDLHAWHGDWPLPVK	0.57 : 1
	C#C#SDVFNQVVK	0.48 : 1
Adh2	YSGVC#HTDLHAWHGDWPLPTK	> 200 : 1
	C#SSDVFNHVVK	> 200 : 1
Ald4	TFEVINPSTEEEIC#HIYEGR	> 100 : 1
Bmh1	SEHQVELIC#SYR	0.95 : 1
Cdc19	YRPNC#PIILVTR	0.49 : 1
	NC#TPKPTSTTETVAASAVAAVFEQK	0.65 : 1
	AC#DDK	0.67 : 1
Fba1	SIAPAYGIPWLHSDHC#AK	0.60 : 1
	EQVGC#K	0.63 : 1
Gal1	LTGAGWGGC#TVHLVPGGPNGNIEK	1 : > 200
Gal10	HHIPFYEVDLC#DR	1 : > 200
	DC#VTLK	1 : > 200

Source: Gygi, et al. 1999. *Nature Biotechnology.* 17: 997, Table 2 (partial).

How well did ICAT work? It appeared to perform well for those proteins it was able to identify. For example, the protein Pck1 had an ethanol/galactose expression ratio of 1.57 ± 0.15. This value is in agreement with previously published reports and 2D gel electrophoresis analysis. In addition to identification and relative quantifications, ICAT can perform analyses that other methods cannot. When the isozymes of alcohol dehydrogenase (Adh1 and Adh2) were considered, ICAT had better resolution than DNA microarrays because the two genes have very similar DNA sequences. Adh1 converts acetaldehyde to ethanol, while Adh2 converts ethanol to acetaldehyde (Figure 6.30, page 194).

DISCOVERY QUESTIONS

56. When ethanol is the only source of food, yeast cells rapidly convert it to acetaldehyde. Can you predict which form of ADH would be more abundant when cells were grown on either galactose or ethanol? Support your answers using Figure 6.30.

57. Since the ICAT method relies upon pairs of proteins (heavy and light), it does not measure any proteins that lack a partner peak. What is the problem with this limitation? If you were to use MS/MS to identify nonpaired proteins, what information would be lacking that ICAT is designed to produce?

ICAT was able to measure how yeast cells respond at the protein level when the food source varies. When grown on galactose, Adh1 was present at twice the level of ethanol growth (see Table 6.3 for summary and Figure 6.31, page 194, for details). Conversely, when ethanol was the only source of food, Adh2 was present at more than 200 times the level of galactose growth. This is particularly impressive from a technical standpoint, since the two forms of ADH are 93% identical at the amino acid level and yet ICAT could distinguish between them. DNA microarrays often fail to distinguish the two ADH mRNAs, which are 88% identical in the coding region.

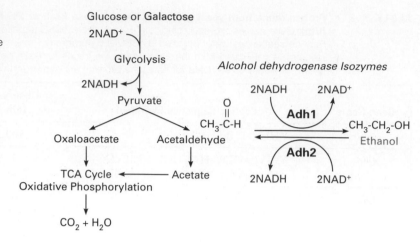

FIGURE 6.30 • **Reversible metabolic pathway for ethanol and galactose utilization.** The major proteins and products are shown. The two alcohol dehydrogenases are indicated as Adh1 and Adh2.

ICAT works in complex proteomes. It identified many of the proteins and determined relative amounts expressed in two different populations of cells. This is the beginning of a more mature form of proteomics that can examine dynamic cellular responses at the protein level as cells respond to changing environments. Using the ICAT method, it should be possible to examine variations among individuals of a species to understand the functional consequences of SNPs (see Chapter 3). We should be able to compare DNA microarray expression profiles with ICAT proteomics data and see how well they correlate (Chapter 9). The utility of this method is impressive and will increase as it becomes more refined by Aebersold's lab and others around the world.

The proteome story might appear complete except for a few details, but scientists always discover new questions to ask after data are collected. At this point, take a few minutes to dream where you would like to see the field of proteomics go. Think wildly and don't worry about how to do it experimentally.

Here's What We Can Do Today

- Use sequence information to predict functions and interacting proteins.

- Use yeast two-hybrid technology to determine protein-protein interactions.

- Identify many proteins in a proteome, though proteins expressed at very few copies per cell are still elusive.

- Use stable isotope methods to quantify relative amounts of proteins in two populations of cells.

What can't we do that you think would help us better understand what happens inside cells? Take a few minutes to write down your ideas.

The Last Unexplored Ecosystem on Earth Is Inside the Cell

You may have written down some areas for exploring that are not included in this section. If so, keep thinking about your ideas, and you may be able to figure out ways to

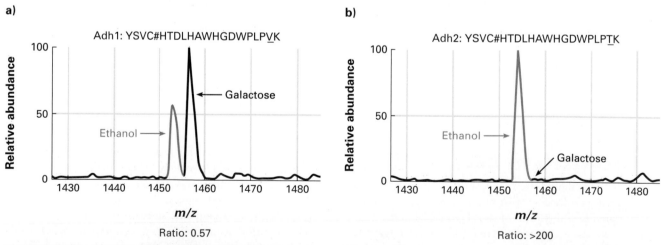

FIGURE 6.31 • **Reversible switch in metabolism.** ICAT detection of **a)** Adh1 and **b)** Adh2 from yeast cells grown in the presence of ethanol (gray) or galactose (blue). Relative amounts of these two proteins are indicated with ratios at the bottom.

substantially improve our understanding of cells. Although your list may have contained more than three new goals to pursue, we will briefly examine three that a few labs have begun to address: (1) protein chips, (2) single-cell analysis, and (3) metabolomics.

Can We Make *Protein* Chips?

DNA microarrays have been so popular and successful that many people are trying to create protein chips. So far, protein chips have been designed for three basic purposes: to identify and quantify proteins in a complex sample, to determine protein interactions, and to determine protein functions.

ID and Quantify Pat Brown, who popularized DNA microarrays, has tried to do the same with protein chips. Brian Haab, Maitreya Dunham, and Pat Brown published their first protein chip paper in January 2001 in *Genome Biology.* (*Genome Biology* is a free online journal, so you can go to its web site and download the PDF file for this paper.) Brown's team used monoclonal antibodies to detect specific proteins. For some protein microarrays, they spotted the antibodies and detected proteins in a mixed sample. On others, they spotted the proteins and detected antibodies in a mixed solution. Figure 6.32 on your CD-ROM shows the results of an **antibody microarray** that detected proteins labeled with the fluorescent dyes **Cy3** (green) and **Cy5** (red). For their proof of concept experiment, they worked with 115 antibody/**antigen** pairs (antigens are the proteins that antibodies recognize), creating six different cocktails of the 115 antigens, each with different concentrations of the various antigens. Each of these six cocktails was labeled red, while a single "reference" cocktail with a fixed concentration of the same 115 antigens was labeled green. Equal volumes of the experimental and reference proteins were mixed and then added to the antibody microarray and the ratio of the two colors was determined for each feature.

> **FIGURE 6.32 • Antibody arrays detected fluorescently labeled antigens.**
> Go to the CD-ROM to view this figure.

In an ideal protein microarray world, we could create arrays with monoclonal antibodies for each protein in the proteome and determine which proteins were present and in what amounts. Two protein samples would be applied to the microarray at the same time so we could compare two proteomes (e.g., healthy vs. diseased). Unfortunately, protein chips are not yet able to perform a two-proteome comparison. Furthermore, Brown's protein chip worked better when the antigens were spotted on the glass and used to identify which antibodies were present in a mixed solution

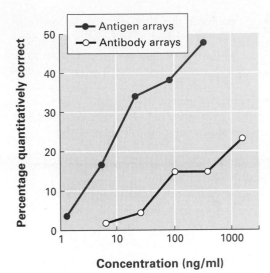

FIGURE 6.33 • Comparison of two protein microarrays. Percentages of antibodies and antigens yielding quantitatively correct results when known protein mixtures were incubated on protein microarrays.

(Figure 6.33). Antigen microarrays have limited utility except when studying immune responses to known antigens. So, this attempt at using antibodies was nice, but the search continues for better methods.

LINKS

Pat Brown
Protein Chip Paper
Antibody Resource
Gavin MacBeath
Stuart Schreiber

DISCOVERY QUESTIONS

58. What reagent is the limiting factor for the success of Brown's method? Does this limitation make the method impractical?

59. Go to the Antibody Resource Page and choose a couple of links. How many monoclonal antibodies are commercially available? Do these sites change your evaluation of the practicality of Brown's antibody microarray method?

60. Of the two protein microarrays produced by Brown's lab, which would be more useful for defining a proteome? What is the limitation for this type of array?

Proteome Interactions Gavin MacBeath and Stuart Schreiber from Harvard University have created useful protein microarrays for defining protein-protein interactions. They covalently attached proteins known to be one-half of interacting pairs of proteins to glass microscope slides. The spotted proteins created a microarray of in vitro "bait" analogous to the bait proteins used in yeast two-hybrids. In their first experiment, they demonstrated the specificity of their method as three pairs of proteins were shown to bind to the appropriate partners but not the other bait proteins (Figure 6.34a on your CD-ROM). In a rather dramatic way, they also showed that bait and prey proteins could discriminate even when only 1 feature out of 10,800 contained the appropriate bait protein (Figure 6.34b).

FIGURE 6.34 • **Detection of protein-protein interactions on protein microarrays.**
Go to the CD-ROM to view this figure.

In addition to determining protein-protein interactions, MacBeath and Schreiber wanted to determine whether they could measure the protein target for small molecules (i.e., ligands and metabolites). They attempted to bind dye-labeled ligands to the appropriate receptor proteins that had been spotted onto a protein microarray (Figure 6.35 on your CD-ROM). This experiment was a significant step forward into the next generation of "-omics"—metabolomics (see pages 200–202).

FIGURE 6.35 • **Detecting the targets of small molecules on protein arrays.**
Go to the CD-ROM to view this figure.

DISCOVERY QUESTIONS

61. In Figure 6.34b, only two bait proteins were spotted on the microarray. Think of a better experimental design for testing specificity than the one shown here. What negative control would you like to have seen in Figure 6.35?

62. In these protein microarray experiments, three different dyes were used. What properties would you expect these dyes to have so you could be sure that the bottom panels of Figures 6.34 and 6.35 accurately represent the specificity of binding?

63. Can you think of a way to modify the ligand microarray experiment (Figure 6.35) to identify in serum samples the number of metabolic intermediates and how much of each is present? (Hint: Combine Pat Brown's antibody microarray with the method used by MacBeath and Schreiber.)

Kinase Functionality Mike Snyder's lab (the team that created the mTn-tagged functional mutants described on pages 162–166) developed a high-throughput method to survey the functions of kinases. Their first innovation was the production of the protein array itself. Instead of spotting onto glass, Snyder's group developed an inexpensive way to produce silicon polymer microwell arrays, analogous to 96 well plates, only smaller (Figure 6.36). The microwells have rounded floors, are 300 μm deep, and hold approximately 300 nl of liquid. The microwell arrays are placed on microscope slides to facilitate handling, with two of them fitting on a single slide. For this work, the investigators pipetted all reagents by hand, but the technology is amenable to high-throughput automated liquid handling methods already used in many labs.

The investigators produced 17 different microwell arrays, one for each kinase substrate (Figure 6.37). In the yeast genome, there are 122 kinase genes; Snyder's lab produced large amounts of 119 of the 122 kinase enzymes and incubated each one in one microwell for each of the 17 microwell arrays. By using microwells instead of traditional microarrays where the substrates would have been spotted onto a single glass microscope slide, the reaction conditions in each well

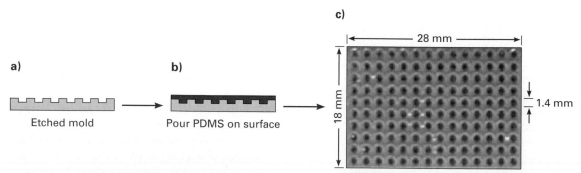

FIGURE 6.36 • **Microwell array manufacturing. a)** Molds were etched, and then **b)** monomeric elastomer was used to form a silicon-based polymer (blue). **c)** When removed from the mold, an array of microwells was produced and ready for the next step.

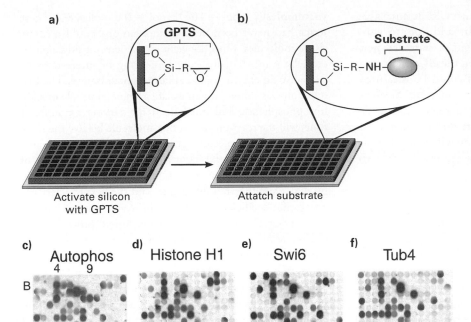

a) The silicon polymer was activated with a crosslinker called 3-glycidox-propyltrimethoxysilane (GPTS).

FIGURE 6.37 • **Coupling substrate to the array of microwells.**
a) The silicon polymer was activated with a crosslinker called 3-glycidox-propyltrimethoxysilane (GPTS).
b) The peptides used as substrates were covalently linked to the silicon.
c) to **f)** Four of the 119 kinase microwell assays are shown in this figure. Each microwell array contained one substrate (labeled at the top) while a different kinase and radioactive ATP (^{33}P-γ-ATP) were added to each microwell. The microwell arrays were imaged with a sensitive device called a phosphoimager.

could be manipulated independently to optimize the reactions.

For their enzymatic assays, radioactive ATP was used to label all polypeptide substrates phosphorylated by a given kinase (Figure 6.37c–f). In addition to testing each kinase with each substrate, they also tested each kinase with itself to see if any were able to **autophosphorylate,** that is phosphorylate themselves. For each microwell array, position I9 was a negative control. You can see that the kinase Mps1p (well B4) was able to phosphorylate all four substrates, including itself (Figure 6.37c).

DISCOVERY QUESTIONS

64. Microwell arrays worked very well with kinases; can you think of other proteomics applications?

65. Could microwell arrays be used with a nonradioactive detection system?

66. Think about the correlation between the primary amino acid sequence, the 3D shape of each kinase, and the substrates that each kinase can phosphorylate. Snyder's group aligned the amino acid sequences for 107 kinases and created a phylogenetic kinase tree. Look at the PDF file (on the Web site) of an unrooted phylogenetic tree that illustrates the relationship of the 107 kinases' amino acid sequences. Green indicates kinases that phosphorylate tyrosines in a poly(tyrosine-glutamate) substrate; "blue kinases" can use the same substrate

weakly; "yellow kinases" could not phosphorylate this substrate.

a. What correlation do you see between primary structure and substrate preference?

b. If you could see the 3D structures for all of these kinases, hypothesize what you would expect to see for all the "green kinases."

c. Hypothesize what you would see when you compared the 3D structure of a "green kinase" and a "blue" one, a "green kinase" and a "yellow" one.

METHODS
Phylogenetic Tree

SEQUENCES
Kinase Tree

It is striking to see so many green-coded kinases scattered all over in the phylogenetic tree. This means their amino acid sequences vary substantially, though subsets of them cluster on particular branches, indicating that kinases within a subset have similar amino acid sequences. Even though the green-coded kinase sequences vary, they all act on the same substrate, which indicates their active sites are conserved. This leads to the hypothesis that the differences in their sequences lie outside the active site. For those kinases that do not phosphorylate the same substrate yet clustered near the green-coded ones, we might expect to see a large degree of amino acid conservation overall but variations within the active site. This could be confirmed if we had access to all the 3D structures.

With the combination of all the different enzymes and all the different substrates in a proteome, Snyder's method will be very useful. Any enzyme that can produce

METHODS

Capillary
Electrophoresis

LINKS

Norm Dovichi

a detectable product should be amenable to this approach. In addition to substrates, enzyme inhibitors, potential new medications, and even the effects of pesticides and herbicides could be measured by the microwell array method. Snyder's method is easy enough to fabricate in any lab and suitable for scaling up for large projects. There do not seem to be many limitations to this method, so you will probably hear more about it in the future.

Are All Cells Equal?

We have been talking about what happens inside a cell, but in reality all our data have been based on populations of cells. Now we have tools sensitive enough to detect individual molecules. Norm Dovichi has recently moved from the University of Alberta to the University of Washington, Seattle. Dovichi is an analytical chemist who "was trained to work with lasers" and has applied his skills to biological molecules. First, he helped invent the capillary electrophoresis technology that has driven the high-throughput DNA sequencing. He chose DNA methods because "compared to proteins, it was easy." Now that the "easy" stuff is done, Dovichi says his lab group is working on the "really cool stuff."

The initial goal was to improve laser-based methods for detecting small amounts of proteins, which his group accomplished, and in the process coined two new terms: **zeptomoles** (1 zmol = 10^{-21} mol = 602 molecules) and

yoctomoles (1 ymol = 10^{-24} mol = 0.6 molecules). These terms had never been needed since no one could measure single molecules. Once Dovichi could detect a single molecule, he wanted to measure something of interest, so he turned to an enzyme called **alkaline phosphatase.** The first question was simple: If you isolate a single molecule of alkaline phosphatase and incubate it for varying amounts of time, will the reaction rate of that molecule behave the same way as a pool of molecules? A single molecule was incubated for one, two, four, and eight minutes, and the amount of product was measured (Figure 6.38). The individual enzyme produced product at a linear rate, and the best fit line has a correlation coefficient of 0.999. This result was not surprising but significant nonetheless. Dovichi had proven his detection system was able to measure the output of a single enzyme molecule.

In science, we need to know our results can be reproduced when the same experiment is performed a second time. When Dovichi repeated the time course experiment, he noticed that each enzyme's rate of product formation was linear but the individual enzymes did not produce product at the same rate. So his lab measured the reaction rates (the slope of the line in Figure 6.38b) for 83 different molecules (Figure 6.39a). As you can see, not all molecules behave the same way under identical reaction conditions. The reaction rates for the 83 individual molecules of alkaline phosphatase varied over a tenfold range. Similarly, when reaction rates for individual enzymes were measured at different temperatures, the individual enzymes did not produce identical amounts of product (Figure 6.39b). Although he is not sure, Dovichi

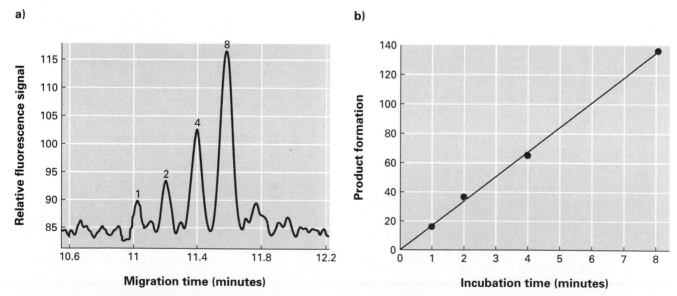

a)

b)

FIGURE 6.38 • **Measuring the rate of product formation from a single enzyme molecule. a)** Product formation generated by reacting a single enzyme for 1, 2, 4, and 8 minutes.
b) Kinetic plot of the data from a). The amounts of product for each incubation time are shown and the straight line is the least-squares fit to the data. Correlation coefficient is 0.999.

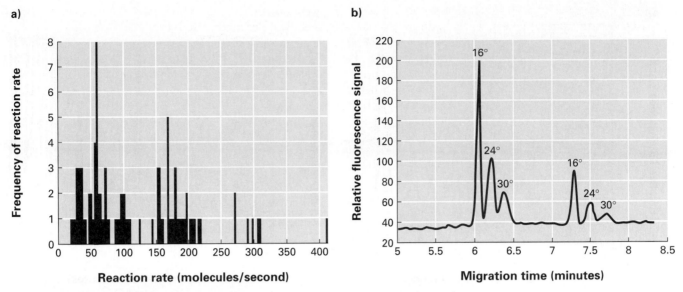

FIGURE 6.39 • Variation between enzymes. a) Histogram of reaction rates observed for 83 different alkaline phosphatase molecules assayed individually. Activity was determined from the incubation time and product formation. **b)** Electropherogram of alkaline phosphatase assay using two different molecules. Each molecule was incubated sequentially at 16°C, 24°C, and 30°C.

believes the degree of glycosylation on a particular enzyme molecule may be the determining factor for variation. Another possibility may be subtle variations in tertiary structure of the folded enzyme. Whatever the reason, we have experimental evidence that multiple copies of a single protein do not function identically.

Next, Dovichi's lab wanted to determine what happens to individual molecules when they are subjected to **heat denaturation.** You can imagine two possible outcomes. The first was that all the enzymes in a population would gradually become denatured, gradually decreasing the productivity of any given molecule. The second was that an individual molecule would denature almost instantly, but each molecule in a population would denature at its own rate. The consequence of the second outcome would be that halfway through a denaturation experiment, individual enzymes would either have their normal level of activity or zero activity. Dovichi's team determined that alkaline phosphatase molecules denatured very rapidly, producing two kinds of activities in a population: zero or normal. It is amazing that proteins behave digitally (toggling either on or off) rather than in a gradual fashion. Proteins acting as toggle switches will be addressed further in Chapter 8 when we examine how proteins can make biochemical decisions.

As important as it is to understand what single molecules do, we ultimately want to know what the entire cell does. Based on Dovichi's results from studying single molecules, you might imagine that two mitotic clones derived from the same parental cell might not be identical. Dovichi's group decided to test this hypothesis by comparing two human colon

adenocarcinoma cells grown in tissue culture. They conducted a one-dimensional analysis, utilizing capillary electrophoresis to separate and quantify the proteomes from two individual cells (Figure 6.40, page 200). They compared these two single-cell proteomes to a composite proteome taken from a pool of mitotic clones of the two individual cells. At least 30 peaks were visible, and all three proteomes were very similar but not identical. In cell a, peaks 14 and 17 were barely detectible and peaks 12, 19, and 22 were reduced in magnitude. Peak 10 is especially interesting since its abundance was similar but its migration time was altered. There are several plausible explanations, but Dovichi's group is especially intrigued that this may represent differences in posttranslational modifications such as the addition of a single phosphate group.

DISCOVERY QUESTIONS

67. How could you test the hypothesis that the differences between peak 10 in cell a and cell b is due to phosphorylation?

68. With the ability to detect single molecules, what do you think the next step for Dovichi's group will be as they develop tools to describe the proteome of single cells?

Genomic and proteomic approaches provide us with new tools, new types of data, but more importantly, new types of questions that reflect new ways of thinking about the cell.

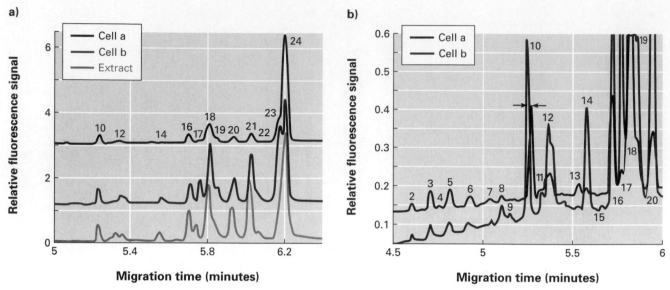

FIGURE 6.40 • Capillary electrophoresis of single-cell proteomes. a) Two individual human colon adenocarcinoma cells (blue and black; cell line HT29) were subjected to one-dimensional separation. The "extract" was produced from approximately 10^6 cells (gray). Only the major peaks are labeled. **b)** Higher resolution of the central portion of a). The arrows mark differences in migration times between the two number 10 peaks.

METHODS

Arabidopsis

LINKS

Oliver Fiehn

Dovichi's group has detected a difference in peak 10 that may represent differences in phosphorylation events as two genetically identical cells responded differently to their shared environment. We could use antibodies that are specific for phosphoproteins or Chait's stable isotope method to determine whether protein 10 was differentially phosphorylated. With time, Dovichi and others will address these technical challenges and further expand our understanding of cells.

For now, Dovichi's lab is busy developing the next generation of gel-free 2D analysis for proteins. His group has developed a capillary electrophoresis procedure to measure the molecular weight and then use the pI of proteins to produce 2D maps of proteomes from single cells. "What would be really cool," Dovichi explained, "would be to couple this single-cell 2D analysis with a method to identify each spot," such as MS/MS.

What Does a Proteome Produce?

Imagine a day when we can define proteomes easily, determine which proteins interact, and measure their functions in vitro. What could be left? How about lipids? sugars? steroids? ion concentrations? metabolic intermediates? The list of cellular components is substantial, and so far we have only discussed proteins and mRNA. There is more to making a cell than just transcription and translation. Although we cannot define complete proteomes yet, researchers around the world are already looking past this temporary limitation to defining **metabolomes,** or the entire metabolic state of cells.

Oliver Fiehn and his colleagues at the Max Planck Institute in Potsdam, Germany, are among the world leaders in profiling the metabolomes of model plants such as potatoes, tobacco, and *Arabidopsis*. *Arabidopsis* is a small flowering plant that goes from seed to flower in about two weeks, can be grown in culture, has been very important in understanding the biology of all flowering plants, and had its genome completely sequenced in 2000.

Fiehn's team wanted to compare the metabolomes of four different strains of *Arabidopsis*. Two strains were considered *wt*, though they differ from each other by carrying an undetermined number of allelic variations. The other two were mutants derived from these parental strains, each with a single gene deleted. Col-2 is a *wt* strain from which the mutant strain Dgd1 was derived; Dgd1 lacked an enzyme used in the production of a particular lipid. Phenotypically, Dgd1 did not photosynthesize well and was hypersensitive to light stress. C24 is the other *wt* strain from which Sdd1-1 was derived; Sdd1-1 lacked a regulatory protein involved in controlling **stomata** development (the pores on the underside of leaves used for gas exchange). The only apparent phenotype in Sdd1-1 was a two- to fourfold increase in the number of stomata per leaf.

Using **gas chromatography** to separate the cytoplasmic nonprotein compounds and then MS to identify them, Fiehn's group was able to determine the identity of 326 distinct compounds from the leaves of all four strains of *Arabidopsis*. In the Dgd1 mutant, 153 of the 326 compounds showed significant concentration differences compared to the parental strain Col-2 (Figure 6.41a). In contrast, in the Sdd1-1 mutant strain, only 41 of the 326 compounds

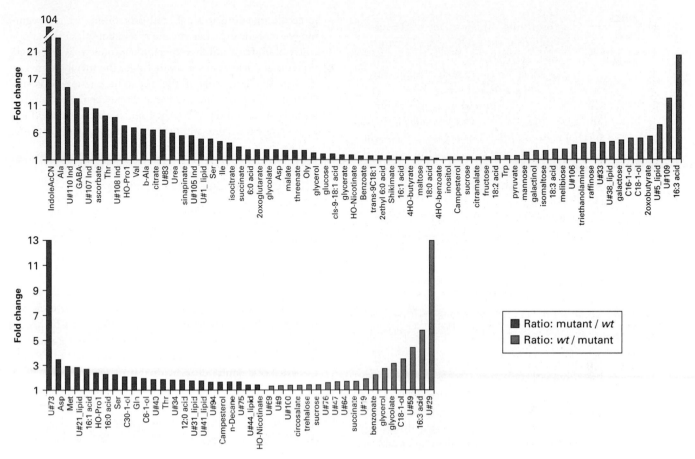

FIGURE 6.41 • **Significant changes in metabolomes when comparing** *wt* **to mutant strains. a)** Alteration in average metabolite levels (*t*-test *p*-value < 0.01) of Dgd1 compared to its parental strain Col-2. Only 67 of the 153 significant differences detected are shown. **b)** When compared to its parental strain (C-24), Sdd1-1 had 41 significant differences in its metabolome; all 41 are shown.

differed significantly from the parental strain of C24 (Figure 6.41b). What was surprising in both cases was that while only one protein was lacking in each proteome, the metabolome responded in complex and unexpected ways. Some metabolites were increased, while others were reduced. In short, the proteome is complex, but the metabolome is exponentially more complex, perplexing, and challenging to study. In order to fully understand cells, we must measure *all* the components under changing conditions. This task will take a lot more effort and collaborations to produce the right technology and experiments.

DISCOVERY QUESTIONS

These three questions are designed to help you discover the wealth of metabolomic information currently available, and the complexity of defining metabolomes.

69. Go to the What Is There? web site. Click on the "View Models" link from the menu at the top of the page. You will get a list of many genomes' worth of information. Find *C. elegans* at the bottom and click on the "CE" link in the first column for that model.

a. Click on "polysaccharide metabolism."

b. Click on "G6PGLG.ANA."

c. Click on "Diagram Picture." You will see a pathway that should look familiar (the beginning of glycolysis). Click on the blue box around α-D-glucose 6-phosphate. You will see the structure for this compound.

d. How many compounds would you need to measure to determine the G6PGLG.ANA portion of a metabolome?

70. Go to the Enzymes and Metabolic Pathways database search page for an amazing project. In the search terms window, enter the word

LINKS

What Is There?

Enzymes and Metabolic Pathways

LINKS

KEGG

Human Proteome Initiative

"orotidine" (this was one of the substrates for the ODCase we examined on page 172).

   a. Click on the "search" button. You will get several hits.

   b. Click on the "SVG" button on the far right of hit "CO_2, NH_3—UMP, CO_2 anabolism (ATP, FAD^+) (cytosol, plasma membrane.)" You will see a diagram of the metabolic pathway to synthesize UMP. Mouse over any of the enzymes and compounds, and you will notice that they are all hyperlinks.

   c. Click on UMP at the bottom of the list. You will get three options. Click on the button next to "UMP." In the right frame, you will see some text describing your choice of UMP. Click on the "depict" button. Soon you will see the structure of UMP.

   d. Return to the pathway and click on "carbamoyl phosphate" and compare its structure with UMP's.

   e. Return to the pathway again. ODCase is the last enzyme (E.C. number 4.1.1.23) in this pathway to produce UMP. Notice carbon dioxide is removed during the final step, but it is also consumed at the top of the pathway. Would this type of recycling present a problem if you wanted to define a metabolome?

71. You can explore the metabolic pathways of many different species at the Kyoto Encyclopedia of Genes and Genomes (KEGG) web site. Go to KEGG, click on any section of the metabolic overview, and gradually zoom in until you see individual enzymes and metabolites listed. This will give you a sense of how much work is involved in characterizing a single metabolome from one cell at one time point.

Summary 6.4

Proteomics is a very difficult field, and the methods have not yet produced the quality or quantity of data we want. 2D gels were the first method to allow us to identify proteins in a proteome, but newer methods that utilize tandem mass spectroscopy enable us to identify more proteins. By incorporating stable isotopes, we can quantify as well as identify proteins in proteomes as they change over time. Protein chips are quickly becoming more sophisticated and no doubt will have a substantial impact on proteomics. As our tools become more precise, we can begin to study single proteins and discover new properties that were masked when we studied them in large pools of many molecules. From single-

molecule functions to single-cell proteomics, we are beginning to recognize the heterogeneity and complex cellular circuitry of proteins and their metabolic products. When single proteins are deleted from a proteome, the metabolome can be very different even if the apparent phenotype is unchanged. The proteome and metabolome seem too daunting to understand, but in 1980, the human genome seemed too big to sequence. Time and new technologies can change the way we view new areas of research.

Chapter 6 Conclusions

The postgenomics era could be called the proteomics era as the research community has begun to tackle this next huge challenge. The Human Proteome Initiative is gaining momentum and, more importantly, increasing its knowledge base. Proteomics has been more difficult than genomics because we need so much more information. We need to understand the many ways in which proteins work, their 3D structure, the protein interaction network, and which proteins and how much of each are present in any given proteome. In addition, we would like to understand more fully the stochastic nature of individual proteins and what effect this unpredictable behavior has on single-cell proteomes and metabolomes. You may have thought that the complete sequencing of the human genome meant all the challenging research had been completed, but hard work and great discoveries are still ahead. In addition to fully describing many different proteomes that change over time, we would like to integrate proteomics with all the other forms of information covered in the previous five chapters.

   The area of proteomics is like the front line of an avalanche where advances protrude from different places along the edge but each advancing edge is followed very quickly by the rest of a fast-paced progression. The methods of proteomics change quickly and are making great progress. In a few years, our perspective on many topics will change. New perspectives will spawn new questions and insights into the complexity of cell webs. In Chapters 7–9, we will examine how proteomes, genomes, and metabolomes interact to produce cellular responses. This will culminate in the greatest challenge to date: synthesizing all the information into comprehensive models.

References

Proteomics References

Functional Proteomics
Caplen, N.J., S. Parrish, et al. 2001. Specific inhibition of gene expression by small double-stranded RNAs in invertebrate and vertebrate systems. *PNAS.* 98(17): 9742–9747.

Elbashir, S.M., J. Harborth, et al. 2001. Duplexes of 21-nucleotide RNAs mediate RNA interference in cultured mammalian cells. *Nature.* 411: 494–498.

Fire, Andrew, SiQun Xu, et al. 1998. Potent and specific genetic interference by double-stranded RNA in *Caenorhabditis elegans*. *Nature*. 391: 806–811.

Fraser, Andrew G., Ravl S. Kamath, et al. 2000. Functional genomic analysis of *C. elegans* chromosome I by systematic RNA interference. *Nature*. 408: 325–330.

Gönczy, Pierre, Christophe Echeverri, et al. 2000. Functional genomic analysis of cell division in *C. elegans* using RNAi of genes on chromosome III. *Nature*. 408: 331–339.

Kumar, Anuj, Kei-Hoi Cheung, et al. 2000. TRIPLES: A database of gene function in *Saccharomyces cerevisiae*. *Nucleic Acids Research*. 28(1): 81–84.

Ross-Macdonald, Petra, Paulo S. R. Coelho, et al. 1999. Large-scale analysis of the yeast genome by transposon tagging and gene disruption. *Nature*. 402: 413–418.

Winzeler, Elizabeth A., Daniel D. Shoemaker, et al. 1999. Functional characterization of the S. cerevisiae genome by gene deletion and parallel analysis. *Science*. 285: 901–909.

Structural Proteomics
Abbott, Alison. 2000. Structure by numbers. *Nature*. 408: 130–132. [Commentary]

Christendat, Dinesh, Adelinda Yee, et al. 2000. Structural proteomics of an archeaon. *Nature Structural Biology*. 7: 903–908.

Martz, Eric. 2001. Protein Explorer. <http://www.umass.edu/microbio/chime/explorer/>. Accessed 17 February 2002.

Murata, Kazuyoshi, Kaoru Mitsuoka, et al. 2000. Structural determinants of water permeation through aquaporin-1. *Nature*. 407: 599–605.

Nienaber, Vicki L., Paul L. Richardson, et al. 2000. Discovering novel ligands for macromolecules using X-ray crystallographic screening. *Nature Biotechnology*. 18: 1105–1108.

Roepe, Paul D. 2001. A peptide needle in a signaling haystack. *Nature Genetics*. 27: 6 8.

True, Heather L., and Susan Lindquist. 2000. A yeast prion provides a mechanism for genetic variation and phenotypic diversity. *Nature*. 407: 477–483.

Wu, Ning, Yirong Moi, et al. 2000. Electrostatic stress in catalysis: Structure and mechanism of the enzyme orotidine monophosphate decarboxylase. *PNAS*. 97: 2017–2022.

Quantitative Proteomics
Andersen, Jens S., and Matthias Mann. 2000. Functional genomics by mass spectroscopy. *FEBS Letters*. 480: 25–31. [Review Article]

Celis, Julio, M. Kruhoffer, et al. 2000. Gene expression profiling: Monitoring transcription and translation products using DNA microarrays and proteomics. *FEBS Letters*. 480: 2–19. [Review Article]

Chalmers, Michael J., and Simon J. Gaskell. 2000. Advances in mass spectroscopy for proteome analysis. *Current Opinion in Biotechnology*. 11: 384–390.

Garin, Jérome, Roberto Diez, et al. 2001. The phagosome proteome: Insight into phagosome functions. *Journal of Cell Biology*. 152(1): 1665–180.

Gygi, Steven P., Beate Rist, et al. 1999. Quantitative analysis of complex protein mixtures using isotope-coded affinity tags. *Nature Biotechnology*. 17: 994–999.

Gygi, Steven P., Beate Rist, and Ruedi Aebersold. 2000. Measuring gene expression by quantitative proteome analysis. *Current Opinion in Biotechnology*. 11: 396–401.

Mann, Matthias. 1999. Quantitative proteomics? *Nature Biotechnology*. 17: 954–955. [Commentary]

Oda, Y., K. Huang, et al. 1999. Accurate quantitation of protein expression and site-specific phosphorylation. *PNAS*. 96: 6591–6599.

Patterson, Scott D. 2000. Mass spectroscopy and proteomics. *Physiological Genomics* 2: 59–65. [Review Article]

ProteomeMetrics. 2001. PROWL for Intranet. <http://prowl1.rockefeller.edu/>. Accessed 28 Feb.

Interactions of the Proteome
Bartel, P. L., J. A. Roecklein, et al. 1999. A protein linkage map of Escherichia coli bacteriophage T7. *Nature Genetics*. 12(1): 72–77.

Enright, Anton J., Ioannis Iliopoulos, et al. 1999. Protein interaction maps for complete genomes based on gene fusion events. *Nature*. 402: 86–90.

Hybrigenics. 2001. PIMRider. <http://pim.hybrigenics.com/>. Accessed 24 February 2002.

Marcotte, Edward M., Matteo Pellegrini, et al. 1999. A combined algorithm for genome-wide prediction of protein function. *Nature*. 402: 83–89.

Marcotte, Edward M., Matteo Pellegrini, et al. 1999. Detecting protein function and protein-protein interactions from genome sequences. *Science*. 285: 751–753.

Mayer, Melanie L., and Philip Hieter. 2000. Protein networks—Built by association. *Nature Biotechnology*. 18: 1242–1243. [Commentary]

Oliver, Stephen. 2000. Guilt-by-association goes global. *Nature*. 403: 601–603. [Commentary]

Rain, Jean-Christophe, L. Selig, et al. 2001. The protein-protein interaction map of *Helicobacter pylori*. *Nature*. 409: 211–215.

Schwikowski, Benno, Peter Uetz, and Stanley Fields. 2000. A network of protein-protein interactions in yeast. *Nature Biotechnology*. 18: 1257–1261.

Uetz, Peter, Loic Grot, et al. 2000. A comprehensive analysis of protein-protein interactions in *Saccharomyces cerevisiae*. *Nature*. 403: 623–627.

Walhout, Albertha J., R. Sordella, et al. 2000. Protein interaction mapping in *C. elegans* using proteins involved in vulval development. *Science*. 287: 116–122.

Protein Arrays
Haab, Brian B., Maitreya J. Dunham, and Patrick O. Brown. 2001. Protein microarrays for highly parallel detection and quantitation of specific proteins and antibodies in complex solutions. *Genome Biology*. 2(2): 0004.1–0004.13

MacBeath, Gavin, and Stuart L. Schreiber. 2000. Printing proteins as microarrays for high-throughput function determination. *Science*. 289: 1760–1763.

Zhu, Heng, James F. Klemic, et al. 2000. Analysis of yeast protein kinases using protein chips. *Nature Genetics*. 26: 283–289.

Zhu, Heng, and Michael Snyder. 2001. Protein arrays and microarrays. *Current Opinion in Chemical Biology*. 5: 40–45. [Review Article]

Metabolomics
Fiehn, Oliver, Joachim Kopka, et al. 2000. Metabolic profiling for plant functional genomics. *Nature Biotechnology* 18: 1157–1161.

Glassbrook, Norm, Chris Beecher, and John Ryals. 2000. Metabolic profiling on the right path. *Nature Biotechnology*. 18: 1142–1143. [Commentary]

Single-Cell Proteomics

Chen, DaYong, and Norman J. Dovichi. 1996. Single-molecule detection in capillary electrophoresis: Molecular shot noise as a fundamental limit to chemical analysis. *Analytical Chemistry.* 68(4): 690–699.

Craig, Douglas, Edgar A. Arriaga, et al. 1999. Studies on single alkaline phosphatase molecules: Reaction rate and activation energy of a reaction catalyzed by a single molecule and the effect of thermal denaturation—The death of an enzyme. *Journal of the American Chemical Society.* 118(22): 5245–5253.

Zhang, Zheru, Sergey Krylov, et al. 2000. One-dimensional protein analysis of an HT29 human colon adenocarcinoma cell. *Analytical Chemistry.* 72(2): 318–322.

Additional References

Adam, Gregory A., Benjamin F. Cravatt, and Erik J. Sorenson. 2001. Profiling the specific reactivity of the proteome with non-directed activity-based probes. *Chemistry and Biology.* 8(1): 81–95.

Gmuender, H., K. Kuratli, et al. 2001. Gene expression changes triggered by exposure of *Haemophilus influenzae* to Novobiocin or Ciprofloxacin: Combined transcription and translation analysis. 2001. *Genome Research.* 11(1): 28–42.

Holt, Lucy J., Carolyn Enever, et al. 2000. The use of recombinant antibodies in proteomics. *Current Opinion in Biotechnology.* 11: 445–449.

Nature. 1999. A series of news briefings about proteomics appeared in this issue of *Nature.* 42: 715–720.

Ren, Bing, F. Robert, et al. 2000. Genome-wide location and function of DNA binding proteins. *Science.* 290: 2306–2309.

UNIT THREE

Whole Genome

Perspective

CHAPTER 7

Genomic Circuits

in Single Genes

Development can be thought of in two halves; the early part is tightly controlled by genes and the late part generates the features that we observe are unique to individuals, even identical twins. We are especially interested in the early part that is tightly controlled by the genome. Every cell in an organism contains the exact same DNA. Genetically, there is no difference between your liver cells, your brain cells, and your big toe cells. How can our bodies produce these different tissues if the genes are all the same? The answer is that each cell only expresses a subset of its genes. The subset of genes expressed is tightly controlled during development, and this control is exerted over *location* (which cells do or don't express the gene), *time* (when during development should the gene be turned on and off), and *amount* (do you need only 2 molecules of the encoded protein per cell or 20,000?).

Section 7.1 focuses on the **circuits** (interaction of proteins and DNA sequences in the promoter and enhancer) of a single gene that serves as a model for genomic control over the expression of genes. The gene we understand better than any other is *Endo16,* which is expressed in the developing gut of a sea urchin embryo. Over several years, Eric Davidson and his collaborators have slowly dissected *Endo16* to reveal an amazing circuitry that regulates its expression. In order to fully appreciate how a genome regulates the expression of individual genes, we will study how a sea urchin embryo regulates *Endo16.* The data are impressive, and the resulting model is elegant. In Section 7.2, we begin to coalesce what is known about the genomic regulation of genes into interactive models that are prototypes of what is needed to model entire genomes.

One final note: Chapter 7 is our window into the future. If you wanted to rewire your home, you would practice on a simple object first—a lamp, for example. Then you would gradually increase the level of complexity until you were ready to take on the fuse box and major appliances. Similarly, if we want to understand how genomes respond to their environment, we should begin with a simple wiring diagram of a single gene. For the next 100 years, we will continue to analyze how genomes and proteomes are interconnected. Discovering how *Endo16* is regulated will help you understand Chapters 8 and 9, which expand the scale of gene and protein interactions.

7.1 Dissecting a Gene's Circuitry

Throughout your biology education, you have been told that promoters control transcription. Later you learned that individual proteins, called transcription factors, bind to particular sites within promoters to facilitate transcription. Repressor proteins bind to other sites to block transcription. For most biologists, this explanation has been sufficient. Luckily, Eric Davidson wanted to understand exactly how genes

are regulated during early development. He chose the sea urchin as his model organism in part because it has been used for many years to study development. Of all the genes in sea urchin, why *Endo16*? It does not really matter why, only that *Endo16* is the best-understood gene in the world. In this section, we examine a lot of detailed information used by Davidson to produce a wiring diagram and a computer algorithm that model the genomic regulation of *Endo16.*

METHODS

Sea Urchin

LINKS

Eric Davidson

How Do Genomes Control Individual Genes?

Every gene needs to be controlled in three ways (location, time, amount), but we do not understand how this control is accomplished. Microarray data (Chapters 4 and 5) help us understand transcriptional output, and proteomics methods (Chapter 6) help us understand protein production. But the initial control mechanisms that regulate whether a gene will be active or silent are still hidden from us inside "black boxes." Davidson describes components at the 5' end of a gene that control transcription as "**cis-regulatory elements.**" Before we dissect the cis-regulatory elements of *Endo16*, we need to appreciate how tightly the expression of *Endo16* is controlled.

Are Genes Hardwired? A gene's cis-regulatory elements are the DNA sequences upstream of the start transcription site that control transcription. *Cis*-regulatory elements are different from *trans*-regulatory elements, which are DNA sequences further away from the coding sequence, often located on a separate chromosome (Figure 7.1). *Cis*-regulatory elements are **modular** in their organization. Modular in this context means the DNA can be divided into functional units, each of which performs a particular job when studied individually by investigators. Each cis-regulatory module is composed of a sequence of DNA to which one or more DNA-binding proteins can bind, either to help

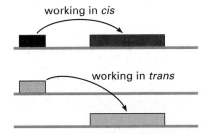

FIGURE 7.1 • Spatial difference between *cis*- and *trans*-regulatory elements. Cis-regulatory elements (dark blue) are adjacent to the coding sequences they regulate (dark gray), while trans-regulatory elements (light blue) regulate coding DNA located further away (light gray).

LINKS

100 Years

initiate transcription (transcription factors) or to repress transcription (repressors).

The best understood cis-regulatory elements are found in the sea urchin *(Strongylocentrotus purpuratus),* a model organism developmental biologists have studied for over 100 years. Sea urchins produce transparent embryos that are easy to produce and grow in the lab by in vitro fertilization. As with all model organisms, it is a useful tool for understanding the fundamentals shared by many species; in this case we can learn a lot about animal development. As with all animals, the new zygote is formed by the fusion of egg and sperm. From this single diploid cell, all the different tissues must arise. How does this mixed bag of proteins, carbohydrates, lipids, and nucleic acids know to dedicate the correct number of cells to form each of the required organs?

In the case of *S. purpuratus,* we know a great deal about how one cell mitotically replicates and commits each subsequent daughter cell to a particular fate. From one cell come two cells, then four, etc. After six rounds of mitosis and cytokinesis, the sphere of cells is composed of 60 cells (Figure 7.2). Biologists have injected embryos with nontoxic dyes and followed the distribution of the dye in older tissues to produce a **fate map,** which indicates what type of tissue each cell will

become. Fate maps have proven useful for understanding the significance of a gene being transcribed in a subset of cells at a particular time during development. Davidson's extensive analysis of the sea urchin model system has contributed substantially to our understanding of animal development.

Developmental biology has its own jargon and specific terminology, but only a few terms are critical for us to understand at this time. The most important are **blastula** and **gastrula.** Blastula refers to a hollow sphere of embryonic cells (Figure 7.2b) that is ready to undergo a process called gastrulation. Gastrula is the term used to describe an embryo during **gastrulation** when a subset of blastula cells begin to **invaginate** or move into the cavity of the blastula/gastrula (the "sea urchin" methods page has details and animations). These invaginating cells are genetically programmed to perform their movement, and without this internalization, the embryo will fail to develop normally. As the cells begin to invaginate, they repress some genes and activate others. The cells that form the elongating tube inside the gastrula will become the **endodermal** cells that line the gut of the future larva. The place where the cells begin to invaginate will become the mouth, and the place where this tube hits the other side of the embryo will become the anus. This elongated tube of cells is called the **archenteron.** As the archenteron continues to elongate, it differentiates into foregut, midgut, and hindgut to denote the eventual fate of becoming the mouth, digestive tract, and anus, respectively. **Ectodermal** cells will form the exterior surface cells of the larva.

Molecular Dissection of Development Eric Davidson and his colleagues have generated an amazing amount of high-quality data that illuminate the details hidden in what some people incorrectly referred to as "**junk DNA.**" As we all know from yard sales, one person's junk is another person's treasure. Davidson's group has sifted through the cis-regulatory elements of the gene *Endo16* and found many hidden treasures.

Although microarray technology (Chapters 4 and 5) has enabled us to simultaneously measure the activity of many genes, we cannot use microarrays to understand the circuitry that controls individual gene transcription. We have to select one gene and then use a **reductionist**'s approach to take it apart. This approach seems obvious at first, but if you have ever taken apart an appliance, clock, or coaster brakes on a one-speed bike, you know that dissection is easier than reconstruction. Therefore, selecting which gene to take apart is the first important decision to be made. Davidson's group chose one gene that exhibits all three levels of transcriptional control—time, location, and amount.

Expression of *Endo16* *Endo16* is as unlikely a hero as you could find anywhere in biology. During development, the expression of *Endo16* is tightly controlled for location, time,

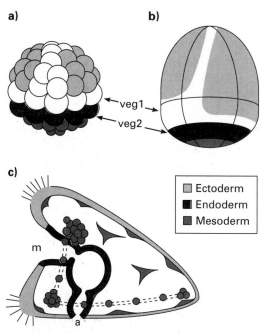

a)

b)

c)

Ectoderm
Endoderm
Mesoderm

FIGURE 7.2 • **Sea urchin *S. purpuratus* embryogenesis. a)** 60-cell embryo undergoing rapid cell division. Cell fates have been determined for many of these cells as shown by the color code for future ectoderm, mesoderm, and endoderm layers. Cells not yet committed to one of the three layers appear white. **b)** A 24-hour-old blastula composed of about 500 cells. **c)** Early larva 65 hours after fertilization composed of about 1,500 cells. The mouth (m) and anus (a) are shown on the larva.

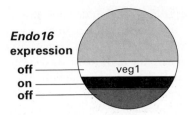

FIGURE 7.3 • Location and timing of *Endo16* expression. Prior to the blastula stage, *Endo16* is expressed in the veg2 cell layer. Only the veg2 layer will become endoderm, and the cell layers above and below veg2 do not express *Endo16*. Light gray indicates ectoderm, blue indicates endoderm, and dark gray indicates mesoderm; white is uncommitted.

and amount of RNA (Figure 7.3). *Endo16* has been the subject of intense research mainly because it is an ideal (i.e., model) gene to understand gene regulation during development. Slightly before gastrulation, cells at the base of the blastula express *Endo16,* and later all cells of the invaginating archenteron express *Endo16*. These cells will become endodermal cells in the larva (Figure 7.4). This expression pattern has made *Endo16* an excellent genetic marker to identify which cells will form the endoderm. Later in gastrulation, the cells of the foregut and hindgut repress *Endo16* so that only midgut cells still express it. By late gastrulation, the midgut cells transcribe *Endo16* at an even higher rate than before.

METHODS

In Situ

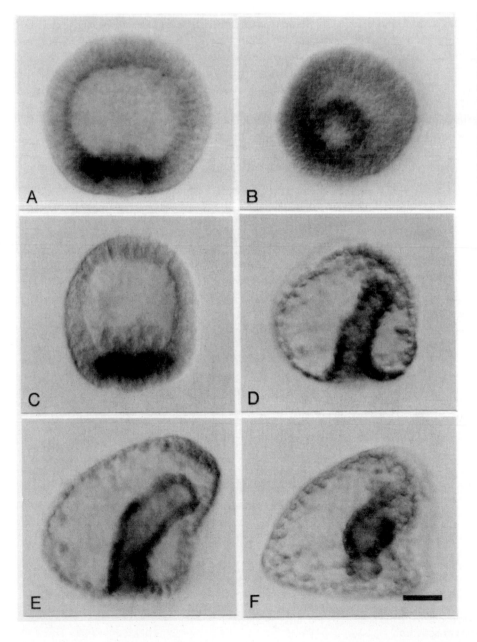

FIGURE 7.4 • *Endo16* transcription during sea urchin development. A) and **B)** Using in situ hybridization, *Endo16* mRNA is visible in 20-hour-old blastulas, **C)** a 30-hour-old blastula at the beginning of gastrulation, **D)** a 48-hour-old gastrula, **E)** a 60-hour-old embryo as it begins to take on the triangular shape of the larva, and **F)** a 72-hour-old larva stained darkly in the bulbous midgut with the *Endo16* probe. The lighter color outside the midgut is due to pigmentation in the embryo.

The *Endo16* protein (Endo16p) is secreted by the cells and localized to the basal extracellular matrix (i.e., cell side away from the hollow tube of the archenteron). Endo16p is a large glycoprotein that binds calcium. A similar gene has been identified in mice, and the two orthologs are similar in sequence to serum albumin, which is known to bind calcium. It appears that Endo16p and the mouse ortholog play active roles in cellular adhesion to the extracellular matrix, and possibly other cells too.

How Does a Gene Control Location, Timing, and Quantity of Transcription?

As a model gene, *Endo16* illustrates how a gene can control its expression. No other gene has been dissected and reassembled as well as *Endo16*. Therefore, we study *Endo16* because its circuitry and logic are wonderful in their design and exquisite in their control. A vast majority of the work in this section was conducted by Chiou-Hwa Yuh, a senior research fellow at Caltech.

The modules of *Endo16*'s cis-regulatory elements are located in the 2,300 bp upstream of the coding DNA (Figure 7.5). Each module (G–A) has a function when studied individually because DNA-binding proteins recognize specific DNA sequences within each module, and these protein-DNA interactions regulate transcription.

> ### DISCOVERY QUESTIONS
>
> 1. Why would a gene need so many different transcription factors to control its expression? Why not just use 5–10 DNA-binding proteins?
>
> 2. Of all the DNA-binding proteins shown in Figure 7.5, which would you want to study first? Why?

The term "watchmaker" can be used to describe anyone who successfully uses the reductionist approach to dissect an object and understand how it works. A related expression is "for every job you have to have the right tools." Davidson's group used molecular methods to build the right tools and became expert "watchmakers" as they dissected the cis-regulatory element modules of *Endo16*. We do not need to understand every aspect of these tools, but we should recognize the strategy involved (Figure 7.6). The "tools" numbered 2–8 illustrate DNA constructs created to fuse each module individually onto the most basic promoter (Bp) that allows RNA polymerase to bind and begin transcription. Construct 1 is the *wt* cis-regulatory element, and construct 9 is the Bp alone placed upstream of the coding DNA. The investigators used a reporter gene rather than the coding portion of *Endo16*. As their reporter, they chose **chloramphenicol acetyl transferase (CAT)**. The CAT gene was cloned from antibiotic-resistant bacteria that neutralize the antibiotic chloramphenicol by adding an acetyl group. CAT is not normally found in sea urchin, and its output (both mRNA and protein) can be monitored easily by investigators. Each of the seven modules was studied individually, but later you will notice that modules D and C are commonly studied jointly.

Location at 48 Hours Constructs 2 through 8 can be divided into two categories: those that expressed CAT in the endoderm (where *Endo16* is normally expressed) and those that expressed CAT outside the endoderm. Table 7.1 summarizes these data, specific examples of which are provided in Figure 7.7, page 212. Notice that constructs 1, 2, 7, and 8 promote the production of CAT in endodermal cells. These constructs include modules G, B, and A. Constructs containing modules C, D, E, and F do not promote the production of CAT in endoderm cells but do permit CAT production in **mesoderm** and ectoderm cells.

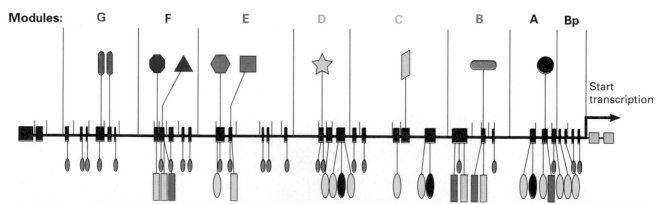

FIGURE 7.5 • *Endo16* **cis-regulatory element (horizontal black line) divided into the eight modules lettered G through A.** Within each module, DNA sequences are recognized by DNA-binding proteins (blue boxes on horizontal black line). The DNA-binding proteins are illustrated as colored shapes above and below the DNA. The larger shapes above the DNA bind to only one module, while those below bind to two or more modules. Bp is the "basic promoter" where the RNA polymerase binds to begin transcription.

TABLE 7.1 • **Expression of CAT from different cis-regulatory modules as promoters (see Figure 7.6) in 48-hour embryos.**
Numbers in the table indicate the percentage of embryos expressing CAT enzyme in the indicated cell types; some constructs expressed CAT in more than one cell type, so the total percentage in a column may exceed 100. Color coding highlights the critical data for the three cell types (see Figure 7.2). "Total # embryos" indicates how many embryos exhibited a detectable signal. Numbers in parentheses indicate no more than three cells per embryo expressed CAT compared to the widespread expression for constructs 1, 2, 7, and 8.

Construct	1 wt	2 G+Bp	7 B+Bp	8 A+Bp	6 C+Bp	5 D+Bp	4 E+Bp	3 F+Bp	9 Bp
Endoderm	97	94	89	95	0	0	0	0	0
Mesoderm	3.9	6	13	5.9	0	63	78	100	82
Ectoderm	5.8	12	19	12.3	0	41	28	4	27
Total # embryos	70	74	72	74	0	(11)	(41)	(45)	(13)

Source: Derived from Yuh and Davidson. 1996. *Development.* 122: 1074, Table 1 (partial).

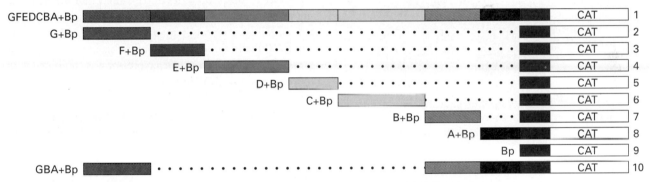

FIGURE 7.6 • **The first ten DNA constructs built by Davidson's lab to dissect *Endo16*'s cis-regulatory elements.** Each colored block represents a module (see Figure 7.5), and the modules are labeled G to A from left to right. Bp (black box) is located immediately upstream of the CAT-coding DNA. Dotted lines indicate which portions of the *wt* cis-regulatory element have been deleted. For example, construct 2 (G+Bp) has module G fused directly onto the Bp, which was fused onto the reporter gene CAT.

DISCOVERY QUESTIONS

3. If modules G, B, and A all promote about the same expression, why would *Endo16* need three redundant modules that don't do much better in combination (GBA+Bp) than they do individually (see Table 7.1)?

4. Hypothesize how modules G, B, and A can be active in endodermal cells but not in other cell types. Refer to Figure 7.5 to formulate your hypothesis.

Why Do Modules F, E, and DC Promote Expression in the Wrong Cells? From the data in Table 7.1, the roles of modules F–C are unclear since they do not promote the production of CAT in endodermal cells where *Endo16* is normally expressed but they do promote its production in mesoderm and ectoderm. It seems counterproductive for portions of a gene's cis-regulatory element to promote transcription in the wrong cell types. Therefore, new constructs were created that combined the best endoderm promoter (GBA+Bp) with each of the remaining modules (Figure 7.8, page 212). The idea was to test whether modules F–C had any influence on the transcription promoted by modules G, B, and A.

When the new constructs were tested, they exhibited different capacities to promote the formation of CAT (Table 7.2, page 213). However, the level of CAT production was essentially unchanged in endodermal cells, but the level of "inappropriate" CAT expression in mesoderm and ectoderm was altered. Modules DC (construct 13) reduced the capacity of the promoter to function in mesoderm cells, while modules F (construct 11) and E (construct 12) each reduced the expression of CAT in ectoderm cells. Therefore, modules F–C appear to function as cell-type specific repressors of transcription, which helps explain why *Endo16* is not expressed in ectoderm or mesoderm cells.

30 hours   48 hours

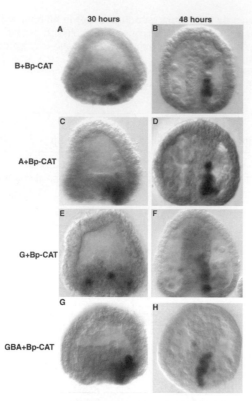

FIGURE 7.7 • Photomicrographs showing the location of CAT mRNA. In situ hybridization was used to detect CAT mRNA produced with different modules from the *Endo16* cis-regulatory element in embryos at 30 (left column) or 48 (right column) hours postfertilization. The constructs used in each case are described to the left with the module controlling the expression written before the "Bp-CAT" core construct. The modular design of the constructs can be seen in Figure 7.6: (A–B) construct 7 = B+Bp-CAT; (C–D) construct 8 = A+Bp-CAT; (E–F) construct 2 = G+Bp-CAT; and (G–H) construct 10 = GBA+Bp-CAT.

Combining the data from Tables 7.1 and 7.2 with Figure 7.7, we can build our first comprehensive model for the function of each module. Modules G, B, and A strongly *promote* "appropriate" expression in the endoderm. Modules F and E *repress* expression in the ectoderm, while modules DC *repress* expression in the mesoderm. This level of promoter dissection is impressive but not unique. Many other research groups had cut and pasted modules before, but they often stopped at this point. Davidson's group wanted to go further, but they needed many more constructs than those we have already seen (Figure 7.9).

Davidson's team wanted to know which activating modules (G, B, or A) were specifically repressed by modules DC, E, and F. For this work, they reported their data in a new format that requires a little explanation. Rather than simply reporting the amount of CAT enzyme activity in a large table, they used bar graphs to display the *relative* amount of activity compared to the lowest level of CAT activity when only the basal promoter (Bp) was placed upstream of the CAT-coding DNA. If there were no difference (e.g., Bp-CAT), no bar graph is visible and you see only the vertical line at the number 1.00 along the logarithmic scale at the top of Figure 7.10, page 214. If a construct led to an increased production of CAT enzyme compared to Bp-CAT, then a bar extends to the right and displays fold induction. Constructs that expressed less CAT have bar graphs extending to the left and displaying fold repression. Figure 7.10 illustrates six classes of constructs. All bar graphs should be compared to the top one within each class.

Notice in the X-Bp series of constructs that the repressor modules can repress expression in some constructs but not others. When a repressor module was inserted between G and BA (GXBA-Bp series), each of the repressors reduces expression by about half. When a repressor was combined with module A alone (XA-Bp series), the repressor effects were even greater, especially when more than one repressor was combined. When repressor modules were combined with G, or B, or GB, the effects were not as uniform or strong compared to the XA-Bp series. Certain combinations of repressor modules did have stronger effects on either G or B, but not as strong as when module A was present. These data led the investigators to conclude that the repressors work primarily by affecting module A's ability to initiate transcription.

DISCOVERY QUESTIONS

5. How do repressor modules exert their effect on *Endo16* expression?

6. Why does *Endo16* need two types of repressor modules (DC vs. F and E)?

7. Draw a picture that shows how repressor modules might be able to block transcription even when inducer modules are present. Review Figure 7.5 and include DNA-binding proteins in your model.

GFBA+Bp   CAT   11
GEBA+Bp   CAT   12
GDCBA+Bp   CAT   13

FIGURE 7.8 • Three DNA constructs to elucidate the roles of modules F, E, and DC. Same colors and symbols as used in Figure 7.6. The numbering of the constructs continues from Figure 7.6.

TABLE 7.2 • **Percentage of embryos expressing CAT enzyme when the best endoderm promoter was combined with the remaining four modules.** Location and frequency of CAT expression in 48-hour embryos containing constructs 10–13 (Figure 7.8). Some constructs expressed CAT in more than one cell type, and thus the total percentage in a column may exceed 100. Color coding (see Figure 7.2) highlights the significant data.

Name	10 GBA + Bp	11 GFBA + Bp	12 GEBA + Bp	13 GDCBA + Bp
Endoderm	94	99	94	94
Mesoderm	9.4	10.4	30.4	1.4
Ectoderm	15	5.7	4.3	26

Source: Derived from Yuh and Davidson. 1996. *Development.* 122: 1075, Table 2.

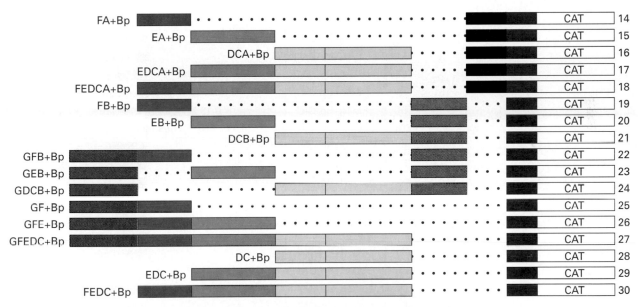

FIGURE 7.9 • **Additional DNA constructs to elucidate the roles of modules F, E, and DC.** Same colors and symbols as used in Figure 7.6. The numbering of the constructs continues from Figure 7.8.

What Happens to Cis-Regulatory Element When Exposed to Lithium?

At this point, we know that modules G, B, and A enhance the basic promoter's ability to drive transcription. Modules DC act to repress expression in mesoderm, while modules E and F repress expression in ectoderm. Davidson and his lab wanted to understand another aspect of *Endo16*'s transcriptional control. It has been known for many years that lithium (Li) is a **teratogen,** which means lithium causes birth defects in developing embryos exposed to it. This explains why it is very important to know whether a woman is pregnant before prescribing Li to control her depression. If she were pregnant, the Li would lead to serious birth defects and probably a miscarriage.

Lithium has many effects inside sea urchin embryos, one of which is the conversion of cells from ectoderm fate into cells that will become endoderm. Since *Endo16* is expressed in endodermal cells, Li-treated embryos express more *Endo16* than untreated control embryos. The investigators wanted to find out which modules were responsive to Li treatment. These experiments used the same DNA constructs used in early studies, and the results are shown in Figure 7.11, page 215. In this figure, the bar graphs show the ratio of CAT enzyme produced in Li-treated embryos to the amount of CAT produced in untreated embryos. For example, if Li treatment had no effect on a construct, the bar graph would be near 1 since the amount of CAT produced in the presence of Li would be equivalent to the amount of CAT produced in the absence of Li. If the ratio of CAT activity is below 1, Li treatment resulted in less CAT activity than in untreated embryos and the construct was not sensitive to Li. However, if the bar graph extends beyond 1, the construct was sensitive to Li and produced more CAT in the presence of Li than in

FIGURE 7.10 • **Repressive functions of modules F, E, and DC on modules G, B, and A in 48-hour embryos.** The base level of CAT enzyme produced in the entire embryo promoted only by the Bp is set at 1.00. The ability of each construct to express more (bar graphs extending to the right) or less (bar graphs extending to the left) CAT enzyme is indicated. The number at the end of each bar graph denotes the fold change from the basal expression, which was 2.5×10^5 active CAT enzyme molecules per embryo. The scale is logarithmic, the error bars indicate standard error, and numbers in parentheses under the construct names denote how many separate experiments were performed.

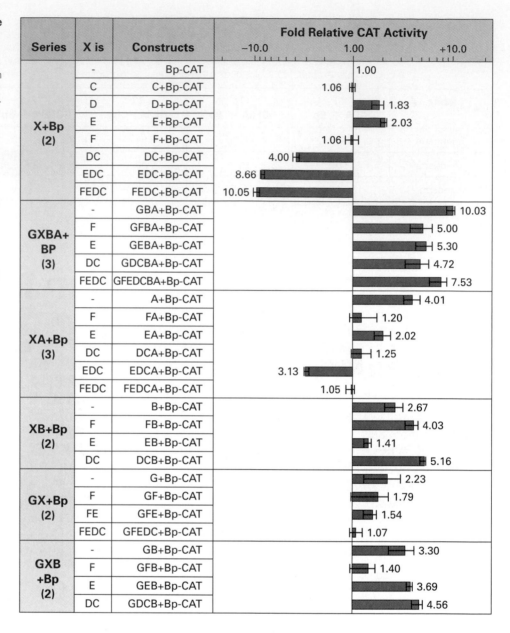

Series	X is	Constructs	Fold Relative CAT Activity
X+Bp (2)	-	Bp-CAT	1.00
	C	C+Bp-CAT	1.06
	D	D+Bp-CAT	1.83
	E	E+Bp-CAT	2.03
	F	F+Bp-CAT	1.06
	DC	DC+Bp-CAT	4.00
	EDC	EDC+Bp-CAT	8.66
	FEDC	FEDC+Bp-CAT	10.05
GXBA+Bp (3)	-	GBA+Bp-CAT	10.03
	F	GFBA+Bp-CAT	5.00
	E	GEBA+Bp-CAT	5.30
	DC	GDCBA+Bp-CAT	4.72
	FEDC	GFEDCBA+Bp-CAT	7.53
XA+Bp (3)	-	A+Bp-CAT	4.01
	F	FA+Bp-CAT	1.20
	E	EA+Bp-CAT	2.02
	DC	DCA+Bp-CAT	1.25
	EDC	EDCA+Bp-CAT	3.13
	FEDC	FEDCA+Bp-CAT	1.05
XB+Bp (2)	-	B+Bp-CAT	2.67
	F	FB+Bp-CAT	4.03
	E	EB+Bp-CAT	1.41
	DC	DCB+Bp-CAT	5.16
GX+Bp (2)	-	G+Bp-CAT	2.23
	F	GF+Bp-CAT	1.79
	FE	GFE+Bp-CAT	1.54
	FEDC	GFEDC+Bp-CAT	1.07
GXB +Bp (2)	-	GB+Bp-CAT	3.30
	F	GFB+Bp-CAT	1.40
	E	GEB+Bp-CAT	3.69
	DC	GDCB+Bp-CAT	4.56

untreated embryos. In this manner, the investigators could test each construct ± Li, and if the ratio was significantly larger than 1, the construct was sensitive to Li.

The results in Figure 7.11 are striking in their clarity. From the X+Bp constructs, we see that Li was unable to produce more CAT when each repressor module was tested individually. However, when module A was present (GXBA+Bp and XA+Bp constructs), Li treatment resulted in a substantial increase in CAT activity, which indicates these constructs were responsive to Li. When module A was combined with one of the repressors, the amount of CAT produced was increased. This experiment was well designed since there were constructs that contained module A but lacked repressor modules, and these constructs were not affected by Li. Additional data confirm this conclusion. The

bottom three class of constructs (all lacking module A) lacked the increased production of CAT when exposed to Li whether repressor modules were present or not. Therefore, Li does not activate modules G, B, or A of the cis-regulatory element; rather it blocks the repression by modules F, E, and DC as long as module A is present.

What Controls *When Endo16* Is Expressed in the Embryo? At this point, the investigators had established which modules promote and repress transcription and which modules address the location component of *Endo16* expression in embryogenesis (Figure 7.10 and Tables 7.1 and 7.2). Building on their success, they turned to the question of timing. Blastula formation occurs at about 24 hours and gastrula at about 48 hours. Sixty-five hours after fertilization,

Series	X is	Constructs	CAT Activity Per Embryo: Li/Untreated			
			1	2	3	4
X+Bp (2)	-	Bp-CAT	0.61			
	F	F+Bp-CAT	0.94			
	E	E+Bp-CAT	0.77			
	DC	DC+Bp-CAT	0.68			
	FEDC	FEDC+Bp-CAT	1.21			
GXBA+ BP (3)	-	GBA+Bp-CAT	1.06			
	F	GFBA+Bp-CAT		2.68		
	E	GEBA+Bp-CAT		2.18		
	DC	GDCBA+Bp-CAT		2.27		
	FEDC	GFEDCBA+Bp-CAT		2.47		
XA+Bp (3)	-	A+Bp-CAT	1.05			
	F	FA+Bp-CAT		2.41		
	E	EA+Bp-CAT		1.65		
	DC	DCA+Bp-CAT				3.88
XB+Bp (2)	-	B+Bp-CAT	0.91			
	F	FB+Bp-CAT	0.84			
	E	EB+Bp-CAT	0.99			
	DC	DCB+Bp-CAT	1.10			
GX+Bp (2)	-	G+Bp-CAT	1.01			
	F	GF+Bp-CAT	1.34			
GXB+ Bp (2)	-	GB+Bp-CAT	1.00			
	F	GFB+Bp-CAT	0.73			
	E	GEB+Bp-CAT	0.99			
	DC	GDCB+Bp-CAT	1.05			

FIGURE 7.11 • **Effect of LiCl on the cis-regulatory element constructs.** The amount of CAT enzyme activity in the presence of Li is divided by the amount of CAT enzyme activity in untreated embryos. In this graph, the further the bar goes to the right, the more CAT produced in the presence of Li and thus the more sensitive the construct was to the endoderm-inducing effects of Li. Li was added at the two-cell stage and CAT enzyme assays were performed 48 hours postfertilization.

the larva stage is achieved (see Figures 7.2 and 7.4 as well as the "sea urchin" methods web page for details).

The data for the timing experiments are plotted as time courses for each DNA construct; CAT enzyme activity was measured at each time point (Figure 7.12, page 216). To provide a baseline for comparison, they measured the CAT activity using constructs 1, 2, 7, 8, and 10 (see Figure 7.6). From these data, we can see that modules G, B, and A each promoted CAT transcription, and constructs that incorporated all three produced the most CAT. However, with the sampling of CAT production over time, we can see that the three inducing modules exhibited different temporal profiles. Module A induced CAT production during the first 48 hours and then dropped off. Module B promoted CAT production primarily at the 60- and 72-hour time points (after module A had already peaked). Module G did not promote much CAT production by itself, though there is a marginal increase around 48 hours. When modules GBA were combined, the expression level was almost equal to the wild-type cis-regulatory element containing all eight modules.

The investigators wanted to take into consideration the fact that the number of cells is changing over the same time period. They normalized their data by dividing the number of CAT molecules produced by the number of cells that typically produce Endo16p in *wt* embryos. To perform this normalization, they had to determine the number of Endo16p-expressing cells from 20–72 hours of development in *wt* sea urchin embryos (Figure 7.12b inset). The ability to count these cells is one advantage of working with the sea urchin embryo. This allowed them to calculate the number of CAT molecules produced in endodermal cells that normally transcribe *Endo16* (Figure 7.12b). After this normalization, the whole embryo data interpretation did not change substantially. During the first 48 hours, most of *Endo16* transcription was under the control of module A. Module B began to exert an influence at 60 hours, when B and A equally contributed to CAT expression. By 72 hours, module A contributed a smaller percentage of the total expression, and module B had taken over control of expression. In fact, module B alone was almost sufficient to supply the entire amount produced by the *wt* cis-regulatory element at 72 hours. The little bit that B alone was lacking could be accounted for by the amount induced by module A. As seen in Figure 7.12a, module G played a trivial role by itself. It is interesting that the shape of the normalized

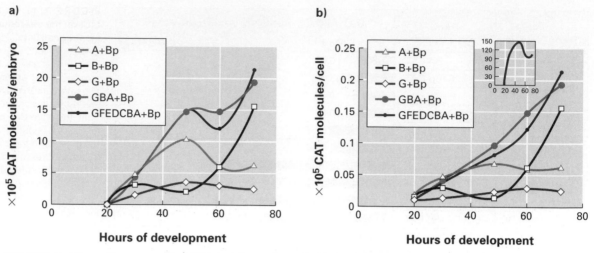

FIGURE 7.12 • **Measurement of CAT protein produced in 20- to 72-hour embryos.**
a) The number of CAT molecules produced *per embryo*. The data points are averages of 100 embryos per time point; the smooth lines make it easier to see trends but are not intended to imply the level of CAT in times when samples were not taken. **b)** The number of CAT molecules produced *per cell* (the number of CAT molecules in the embryo divided by the number of cells that normally express endogenous *Endo16* at each time point). The inset graphs the number of cells (Y-axis) that normally express *Endo16* during development (X-axis).

B+Bp time course was very similar to the time course for the full-length cis-regulatory element.

What Does Module G Do?

To determine the role of module G in *Endo16* transcription, the investigators built DNA constructs 31 through 38 as shown in Figure 7.13a. For these constructs, they removed the *Endo16* Bp and replaced it with a weakened viral promoter (SVp). When each inducing module was placed individually upstream of SVp, each construct produced a different temporal profile (Figure 7.13b) though similar in shape to the analogous ones in Figure 7.12a. These profiles indicated that modules G, B, and A exerted their temporal influence on transcription without any participation by Bp. When the three inducing modules were analyzed in pairs or all together, the amplitudes and shapes of the time courses were altered. The amplitude of CAT production by module A was increased approximately fourfold when module G was added onto A+SVp, though the shape of the curve was essentially unchanged. Interestingly, module G did not alter the amplitude or shape of module B's ability to produce CAT. When modules B and A were combined, the shape and amplitude were very similar to those of GBA+Bp in Figure 7.12a. When module G was added onto BA+SVp, the amplitude is substantially increased at the 48-hour time point, which is when module A is exerting its maximum effect. Furthermore, the addition of module G (GBA+SVp) has increased the output from module B at 60 and 72 hours, when module B becomes active. However, module G increased the output of module A (at 48 hours) to a greater extent than G

increased the output of module B (at 60 and 72 hours). In short, module G acts as an amplifier for module A and to a lesser extent module B, but only if B is combined with A.

DISCOVERY QUESTIONS

8. Summarize the major spatial, temporal, and capacity aspects of modules G, B, and A.

9. Module G acts as an amplifier, but this might seem wasteful. Hypothesize why it might be beneficial to have a separate amplifier module rather than simply boosting the capacity of modules B and A directly.

Why Bother Drawing a Gene's Circuit Diagram? Most biologists collect data until they can build a conceptual model that explains the process they study. A good model allows investigators to make predictions derived from the logical consequences of the model. These predictions can be tested in the lab or field, and one of two things will happen: The predicted behavior will be observed, or it will not. If the prediction is substantiated, the model has passed its first test. If not, the model needs refinement to incorporate the new data. Predicting, experimental testing, and refining the model is a **reiterative** process that continues to improve the model. Once a model has been refined enough, it can provide understanding and insights that would not have been possible without it. These insights can lead to new areas of

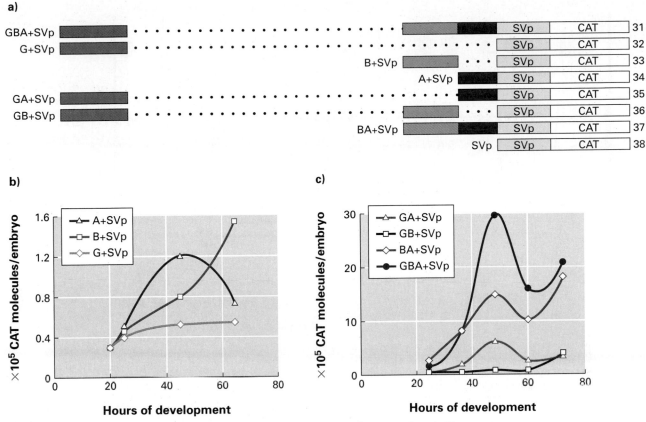

FIGURE 7.13 • **The role of module G. a)** For these constructs, the Bp was replaced with a basic promoter from the virus SV40 (SVp; same colors and symbols as in Figure 7.6). The numbering of the constructs continues from Figure 7.9. Measurement of CAT enzyme molecules produced per embryo from **b)** 20- to 72-hour embryos, or **c)** 22- to 72-hour embryos.

research and discoveries. In short, a good model is the end product of a lot of research and provides the starting place for additional research and understanding.

Circuit diagrams are one form of modeling. The use of circuit diagrams is recent, or, to use current jargon, not mature. We have been using forms of circuit diagrams throughout this book, and biologists have been drawing versions for a few years. However, we lack a standardized set of symbols. The issue of standardized circuit diagrams will be discussed in Chapter 8, but for now, let's stick to the symbols used by Davidson and his group. Figure 7.14, page 218, shows us the "hardwiring" that has evolved in the sea urchin *Endo16* cis-regulatory elements.

Circuit Diagram of *Endo16* Cis-Regulatory Elements

From the data presented so far, we can create our first *Endo16* circuit diagram that explains how the cis-regulatory elements function in early and late development (see Figure 7.14). The purpose for creating a circuit diagram is not to allow electrical engineers to create a copper version of *Endo16*. The real benefit of circuit diagrams such as this is their predictive power. When combined, modules B and A

determine the shape of the profile and module G amplifies the capacity of B and A. Early in development (up to 48 hours), regulation of *Endo16* is controlled by module A, but modules G and B can influence module A. Repressor modules F–C function by negatively affecting module A. Modules DC repress transcription in mesoderm, and modules F and E repress transcription in ectoderm. Finally, Li treatment induces "inappropriate" transcription of *Endo16* in nonendoderm cells by blocking the ability of the repressor modules to suppress transcription via module A.

DISCOVERY QUESTIONS

10. Use Figure 7.14 to predict whether module A would be active in ectoderm cells in the absence of modules F and E. Would module B be active in the absence of A?

11. Use Figure 7.14 to predict what proteins are present or absent in endoderm cells that prevent the repression of *Endo16* by modules F–C.

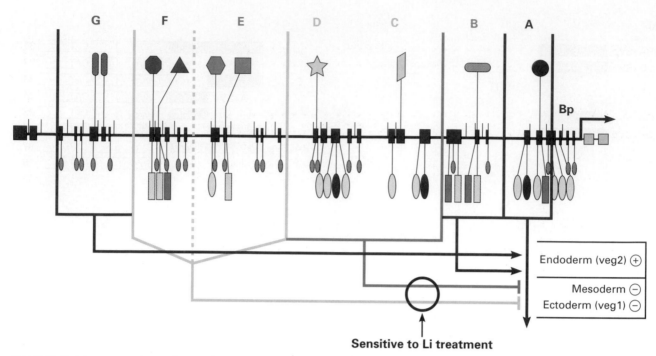

FIGURE 7.14 • **Circuit diagram of the *Endo16* cis-regulatory elements.** The modules of the *Endo16* cis-regulatory element and associated DNA-binding proteins are shown as in Figure 7.5. Modules that induce expression are denoted by blue lines; repressive modules are denoted by gray lines. The three cell types are listed to the right, and the net effect on each is shown: + indicates activation; − indicates repression.

> 12. Use Figure 7.14 to predict what would be needed for module B to become more productive later in development.

So far, the circuit diagram explains the data we have covered, but that is to be expected since it was constructed from the same data. The interesting question is whether the model allows us to make any new predictions. Figure 7.15 shows the first prediction that could lead to new research and subsequent discoveries. Bmer is predicted to be one of the DNA-binding proteins that enables module B to have its 60- to 72-hour location, timing, and amount specificity so that *Endo16* is expressed at very high levels in the midgut but completely shut off in the foregut and hindgut. The gene encoding Bmer is predicted to have its own cis-regulatory elements that permit its expression only in the midgut. The circuit diagram prediction also implies that there are transcription factors that bind to the MG module and repressors that bind to the FG and HG modules. When produced in the appropriate cells, Bmer would activate module B, which can be amplified by module G resulting in the production of *Endo16* at very high levels only in midgut cells beginning 60 hours after fertilization. All these predictions are due to the more comprehensive understanding provided by the *Endo16* cis-regulatory element circuit diagram.

DISCOVERY QUESTION

13. This question is challenging. Use the predictive powers of the circuit diagrams in Figures 7.14 and 7.15a to predict which DNA-binding proteins are present in which cells and at what times during development. Only address the nine unique factors that are drawn above the line in Figures 7.5, 7.14, and 7.15. Once you have made your prediction, write a brief summary of the research projects you would launch to test your predictions. You may use any method with which you are familiar.

As you can imagine, your predictions could lead to new insights for gene regulation, animal development, and genome control. The experiments you devise could be developed into research projects for publication since you are dealing with an area that has not been experimentally tested.

***Endo16* Cis-Regulatory Elements** In Discovery Question 13, you had a chance to test the predictive powers of the *Endo16* circuit diagram. You might be surprised that you could conduct genomic research with pencil, paper, and circuit diagrams. But one aspect of biology unaffected by genomics is

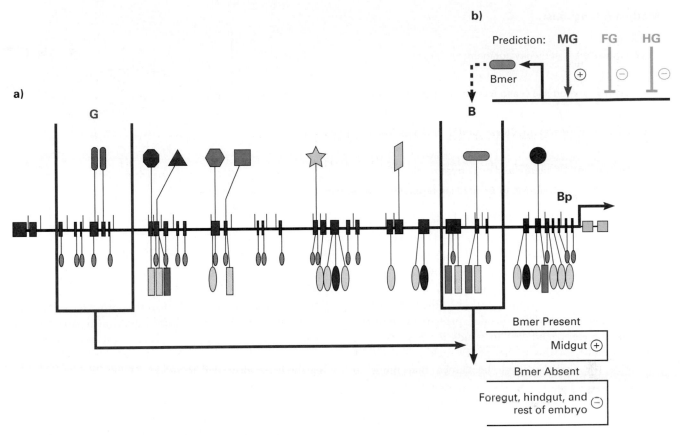

FIGURE 7.15 • **Regulation of module B activity. a)** Modified circuit diagram from Figure 7.14. The DNA-binding protein (Bmer) activates module B beginning about 60 hours after fertilization. Module B is amplified by module G to transcribe *Endo16* at a high rate only in midgut cells of the larva. **b)** Expression of Bmer is controlled by its cis-regulatory elements, which are comprised of three modules: MG is an activator in midgut cells; FG is a repressor in foregut cells; HG is a repressor in hindgut cells. The dotted arrow indicates that Bmer would bind to module B of the *Endo16* cis-regulatory element to control the expression of *Endo16* in midgut cells.

that good research always starts at the desk and not the bench. Good *ideas* drive research. Good technology allows good ideas to be realized, but without the quiet, contemplative time, science would not progress.

It seems that Yuh and Davidson and their colleagues must create time for contemplation because their watchmaker-like dissection of *Endo16* continued. They looked at the prediction about the Bmer protein and decided to determine which DNA-binding proteins did what jobs. There are 15 different DNA-binding proteins that bind with high specificity to the cis-regulatory elements of *Endo16*. However, they could not possibly analyze all 30+ binding sites using 15+ DNA-binding proteins, so they decided to start with only one module. Which module would you start with?

Yuh and Davidson decided to start with module A (Figure 7.16, page 220). Why? As they explained in a paper, "Module A interacts with all of the other *Endo16* cis-regulatory elements and is either absolutely required for their operation or

synergistically enhances their output. Moreover, it serves as a central switching unit, acting according to inputs from the other modules." They reasoned if you are going to study one module, start with the most critical one and one where you are likely to see an effect when the system is perturbed.

What Does Module A Look Like at the DNA Level? Let's zoom in for a closer look at the DNA sequence and protein binding sites for module A (Figure 7.17, page 221). Some of the proteins that bind to and facilitate the actions of the eight modules (Bp, and G–A) have been identified. Of those given a name, some have been better characterized than others. For example, protein "C" is a cAMP-responsive, element-binding (CREB) protein family member, which among other functions has been shown to play a role in memory formation in flies and mammals.

The functions of module A are mediated through interactions at eight different sites where at least four different proteins bind. Only two of these four genes have been

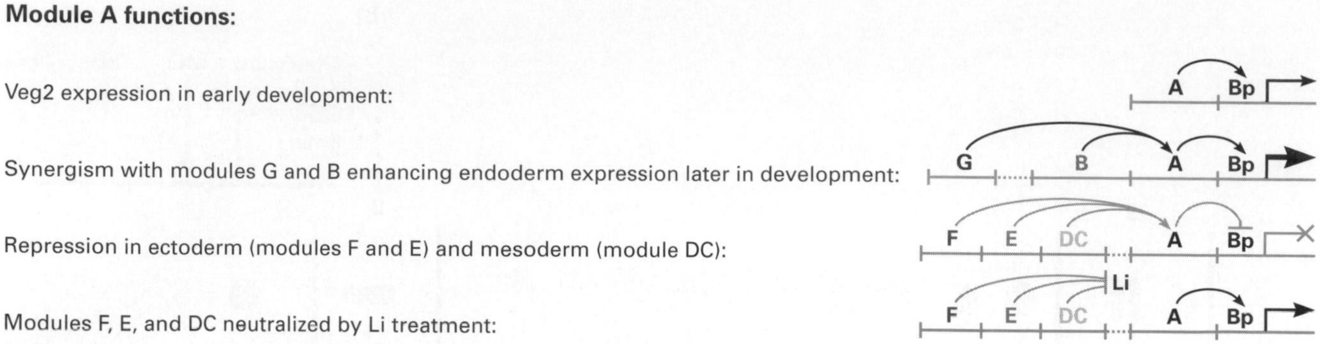

Module A functions:

Veg2 expression in early development:

Synergism with modules G and B enhancing endoderm expression later in development:

Repression in ectoderm (modules F and E) and mesoderm (module DC):

Modules F, E, and DC neutralized by Li treatment:

FIGURE 7.16 • **The roles of *Endo16* cis-regulatory element modules.** Module A communicates the output of all upstream modules to the Bp. Arrowhead denotes activation; ⊥ denotes repression; X denotes lack of transcription.

cloned and sequenced. *Otx* is a member of the orthodenticle transcription family. Otx is a shorthand way of saying orthodenticle (O) transcription (tx). The other cloned gene is called SpGCF1, and its encoded protein binds many places in the eight modules. Sp stands for *S. purpuratus,* the Latin name for this species of sea urchin, and GCF1 represents the first *Factor* to be cloned that binds to *GC*-rich DNA. SpGCF1p binds once in module A, twice in Bp, and appears to weakly stimulate transcription. Because SpGCF1p had been characterized before by this and other labs, Yuh and Davidson decided to study other proteins that bind to module A.

The researchers' challenge was to determine the roles for the remaining DNA-binding proteins in module A. As mentioned above, Otx protein binds to module A and nowhere else in *Endo16*'s cis-regulatory elements. The four binding sites labeled CG1, CG2, CG3 and CG4 (see Figure 7.17) are all bound by the same transcription factor, but the function of each binding site was unknown. Finally, two proteins called P and Z also bind to module A and nowhere else in *Endo16*. Yuh and Davidson decided they needed more molecular tools for this job, so they created a new series of mutant cis-regulatory element DNA constructs and performed experiments similar to those described above.

How Do the Transcription Factors That Bind to Module A Work within *Endo16*? Like all good scientists, this group began their next steps where they ended previously. They spliced together different portions of the cis-regulatory elements and placed them upstream of the Bp and the CAT reporter gene. These constructs were injected into developing embryos and CAT enzyme activity was measured. The activity is reported as the number of CAT enzyme molecules produced per embryo. As you can see in Figure 7.18a, page 222, when modules B and A work together, their output is a synergistic product of modules B and A assayed separately. Interestingly, the output of B+A is exactly 4.2 times the output of module B alone! This numerical insight provided

the investigators with a new perspective and new questions. (1) Why does the combined B+A output have the same shape but greater magnitude than the output from B alone? Why doesn't B+A equal the sum of individual parts B and A when measured in isolation? (Historical side note: This should sound familiar. "Why do wrinkled and smooth parental pea plants fail to produce mildly wrinkled peas?" Though Gregor Mendel's and Davidson's questions were addressing very different aspects of genetics, it is interesting to see that the best questions never get too old to be asked again and again in new situations.) (2) What specific DNA-binding sites in module A are responsible for the 4.2-fold increase in expression of module B as seen in Figure 7.18a?

To determine the answer to their questions, the investigators focused on the CG1 and P sites in module A (see Figure 7.17b). When either CG1 or P were mutated, the output dropped by about 50%, and it no longer had the 60- to 72-hour rise in output (Figure 7.18b). In fact, the BA constructs with either CG1 or P mutated looked very similar to module A alone (Figure 7.18c). Module A, however, was not affected by either mutation when compared directly (Figure 7.18d). So what does all this mean?

The investigators drew several conclusions from these data, as well as additional data not shown here:

1. Both CG1 and P sites are needed to provide the full 4.2X module B production of CAT (Figure 7.18b).

2. In module A, the CG1 and P sites have no other function since transcription from module A is unaffected by these mutations (Figure 7.18d).

3. In experiments not shown here, they tested all other binding sites in module A and found that no other sites in module A had any effect on the output of module A when studied in isolation (similar to Figure 7.18d).

4. Module B does not communicate directly with Bp, but must interact with CG1 and P of module A. This conclusion is based on the fact that none of the BA constructs with mutated CG1 or P have curves that rise late

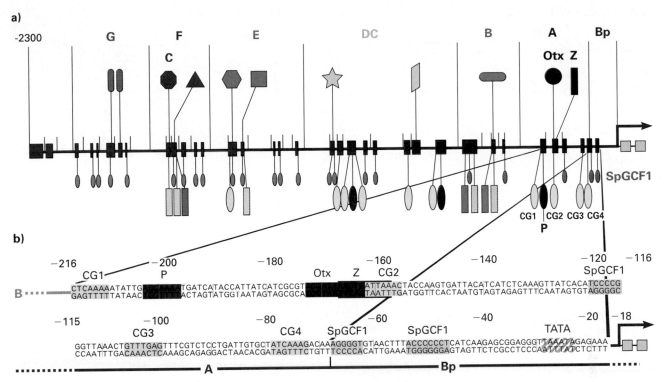

FIGURE 7.17 • **Module A and Bp showing protein binding sites. a)** Several of the module-specific DNA-binding proteins have been named. CG1–4 indicate the same protein binds to four different sites. Note that all the small blue ovals below the line are copies of SpGCF1.
b) The nucleotide sequence of module A and Bp. The sequence shown for module A starts at base −216, which is the 5' edge of the CG1 binding site. Module A sequence continues onto the next line and terminates with the black vertical line at base −68 (5' edge of a SpGCF1 binding site). Bp begins at base −67 and extends to base −1, though the sequence in this figure stops at base −18. The sequences to which DNA-binding proteins bind are enclosed in colored boxes.

the way module B does when studied in isolation (Figure 7.18a–c).

In short, module A functions as a **bimodal toggle switch.** Either module A alone determines the output of *Endo16* (first 48 hours), or there is input from module B,

which acts on the CG1 and P sites (beginning at 60 hours). In other words, module A toggles from complete control of gene output to complete submission of control to module B. When module B is activated (by the predicted Bmer protein; see Figure 7.15), module A acts as a multiplier of module B's output. These interpretations beg the question: Which binding sites in module A determine its output in the absence of module B input (i.e., first 48 hours of development)? Can you design experiments to answer this question?

Many more constructs were needed, but only those involving Otx and Z sites are presented here. Notice from Figure 7.17 that the Otx and Z sites abut one another. For these experiments, the investigators produced various mutations in the Otx and Z DNA sequences to block the binding of the appropriate transcription factors. The different DNA constructs were placed into sea urchin embryos, and the amount of CAT enzyme produced was measured for each time point.

For comparison, the ability of module A alone to produce CAT is graphed (Figure 7.19a, page 223). When the Otx binding site was mutated, module A is unable to promote CAT production. Interestingly, when only an 18 bp section of module A was used *(OtxZ),* the ability to promote CAT

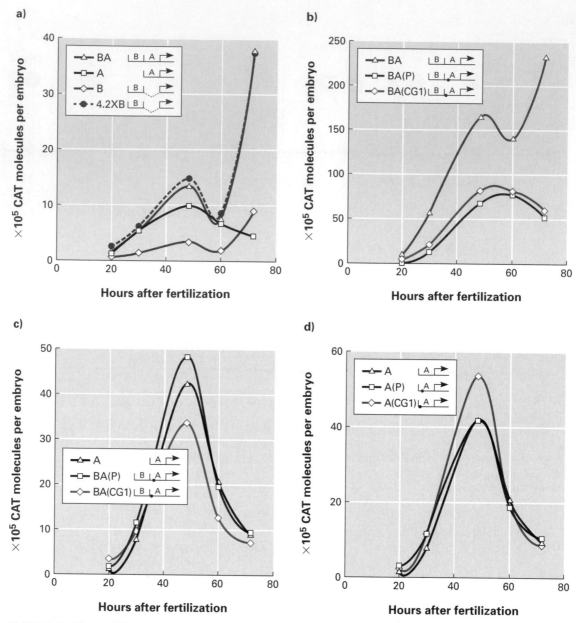

FIGURE 7.18 • **CAT output (molecules per embryo) with different *Endo16* cis-regulatory element DNA constructs.** In the panel legends, dots within a module indicate the location of mutated P or CG1 sites. The name of each construct contains the mutated site in parentheses. The dotted line in a) was produced by taking the time course of module B alone and multiplying it by 4.2; thus it is a theoretical time course. All other time courses were experimentally derived.

production was recovered substantially. It is surprising that a mere 18 bp fragment, which represents only 10% of module A, is able to produce a *wt* temporal profile with amplitude about half that of full-length module A. Using the results shown in Figures 7.18 and 7.19 and the circuit diagram in Figure 7.14, the investigators made a prediction and designed an experiment to test it.

Prediction: If module B works through module A via the CG1 and P sites, then mutating Otx should not affect the ability of A to function as a conduit for module B.

When B was combined with A or A + mutated Otx, there was no difference in output (Figure 7.19b). These data confirm the prediction that module A uses only sites CG1 and P to amplify B; Otx does not play a role in this amplification. However, Otx is the critical site to determine the output of module A in the absence of module B (i.e., during the first 48 hours of development). The summary table in Figure 7.19c shows how important Otx is for module A to function properly. Especially noteworthy are the experiments performed on the *OtxZ* 18 bp fragment. This fragment was able to produce about half as much CAT as full-length

a)

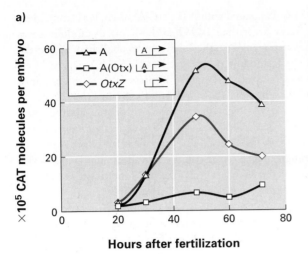

b)

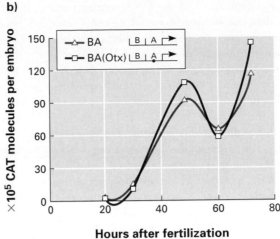

c)

Constructs	Endoderm
Endo16	97.0%
A	78.9%
A(Otx)	5.1%
OtxZ	56.5%
(Otx)Z	3.6%
Otx(Z)	72.0%

FIGURE 7.19 • Role of Otx site in the output of module A. a) Time course expression of CAT protein using different versions of module A as the promoter. Panel legend uses dots to indicate the location of each mutated site, and the site inside the parentheses has been mutated *OtxZ* (in italics) indicates an 18 bp fragment of Otx and Z DNA was the only portion of module A upstream of Bp. **b)** Time course expression of CAT protein using modules B+A or B + mutated A as the promoters. **c)** The percentage of embryos that exhibited CAT expression in the endoderm for each construct. Parentheses denote sites that were inactivated. To inactivate the Otx site, the sequence ATTA was changed to GCCG; Z was inactivated by changing TGATTAA to CAGCCGG (see Figure 7.17b). *OtxZ* indicates an 18 bp fragment was used as in panel a).

module A, but when module A had an inactivated Otx, CAT production dropped 15-fold. In contrast, when Z is inactivated, the productivity of the 18 bp fragment is enhanced so that it promotes better than the *OtxZ wt* fragment did.

DISCOVERY QUESTIONS

16. Look at Figure 7.17b and locate the Otx and Z binding sites. Formulate a hypothesis to explain why the *Otx(Z)* fragment promoted better than the *OtxZ* fragment.

17. Look at Figure 7.4 and predict the expression pattern for *Endo16* in an embryo that has a mutated Otx site. Remember that blastula are formed in 24 hours, gastrulation occurs at about 48–60 hours, and a larva is about 72 hours old.

Which Sites Within Module A Are Responsive to Repression by Modules DC, E, and F? Determining how a repressor interacts with module A required the use of Li again. In normal sea urchin embryos, Li caused the loss of repression in ectoderm and mesoderm cells that normally repressed *Endo16*. With Li present, cells that normally repressed *Endo16* transcribed it. A large set of experiments was needed to tease out the parts of module A important for repression. To minimize the number of variables, the investigators decided to use only module F as their test case for a repressor module—F normally represses *Endo16* in ectodermal cells. For these experiments, some new DNA constructs were built, and some constructs were used from earlier studies. For each construct tested, the amount of CAT enzyme produced in Li-treated embryos was determined and divided by the amount of CAT enzyme produced in untreated embryos. The ratios (Li-treated : untreated) were displayed as bar graphs. If the Li-treated embryos produced the same amount of CAT enzyme as untreated embryos, then the ratio would be 1 (Figure 7.20). If the DNA construct was sensitive to Li, then the bar graph would exceed 1 since more CAT would have been produced per embryo when exposed to Li. Again, the goal was to determine which parts of module A interacted with the repressor modules. DNA constructs whose ratios exceed 1 contain functional repressors, while those with ratios close to 1 do not.

The clarity of the results is impressive and should help you appreciate the quality of this research. The first constructs (GBA and GFBA) show how the system works when repressor F is interacting normally with module A and Li is present. Note that the ratio of CAT expression increases dramatically when modules F and A are in the same construct. The next six constructs show that it was module F interacting with module A, and not modules G or B, that produced the Li response.

The final seven constructs in Figure 7.20 focused on smaller subregions within modules A and F. The nomenclature for the next set of constructs needs a bit of clarification. The letter *C* stands for the CREB binding site within module F (see Figure 7.17a). When only this portion was used upstream of Bp+CAT, an italic *C* was used in Figure 7.20.

CAT Activity Per Embryo: Li/Untreated

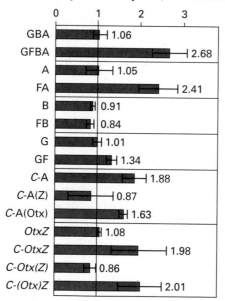

FIGURE 7.20 • **Repressor-sensitive portion of module A.** Relative amounts of CAT produced are displayed as bar graphs. Each bar graph represents a ratio of CAT enzyme produced in the presence of Li divided by CAT enzyme produced in the absence of Li. Bars that extends above 1 indicate that more CAT was produced in the presence of Li and therefore repressor module F was working properly. 48-hour-old embryos containing each construct were used to detect Li sensitivity. Otx and Z constructs use the same nomenclature as in Figure 7.19. *C* is a binding site in module F and stands for *C*REB, which is a DNA-binding protein.

When *C* was paired with A (e.g., *C*-A), there is a strong Li effect (ratio larger than 1). When the Z site in module A was inactivated (*C*-A[Z]), the Li effect was abolished. When the Otx site in module A was inactivated (*C*-A[Otx]), the Li effect was retained. Even more impressive were the last four constructs, in which both the A and F modules were reduced to short DNA fragments and the rest of the modules were omitted. *Otx*Z alone was insensitive to Li treatment, as was the entire A module. *C*-*Otx*Z produced an effect similar to the much larger FA construct. When the Z site (*Otx*[Z]) was mutated, the Li effect was lost, but mutating the Otx site (*[Otx]Z*) did not reduce the Li effect at all.

It is worth noting the length of the DNA pieces used for these experiments. The combination of Otx and Z sites from module A consisted of 18 bp (TCGCGTAGG*ATTA*AG*T-GATTAA*). When Otx was mutated, ATTA was changed to GCCG, while the Z site was mutated from TGATTAA to CAGCCGG (see Figure 7.17b to locate Otx and Z sites). Amazing how simple changes in "junk DNA" produce dramatic consequences.

DISCOVERY QUESTIONS

18. Summarize the data from Figure 7.20.

19. What assumptions were made by using only module F for these experiments? What were the practical reasons for making these assumptions?

20. Are there any parts of module A that remain to be dissected?

Which Portion of Module A Interacts with the Basic Promoter (Bp)? If you compare the output in Figures 7.12a and 7.13a, you can see that the different basic promoters (Bp vs. SVp) did influence transcription. We also know that module A is the conduit through which all others communicate with Bp. Therefore, Davidson's lab wanted to determine which portions of module A were responsible for communicating with Bp. In Figure 7.17, you can see that most of the module A binding sites have been experimentally tested. The only three remaining are CG2, CG3, and CG4. Therefore, it is not surprising that their final experiments focused on these three sites.

The investigators conducted experiments where different cis-regulatory element constructs were placed upstream of the CAT-coding DNA and injected into sea urchin embryos. The amount of CAT enzyme was measured at different time points. When modules B+A were used, we see the standard output. The deletion of CG2 binding site, however, reduced the amplitude by about 50%, while the overall shape of the output was retained (Figure 7.21a). Thus, the CG2 site plays an important scaling role; that is, it influences amplitude of the output. When each of the other two CG sites were mutated and assayed in conjunction with module A but not

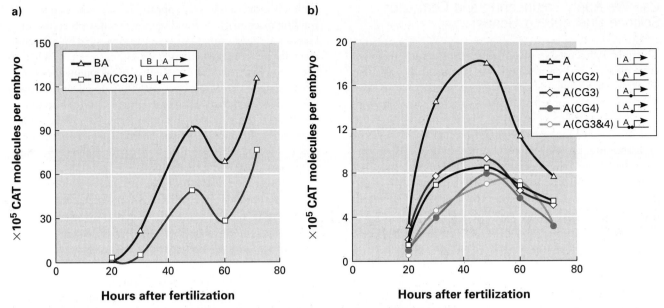

a)

b)

FIGURE 7.21 • **Role of CG2, CG3, and CG4 on interaction between module A and Bp.**
To perform these experiments, TCTAGA was inserted in place of the following nucleotides:
CG2 = TGATTAAACT; CG3 = TGTTTGAGTTT; CG4 = ATCAAAGACAAAGG. CAT activity was
used to determine the number of molecules per embryo. In the legend, dots denote the location
of mutated sites, which are shown within parentheses in the construct names.

module B, all constructs retained the same shape but the amplitude was always reduced by half (Figure 7.21b). Binding sites CG2, CG3, and CG4 are important for module A's ability to multiply the output by a factor of 2. This completes the dissection of module A, since we have already discovered the roles played by binding sites Z, Otx, CG1, and P.

DISCOVERY QUESTIONS

21. Look carefully at Figure 7.17 and find the sequences altered in Figure 7.21. Describe fully the mutations generated in these constructs. Are any other binding sites altered in addition to the CG sites? If so, what impact should this have on the output of A+Bp constructs?

22. Do the locations of the CG binding sites relative to other binding sites in module A suggest any models for controlling transcription? Could these models lead to testable predictions?

23. Predict what would happen when the entire *wt Endo16* cis-regulatory element was placed upstream of the CAT sequence but CG2 was nonfunctional. Draw a graph (use Figure 7.12a as a template).

24. Create a table that summarizes the roles of binding sites Z, Otx, CG1, CG2-4, and P.

We have spent a significant amount of time and effort discovering how cis-regulatory elements work to control the expression of a single gene in the gut of a developing sea urchin. You might be wondering why. Remember, model organisms tell us a lot about all species when we study fundamental processes. Yeast taught us about cancer, and worms taught us about genomic trade-offs in aging. The sea urchin has taught us about genomic regulation of development. Genes needed early in development are tightly regulated, and what appears to be redundant components (e.g., 4 CG sites) can play unique roles. Initially we knew that *Endo16* contained three activator modules (G, B, and A) and three repressor modules (F, D, and C). Based on the experimental evidence, each DNA-binding site has its own function. When the right combination of proteins bind to their appropriate sites, the downstream gene will be expressed *when* it is needed, *where* it is needed, and in the appropriate *amount*. Multiply this level of gene control by 35,000, and you begin to understand the level of control needed in each of your trillion cells to produce dynamic proteomes in each cell. Furthermore, cis-regulatory modules raise many evolutionary questions about their origins and how they might be duplicated and used by other genes that require similar regulation. As more genomes become sequenced, we will be able to trace the evolution of individual cis-regulatory elements. Therefore, during the **postgenomic era** we will continue to mine genome sequences to understand genes and evolution.

Can We Apply Engineering and Computer Science Concepts to Genes?

The results presented in this chapter were the culmination of years of work and on a scale never attempted by any other lab. It is worth remembering that these experiments used DNA constructs not found in sea urchin but that were engineered and placed in *wt* embryos. The investigators assumed that the injected embryos expressed normal amounts of the DNA-binding proteins, and the concentration of these proteins determined the regulation of gene expression. Given the need to perturb a system in order to understand it, all scientists must live with similar assumptions, though it is worth remembering how the data were produced.

After thousands of experiments, we now have enough information to assign a role to each of the eight modules (G–A and Bp) as well as some binding sites within module A. Module A is the central hub that performs a type of computational logic in order to determine where, when, and how much a gene should be transcribed (see Figure 7.16). Some functions of the genome are "hardwired" into the DNA with some of the instructions located in the noncoding regions. Calling noncoding DNA "junk DNA" is equivalent to considering the keyboard and monitor the only important parts of a computer.

Throughout this book, we have seen that the ability to visualize data is nearly as important as collecting the data. By using the term "hardwired," Davidson was creating a mental image that everyone can understand. A vending machine that dispenses cold sodas is hardwired; put the money in,

push a button, and a soda appears. If you were to open the vending machine, you would see a series of wires connecting functional units (money counter, buttons, switching mechanism that chooses which soda to dispense, etc.). When a hardwired machine is built, the electrical engineer draws a circuit diagram showing how each functional unit is connected in order to accomplish the desired output. Each functional unit contains symbols that indicate how it works. Davidson wanted to use similar electrical engineering principles to capture all the information learned about the transcriptional control of *Endo16* (Figure 7.22).

To understand *Endo16*'s circuit diagram, think of each module or binding site as a source from which wires extend. When two wires meet at a node, a small transistor is located to perform a simple calculation. The overall flow of information in the circuit leads from left to right. The outcome of the cis-regulatory elements at the far right is analogous to a soda appearing after you have made your selection. For *Endo16*, the outcome of this circuit is the initiation of transcription at the right time, in the appropriate cells, and to the correct level of RNA production.

Each node performs a calculation, and we need to understand each input and how it affects the output of each node. Node α represents the repressive action of modules F, E, and DC mediated through site Z. Node β receives input from binding sites P and CG1 to amplify the output of module B by twofold. Node γ receives input from binding sites CG2–CG4 to produce a twofold amplification signal. Node δ receives positive input from modules B and G and produces a positive signal. The input from modules G and B are time

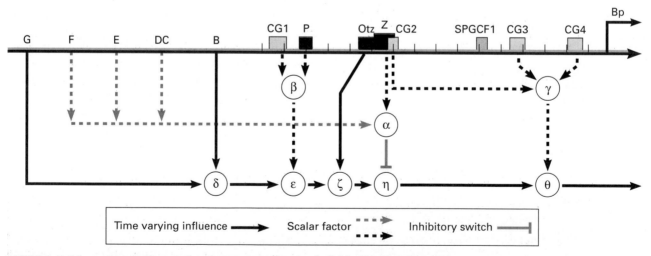

FIGURE 7.22 • **Circuit diagram for *Endo16* transcription.** Individual binding sites within module A are indicated by labeled boxes above the double line representing the DNA. The interactions of the upstream modules with elements of module A are indicated by circles and arrows beneath the DNA. Each labeled circle, or node, represents a specific regulatory interaction. Two types of regulatory influence are indicated: time-varying interactions (solid lines) that determine the temporal pattern of *Endo16* expression, and location/amount interactions that affect the level of *Endo16* transcription by constant scaling factors. Note that the scaling factor of the repressor modules determines which cells will repress *Endo16*.

dependent in that they become active later in development. Node ε receives input from nodes δ and β and produces a product of the two input nodes. Node ζ (zeta) receives input from Otx that is critical to the ability of module A to communicate with Bp early in development. Node ζ is also affected by node ε, which is regulated by the stage of development. Therefore, node ζ is capable of receiving information early (Otx) and late (G and B) in development to produce a signal that is a toggle switch. Node ζ will produce an outgoing signal that is driven either by module A (via Otx) or module B (and module G amplifies module B). The toggle switch of node ζ is controlled by the timing of protein production (e.g., the hypothesized "Bmer," which activates module B).

Once the developmental timing toggle switch ζ has determined whether module A or B is in control of *Endo16* transcription, a few more inputs are needed before the final level of transcription can be determined. One important input is whether a particular cell will transcribe *Endo16* or not. Node α receives its input of repression and sends a signal to node η (eta). Node η also receives a signal from node ζ, promoting transcription of *Endo16*. The decision to make at node η is whether to pass along the positive signal, leading to transcription, or the negative signal, causing transcription to be blocked. Each cell must make its own determination whether to transcribe *Endo16* or not, and the decision is determined by the location of the cell. If the cell is located in the endoderm, then no repression signal reaches node η and *Endo16* is transcribed. If the cell is located in the mesoderm (modules DC) or ectoderm (modules F and E), then node α produces an inhibitory signal that negates the input from node ζ, the same way that pushing the money return button on a vending machine stops the transaction. Endodermal cells do not inhibit node η, and thus the original input from either module A or B is passed on to node θ (theta) which also receives input from node γ. If a positive signal is received from η, then the signal will be amplified by node γ. The final signal transmitted to the basic promoter (Bp) is determined by node θ. The final outcome from θ determines the timing, location, and amount of *Endo16* mRNA produced in each cell and thus helps shape the outcome of the developing embryo.

Models (e.g., circuit diagrams) are ways of displaying information so it can be more readily understood. Compare Figure 7.22 to the table you produced (Discovery Question 24) that lists the functions for the DNA binding sites. Your table contains the same information but is too limited to understand all the intricacies that are a part of *Endo16* gene regulation. For example, what would happen if CG1 and P were nonfunctional? Your table probably indicates that the production of *Endo16* would be reduced by half, but from the circuit diagram we see that this is only true later in development, since node β does not influence node ζ if the signal from node δ is lacking. From the circuit diagram, we can better understand the interactions of the different modules and binding sites, but the circuit diagram visualization

of *Endo16* regulation is not quantitative, so Davidson produced a different model to supplement the circuit diagram.

Calculating Gene Output—The Power of Predictions

Most students of biology like to think of genes as small factories that produce RNA, but we often gloss over the issue of quantity. How does it know how much RNA to produce in each cell? How does the gene know when to increase its productivity when the cell needs it? As with any factory, there must be a way to regulate the rate of production. Davidson's lab has dissected *Endo16*, and we know more about its ability to regulate the amount of RNA than any other gene's. However, the circuit diagram lacked any quantitative information, so let's look at a computational model that allows you to make quantitative predictions (Table 7.3, page 228). In short, Davidson has written a simple computer program that summarizes in seven steps what we know about *Endo16* gene regulation.

If you have ever written any computer code, you will be familiar with the style and logic of this quantitative model. If writing code is new to you, you might not recognize immediately the logic within the code. In fact, the two sentences you just read illustrate the logic in Davidson's model. Notice in Table 7.3 the pattern of several lines as "if . . . else . . ." conditional statements. These conditional statements are saying that two possible options exist. For this paragraph, either you have written computer code or you have not, and there is a consequence for each of the two possibilities (familiarity or no familiarity). As we walk through *Endo16*'s computational model, we will examine several conditional statements and discover the consequences for each of the two possibilities. The model consists of seven parts and applies the following logic.

1. If any of the three repressor modules is activated (in mesoderm or ectoderm) and binding site Z is functional, then the consequence will be an activated (α = 1) node and the transmission of a repressor signal to η, which will result in no input at θ—i.e., *Endo16* will be repressed. Alternatively, α will be inactive (α = 0), the inducing signal from ζ will be transmitted to η and θ, and *Endo16* will be transcribed.

2. If both amplifiers CG1 and P are on, then β is 2, which represents its twofold enhancement of output from modules G+B. Alternatively, β is turned off if either CG1 or P is nonfunctional, and module B is unable to transmit its signal on to Bp. In this case, module A resumes its control over *Endo16* expression.

3. If the amplifiers CG2–CG4 are functioning, then γ is 2, which represents its twofold enhancement of transcription. Alternatively, γ is neutral and there is no additional enhancement of *Endo16* transcription.

4. Time of development controls module B activation. Node δ output depends upon the time-dependent activation of modules G and B, which are regulated by

TABLE 7.3 · *Endo16* **regulatory algorithm.** Boolean logic applied to the circuit diagram in Figure 7.22. The logic is in the left column and explanations are in the right column. Time, location, and amount of *Endo16* transcription are described in these seven steps.

1. if ($F = 1$ or $E = 1$ or $DC = 1$) and ($Z = 1$) 　　　$\alpha = 1$ 　else　$\alpha = 0$	Repression functions of modules F, E, and DC mediated by Z site
2. if ($CG1 = 1$ and $P = 1$) 　　　$\beta = 2$ 　else　$\beta = 0$	Both CG1 and P are needed for synergistic link with module B
3. if ($CG2 = 1$ and $CG3 = 1$ and $CG4 = 1$) 　　　$\gamma = 2$ 　else　$\gamma = 1$	Final step up of system output
4. $\delta(t) = B(t) + G(t)$ 　$\epsilon(t) = \beta \times \delta(t)$	Positive input from modules B and G Synergistic amplification of module B output by CG1-P subsystem
5. if ($\epsilon(t) = 0$) 　　　$\zeta(t) = Otx(t)$ 　else　$\zeta(t) = \epsilon(t)$	Toggle switch determining whether Otx site in module A, or module B will control level of activity
6. if ($\alpha = 1$) 　　　$\eta(t) = 0$ 　else　$\eta(t) = \zeta(t)$	Repression function inoperative in endoderm but active elsewhere
7. $\theta(t) = \gamma \times \eta(t)$	Final output communicated to Bp

Source: Yuh et al. 1998. *Science.* 279: 1901, Figure 6B only.

LINKS

Susan Ernst

upstream proteins such as Bmer described in Figure 7.15b. Module B assumes control during gastrulation about 60 hours after fertilization. Node ϵ is the product of δ and either 2 or 0 (see step 2) depending on the state of CG1 and P as described in the right column.

5. The toggle switch determines whether module A or module B governs the transcription of *Endo16*. If ϵ is not yet activated (i.e., $\beta = 0$ in step 4 because it is too early in development), then ζ equals the time-dependent output from the Otx binding site in module A. Alternatively, ζ equals the output from ϵ (i.e., $\beta = 2$ in step 4), which is governed by module B.

6. Transcription response of the entire gene is governed by the repressors described in step 1. If the repressors are activated (i.e., $\alpha = 1$), then *Endo16* is not transcribed. Alternatively, the repressors are silent (i.e., $\alpha = 0$) and the signal for transcription at node η equals the input from the toggle switch at node ζ (calculated in step 5).

7. At any given time, the instructions produced by θ and passed on to the basal promoter is the product of γ (either 2 or 1 from step 3) and η (calculated in step 6).

In order to predict the amount of *Endo16* transcribed, all we need to know is the experimental values for G+B in step

4 and the output of Otx in step 5. Steps 1–3 determine the amplitude and location of transcription; step 5 determines which module is governing transcription, depending upon timing. Given these numerical values, the final outcome for *Endo16* is as certain as putting money in a vending machine and pushing the button.

Davidson's lab has extensively dissected *Endo16*'s cis-regulatory elements (2,300 bp) that controls when, where, and how much RNA will be transcribed. (The amount of *m*RNA is not controlled by the cis-regulatory elements; *m*RNA processing has its own regulation.) You should appreciate that this 2.3 kb piece of DNA is but one part of one gene within one genome. There are cis-regulatory elements for every gene in every genome. The degree of transcriptional control in *Endo16* may appear to be sufficiently complex to explain all gene regulation, but we have not dissected the genome's full complexity yet. In a paper by Susan Ernst and her colleagues at Tufts University, the investigators reported that *Endo16* RNA can be alternatively spliced in three different ways to produce three different mRNAs and two different proteins. Two of the three mRNAs encode for the same amino acid sequence but have different 3' untranslated regions. The longer form of the protein has additional calcium binding sites, which produces a different conformation of the total protein and presumably a different function. All this variation means that understanding the circuitry of cis-regulatory elements is only the beginning to understanding the production of a protein.

Summary 7.1

In order to survive, an embryo must control when, where, and by how much a gene will be transcribed. *Endo16* is regulated by the activity of its cis-regulatory elements, which work as if they were hardwired like a vending machine. If you insert money and push the button, a soda of your choice will appear. Your ability to control the vending machine is a consequence of its hardwiring; every time you insert money you can predict the outcome. It is almost shocking to think of your genes, perhaps all genes, as being regulated by hardwiring that can be diagrammed as electrical circuits or programmed like simple computer algorithms. We studied *Endo16* so you can begin to appreciate the complexity and predictability inherent in the transcriptional control of any genome (Chapters 8 and 9). Imagine what the circuit diagram would look like for 35,000 human genes. In Section 7.2, we will interact with two Java-based models that are prototypes for visualizing genome regulation.

DISCOVERY QUESTIONS

25. In Chapter 6, we discovered how proteins can behave stochastically. Is stochastic behavior possible in genes such as *Endo16,* or is its transcription identical in every endoderm cell?

26. If you isolated a mutant strain of sea urchin that lacked the Bmer gene, what effect would this have on *Endo16* transcription? What would the consequences be if a fertilized egg lacked a Z binding site? Refer to Figure 7.22 and Table 7.3 to help you answer these two questions.

7.2 Integrating Single-Gene Circuits

Endo16 is an excellent model gene and has been very informative, but the complexity of whole-genome regulation is too overwhelming to diagram as simple circuits. We need to revisit the problem we encountered when considering protein-protein interaction maps. How can we best display complex circuits? It might seem a trivial problem, but imagine looking at a circuit diagram for 35,000 genes or a computer program with enough lines of code to describe the development of a human embryo. Genomic information is accumulating faster than

ever, and we need new tools to visualize all of it simultaneously. A group in Russia has produced some interactive diagrams that we will use as a first step toward comprehensive models of genomic circuitry.

METHODS

Gene Circuits

LINKS

Lee Hartwell

How Can We Describe to Others What We Know About a Genome Circuit?

We want to develop computer models that will enhance our understanding and lead to new discoveries. With a good computer simulation, we could predict what kinds of changes we can make to a system to promote the maximum desired outcome and the least negative consequence. For example, what if you knew every gene that was activated during the progression of cancer (see Chapter 5)? You would like to develop a drug that could silence a critical gene or protein, but you don't know which one to choose. If you could predict the consequences of silencing a particular gene, you could choose a gene that killed the cancer cells but did the least harm to healthy cells. That's what a good understanding of genome-wide circuits could provide.

People have been creating computer simulations for many years—ever since the vacuum tube was invented. In 1952, Hodgkin and Huxley used mathematics to accurately model the action potential of squid neurons, and their model still stands as a monument to experimental design, keen observations, and computational modeling. As computers improve, we can evaluate models faster. However, speed is not the limiting factor—biological understanding is. In a 1999 commentary, Lee Hartwell and his colleagues predicted "the next generation of students should learn how to look for [biological] amplifiers and logic circuits, as well as to describe and look for molecules and genes." Hartwell was the principal investigator (and 2001 Nobel Laureate) who helped discover proteins called cyclins, which control the cell cycle in every plant and animal with a circadian rhythm.

Let's look at a simple interactive network for sea urchin. This model was produced in Russia and is the best public domain model available online. While viewing the model, think about whether it is more or less useful when thinking of networks.

Interacting Gene Networks Go to gene circuits web page and let the Java applet load completely before you click on any buttons (Figure 7.23 on your CD-ROM).

FIGURE 7.23 • Screen shots from a Java program.
Go to the CD-ROM to view this figure.

230 CHAPTER 7 • *Genomic Circuits in Single Genes*

METHODS

Protein-DNA
Circuits

Visual Language

If you mouse over the window without clicking, it speeds up the appearance of the images. Green arrows and boxes are those that you have not yet selected. Once selected, genes will appear either red or blue, according to this key:

Key to Symbols

(upstream genes)

Filled red arrows point to genes that become activated.

Hollow red arrows point to genes that become repressed.

(downstream targets)

Filled blue arrows point to genes that the highlighted gene activates.

Hollow blue arrows point to genes that the highlighted gene represses.

Technical Hints

1. For clarity, you can click once each on a series of genes while holding down the shift key, turning them gray. Then click on the "Pathway" button to eliminate all other genes from view.

2. You can find out a lot about any gene by holding down the control button on your keyboard and clicking once on the gene. You will get a new window with information about the gene.

DISCOVERY QUESTIONS

27. Find and then click on "Endo16." What can you conclude about this model's completeness with regard to *Endo16* gene regulation?

28. Now click on "CyIIIa." What kind of control is exerted on this gene? What type of protein does this gene encode? Use technical hint 2 to find out.

Can We Visualize Circuits for Protein Interaction and DNA Binding?

A Java applet to view protein-DNA circuits was created as an interactive means to show protein-protein interactions as well as protein-DNA interactions (Figure 7.24). An interaction is shown as a black node, with DNA drawn as boxes and proteins as ovals. Arrows leading toward a node indicate which two members are interacting. If a gene and a protein interact, the protein is binding to one of the gene's cis-regulatory elements.

FIGURE 7.24 • Screen shot of a Java web page that shows protein-protein and DNA-protein interactions. Go to the CD-ROM to view this figure.

Arrows leading away from the nodes always point to the protein produced by the gene. If you hold down the shift key and double-click on a node, it reveals the interaction with the regulatory protein and the target DNA. Choose a node of interaction (e.g., *Endo16* DNA and SpOtx protein) and look back at Figures 7.17 and 7.22 to compare the different types of models.

DISCOVERY QUESTIONS

29. Does this protein-DNA interaction applet help you see the complex interactions better than the gene circuits applet?

30. Identify which gene has the largest number of DNA-binding proteins that govern its expression. List each of the proteins that regulate this gene.

31. Find which DNA-binding protein regulates the highest number of genes. List those genes.

You have seen two ways to illustrate a number of the pathways related to *Endo16*. The field of creating gene and protein networks is new. Each group of genomic and proteomic investigators is creating their own set of tools to communicate their view of a network, and there are no universal standards in place. In October 2000, Isabelle Pirson, Jacques Dumont, and their colleagues at the Free University in Brussels, Belgium, proposed a simple set of symbols to illustrate on paper any network combination. However, since there are many competing visual languages in existence, it will take a lot of work and cooperation to produce a standardized way to communicate.

Summary 7.2

A common theme in genomics and proteomics is that simple unifying rules are usually too simplistic. Before we can fully understand how an embryo can develop into an adult, we will need to dissect more genes to discover which mechanisms are common and which are unusual. As we accumulate data, we will build comprehensive models but they need to be comprehensible, too. Interactive models, 3D graphics, layered images—who knows what will work best? However, the two examples studied in this section illustrate why 2D diagrams need substantial improvements to facilitate our understanding of how genomes control gene expression.

Chapter 7 Conclusions

At first, it may have seemed that studying *Endo16* in depth was inappropriate for a genomics textbook. However, no other gene has been dissected as completely, so *Endo16* is the perfect model to understand how genomes are converted from silent code to dynamic cells. Genomics and proteomics are fields that can collect large amounts of data, and the two fields interact when we study how an individual cell produces a dynamic proteome that is altered over time and in response to environmental changes. The best way to understand a machine is to take it apart and reassemble it. By taking apart *Endo16,* you understand the logic behind one gene's wiring diagram and computer algorithm that illuminate how *Endo16* is expressed in the developing gut of a sea urchin embryo. The tools Davidson used were modular, but they need to be applied on a genome-wide scale, which raises a troubling question. How can we possibly understand how *every* gene in a genome is regulated if it took Davidson many years to model *Endo16*? To model entire genomes, we will need new tools and different types of questions. Chapter 8 will focus on genomic circuits composed of fewer than 50 genes. In Chapter 9, you will see what has been done to integrate many forms of data to create a "systems" approach to understanding complete genomes.

References

Specific Pathways

Bhalerao, J., P. Tylzanowski, et al. 1995. Molecular cloning, characterization, and genetic mapping of the cDNA coding for a novel secretory protein of mouse: Demonstration of alternative splicing in skin and cartilage. *Journal of Biological Chemistry.* 270(27): 16385–16394.

Boos, Winfried, and Alex Böhn. 2000. Learning new tricks from an old dog. MalT of Escherichia coli maltose system is part of a complex regulatory network. *Trends in Genetics.* 16: 404–409.

Davidson, E. H., R. A. Cameron, and A. Ransick. 1998. Specification of cell fate in the sea urchin embryo: Summary and some proposed mechanisms. *Development.* 125(17): 3269–3290.

Gladstone Institutes. 1999. G protein signaling pathways. <http://gladstone-genome.ucsf.edu/public/Sample_genmapp.asp>. Accessed 11 October 2000.

Godin, R. E., L. A. Urry, and S. G. Ernst. 1996. Alternative splicing of the Endo16 transcript produces differentially expressed mRNAs during sea urchin gastrulation. *Developmental Biology.* 179(1): 148–159.

Kirchhamer, C. V., C. H. Yuh, and E. H. Davidson. 1996. Modular cis-regulatory organization of developmentally expressed genes: Two genes transcribed territorially in the sea urchin embryo, and additional examples. *PNAS* U.S.A. 93(18): 9322–9328.

Li, Qiliang, Susanna Harju, and Kenneth R. Peterson. 1999. Locus control regions coming of age at a decade plus. *Trends in Genetics.* 15: 403–408.

Ransick, A., S. Ernst, et al. 1993. Whole mount in situ hybridization shows Endo 16 to be a marker for the vegetal plate territory in sea urchin embryos. *Mechanisms of Development.* 42(3): 117–124.

Yuh, C. H., A. Ransick, et al. 1994. Complexity and organization of DNA-protein interactions in the 5'-regulatory region of an endoderm-specific marker gene in the sea urchin embryo. *Mechanisms of Development.* 47(2): 165–86.

Yuh, Chiou-Hwa, and Eric Davidson. 1996. Modular cis-regulatory organization of Endo16, a gut-specific gene of the sea urchin embryo. *Development.* 122: 1069–1082.

Yuh, C. H., H. Bolouri, and E. H. Davidson. 1998. Genomic cis-regulatory logic: Experimental and computational analysis of a sea urchin gene. *Science.* 279: 1896–1902.

Yuh, C. H., H. Bolouri, and E. H. Davidson. 2001. Cis-regulatory logic in the endo16 gene: Switching from a specification to a differentiation mode of control. *Development.* 128: 617–629.

Yuh, C. H., J. G. Moore, and E. H. Davidson. 1996. Quantitative functional interrelations within the cis-regulatory system of the S. purpuratus Endo16 gene. *Development.* 122(12): 4045–4056.

Review Articles

Bonifer, Constanze. 2000. Developmental regulation of eukaryotic gene loci. *Trends in Genetics.* 16: 310–315.

Davidson, Eric H. 1999. A view from the genome: Spatial control of transcription in sea urchin development. *Current Opinion in Genetics & Development.* 9: 530–541.

Endy, Drew, and Roger Brent. 2001. Modelling cellular behavior. *Nature.* 409: 391–395.

Wray, G. A. 1998. Promoter logic. *Science.* 279:1871–1872.

General Reference

Brent, Roger. 2000. Genomic biology. <http://www.molsci.org/htdocs/publications/omicbio/brentcell.html>. Accessed 11 October 2000.

Cho, Raymond J., and Michael J. Campbell. 2000. Transcription, genomes, function. *Trends in Genetics.* 16: 409–415.

Clayton, David F. 2000. The genomic action potential. *Neurobiology of Learning and Memory.* 74: 185–216.

Davidson, Eric H. 2001. *Genomic Regulatory Systems: Development and Evolution.* San Diego: Academic Press.

Gladwell, Malcolm. 2000. *The Tipping Point: How Little Things Can Make a Big Difference.* Boston: Little, Brown and Co.

Hartwell, Leland H., John J. Hopfield, et al. 1999. From molecular to modular cell biology. *Nature.* 402 Supplement: C47–C52.

Legrain, Pierre, Jean-Luc Jestin, and Vincent Schächter. 2000. From the analysis of protein complexes to proteome-wide linkage maps. *Current Opinion in Biotechnology.* 11: 402–407.

Pirson, Isabelle, Nathalie Fortemaison, et al. 2000. The visual display of regulatory information and networks. *Trends in Cell Biology.* 10: 404–408.

CHAPTER 8

Integrated

Genomic

Circuits

In biology, investigators must balance the utility of creating models against the danger of believing their models accurately represent a living system. Models of biological processes are never perfect, but they can help us make new discoveries. In Section 8.1, we examine gene regulation from a different level of control. Rather than determining exactly which DNA sequences control each aspect of a gene's overall productivity (Chapter 7), we will study gene regulation at the level of protein production. How can one gene influence another? Can genes work together to toggle between two alternative outcomes? Is it possible to apply our understanding of genetic toggle switches to design gene circuits in vitro that function in vivo? To answer these questions, we will explore a series of case studies that have led the way in modeling genomic responses on a small scale. Once we understand these types of **integrated circuits** (multigene interactions), can we calculate their reliability and ask why we are diploids and why we have apparently redundant genes?

When creating a model, it is best to start small and simple before building more complex models. In Section 8.2, we consider three integrated circuits. In order to understand how we learn new information, we will integrate a series of small circuits into a larger network. From this complex integrated circuit, we hope to discover new properties that were not apparent when each circuit was studied in isolation. Similar principles are applied to understand cancer and even a functioning organism—a virus called T7. To understand how the simple genome of T7 functions, we will test different arrangements of genes that can lead to different phenotypes. Ultimately, we want to understand how organisms function by understanding how proteins work alone and as a part of integrated circuits.

8.1 Simple Integrated Circuits

We know that our genes can be regulated so they are activated in some cells and repressed in others. We also know that proteomes are dynamic, changing in response to environmental influences and aging. How can a single cell alter its proteome if its **cis-regulatory elements** do not change? Cells need a mechanism to switch from on to off and vice versa. Genes need to sense their intracellular environment and respond accordingly. However, we don't want our cells to change so rapidly that genes are turned on and off every second of every minute. It would be a disaster for our brain cells to sense a drop in glucose and respond by converting themselves into liver cells that can store sugar. Therefore our genes need to be tolerant of some cellular variations. Furthermore, vital cell functions need to have alternative means for accomplishing a necessary task. Our genomes must be prepared for circumstances that might block one circuit from performing its cellular role. For example, human cells normally consume oxygen in order to produce adenosine

triphosphate (ATP). This is a good strategy until you are being chased by a bear. At times like this, it is good to have an alternative means to produce enough ATP to continue running. Knowing how genomic circuits function allows us to calculate the reliability of each component in the circuit, which can further our understanding of genomes as they are regulated in living cells.

LINKS

Harley McAdams

Adam Arkin

Can Genes Form Toggle Switches and Make Choices?

Let's look at one universal issue related to networks—**bistable toggle switches.** You know what a toggle switch is; it turns on your lights, computer, DVD player, etc. A bistable toggle switch will remain in one position (on or off) until the circuit determines the switch should be toggled to the other position. This is easy to understand in electrical engineering terms, but how can genes toggle between on and off? In Chapter 7, we saw how transcription factors regulate whether a gene will be "on" or "off," but what controls the transcription factors? And what controls the proteins that control them? Part of the answer is that the egg is not an empty bag of water but is filled, thanks to mom, with many lipids, carbohydrates, nucleic acids (including mRNA ready for translation), and proteins, including transcription factors. Developmental biologists have revealed to us what causes an egg to enter mitosis and cytokenesis and the beginnings of a new organism. Nonembryonic cell division repeats itself according to some internal regulatory mechanism. Normally, our cells can control their cell division toggle switch, but when they lose control of this switch, we develop cancer. The answers to these questions about genetic toggle switches are neither esoteric nor insignificant.

How Do Toggle Switches Work? There are two ways to start answering this question: Start with data and build a model, or start with a model using engineering principles and improve the model with experimental data. Harley McAdams of Stanford University's School of Medicine and Adam Arkin in the Physical Biosciences Division of the Lawrence Berkeley National Laboratory combined the best of both approaches in an elegant analysis of genetic toggle switches. The first issue they had to address was noise.

Noise in a regulatory system such as a toggle switch means that, unlike your computer, genetic switches have to deal with a degree of uncertainty. We know that gene activation occurs when transcription factors bind to cis-regulatory elements. But when a cell undergoes mitosis and cytokenesis (eukaryotes) or cell division (bacteria), the first source of noise is introduced—which daughter cell will receive more of the existing transcription factors? Of course, if cells were as wise as Solomon, the pool of transcription factors would be split right down the middle, 50:50. However, cells are not "wise" and to some extent the partitioning process during cell division is random, or **stochastic.** For example, if a cell had 50 copies of the Otx transcription factor, 6% of the time a particular daughter cell might get 19 or fewer copies, while 6% of the time it might get at least 31 copies (Math Minute 8.1). That could have a profound impact on the subsequent regulation of *Endo16* expression.

Another component of genetic noise is that there are only a few binding sites for each protein, and binding occurs at a slow rate. For example, Otx may be able to bind to only a few cis-regulatory elements in the entire genome, and it has to find these sites. Each cis-regulatory element must be found by a small number of DNA-binding proteins. The limited number of transcription factors and binding sites results in a random distribution of times when all the transcription factors are in the right places for any given gene. Another example of slow reaction rate is that once the cis-regulatory element is fully occupied and ready to initiate

transcription, the first RNA will be produced a variable number of seconds later due to noise in the initiation of the transcription machinery. Transcription takes an *average* of several seconds to begin, but again this is an average, with a distribution of times shorter and longer than the average.

What effect do noise and stochastic behavior have on a cell? In prokaryotes and eukaryotes, proteins are produced in bursts of varying durations and with varying outputs. Therefore, the total number of proteins produced from any one gene is not the same each time but rather an average with a **normal distribution** (see Math Minute 8.1). By producing proteins in bursts rather than at a constant rate, it provides proteins a higher probability of forming dimers that may be required for full function. The bottom line is, most students learn "gene Y is activated and produces X proteins per minute," but this summary statement is an oversimplification of a messy and mildly chaotic world inside each of your cells.

MATH MINUTE 8.1 • HOW ARE STOCHASTIC MODELS APPLIED TO CELLULAR PROCESSES?

At first, it is hard to imagine that some cellular processes are random. But random doesn't necessarily mean chaotic; it is just a way of saying that the outcome is not exactly the same every time the process is repeated. Even sophisticated machinery designed to manufacture thousands of identical automobile parts produces parts that are nearly the same, but not 100% identical. The field of *probability theory* provides stochastic models for random processes. We have already seen one example in Math Minute 6.1: a model for sampling from a finite population using the hypergeometric frequency function. Now we will explore two stochastic models for cellular processes.

The Binomial Model

If 50 molecules of Otx (see Chapter 7) are floating around inside a nucleus prior to cell division, it seems likely that the two daughter cells will not always inherit exactly 25 molecules each. In this situation, randomness captures the idea that if a large number of identical cells divided, the outcome (i.e., the number of molecules inherited by each daughter) would vary. Some outcomes would occur quite often, while others would be rare. The fraction of the time that each possible outcome occurs in the long run (i.e., in a large number of cells) is an estimate of the outcome's probability.

The standard stochastic model for situations like the allocation of Otx molecules between two daughter cells uses the *binomial frequency function.* This model assumes that a particular "experiment" is repeated n times, where each repetition, or trial, is independent of all the others. In the example of Otx, a trial consists of determining which daughter cell gets a particular molecule. We assume that the fate of each molecule is independent of the other 49, a reasonable assumption if each molecule of Otx has randomly selected a location inside the nucleus. Since all 50 molecules must wind up in one of the two daughter cells, there will be 50 trials ($n = 50$). Each trial results in a "success" with probability p. In this example, a trial is counted as a success if a particular daughter cell gets the Otx molecule in question. To keep track of how many molecules go to each daughter, it is helpful to distinguish the cells by their relative positions after cell division: daughter L (cell on the left) and daughter R (cell on the right). Let's arbitrarily pick daughter L as the one we follow. In other words, the number of successes in 50 trials is the number of Otx molecules that go to daughter L. Since daughter L is just as likely to get each molecule as is daughter R, $p = 0.5$.

Under the binomial model, you can compute the probability of achieving k successes out of n independent trials with the binomial formula:

$$\binom{n}{k} p^k (1-p)^{n-k},$$

where $\binom{n}{k}$ is the binomial coefficient defined in Math Minute 6.1 (page 168). Therefore, the probability that daughter L receives 25 molecules of Otx is

$$\binom{50}{25}(.5)^{25} (1 - .5)^{50-25} \approx 0.112.$$

You can find the probability that daughter L receives 19 or fewer molecules (meaning daughter R receives 31 or more molecules) by computing the probability of each outcome satisfying this criterion (there are 20 such outcomes), and adding the 20 probabilities to get 0.06. Similarly, you can determine the probability daughter L receives 31 or more molecules (meaning daughter R receives 19 or fewer molecules) to be 0.06. Thus, with probability 0.12, each daughter will be 6 or more molecules away from the average value of 25.

The Normal Model

Many random factors influence the amount of protein produced by a gene at a particular time, including the number, location, and timing of all proteins needed to transcribe and translate the gene. In this situation, randomness means that if you measure the amount of protein produced by the same gene in thousands of identical cells (or in a single cell at thousands of time points), the outcome (i.e., number of protein molecules produced) will vary. Some outcomes will occur more frequently than others.

The standard stochastic model for a random quantity that represents the accumulation of many small random effects (e.g., protein production) is the normal distribution (also called the Gaussian distribution, or bell curve). The use of the normal distribution model is justified by one of the most powerful results in probability theory, the Central Limit Theorem.

Let X be the number of molecules of protein produced by a gene. You can compute the probability that the value of X is in a certain interval by finding the appropriate area under the curve given by the normal probability density function:

$$f(x) = \frac{1}{\sigma\sqrt{2\pi}}\, e^{-\frac{(x-\mu)^2}{2\sigma^2}}.$$

In this function, μ is the mean of the distribution (the average, or expected, value of X) and σ is the standard deviation of the distribution (a measure of the variation in values of X). The values of μ and σ can be estimated by taking a random sample of measurements (i.e., measuring the quantity of protein produced at several randomly chosen times), and calculating the mean and standard deviation of these measurements.

For example, if $\mu = 300$ and $\sigma = 20$, the probability that X is between 310 and 330 is given by the shaded area below:

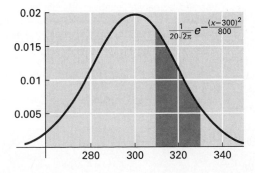

You can look up the numerical value of this area (approximately 0.2417) in a table of normal probabilities, or you can use numerical integration to estimate the area. In addition, many mathematical, statistical, and spreadsheet programs provide a function for computing probabilities using the normal probability density function.

A handy property of the normal probability distribution is that X is in the interval $(\mu \pm \sigma)$ 67% of the time; X is in the interval $(\mu \pm 2\sigma)$ 95% of the time; and X is in the interval $(\mu \pm 3\sigma)$ 99% of the time. For example, with $\mu = 300$ and $\sigma = 20$, we know that X is between 260 and 340 ($300 \pm 2 \times 20$) with probability 0.95. We used this property in Math Minute 6.2 (page 179) to determine whether a particular node in a graph had an unusually large degree. Because the normal distribution is so often a reasonable approximation for random quantities, we can use this property any time we look at data with error bars to get a rough estimate of the probability that the measured quantity is within the interval denoted by the error bars.

Standard stochastic models are excellent starting points for understanding random cellular processes. However, these models rely on certain assumptions, which may or may not hold. Like all models, stochastic models can be refined after gathering experimental data.

Genes are noisy, but what does this have to do with a genetic toggle switch? Everything. Let's imagine two proteins that each bind to different but overlapping binding sites and these sites have competing roles. For example, look back at the *Endo16* cis-regulatory element in Figure 7.17 on page 221 and find the Z and CG2 binding sites. Here is a small segment of DNA that can accommodate two different proteins, but only one at a time. Either Z can bind, or CG2, but not both. The Z site is responsible for repressing, and CG2 for amplifying, the output of module A. Given that there is noise within the system, two genetically identical cells (descendants from the same fertilized sea urchin egg) may have exactly opposite developmental fates. The same mechanism helps explain why even "identical" human twins have different fingerprints.

As a result of this noise, genetic toggle switches can be regulated by DNA binding site competition and stochastic production of transcription factors. Evolution has produced many toggle switches. The only component missing in order to stabilize the genetic choice is a feedback loop that reinforces what was initially a random "decision" (Figure 8.1a). Let's look at this diagram and understand a theoretical switch before we study a naturally occurring toggle switch. Protein A can bind to the cis-regulatory elements of genes b and c to initiate transcription for both genes. Protein B has three possible fates: it can be degraded by the cell; it can go and perform other functions; and most importantly for us, it can repress the expression of gene c. Conversely, protein C has three fates, one of which is to repress gene b. Will protein A bind to b and suppress c? Or will it bind to c and suppress b? Either outcome is possible; the determining factors are stochastic—the amount of A and its ability to find a limited number of binding sites. Once the decision is made, a genetically identical population of cells can be split into two subtypes as shown in Figure 8.1b. An individual cell will make only B or C. No cells will make both, nor are there any genetic hardwiring instructions that allow us to predict which path any particular cell will choose.

a)

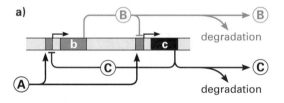

b)

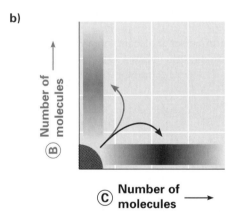

FIGURE 8.1 • **Toggle switch circuit. a)** Two promoters (dark gray boxes) upstream of two genes (boxes with lowercase letters) are controlled by protein A (proteins are represented by capital letters inside circles). Protein B represses gene c and protein C represses gene b. **b)** The shading displays the number of genetically identical cells containing different numbers of molecules (B or C). Initially, cells contain A but neither B nor C (mass of cells at the origin). More cells are expressing C as indicated by the darker shading gradation along the X-axis. Within each shaded gradation, the number of B or C molecules varies around a mean value since the number of proteins produced from any gene is stochastic. Arrows indicate the choice made by cells to express either B or C.

Theory Is Nice, but Do Toggle Switches Really Exist?

Theoretical models help us comprehend general principles, but they are useful only if they approximate reality. We will study two examples of genetic toggle switches: the first occurs in nature and the second was bioengineered. Many pathogens have the ability to evade our immune systems by changing their protein exteriors. How can genetically identical pathogens present different exteriors? They take advantage of noise and toggle switches. As your immune system learns to search and destroy, the pathogen changes its appearance. For the pathogen, it is easy to see that there are evolutionary forces at work to maintain noise and toggle switches, but mechanistically, how does it work?

Let's take a closer look at a real toggle switch that controls the behavior of the bacterial virus called λ phage. λ has two behaviors from which to "choose." It can either live quietly within its *Escherichia coli* host (**lysogenic** lifestyle), or it can replicate rapidly and blow up its host as the progeny are launched to infect new hosts (**lytic** lifestyle). The choice be-

tween peaceful coexistence and lethal parasitism is made by a single protein with the inconspicuous name of CII.

The λ phage toggle switch (Figure 8.2) and the theoretical switch in Figure 8.1 are very similar. CII is the equivalent to protein A. The amount of CII is the critical parameter, and one of two outcomes is possible. If CII finds the promoter P_{RE}, it will initiate transcription toward the left and result in the production of CI. CI can bind to the promoter P_L upstream of *cIII* and lead to the production of CIII. CIII prevents the destruction of CII, thus CIII indirectly reinforces its own production in a positive feedback loop. CI can reinforce its own production directly by binding to sites labeled O_{R1} and O_{R2} to repress the production of Cro protein (CI acting as a repressor of *cro*). CI can also promote its own production in a positive feedback loop when acting as a transcription factor for its own gene, *cI*. Once this toggle choice is made, λ is locked into peaceful lysogenic coexistence with its host *E. coli* unless new environmental forces perturb the system (e.g., UV light, change in nutrient availability). However, the choice could have gone the other way. CII protein could have been degraded if it took too long to find P_{RE}, since *E. coli* makes a protease that can destroy CII. If the protease finds CII before CII finds P_{RE}, the lytic lifestyle is chosen. In the absence of CII, the promoter labeled P_R is weakly active and begins transcribing to the right, resulting in the production of Cro protein. Cro binds to O_{R3} and O_{R2}, which leads to the repression of *cI* and the further activation of *cro*. The genetic choice of Cro and a lytic lifestyle eventually leads to the production of hundreds of fully mature viruses that swell the *E. coli* and lead to its demise.

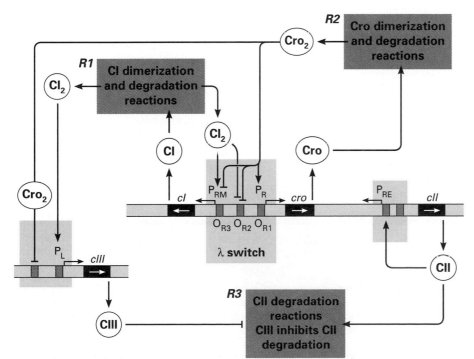

FIGURE 8.2 • A toggle switch that chooses between coexistence and murder. DNA is shown as blue bands, promoters as light gray boxes with arrows pointing to their genes. Genes are named black boxes with white arrows indicating the direction RNA polymerase travels to transcribe the genes. Genes are induced (arrows) or repressed (⊥) as indicated. Three regulatory regions (blue boxes labeled *R1*, *R2*, and *R3*) determine the lifestyle "decision" for λ phage. Named circles are proteins with subscript 2 indicating dimerization. Arrows into and out of regulatory regions represent a flow of information.

LINKS

Timothy Gardner

Jim Collins

Charles Cantor

There are several noisy factors in the choice made by λ phage, such as the limited number of proteins and binding sites as well as the unpredictable amount of time it takes to initiate transcription. Another factor is the burst of protein production. Notice in Figure 8.2 that both Cro and CI must form homodimers in order to be functional. Dimerization is more likely to happen when proteins are produced in bursts than when the same number of proteins are made at a slow but steady rate. A final component worth noting is that environmental influences can skew this decision. For example, if the bacterium host happens to be growing in a nutrient-rich environment (e.g., in a flask with lots of glucose), the bacterium produces more protease, resulting in faster destruction of CII and the production of many new λ phage (lytic lifestyle). Conversely, if the bacterium happened to be in a nutrient-poor environment (e.g., on the bottom of your shoe), there are fewer (but not zero) protease molecules, so CII has a higher probability of finding its binding site on P_{RE} before being destroyed. A longer half-life for CII leads to peaceful coexistence (lysogenic lifestyle), which makes good sense for the virus. Why reproduce rapidly if the environment is not conducive to making more potential hosts? Why not wait for the nutrients to arrive (e.g., when you step in something yucky) so the bacteria can grow? When the nutrients arrive, bacteria will grow faster, proteases will be more numerous, CII will be destroyed more readily, more viruses will form, more bacteria will lyse, and viruses will infect more hosts. The selective advantage for a noise-tolerant toggle switch is impressive.

DISCOVERY QUESTIONS

3. What would be the consequences if CI degradation were more prevalent than CI dimerization?

4. If the P_L promoter were inactivated, would this change the outcome of the toggle switch for lysogenic vs. lytic lifestyles? Explain your answer.

5. How does Cro affect the ability of CII to switch λ from lytic to lysogenic?

Can Humans Engineer a Genetic Toggle Switch?

Often there is a division among biologists who like to argue about which experiments are "real science"—those working with real organisms and molecules vs. those working with theoretical models to unify experimental observations. Both approaches are necessary for a complete understanding, but it is important to remember that predictions based on theoretical models must be tested with real experiments. In the 20 January 2000 issue of *Nature,* two papers described genetic toggle switches that had been genetically engineered in the lab. Both groups used *E. coli* as the host for their switches, but they made two different types of switches. One group built a toggle switch very similar to the one used by λ phage to choose its lifestyle. We examine this switch first. Later we examine a switch that oscillates like a circadian clock.

How to Build a Toggle Switch. Timothy Gardner and his colleagues Jim Collins and Charles Cantor at Boston University built a toggle switch that can choose one of two states. However, rather than letting *E. coli* decide which direction to go, they designed a toggle switch that could be regulated by the investigators (Figure 8.3). In this switch, they needed two **constitutive** (always activated) promoters (colored black and gray) and two repressor genes (also colored black and gray). The black repressor protein inactivates the gray promoter, which drives the production of the gray repressor protein. Conversely, the gray repressor protein inactivates the black promoter,

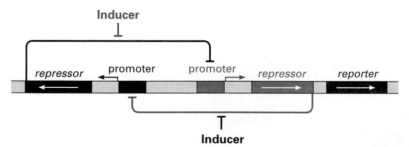

FIGURE 8.3 • **Theoretical two-gene bistable toggle switch.** The black gene *"repressor"* is transcribed from its black promoter. The black repressor protein binds to the gray promoter DNA sequence to block the production of the gray repressor protein. The gray repressor protein blocks the production of the black repressor protein when the gray repressor protein binds to the black promoter. To detect which state the toggle switch in in, GFP was placed downstream of the gray promoter so the gray repressor and GFP are produced simultaneously.

which drives the production of the black repressor protein. Thus, if the black repressor protein were produced, the gray repressor protein could not be produced and vice versa. This is an example of a bioengineered bistable toggle switch. To control this switch, the investigators utilized two inducer drugs, each of which incapacitates one of the repressor proteins.

The group from Boston University built and tested several different plasmids with a variety of promoters to see which would successfully produce a bistable toggle switch (Figure 8.4). To measure switching, the **reporter gene** GFP was added to the system to produce a glow-in-the-dark protein when the gray repressor protein was produced. In Figure 8.4a, the investigators compared two plasmids, pIKE107 and pIKE105. The only difference between these two was the use of two different ribosomal binding sites that affected the efficiency of translation of the RNA into protein. Both 107 and 105 were capable of producing GFP (and thus the gray repressor as well) using an inducible promoter and a drug called IPTG. Note that 105 did not produce as much GFP as 107. When the inducer IPTG was removed, 105 was incapable of sustaining its output, therefore it was not a bi*stable* switch. However, 107 was capable of sustaining the production of the gray repressor and GFP even after IPTG was removed—thus it was stable. When 107 was exposed to the second inducer drug called anhydrotetracycline (aTc), the production of GFP (and thus the gray repressor) was eliminated within a couple of hours due to the production of the black repressor. The experiment was well controlled (Figure 8.4b–c) since each half of the toggle switch was able to induce GFP production but neither of these "half-switches" was stable. Gardner's research demonstrated it was possible to take a theory (see Figure 8.3) and build a biologically functional switch in a test tube that worked inside living cells—the model was supported with experimental evidence (see Figure 8.4).

DISCOVERY QUESTIONS

6. Explain why pIKE105 was considered a failure while the control plasmid pTAK102 was deemed a success?

7. Based on the data in Figure 8.4, determine which drug (IPTG or aTc) was the gray inducer and which was the black inducer in Figure 8.3.

8. Design a bistable toggle switch that could be used in gene therapy to produce a protein on demand (e.g., blood clotting factor). Include in your design how the protein of interest could be turned on and off.

METHODS

GFP

Inducible Promoter

LINKS

Indicated Plasmids

The success of Gardner's work is encouraging to those who want to understand genomic circuits. Theoretical models can be designed and validated experimentally. It also represents a success in "forward engineering" in which simple genetic circuits serve as models for more complex systems. In a more applied sense, the bistable toggle switch may prove useful in gene therapy and other biotechnology methods that require a silent gene to be activated at a certain point and then sustained. Finally, Gardner's team produced a bistable toggle switch, which represented the first "**genetic applet,**" a term derived from computer programs called Java applets. Applets are self-contained programs, and this genetic applet is capable of being programmed (turned on or off similar to the digital version of 1 or 0). Although the genetic applet is a long way from becoming a biological computer, the potential to store information in DNA is intriguing.

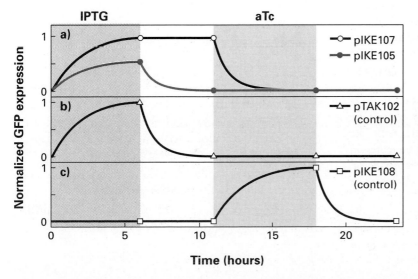

FIGURE 8.4 • Experimental two-gene bistable toggle switch. *E. coli* were transformed with the indicated plasmids and exposed to the inducer drugs as indicated by the shading. **a)** The only difference between pIKE107 and pIKE105 was different ribosomal binding sites that affected the rate of translating the encoded proteins. **b)** pTAK102 control plasmid contained only the IPTG-inducible promoter upstream of the GFP gene. **c)** pIKE108 control plasmid contained only the aTc-inducible promoter upstream of the GFP gene.

FIGURE 8.5 • **Two plasmids needed to make and see the output of the "repressilator" genetic circuit.** The repressilator plasmid contains a cyclic negative feedback loop composed of three repressor genes and their corresponding promoters. $P_L$lac01 and $P_L$tet01 are strong promoters that can be tightly suppressed by LacI and tetracycline, respectively. The third promoter, P_R, is repressed by CI. The three repressor genes are appended with the suffix "lite" to indicate the encoded proteins degrade rapidly. The gfp-aav protein encoded on the reporter plasmid is also degraded rapidly, and its production is regulated by the $P_L$tet01 promoter.

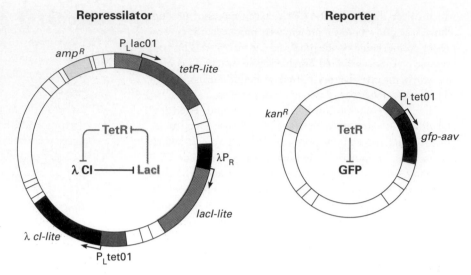

Can Humans Build a Synthetic Circadian Clock from a Toggle Switch Design?

LINKS
Stan Leibler

One lesson from genomics is that interdisciplinary collaborations are the rule rather than the exception. The fields of genomics and proteomics need biologists, physicists, mathematicians, chemists, even graphic artists. In order to build a synthetic biological clock, Michael Elowitz, a graduate student in Princeton University's Molecular Biology Department, teamed up with Stan Leibler in the Physics Department. To create a self-perpetuating cycling toggle switch required more genes than a two-gene genetic applet (Figure 8.5). First, you will notice that two plasmids were used. The larger repressilator plasmid controlled the cyclical nature of the output, while the smaller reporter plasmid produced GFP. Three promoters and three protein repressors were used in the repressilator plasmid: the λ CI protein represses the production of LacI; the LacI protein represses the production of TetR protein; TetR represses the production of CI. You can see that each gene was repressed by one of the three repressors encoded on the repressilator plasmid. Therefore, the repressilator was a genetic closed circuit or cyclical negative feedback loop. Since it was impossible to directly observe the repressilator circuit in action, the reporter plasmid encoding GFP was constructed. The production of GFP was constitutive but could be repressed by TetR. When TetR was produced from the repressilator, GFP production was inhibited and the *E. coli* cells lost their fluorescence.

Figure 8.5 is a nice theoretical model, but does it actually work? Figure 8.6 dramatically illustrates how well the theory worked inside growing *E. coli*. The photos allow us to follow a single cell (Figure 8.6b) through its oscillations of GFP production (Figure 8.6a). The amount of GFP was measured for the entire population of cells, which revealed a fluorescence periodicity of about 150 minutes. What is so striking about this time period is that it is *longer* than the cell cycle is for *E. coli*. The graph indicates that cells divide more rapidly than the repressilator cycles, and thus the investigators had produced a cyclical circuit that outlived any single cell cycle.

The very regular periodicity of the repressilator helps you appreciate your own circadian rhythm and how it too can be controlled by a small number of genes that produce cyclical amounts of protein. But there is a fly in this ointment of perfect timing. Noise and stochastic behavior of proteins are a fundamental property of gene regulation and protein production, so you might expect some problems with the regularity of the repressilator.

DISCOVERY QUESTIONS

9. How can a biological clock outlive its host cell?

10. Graph the production of CI, LacI, and TetR proteins on top of the graph in Figure 8.6c.

11. Why did the investigators choose proteins that are rapidly degraded by cells? What would have happened if the proteins were long-lived?

12. Predict what might happen to the repressilator periodicity inside sister cells after division.

13. What would be required for sister cells to maintain the exact same periodicity of GFP production?

a)
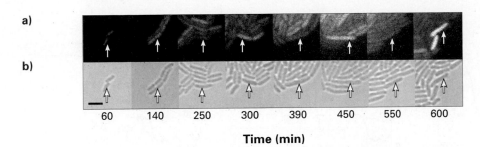

b)

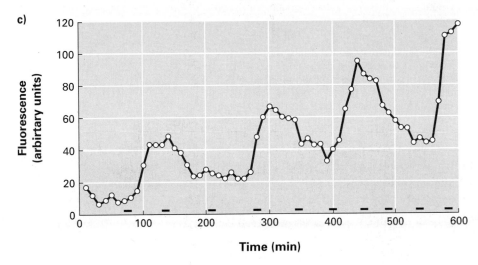

FIGURE 8.6 • **Cyclical toggle switching in live bacteria. a)** Fluorescence and **b)** phase contrast microscopy images of cells, revealing the time course of GFP expression and cell growth beginning with a single bacterium containing the repressilator and reporter plasmids (see Figure 8.5). Scale bar in b) indicates 4 μm in the photographs. **c)** The pictures in a) and b) correspond to peaks and troughs in the time course of GFP fluorescence intensity of the selected cell. Bars at the bottom of c) indicate the timing of cell division, as estimated from the phase contrast photomicrographs in b).

Given the clarity of the data in Figure 8.6c, you might think humans can build a clock that is unaffected by the inherent noise of gene regulation. But a clock can only be as consistent as its component parts (Figure 8.7, page 242). We can follow one cell (blue trace in Figure 8.7a–c) and its sisters from two rounds of division (in gray and black) and see that an oscillation has been retained in all three cells, but its timing and amplitude are not retained through the generations. Note, however, that the original cell maintains its 150-minute cycle. Interestingly, when cells stopped dividing due to nutritional limitations (**stationary phase**), the repressilator stopped working. Therefore, cell growth is required for the proper function of the repressilator.

DISCOVERY QUESTIONS

14. Predict what would happen if IPTG were added to the repressilator since IPTG disrupts the function of LacI.

15. Hypothesize why some sibling cells altered periodicity while others altered amplitude.

If Toggle Switches Are So Noisy, How Can Multicellular Organisms Develop?

It seems that we have just described all genetic toggle switches as overwhelmed with noise and impossible to coordinate—the genomic equivalent to herding cats. But we know from our own experiences that life is not completely chaotic. You do not have brain cells trying to become liver cells. Every human went through gastrulation at the exact same time during gestation. How can cell populations with stochastic toggle switches work collectively toward a common goal? The team analogy is a good one because coordinating genes in cells is similar to coordinating 11 football players on the field. Picture the offense with a quarterback (QB) who throws the ball, linemen who block defenders, and receivers who run downfield hoping to catch the ball, save the game, become a hero, etc. There are three keys to winning a football game, just as there are three keys to coordinating cell populations with noisy toggle switches.

1. Each player does not have to ensure that all the other players are in the right place. The QB and the two behind him can survey all other players and yell reminders to those who have lined up in the wrong place. This is called **cooperation through communication.**

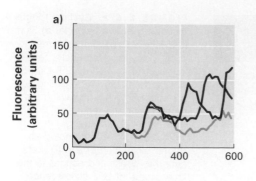
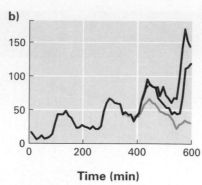
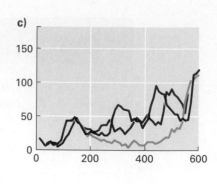

Time (min)

FIGURE 8.7 • **Examples of sister cells that maintain periodicity but not synchronic-
ity. a)** to **c)** In each case, the fluorescence time course of the cell depicted in Figure 8.6c is
redrawn in blue as a reference, and two of its siblings are shown in black and gray. **a)** Siblings exhibiting post–cell division phase delays relative to the reference cell. **b)** Phase is
approximately maintained but amplitude varies significantly after division. **c)** Reduced period
(black) and long delay (gray).

2. At various times, the QB can consult a list of points to make sure everyone has done the right move. Watch how the QB will shout and sometimes raise and lower one leg to signal others to move a bit to the left or right. And what happens if everyone is confused? The QB can call a time-out to give the players a chance to get coordinated again. Each of these points prior to starting the play is called a **checkpoint.**

3. Any team that really wants to win has a contingency plan. Bill Cosby has a great comedy routine where he re-lives his childhood football games and everyone is given very complex directions on where to go so the QB can throw the ball to someone. On real teams, the QB has two to four players running around so if one is not a good target, the QB can look for other options. This du-plication of options to accomplish a goal (winning the game) is **redundant.**

Cells can use the same three keys to achieve coordination.

1. A subset of cells can secrete a product that will commu-nicate a message to keep all cells in synch.

2. Cellular proteins establish a quality control at various checkpoints, such as DNA replication and the stages of mitosis. Checkpoints ensure the quality of the eventual outcome, but the exact timing for any given cell can vary due to stochastic protein function and noise.

3. Critical processes in cells have redundant circuits to create fail-safe approaches to vital processes such as re-sponding to environmental signals. Some pathways have multiple ways of becoming activated and/or mul-tiple ways of producing a cellular response. Redun-dancy can also be achieved by having isozymes that can perform essentially the same job, though they may have slightly different tolerances to environmental perturbations.

Redundancy: Is It Really Beneficial to Have More Than One Copy of a Gene?

Measuring **reliability** is essentially an engineering question. Is it really necessary to have more than one way to stop your parked car from rolling down a hill? In this context, the answer seems obvious, but genetic redundancy is less intuitive. For many years, it has been argued that having more than one locus en-coding a particular function imparts a selective advantage. It might be a backup in case one gene is mutated and loses its function. The duplicated gene can mutate over time and pro-duce new functions for the cell. Freshly evolved duplications can provide a wider range of tolerance to environmental condi-tions to ensure the common function is accomplished (e.g., one may work better in cold temperatures and the other in hot). Bi-ologists should not reinvent the wheel to measure the value of redundancy. Why not borrow from well-established engineer-ing methods for reliability analysis to assess the likelihood that a particular function will be performed successfully (Figure 8.8)?

In equation a and Figure 8.8, we are interested in pro-ducing protein B. The reliability of any given gene being produced is arbitrarily set at 0.9 or 90% and is represented by the line segment with the gene labeled "*x.*" In this relia-bility analysis, we will follow what happens when the DNA required to produce B is altered in different ways.

a. Reliability (R) = P
$$= \text{probability of B being produced}$$
$$= 0.9 \text{ or } 9{,}000 \text{ out of } 10{,}000 \text{ success rate}$$

If producing B required two genes (Figure 8.9), the relia-bility of this step drops from 0.9 to 0.81 because both x and y have to be functional so the rule of multiplication applies.

b. x and y must work:
$$R = P^2$$
$$= 0.9 \times 0.9$$
$$= 0.81 \text{ or } 8{,}100 \text{ out of } 10{,}000 \text{ success rate}$$

If x were duplicated (Figure 8.10), the reliability of producing B increases substantially to 0.99. To determine this reliability, first we must calculate the probability of failure (Q) by taking the probability of either failure or success (total of 1) minus the probability of success (P). For a single gene x in a haploid genome (see Figure 8.8) the probability of failure is:

c.
$$Q = 1 - P$$
$$= 1 - 0.9$$
$$= 0.1$$

In Figure 8.10, reliability equals all possible outcomes (1) minus the probability of failure $(Q^2; Q \times Q)$ because both the upper x and the lower x have to fail (multiplication rule again) in order for the production of B to be unsuccessful.

$$R = 1 - (0.1 \times 0.1)$$
$$= 1 - Q^2$$
$$= 1 - 0.01$$
$$= 0.99 \text{ or } 9,900 \text{ out of } 10,000 \text{ success rate}$$

By creating a diploid genome, we have increased the reliability from 90% to 99%. This increase in reliability would also be true for haploids that duplicated a single gene. What is the reliability consequence if a diploid organism duplicates a single gene (Figure 8.11)? Using the multiplication rule, the probability of failure for each locus (x and x') is multiplied, and we see another substantial increase in reliability.

d.
$$R = 1 - (Q^2 \times Q^2)$$
$$= 1 - Q^4$$
$$= 1 - (0.1^4)$$
$$= 1 - 0.0001$$
$$= 0.9999 \text{ or } 9,999 \text{ out of } 10,000 \text{ success rate}$$

We have determined that two genes in a haploid genome were less reliable (equation b) than when only one gene was required (equation a). This makes sense because with two genes there are two ways to fail instead of just one. Redundancy should help ameliorate the weakness of two genes. Let's determine the reliability in diploids with duplicated genes when two genes are required to complete the function (Figure 8.12). To calculate this reliability, we need to combine equations b and d.

e.
$$R = (1 - Q^4)^2$$
$$= (1 - 0.0001)^2$$
$$= (0.9999)^2$$
$$= (0.9998) \text{ or } 9,998 \text{ out of } 10,000 \text{ success rate}$$

Note that the reliability for two genes that have been duplicated in a diploid (equation e and Figure 8.12) was not quite as high as for one duplicated gene in a diploid (equation d and Figure 8.11). For half the number of alleles (four instead of eight), the organism has a slightly higher reliability. The value in redundancy is that if one allele is mutated, the function will still be completed. If the individual in Figure 8.12 carried a nonfunctional x allele and a nonfunc-

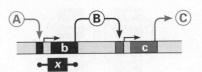

FIGURE 8.8 • **A three-gene pathway for the production of protein C.** The genetic unit represented by the line segment marked with "x" indicates one or more genes that accomplish the task of producing B. In Figures 8.8–8.13, more than one gene will be included in the line segment x, but these genes are assumed to be *un*linked.

FIGURE 8.9 • **A two-gene model to produce B.** These two genes are from unlinked loci, though they have been diagrammed as adjacent for simplicity.

FIGURE 8.10 • **A diploid model to produce B.** The two alleles are on homologous chromosomes.

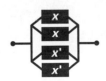

FIGURE 8.11 • **Diploid genome with a duplicated gene to produce B.** The two genes (x and x') are unlinked, while alleles for a given gene are located on homologous chromosomes.

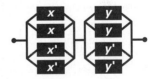

FIGURE 8.12 • **Diploid genome requiring two genes (x and y) to produce B and both genes have been duplicated (x' and y').** None of the genes (x, x', y, and y') are linked, though the pair of alleles for each locus are located on homologous chromosomes.

tional y allele, the redundancy of the system maintains a high degree of reliability (Figure 8.13, page 244).

f. $R =$ modified equation e to take into account that only three viable alleles exist for each locus
$$= (1 - Q^3)^2$$
$$= (1 - 0.001)^2$$
$$= (0.999)^2$$
$$= (0.998) \text{ or } 9,980 \text{ out of } 10,000 \text{ success rate.}$$

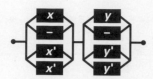

FIGURE 8.13 • **Mutant alleles affect genome's reliability.** Genome from Figure 8.12, but one *x* allele and one *y* allele are nonfunctional as indicated by the "−" sign.

LINKS
Upinder Bhalla
Ravi Iyengar

DISCOVERY QUESTIONS

16. Does the reliability in Figure 8.13 surpass the reliability in Figure 8.11 if you assume one allele of the four in Figure 8.11 were nonfunctional? Explain your answer and support it mathematically.

17. Based on reliability calculations, would a tetraploid be more or less reliable? If tetraploids are more reliable, why aren't more organisms tetraploid?

18. If you were designing a metabolic pathway to be as reliable as possible, would you design:
    a. more components, each with a specialized function, or
    b. fewer components that could "multitask" (each one performing multiple roles)? Explain your answer.

Summary 8.1

We have seen that genetic toggle switches can be formed with very few genes. Genomes contain hundreds or thousands of toggle switches in order to produce proteomes and metabolomes in response to environmental changes. The "choice" between two responses can be driven by particular factors (e.g., nutrient concentration for λ phage lysogenic vs. lytic lifestyles) or by stochastic behavior of proteins (e.g., the repressilator in *E. coli*). Given our understanding of how bistable genetic switches operate, we can design and produce bistable "genetic applets" that can be tested experimentally and might be developed for gene therapy. By applying engineering principles to toggle switch design, we can create biologically functional genetic switches and calculate their reliability. If you were designing a bistable switch for gene therapy, you would want to incorporate a level of redundancy that provided the patient with an acceptable degree of reliability. It might seem uncomfortable to treat genes like machines with reliability that can be calculated, but this approach to genome analysis is valuable. Applying engineering principles to genomics helps us understand why diploids evolved the apparent waste in complex circuits with redundancies. In Section 8.2, we consider how more complex genomic circuits use toggle switches and redundancy to accomplish vital functions.

8.2 Complex Integrated Circuits

Every time we build a model, we are trying to create a simple version of a complex system. From the lessons of simple circuits in Section 8.1, is it possible to understand complex circuits? If we can understand complex circuits, will we discover new properties that were not present in the dissected circuits? We will study three complex genomic circuits that might reveal emergent properties that could not be detected any other way. How do we convert environmental stimulation into memories? Is it possible to understand cancer formation by studying protein circuits? Can the reproductive success of a virus be influenced by the order of its genes? It is worth noting that the following three case studies were conducted by mining data available in public databases and the literature. All of these studies combine experimental data with in silico research to form new models and make predictions to refine the models. Over time, we hope these models will increasingly resemble biological reality.

Are Circuits the Key to Learning?

You may be familiar with the saying "easier said than done." In other words, to say that there is a way to coordinate all these genomic circuits sounds simple and is intuitively appealing, but where is the proof? To paraphrase the movie *Jerry Maguire,* "Show me the data!"

Hopefully by this point you are gaining an appreciation for the complexity of the problems ahead. Only with the benefit of genomic data analysis—sequences, variations, expression profiles, proteomics, biochemistry, computer science, mathematics, etc.—can we begin to piece together the necessary information to see coherent patterns and circuits. Upinder Bhalla (the National Center for Biological Sciences, Bangalore, India) and Ravi Iyengar (Mount Sinai School of Medicine) analyzed many years of data, made some insightful discoveries, and set the pace for others to follow. They used an engineering approach to understand four cell signaling circuits and discovered some interesting cross talk between the circuits. They used data in the public domain to create a complex computer model that accurately simulates the neurocircuitry necessary for learning.

What Is the Goal of Making Complex Models? As we have seen before, in order to comprehend complexity, we need to simplify. That sounds like an oxymoron, but in fact, we do this all the time. No one says "compact disc read only memory"; we just say CD-ROM. In that simplification, you have encapsulated a lot of information into five letters. Bhalla and

Iyengar decided to make a few simplifying assumptions first, rather than beginning with the most complex model possible. This simplification is the principle known as **Occam's Razor**—start with the simplest possible explanation first, rather than more complex ones. As the simple model is compiled and analyzed, the need for specific experiments becomes apparent and the model can become more complex as necessary.

It is interesting to note how certain pathways are more popular than others. For example, every introductory biology textbook discusses how an adrenaline rush can stimulate the production of cyclic adenosine monophosphate (cAMP). Bhalla and Iyengar decided to model a different aspect of cAMP signaling and made a couple of simplifying assumptions. They assumed that the cytoplasm was a well-stirred bag of liquid where all components have equal access to each other. Because of the uniform distribution within the cytoplasm, it is important that each component has a mechanism for delivering its message only to the correct target molecule and only in one direction. If this were not the case, it would be like having wires in your CD player with no insulation on them. The electrical currents would go all over the place, and you would never hear any music. To generate a computer model of their uniformly mixed cell, the investigators needed to know the reaction rates of every enzyme and the concentration of every component in the four signaling circuits. However, easier said than done . . . (Figure 8.14). The level of complexity in ge-

nomic circuits can grow to staggering proportions. A 3-molecule system only requires 7 measurements, but an 18-molecule system requires 162 measurements!

It became too difficult to use standard computational approaches for this level of complexity, so Bhalla and Iyengar utilized a program called GENESIS, a neural network simulator to analyze the four interacting pathways. Even with the help of very sophisticated computation, two simplifying assumptions were made that do not reflect reality. First, they ignored **compartmentalization.** For example, some components may be embedded in phospholipid bilayers and thus not freely available to all other components. In fact, we know this is true of many components in their analysis, but it was impossible to quantify this aspect. Second, they ignored **regional organization of components.** By clustering some components near each other rather than letting them drift by **Brownian motion** (random motion), there can be a significant increase in efficiency. The mitochondrion is an excellent example; metabolism is more efficient due to the clustering of molecules used in the electron transport pathways. Microorganisms also use gene clustering for the production of antibiotics.

Do Complex Models Reveal New Insights and Emergent Properties? The fundamental assumption used by most biologists is that interconnected circuits work synergistically.

STRUCTURES

cAMP

LINKS

GENESIS

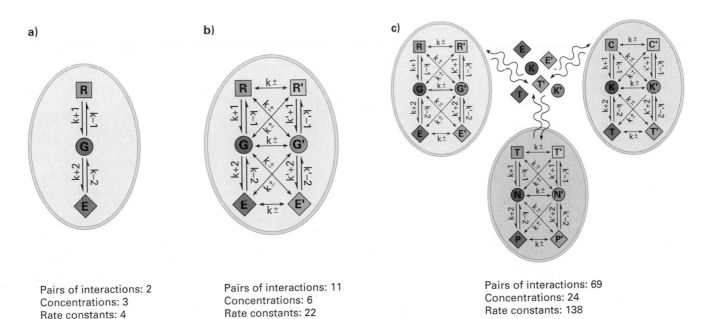

a)

Pairs of interactions: 2
Concentrations: 3
Rate constants: 4

b)

Pairs of interactions: 11
Concentrations: 6
Rate constants: 22

c)

Pairs of interactions: 69
Concentrations: 24
Rate constants: 138

FIGURE 8.14 • Increased need for information as circuits become more complex.
a) to **c)** k + x represents the rate constant for the forward reaction and k − x the reverse reaction; 1 and 2 refer to pathways 1 and 2. In this simplified model, each component in a pathway can communicate only with its nearest neighbors. The impact of this simplifying assumption is most apparent in panel c. The symbols used are: R = receptor; G = G protein; E = effector; T = transcription factor; N = nucleic acids; and P = proteins in the nucleus.

METHODS

Brain Anatomy

LINKS

Long-Term
Potentiation

The existence of **synergy** is why many in field biology dislike the reductionist approach used by molecular biologists. However, you will see the reductionists' goal is not merely to disassemble a cell to look at the parts but to understand the parts well enough to explain synergistic interactions and move beyond descriptions toward testable predictions. The particular case under investigation was one that philosophers and biologists have pondered for centuries: *How do we learn?*

Neurobiologists have given us great insights into the mechanisms utilized whenever an animal (e.g., flies, humans) learns something new. Intuitively we know that our neurons must undergo some sort of change in order for us to retain information. There is a genetic component to this trait, so we know proteins are involved. Somehow, proteins must alter what a neuron does—become depolarized when stimulated and release neurotransmitters to relay this information. Sounds simple, huh? This change in function by a neuron is called **long-term potentiation (LTP),** which means a neuron can be stimulated and the consequence of the stimulation is maintained after the original stimulus is gone. To see a simple example of this, stare at a bright light and then turn away. Even though the light is no longer hitting your neurons, you still "see" it. Your neuron's function was changed so that it performed differently after the stimulus was removed. Bhalla and Iyengar decided this was the complex circuitry they wanted to model—the hardwiring of your brain.

The last piece of background we need is brain anatomy to provide the context of learning. Where in the brain are the processes taking place? Deep within your cerebrum is a collection of neurons called the **hippocampus.** For at least 30 years, memory research has focused on the hippocampus as the center of learning/memory. Inside the hippocampus are layers of neurons and each layer has a name. We will focus on the CA1 layer. On the cell body and dendrite of the CA1 neurons are bumps called spines. Embedded in these spines are integral membrane proteins, three of which are of particular interest to us. The neurotransmitter glutamate binds to its receptor called mGluR (for *mouse glu*tamate *r*eceptor). As with all receptors, it facilitates **signal transduction,** which means it transmits the extracellular signal across the plasma membrane. NMDAR (*N*-methyl-*D*-*as*partate *r*eceptor) is a voltage-sensitive calcium ion channel. The final plasma membrane component is AMPAR (α-amino-3-hydroxy-5-methyl-4-isoxazolepropionate receptor), another glutamate receptor that acts as an ion channel when stimulated by glutamate. LTP is initiated when mGluR and NMDAR are stimulated by a certain amount and frequency of stimuli. Experimentally, LTP can be induced in mouse neurons when stimulated with 3 mild electrical inputs of 100 Hz pulses, 1 second each and separated by 10 minutes. Bhalla and Iyengar set out to construct a computer model of the complex series of events from stimulation to LTP.

How Much Math Do I Need to Know to Understand These Calculations?

Molecular biology is becoming mature enough to need the assistance of many other disciplines, especially mathematics. However, the math used in this study is quite simple. To start their in silico research, Bhalla and Iyengar considered only two types of connections: protein-protein interactions and **second messengers.** Two additional aspects they considered were: (1) proteins **degrade,** that is, they are destroyed by the cell over time; and (2) enzymes have reaction rates—for example, there is an average amount of time it takes a kinase to consume ATP and add a phosphate onto its substrate. In the reaction below, A and B become joined into AB. This illustrates how two proteins can bind to each other, such as a kinase and its protein substrate. The binding has a forward reaction rate (k_f) and a backward reaction rate (k_b). The second reaction shows the conversion of A and B into C and D. For example, we could measure the amount of Na^+ inside a cell (A) and the amount of K^+ outside a cell (B) as they are changed into Na^+ outside a cell (C) and K^+ inside a cell (D). The forward rate of this conversion (k_f) is the rate of the Na/K pump, and the backward rate (k_b) is the rate of ions passing through ion channels. This second reaction can be written as an equation, which says that the change (d) in the concentration of A ([A]) over a change in time (/dt) is equal to the production of A (the backward rate times the concentrations of C and D, which is written: k_b[C][D]) minus the amount of A lost (due to the forward reaction that consumes A, which is written: k_f[A][B]). Since you have studied these types of interactions and rates before, this level of circuitry should be okay so far. The math is only multiplication, division, and subtraction. To determine LTP, all we need is numbers to replace the variables.

$$A + B \underset{k_b}{\overset{k_f}{\rightleftharpoons}} AB$$

$$A + B \underset{k_b}{\overset{k_f}{\rightleftharpoons}} C + D \quad d[A]/dt = k_b[C][D] - k_f[A][B]$$

One more enzymatic interaction is needed to analyze the learning circuit. The equation on page 247 states that an enzyme (E) binds to its substrate (S) to form a complex of the two (ES). This step can go forward (k_1) or backward (k_2), meaning it is reversible. However the second step is irreversible as indicated by the forward arrow and only one rate constant (k_3). Therefore the ES complex can be converted into the original enzyme (E; an enzyme is never consumed in a reaction) plus a new product (P). Each step occurs at a measurable rate, and these values are called **rate constants** (the *k* values). For example, the enzyme adenylyl cyclase produces cAMP from the substrate ATP. Adenylyl cyclase (E) binds to ATP (S) and forms (k_1) a complex of the two (ES). ATP and adenylyl cyclase can fall apart (k_2) or proceed (k_3) toward an irreversible production of adenylyl cyclase plus cAMP (P).

$$E + S \underset{k_2}{\overset{k_1}{\rightleftharpoons}} ES \overset{k_3}{\rightarrow} E + P$$

That's all the math and chemistry you need to understand in order to grasp this very sophisticated in silico analysis of learning! We still have one problem, though. We don't have any real numbers to put into these equations. Luckily, biochemists and cell biologists have published all the values needed to initiate the GENESIS software analysis of a nerve's ability to learn. So we are ready to start the simulation.

What Are the Individual Circuits Used to Learn?
Fifteen well-studied circuits were needed to produce Bhalla and Iyengar's LTP simulation (Figure 8.15, page 248). They started by building computer models of each of the 15 circuits individually. At each step, they made sure that the models matched experimental data. This alone was an impressive task, but here is where the fun starts. They began to build integrated circuits by gradually combining each of the 15 individual circuits. Initially, they only allowed two types of connections between circuits: (1) secondary messengers arachidonic acid (AA) and diacylglycerol (DAG), and (2) an enzyme in one circuit bound to its substrate produced in another circuit.

Permitting these two types of interactions allowed the investigators to combine the circuits labeled a, b, e, f, h, k, and l in Figure 8.15 into one integrated circuit (Figure 8.16, page 249). Notice that epidermal growth factor (EGF) leads to the activation of two enzymes: mitogen-associated protein kinase (MAPK) (in circuit a) and PLCγ (in circuit f). MAPK has been studied intensively for many years. It is associated with substances that stimulate mitosis, and it phosphorylates proteins. Loss of control of MAPK can lead to the formation of cancerous cells. PLCγ is one **isoform** (there are several PLC genes in each person, and this version, or isoform, is called γ) of phospholipase-C that cuts the phospholipid called phosphotidyl inositol bisphosphate (PIP2) into inositol trisphosphate (IP3) and DAG. By connecting these seven circuits, a new layer of complexity became necessary—feedback loops. A **feedback loop** occurs when the product has a stimulatory or inhibitory effect on one of the upstream components, such as an enzyme that leads to product formation.

Is It Possible to Generate Persistent Kinase Activation After a Transient Stimulus?
Do individual circuits perform synergistically when they are integrated into a larger pathway? Can a computer model accurately simulate LTP? Can we use this model to make predictions that would lead to

improved understanding about the way we learn new information? We know that these individual components and circuits are involved in learning, and we know that LTP is the result of a long-term activation of a kinase. But is it really possible to simulate something as complex as learning on a few megabytes of computer microcircuits?

LINKS
15 Circuits

How did the computer model compare with real data? In Figure 8.17a, page 249, you can see the simulation was very similar to the experimental data. In this graph, the MAPK activity was plotted as a function of time (d[A]/dt). The activation of MAPK was transient due to the normal degradation mechanisms in the cell. Equally impressive is Figure 8.17b, where the investigators compared simulated PLCγ activity (dashed lines) with experimental data (solid lines) in the presence (triangles) or absence (squares) of EGF over a 10,000-fold range of calcium concentrations. Given the close agreement between experimental data and the computer model, these two investigators had a good foundation on which to build more complex integrated circuits.

Can the Model Make Predictions that Determine the Amount of Stimulus Needed to Sustain Kinase Activity?
In Figure 8.16, you can see two areas that provide feedback—PKC and MAPK. PKC activates Raf, which activates another kinase MEK, which activates MAPK (two isoforms of MAPK were used in this simulation, numbers 1 and 2). MAPK activates cytosolic phospholipase A-2 (PLA$_2$), to produce AA, which is half of the stimulus needed to activate PKC. So what concentration and duration of stimulus are needed to produce a long-lasting activation of PKC and MAPK? Although Figure 8.18, page 250 is data intensive, it is worth dissecting carefully. First, let's look at the PKC data (open symbols) in the figure. Neither 10 minutes at 5 nM EGF (open circle) nor 100 minutes at 2 nM EGF (open square) led to any significant activation of PKC. However, 100 minutes at 5 nM EGF (open triangle) was sufficient to lead to a protracted activation of PKC. Note how PKC was activated during the 100 minutes and did not rise further, but when the 100 minutes of stimulation was removed, PKC activity remained unchanged. A change in function after removal of stimulus is a hallmark of LTP. MAPK activity (closed symbols) was a bit more complicated, but the take-home lesson was similar. Stimulation for 10 minutes at 5 nM produced a transient activation, while 100 minutes at 2 nM produced a more prolonged activation but at a lower amplitude, though the activation did not extend much beyond the 100 minutes of stimulation. However, 100 minutes of 5 nM stimulation produces a tenfold increase in activity (from 0.01 to 0.1 μM) as well as a prolonged activation (i.e., a bistable toggle switch). Notice how MAPK was activated during the 100 minutes and what happened at about 50 minutes—it jumped to a higher level. Something extraordinary (synergistic?) was happening.

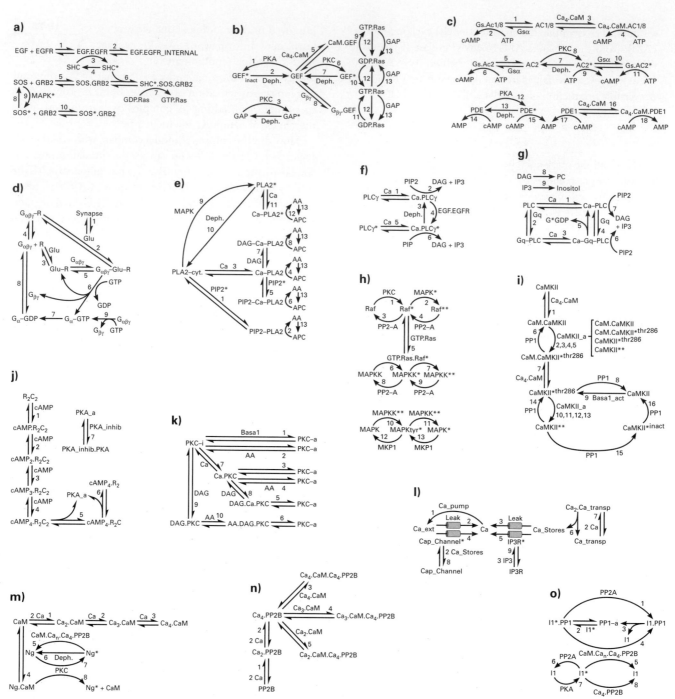

FIGURE 8.15 • **The 15 circuits that were modeled individually before integrating them into more complex networks.** Reversible reactions are indicated by double-headed arrows. Enzyme reactions are drawn as curved arrows, with the enzyme listed in the middle. For visual clarity, the same reactant may be present more than once in a single circuit.

DISCOVERY QUESTION

19. Look at Figure 8.18 and hypothesize what may be the cause of the tenfold increase in MAPK activity between 30 and 50 minutes for the 100 minutes at 5 nM stimulation.

Why Does It Take 100 Minutes of 5 nM EGF to Achieve Long-Term Activation? For the toggle switch of long-term enzyme activation, determining the importance of a particular combination of EGF concentration and duration of exposure is more complex than you might imagine at first. If MAPK could be transiently activated after 10 minutes at

5 nM or 100 minutes at 2 nM, then why was this level of MAPK activation insufficient stimulus to create long–term activation? To determine the answer, we need to analyze Figure 8.19, page 250. These two activity plots were generated by the concentration-effect curves for MAPK activation of PKC (blue line) and PKC activation of MAPK (black line), plotted on the same axes. The curves for PKC vs. MAPK and MAPK vs. PKC intersect at three points: A, B, and T. Point A represents high activity for both PKC and MAPK, whereas point B represents low activity for both. A and B represent distinct steady-state levels of PKC and MAPK activities. A system with two distinct steady states is a **bistable circuit.** The bifurcation point T is important because it defines the **threshold stimulation,** which can be thought of as a toggle switch. If the initial stimulation of EGF (amplitude and duration) is sufficient to activate *either* PKC or MAPK above the level T, then both enzymes will reach steady state at point A. In contrast, if the initial stimulation is below T for *both* PKC and MAPK, then upon removal of the stimulus, both enzymes will relax to B, which would be the basal level of activity.

We have returned to the concept of a bistable toggle switch, but this switch is produced by many more components than the simpler switches we studied in Section 8.1. Remember in λ phage, the choice between lytic and lysogenic lifestyles was determined by the amount and duration of CII protein present (see Figure 8.2). In the bioengineered bistable toggle switch created inside *E. coli,* we saw that a prolonged effect can be achieved even after the stimulus is removed (see Figure 8.4). For EGF stimulation of simulated LTP, it is a critical level of kinase activity in a positive feedback loop that determines whether the stimulus leads to LTP (i.e., learning) or not.

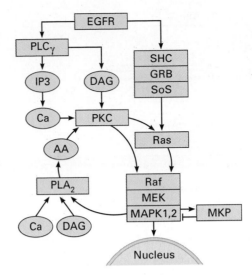

FIGURE 8.16 • **Circuit diagram of signaling pathway beginning with EGF and ending with new gene activation inside the nucleus.** Rectangles represent enzymes, and circles represent messenger molecules. This integrated circuit utilized pathways a, b, e, f, h, k, and l from Figure 8.15.

Bhalla and Iyengar made an interesting analogy that was very appropriate given that they were simulating learning. They reminded us that a bistable (i.e., digital) system can store information the same way that RAM stores information on a computer. It is also worth noting that the simulated enzymatic bistable system was very reliable due to the redundant mechanisms for achieving steady-state level A. High activity of either PKC *or* MAPK was sufficient to flip the system from off (steady-state B) to on (steady-state A). In

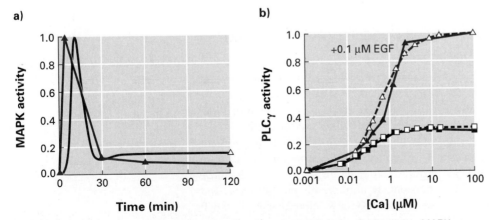

FIGURE 8.17 • **Computer model matches experimental data. a)** Measuring MAPK activity as a function of time. Simulation (open triangle) and real data (filled triangle) are very similar. The stimulus in both cases was a steady supply of 100 nM EGF. **b)** Measuring PLCγ activity as a function of calcium concentration, where dashed lines represent computer simulation and solid lines represent experimental data. Triangles indicate the presence of 100 nM EGF, while squares indicate the absence of added EGF.

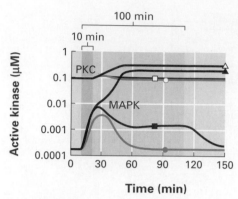

FIGURE 8.18 • **Activation of the feedback loop.**
PKC (open symbols) and MAPK (closed symbols) activities
were graphed to show impact of a positive feedback loop.
Three stimulus conditions are represented: 10 min at 5 nM
EGF (circles), 100 min at 2 nM EGF (squares), and 100 min
at 5 nM EGF (triangles). Dark and light gray shading in the
graph represents the 10 and 100 minutes of EGF exposure,
respectively.

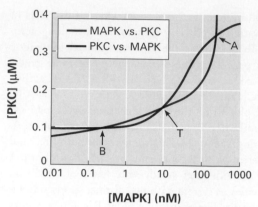

FIGURE 8.19 • **Bistability plot for feedback loop in
Figure 8.18.** The PKC vs. MAPK activities plot (blue line) was
constructed by holding the level of active MAPK constant,
running the simulation until steady state, and reading the
value for active PKC. This process was continued for a series
of MAPK levels spanning the range of interest. A similar
process was repeated for MAPK vs. PKC activities plot (black
line) by holding the level of active PKC constant and calculating MAPK activity. Both plots were drawn with the concentrations of PKC on the Y-axis and MAPK on the X-axis. The
curves intersect at three points: B (basal), T (threshold), and
A (active). A and B are stable points, but T is the toggle
switch point for the two steady state "choices" of A and B.

Section 8.1, we describe biological systems in engineering
terms, including "redundant," "reliable," and "fail-safe." Another engineering term is "**robust,**" meaning the circuit tol-
erates a wide range of environmental conditions. In data not
shown here, five different enzyme concentrations were varied
in their activities (PKC, MAPK, Raf, PLA$_2$, and MAPKK,
which phosphorylates MAPK to activate it), and the simulated bistable circuit was still functional. An interesting consequence of this analysis was that most of the permutations
were equally effective in achieving steady-state A, but PKC
was the least tolerant to change, meaning it was the least robust enzyme of the five tested. The investigators hypothesized that its lack of robustness was due to a limited number
of PKC isoforms in their computer simulation.

MATH MINUTE 8.2 • IS IT POSSIBLE TO PREDICT STEADY-STATE BEHAVIOR?

Figure 8.19 depicts the steady-state MAPK concentration when [PKC] is held constant
(black line), and the steady-state PKC concentration when [MAPK] is held constant (blue
line). If we want to know the response of MAPK to a particular concentration of PKC,
we begin at that value on the PKC axis, move horizontally until we hit the black line,
move vertically until we hit the MAPK axis, and read the concentration at the point
where we hit the MAPK axis. Conversely, if we want to know the response of PKC to
a particular concentration of MAPK, we begin at that value on the MAPK axis, move
vertically until we hit the blue line, move horizontally until we hit the PKC axis, and
read the concentration at the point where we hit the PKC axis. Note that A, T, and B
are stable points, since if [PKC] or [MAPK] is at one of these three points, the response
of the other kinase is at that same point.

You can analyze the stability of the feedback loop between MAPK and PKC by iter-
atively determining the response of [MAPK] to [PKC] and the response of [PKC] to
[MAPK]. The assumption behind this iterative process is that [MAPK] at a particular
time depends on [PKC] at an earlier time. Likewise, [PKC] at a particular time depends

on [MAPK] at an earlier time. A graphical technique known as a cobweb diagram is helpful in following the iterative process.

Let's construct a cobweb diagram for an initial value of [MAPK] just above the threshold T, say halfway between 10 and 100 nM ([MAPK] $\approx 10^{1.5} \approx$ 32 nM). From this point on the MAPK axis, move vertically to the blue line and horizontally to the PKC axis. As described above, the resulting point (approximately 0.23 μM) is the response of PKC to the initial MAPK concentration. MAPK will now respond to this value (0.23 μM) of [PKC]. The resulting [MAPK] (approximately 100 nM) is found by moving horizontally from 0.23 μM on the PKC axis to the black line, and vertically to the MAPK axis. Repeat the process to find the following successive concentration responses: [PKC] = 0.31 μM; [MAPK] = 200 nM (point A); [PKC] = 0.35 μM (point A). The diagram, and thus the concentration of both kinases, has converged to A, as was observed in Figure 8.18.

You may have noticed that the lines from the curves to the axes were merely for keeping track of the concentrations of each kinase, and are always backtracked in the next kinase response determination. Therefore, the cobweb diagram can be constructed by beginning at [MAPK] = 32, moving vertically to the blue line, horizontally to the black line, vertically to the blue line, and so on, until the diagram converges to A.

You can construct cobweb diagrams starting at various values of [MAPK]. If you begin at [MAPK] > T, the diagram should converge to A, but if you begin at [MAPK] < T, it should converge to B. This shows that T is a threshold stimulation level. Given concentration-effect curves like those in Figure 8.19, cobweb diagrams allow us to predict the behavior of systems like this feedback loop.

DISCOVERY QUESTIONS

20. Design an experiment that allows you to test the prediction that robust LTP (i.e., learning) requires more than one isoform of PKC.

21. Describe how the variation in different isoforms of PKC relate to the equations we used to calculate reliability (see pages 242–243).

22. How do the terms "robust" and "reliability" relate to each other in Figure 8.16?

Can the Computer Simulation Accommodate Forgetting as Well as Learning?

Keep in mind that we are oversimplifying what is required to form a memory in your brain, but LTP is one critical step. If LTP requires the long-term activation of either PKC or MAPK, is it possible to ever inactivate these kinases and turn off LTP (i.e., forget a memory)? Bhalla and Iyengar added this aspect of learning to their already successful model. MAPK must be phosphorylated (by MAPKK) in order to be activated, and to inactivate MAPK, there is a phosphatase (MAPK phosphatase [MKP]) that removes the activating phosphate. When MKP was added to their model, the investigators discovered more about the bistable circuit. Figure 8.20, page 252, is another data-rich figure worth careful dissection. PKC activity (Figure 8.20a, open symbols) was stimulated and then exposed to three levels of MKP activity: 10 minutes at 8 nM (open circle), 20 minutes at

4 nM (open square), or 20 minutes at 8 nM (open triangle). Only the greatest exposure to MKP was sufficient to fully inactivate PKC. Likewise, MAPK was treated with 10 minutes of 8 nM (closed circle), 20 minutes of 4 nM (closed square), or 20 minutes of 8 nM (closed triangle) MKP activity, achieving long-term inactivation (steady-state B in Figure 8.19) only with the greatest MKP exposure.

Only the most intense MKP activity was able to inactivate the feedback loop. The rebound in PKC and MAPK activities after the two lower exposures of MKP was due to a couple of factors: the persistence of AA due to a relatively slow time course of degradation, and the time required to dephosphorylate the activated kinases in the MAPK circuit. The investigators simulated a wide range of MKP concentrations and durations of exposure to the bistable circuit and plotted the combinations needed to switch the circuit off (Figure 8.20b). At high MKP concentrations, inactivation occurs quickly, but there is a minimum threshold of nearly 10 minutes exposure. Conversely, when MKP was applied for very long times, at least 2 nM MKP was required to inactivate the feedback loop.

Bhalla and Iyengar created an integrated circuit simulation that exhibited feedback loop properties, a model that has improved our understanding of LTP. The model and real neurons produce prolonged and elevated levels of activation that were initiated by an external signal (EGF binding to its receptor) even after the initial stimulation was removed. Their simulation has quantified what is required to turn off LTP by breaking the feedback loop.

FIGURE 8.20 • **Inactivation of LTP by MKP. a)** Activities of PKC (open symbols) and MAPK (filled symbols) were simulated to show status of feedback. The feedback loop was initially activated by a stimulus above point T in Figure 8.19, and then one of three inhibitory inputs was applied: 10 minutes of 8 nM (circles), 20 minutes of 4 nM (squares), and 20 minutes of 8 nM (triangles) MKP activity. Dark and light gray shading denote 10- and 20-minute exposure times of MKP, respectively. **b)** MKP was applied for varying durations and concentrations to determine thresholds for inactivation of the feedback loop.

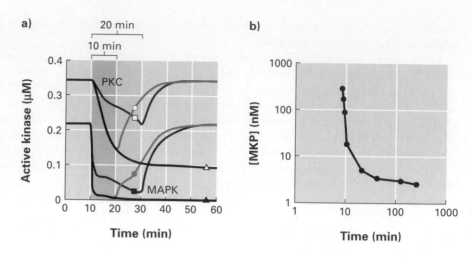

DISCOVERY QUESTIONS

23. Look at Figures 8.16 and 8.20b and hypothesize which enzyme(s) were responsible for the required ten minutes of MKP exposure to inactivate the feedback loop even when the concentration of MKP was increased tenfold.

24. Explain what was happening to the level of phosphorylated MAPK when the amount of MKP was below 2 nM for two hours.

What Roles Do Other Integrated Circuits Play in LTP?

Having enjoyed this much success, the investigators decided to add another layer of circuitry that interacts with the MAPK circuit we just studied. But first, we need to understand a little bit more about LTP. When calcium floods the stimulated neuron, these ions bind to a protein called calmodulin (CMD) to form a calcium/calmodulin (CaM) complex. CaM activates adenylyl cyclase isoforms 1 and 8 (AC1/8), calcineurin (CaN), and a kinase called Ca^{2+}/calmodulin-dependent protein kinase II (CaMKII). LTP is stimulated when CaMKII is activated for extended periods of time. To be activated, CaMKII needs to be phosphorylated (by itself or another kinase). Bhalla and Iyengar hypothesized that protein kinase A (PKA) (a cAMP-dependent kinase) played a critical indirect role for the prolonged activation of CaMKII, and they wanted to use their computer simulation to test their prediction. They called this cAMP/PKA regulation of CaMKII a "gating" control, which can be thought of as another toggle switch. If enough PKA were activated, then CaMKII would become activated (a second bistable switch). The connection between CaM and CaMKII produced a new hardwired connection that integrated smaller individual circuits from Figures 8.15 c, i, j, m, n, o. The investigators produced a new integrated circuit (Figure 8.21) to determine whether interactions between CaMKII, cAMP, and CaN were sufficient to produce prolonged

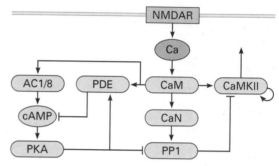

FIGURE 8.21 • **Circuit diagram examining the role of cAMP in LTP.** PDE is a phosphodiesterase that destroys cAMP. PP1 is a protein phosphatase that removes a phosphate (and thus inactivates) CaMKII. CaM is calmodulin that is activated when it binds calcium, and CaMKII is a kinase activated by CaM. CaMKII can autophosphorylate itself (small looping arrow). CaN, calcineurin, is a kinase that becomes activated when it binds CaM. PKA inhibits PP1 by adding a deactivating phosphate onto PP1. NMDAR is a voltage-gated calcium ion channel in the plasma membrane of neurons.

activation of CaMKII even after the amount of cytoplasmic Ca^{2+} inside the neuron returned to resting concentrations. The critical switch point in this circuit is CaM, which activates two competing signals that determine whether CaMKII will be activated or not. As was the case in Section 8.1, stochastic behavior of proteins and noise in the switching mechanism will influence the outcome.

DISCOVERY QUESTIONS

25. How can one kinase (e.g., CaMKII) phosphorylate so many different proteins? In other words, predict how so many different substrates can bind to a single active site.

26. How can a kinase (e.g., PKA) activate one enzyme (PDE) and inhibit another (PP1) even though it adds a single phosphate in both cases?

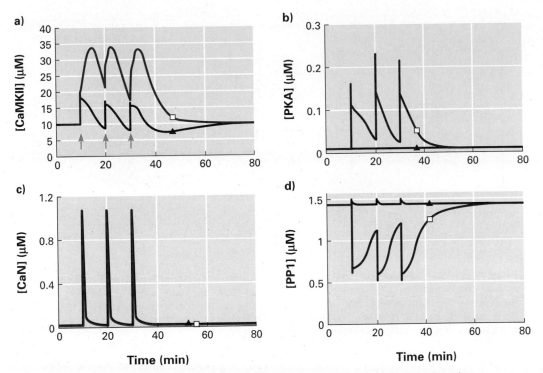

FIGURE 8.22 • Short-term activation of four key enzymes for LTP. Graphs show the simulated activities of **a)** CaMKII, **b)** PKA, **c)** CaN, and **d)** PP1 after stimulation with three 100 Hz pulses lasting 1 second each and separated by 10 minutes (required to produce LTP in real neurons). Open squares indicate that cAMP concentrations were allowed to rise, while filled triangles indicate that cAMP levels were artificially maintained at resting concentrations. The arrows in panel a indicate when the three pluses were given to the system at the 10-, 20-, and 30-minute time points.

27. How many pairs of proteins can you find in Figure 8.21 that are competing to produce opposite outcomes? What role would stochastic bursts of enzymatic activity play in the competitions you found in this figure?

When the new integrated circuit was stimulated sufficiently to produce LTP in a real neuron, the simulated activities of four enzymes were graphed (Figure 8.22). When the concentration of cAMP was maintained at basal levels in the computer simulation, the external stimulation produced three transient peaks of activities for CaMKII, PKA, and CaN, as well as the accompanying loss in PP1 activity. When the concentration of cAMP was allowed to rise as directed by their model, all enzymes exhibited significant changes in activities that were not present when cAMP concentration was fixed at resting levels. PKA needed elevated cAMP in order to become activated. Since we are particularly interested in the system's ability to initiate LTP via CaMKII, CaMKII's response to cAMP concentration changes was especially noteworthy. The amplitude of CaMKII activity was increased by about twofold, and the time it took for the activity to return to resting levels also doubled to about

20 minutes. The prolonged activation of CaMKII was due in large part to the significant inhibition of PP1, which otherwise would have inactivated CaMKII quickly. Nevertheless, CaMKII was *not* activated as long as we would expect in a bistable toggle switch once the stimulus was removed. Thus the bistable toggle switch under these simulated conditions was still in the off position (equivalent to steady-state level B in Figure 8.19).

DISCOVERY QUESTIONS

28. Predict what is needed in order to stimulate CaMKII to become activated long-term. Design an animal experiment to test your hypothesis.

29. Would cAMP levels rise or fall when given the "freedom" to vary after stimulation?

30. Which enzyme(s) exhibited the greatest duration of activity after the stimulus was removed?

Full Implementation Finally, Bhalla and Iyengar were ready to create a fully integrated circuit. This model integrates four individual signaling circuits (Figure 8.23, page 254). When

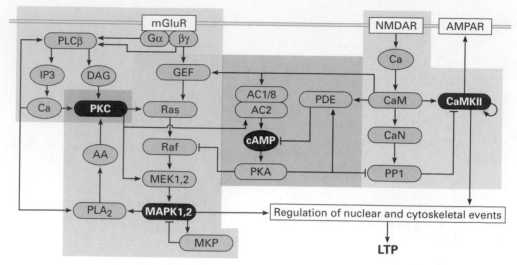

FIGURE 8.23 · **Full-circuit diagram with feedback loop, synaptic output, and CaMKII activity and regulation.** Two possible end points of the model are represented as AMPAR regulation and stimulation of nuclear and cytoskeletal events, which eventually leads to LTP. Note the feedback loop between PKC and MAPK with PLA$_2$ as the connecting enzyme. The four shaded regions highlight the four individual signaling circuits.

electrical engineers view a circuit diagram such as this, they do not examine each piece individually. They look for functional units of circuitry and the connections between them. Figure 8.23 contains four functional units: PKC, MAPK, CaMKII, and the cAMP circuits. We have seen how PKC and MAPK circuits were connected (see Figure 8.16) and how the CaMKII and cAMP circuits were connected (see Figure 8.21). PKC was connected to the cAMP circuit via AC2. These connections enabled Bhalla and Iyengar to produce their final integrated circuit composed of four smaller circuits.

Can the Model Determine Whether Sustained CaMKII Activity Is Possible? How powerful a research tool is the computer model that simulates integrated circuits? Can it elucidate critical features that result in LTP? Does the feedback loop between PKC and MAPK play a critical role in LTP? To build their integrated circuit (see Figure 8.23), all the individual circuits shown in Figure 8.15 were incorporated. This in silico neuron was stimulated with glutamate as well as depolarization at the plasma membrane, which opens the NMDAR calcium ion channel. How did the full integrated circuit respond? Were there any unexpected activities?

The first pair of enzymes examined were PKC and MAPK (Figure 8.24), which are indirectly connected to each other. PKC exhibited a strong on/off response to the three electrical stimulations in the absence of the feedback loop (AA held at basal levels). When the feedback loop was allowed to function, notice how PKC activity is substantially increased, and the duration of the activation is maintained after the stimulation was stopped. Compare the PKC activity to the activity of MAPK. In the absence of the feedback

loop, MAPK activity was undetectable, but in the presence of the feedback loop, MAPK is strongly activated and the activation is sustained. Notice that MAPK activity began at about 30 minutes.

DISCOVERY QUESTIONS

31. Why does MAPK have to "wait" for 30 minutes before it gets activated? Explain your answer in terms of the full circuit in Figure 8.23.

32. Using the full circuit, follow the connections between NMDAR and PKC to explain the stepwise increase in PKC activation to its final level of activity with the feedback loop.

Four additional enzymes were examined in this simulation (Figure 8.25). PKA was activated +/− the feedback loop, as was CaMKII. But note that both of these enzymes exhibited a higher amplitude and sustained activity in the presence of feedback vs. its absence. CaN was activated, but there was no difference +/− feedback loop. PP1 had its activity reduced as one would expect, and its inhibition was of greater amplitude after the stimuli were removed. We are seeing the results of another bistable toggle switch that does not require any transcription or translation. PP1 inhibits CaMKII, but PKA inhibits PP1 while also stimulating its own activity. In this toggle switch, CaN is not involved at all, so it is no surprise that CaN activity was not sustained when the feedback loop was functioning.

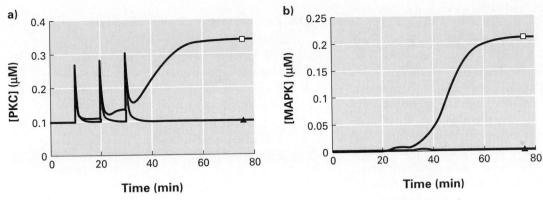

FIGURE 8.24 • Activity profile of **a)** PKC and **b)** MAPK in the fully functional integrated circuit (open squares) and with the feedback loop blocked by holding AA at resting concentrations (filled triangles). Electrical pulses of 100 Hz for 1 second were applied at 10, 20, and 30 minutes. Used with permission from Ravi Iyengar.

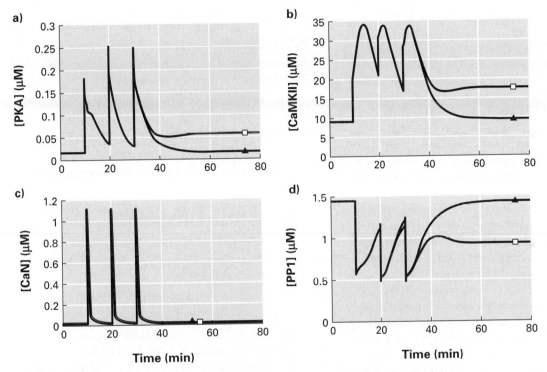

FIGURE 8.25 • Activity profiles of PKA, CaMKII, CaN, and PP1 with the feedback loop functioning (open squares) and with feedback blocked by holding AA fixed at resting concentrations (filled triangles). **a)** The Ca-stimulated PKA waveforms are almost identical, but the baseline rises when feedback is on, because PKC stimulates AC2 to produce additional cAMP. **b)** Activity profile of CaMKII. **c)** Activity profile of CaN is unaffected by the presence or absence of feedback. **d)** Activity profile of PP1 in the presence and absence of feedback loop.

The most striking outcome we discovered was that complex circuits exhibit new properties without the need for new mRNA or protein production! What impact does this have for microarray and proteomics research? This is good news and bad news for students who like to cram in material minutes before a test. The good news is you can learn new information (i.e., LTP is produced) very quickly in the absence of new protein production, which would take about an hour to accumulate. The bad news is that LTP has two phases just like your memory: short-term and long-term. MKP can be thought of as the "off switch" to learning. MKP inactivates MAPK, which is needed to initiate transcription in the nucleus. MAPK indirectly controls the activity of CaMKII (via the feedback loop that includes

PLA2, PKC, AC2, PKA, and PP1), which is also needed to produce LTP. Therefore, if your MKP activity exhibits a stochastic burst, what you "learned" while cramming will be lost.

MKP has the ability to determine which line (open squares vs. filled triangles in Figure 8.25) better represents your ability to remember something. The toggle switch of memory is controlled by the concentration of a few molecules to produce a system that is bistable. Therefore, MKP may act as a timer to bring to an end the early phase of LTP. Of course, you want to be able to retain information for a longer time, but occasionally you forget. MKP becomes the first checkpoint at which some people will not be able to commit a new memory to long-term storage. In Bhalla and Iyengar's full integrated circuit, MKP is not connected to any other proteins, but this is unlikely to accurately reflect reality. What circuits control MKP activity? How do our neurons "decide" which way to flip the toggle switch for LTP?

What Is Long-term Memory and How Can LTP Play a Role?

It has been known for years that LTP is divided into two phases—slow and fast. You have just studied the fast phase, which was composed of four integrated circuits. In a real neuron, the slow phase of LTP changes the structure of synapses between neurons, requiring the production and intracellular transportation of new proteins (i.e., a dynamic proteome). The goal in learning is to change one synapse (to store a new memory) without altering other synapses (which would be like reformatting a floppy disk where the disk is intact but no longer has the instructions to access older memories). MAPK and CaMKII lead to the slow production of new proteins that must be sent to the correct synapses. Bhalla and Iyengar hypothesize that the feedback loop described in their simulation controls the localization of new proteins via the cytoskeleton so they wind up at the correct synapse. Wow! That is a huge conclusion from this work, and for the sake of science, it does not matter if they are right or wrong. Either way, they have made a prediction based on their model that is testable in the lab. If they are right, they may need to rent tuxedos and fly to Stockholm. If they are wrong, someone will demonstrate it and advance the field by eliminating one incorrect possibility. Therefore, a completely computer-based research project has taught us a lot about integrated circuits, provided us with some new insights about LTP, and generated a new idea to explain long-term memory.

DISCOVERY QUESTIONS

33. What difference do you see when comparing the activation of CaMKII in Figures 8.25b and 8.22a? Propose a mechanism for this difference.

34. What is the consequence for LTP when PP1 is more inhibited in the presence of the feedback loop?

35. Explain to DNA-chip experts and proteomics experts why this integrated circuit has a profound impact on their research.

36. Hypothesize how a person who has a photographic memory could have a genotype that permits him or her to never lose fast LTP?

37. Hypothesize how an older person could retain his or her long-term memories but not be able to learn new things.

What Have We Learned (How Much LTP Have We Generated)?
Bhalla and Iyengar's paper had two major goals. Its primary goal was to understand how protein circuits work and whether they can exhibit synergistic properties. From this computer-based simulation, we now understand that integrated protein circuits can explain some long-held principles:

- It is possible to produce a prolonged signaling effect even after the original stimulus is removed.
- Protein circuits can activate feedback loops, which can provide the mechanism for synergistic properties.
- It is possible to understand the signaling threshold required to control a bistable toggle switch.
- A single stimulus can lead to multiple output pathways.
- Complex networks provide the mechanism for critical aspects of "biological design" that can provide the selective pressure for evolutionary steps. These design principles include redundancy, robustness, and fail-safe.
- All of these features can be accomplished in the absence of new protein synthesis.
- Genetic variation in the population will result in certain genotypes that will respond differently to the same stimuli.

The second goal of their research was to gain a better understanding of the process of LTP, which leads to the formation of a new memory. In this area, the investigators predicted:

1. CaN does not play a major role in the LTP activation of CaMKII, which most believe is a critical protein in learning.

2. PKA is downstream of both PKC and MPAK and upstream of CaMKII.

3. PKC can be activated by more than one input, each with critical threshold concentrations to flip the bistable toggle switch to the on position.

4. The adenylyl cyclase isoforms play separate roles in LTP. AC1 is activated by CaM and allows the system to achieve a rapid inhibition of PP1 after Ca^{2+} influx. AC2 is activated by PKC and provides the sustained

inhibition of PP1, which leads to the prolonged activation of CaMKII and thus LTP. AC2 provides the necessary step for the feedback loop, which is critical for the synergistic response.

We have spent a substantial amount of time and energy studying circuits. The work by Bhalla and Iyengar will lead to very focused and efficient research in the lab due to computer-guided hypothesis testing. These two investigators are the first to acknowledge that their model has limitations such as compartmental constraints and changes in protein levels due to translation of new proteins. They said, "models such as these should not be considered as definitive descriptions of networks within the cell, but rather as one approach that allows us to understand the capabilities of complex systems and devise experiments to test these capabilities." They also leave us with an interesting question. Is the LTP integrated circuitry related to immunological "memory," in which memory white blood cells can "remember" pathogens years after initial "learning"? In silico circuit analysis may help us understand immunology as well as many other complex systems.

Can We Understand Cancer Better by Understanding Its Circuitry?

The human genome project is not an end point but rather a beginning. Knowing all the DNA content of humans, even if we knew every single nucleotide polymorphism (SNP) in the gene pool, would only give us the raw material to ask interesting questions. How do we learn? What drives our sexuality? What makes some people early risers or night owls? Why do some people live to be over 100 years old?

We cannot answer these questions yet because genomics is a work in progress. However, each year we have better tools and more data, allowing us to assemble more complex and comprehensive models that better describe biological processes. For example, Kurt Kohn from the **National Cancer Institute (NCI)** produced two huge circuit diagrams of cell cycle control and DNA repair. The first question most nonbiologists ask at this point is "why would the NCI want to study cell cycle and DNA repair?" Cancer is the loss of cell cycle control and the inability to fix damaged DNA. A lot was already known about cell cycle and DNA repair (see Kohn's references), and Kohn summarized all this information in comprehensive circuit diagrams. But first he had to invent a language (Figure 8.26, page 258). Unlike Bhalla and Iyengar, Kohn permitted many more interactions to take place in his circuit diagram, and therefore he needed a larger vocabulary. You should note the different types of interactions needed to complete Kohn's circuits. Some symbols are more intuitive than others, but each interaction is found inside your cells. Do not memorize the symbols, but make sure you understand what each symbol means in the vocabulary.

DISCOVERY QUESTIONS

38. Use Kohn's symbolic language to diagram how *Endo16* is regulated (see Figure 7.22).

39. Use Kohn's symbolic language to "translate" the PKC/MAPK feedback loop (see Figure 8.16).

SEQUENCES
Cell Cycle Control
DNA Repair

LINKS
Kurt Kohn
Kohn's References

Look at the electronic versions of the two circuit diagrams: the cell cycle and the DNA repair circuits. The first things you should notice are their sheer size and complexity. The circuit diagrams facilitate "big picture" comprehension, which is appropriate since most biologists do *not* need to know all the details. The key to understanding complex circuit diagrams is to find functional units that are connected to other functional units. For example, look at the *myc* functional unit that is magnified in Figure 8.27, page 258.

Myc is a known **oncogene**, and here is some information as it appeared in Kohn's annotations appendix:

C35: Cdc25A may be transcriptionally activated by c-Myc: the Myc:Max heterodimer binds to elements in the Cdc25A gene and activates its transcription.

M1: c-Myc and pRb compete for binding to AP2.

M2: AP2 and Max compete for binding to c-Myc. AP2 and Myc associate in vivo via their C-terminal domains.

M3: The E-cadherin promoter is regulated via AP2 recognition elements.

M4: c-Myc and pRb enhance transcription from the E-cadherin promoter in an AP2-dependent manner in epithelial cells (mechanism unknown). Activation by pRb and c-Myc is not additive, suggesting that they act upon the same site, thereby perhaps blocking the binding of an unidentified inhibitor. No c-Myc recognition element is required for activation of the E-cadherin promoter by c-Myc. Max blocks transcriptional activation from the E-cadherin promoter by c-Myc, presumably because it blocks the binding between c-Myc and AP2.

The proteins **retinoblastoma (pRb,** another oncogene) and c-Myc compete to bind to the transcription factor AP2. However, c-Myc can also bind to Max. The two possible combinations c-Myc/Max and pRb/AP2 can each bind to cis-regulatory elements to initiate transcription (shown as two genes being transcribed). c-Myc/Max leads to the production of Cdc25A, a phosphatase that regulates the activity of a number of proteins in other functional domains. However, note the regulation of c-Myc in this functional unit is influenced by a different functional unit as indicated by the arrow to the left of c-Myc.

From our examination of learning circuits in Section 8.2, we know that feedback loops are very important. Let's look

Noncovalent binding, for example between proteins A and B. A filled circle or "node" can be placed on the connecting line to represent the A:B complex itself.

Asymmetric binding where protein A contributes a peptide that binds to a receptor site or pocket on protein B.

Representation of multimolecular complexes: x is A:B; y is (A:B):C. This notation is extensible to any number of components in a complex.

Covalent modification of protein A. The single-arrowed line indicates that A can exist in a phosphorylated state. The node represents the phosphorylated species.

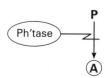

Cleavage of a covalent bond: dephosphorylation of A by a phosphatase.

Proteolytic cleavage at a specific site within a protein.

Stoichiometric conversion of A into B.

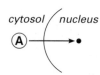

Transport of A from cytosol to nucleus. The filled circle represents A after it has been transported into the nucleus (the node functions like a ditto mark).

Formation of a homodimer. Filled circle on the right represents another copy of A. The filled circle on the binding line represent the homodimer A:A.

z is the combination of states defined by x and y.

Enymatic stimulation of a reaction.

General symbol for stimulation.

General symbol for inhibition.

Shorthand symbol for transcriptional activation.

Shorthand symbol for transcriptional inhibition.

 Degradation products.

One or more genes being transcribed.

FIGURE 8.26 · **Symbolic language created for cell cycle and DNA repair circuit diagrams.** Black lines indicate binding interactions and stoichiometric conversions; red are covalent modifications and gene expression; green are enzyme actions; blue are stimulations and inhibitions. Colors are present only in electronic versions of circuit diagrams.

at one feedback loop in each of these circuit diagrams. In the cell-cycle diagram, locate Cdc25C and CycB:Cdk1, which are located in area H4.

1. Cdc25c is activated by phosphorylation (blue line with green C18 label).

2. Cdc25 removes phosphates from Cdk1 at amino acids Thr14 and Tyr15 (shown as a green line coming from Cdc25c). Removal of the phosphates removes the inhibition from Cdk1.

3. Cdk1 interacts with CycA (black C5 label), and this interaction stimulates the phosphorylation of Cdc25 (green C36). Although this positive feedback loop is difficult to recognize at first glance, with practice and familiarity with the system, you would be able to see other examples.

The second feedback loop is taken from the DNA repair circuit diagram. First, locate **p53** in area E7–9 on the right side of the circuit. Immediately you can tell that p53 is a critical component since there are so many interactions emanating from it. If you look below p53, you will see three black dots (black P18), which indicates that p53 (the protein) can form homotetramers. The homotetramer can initiate transcription (black P17) from seven different genes (one box with seven red lines emanating from it). One of the activated genes is *Mdm2* (red P39), and the protein MDM2 can form

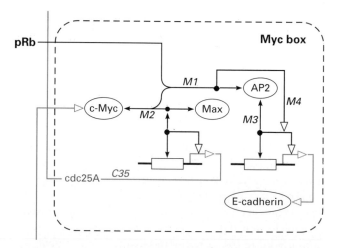

FIGURE 8.27 · **Myc subsection from Kohn's cell cycle circuit.** The Myc box can be found in area D10, in the top right corner of the DNA repair circuit diagram. Symbols are defined in Figure 8.26.

a heterodimer with p53 (black P28), which inhibits the transcription of *Mdm2* gene (blue P29). This represents a negative feedback loop. However, in his paper, Kohn reported that MDM2/p53 formation leads to the degradation of p53 (black P30) as indicated by the blue arrow labeled P31.

All these interactions were discernible from circuit diagrams with very few words, and they allow us to comprehend interactions and make predictions. Kohn's circuit diagrams are works in progress since he is revising the color coding and adding more components to his circuits. Don't make the mistake of assuming that if the component is shown in the diagram all of its interactions are known. Our knowledge is limited, and studying circuit diagrams will focus our research more efficiently. So we are beginning to make headway on larger-scale models, but what about entire organisms?

DISCOVERY QUESTIONS

40. In Kohn's circuit diagrams, try to locate one small functional subcircuit (similar to the *myc* subcircuit), and describe in words what you see. Which components are linked to other functional units?

41. Find BRCA1, the "breast cancer gene," located in area F10 of the DNA repair circuit. Describe in words how BRCA1 interacts with p53.

42. One of the blue lines below *p53* is labeled P31. Notice that two different repressors can block the degradation of p53. Given what you know about circuits, is blocking this degradation an important process or not? What effect would two repressive interactions have on the reliability of blocking the degradation of p53? Can you find any other pathways that also block the degradation of p53?

43. If you became fluent in this symbolic language, do you think you should get a foreign language credit?

If Circuits Are Interconnected, Does Gene Order Matter?

We have seen that cis-regulatory elements are integrated (Chapter 7) and that proteins and cellular messengers are integrated (Section 8.1) to form beautiful biological circuits. But the raw material for evolutionary changes is DNA. The functional (coding and noncoding) aspects of DNA provide the foundation for natural selection. This brings us to a new question. Does the location of a gene within its genome have any influence on its role? This seems like a straightforward question, but it is actually much more complex than you might think. For example, how do you design an experiment to answer this question? We cannot go back in time and genetically engineer some species and see which version survived. What can we do? There are two approaches to this question: the observational approach and the computational approach.

Observational Approach The first approach is the classic evolutionary biologist's approach of comparing what you find in nature. The rationale behind this is very sound. The best way to examine fitness is to see what has evolved over millions of years of selective pressure. Imagine some original cell on earth, some tiny robust prokaryote that was capable of surviving mutations to its own genome (RNA or DNA) and giving rise to a second species on earth. These two species continued to diversify and new genes evolved, became duplicated, modified, duplicated again, etc. All of this genetic experimentation led to new arrangements of genes. For example, species 4 might have genes A, B, C, D, E, F, G, H, I, J, while species 5 might have evolved with a gene order of A, H, G, F, E, D, C, B, B', I, J. At some point during this time of species diversification, sex evolved and the order of genes became scrambled even more with chromosomal rearrangements.

DISCOVERY QUESTIONS

44. What kind of changes occurred to the DNA of species 5 compared to 4?

45. What new opportunities does species 5 have since it has two copies of gene B?

46. What would you predict about the degree of reliability (see pages 242–244) for the process encoded by gene B and B'?

Evolutionary biologists assume that what works best must be what is still in existence today. Some look at morphology, some behavior, and others DNA sequences. As a result of genome comparisons, investigators have identified many gene complexes that are highly ordered. The most famous are the *Hox* genes that are conserved not only in DNA sequence but also in gene order on chromosomes. DNA sequence and gene order are conserved from flies to humans, with genes arranged from head to tail along the length of the chromosomes in humans, mice, worms, and flies. To molecular biologists, however, there is something unsatisfying about this kind of research—it leads to correlations that have not been experimentally tested. Every molecular biologist accepts that the arrangement of *Hox* genes is evolutionarily advantageous, but it is impossible to test this by altering history.

Recent work by many researchers who analyze DNA sequences has produced some interesting observations. Elizabeth Williams and Laurence Hurst found that linked genes often evolve at similar rates. This might seem obvious, but remember that after a few billion years of recombination and mutations, the probability that two genes would stay associated together and mutate at similar rates is very small. The caveat to this type of analysis is you never can be sure if the conserved gene order is due to chance or has been subjected to selective pressures. To address this concern, a group from the European Molecular Biology Laboratory (EMBL) in Germany analyzed the genomic sequences of nine different taxa of prokaryotes (proteobacteria, gram-positive bacteria, and Archaea). They found in many cases, gene pairs were conserved in their order on chromosomes across diverse taxa. Clusters of two to seven genes were conserved in order and orientation for transcription. It is hard to imagine that chance alone could explain conserved gene order and orientation. Nevertheless, this observation is a correlation and not a causative analysis and thus not completely satisfying.

Computational Approach There is a second approach that also has weaknesses but is more satisfying to a growing group of biologists exploring genomics. You will recognize the thought behind this approach. If we know all the genes by sequence, and we know the roles for most of these genes, then why not build a computer simulation, make predictions, test them out, refine the model, etc.? This circuit diagram approach to evaluating gene order for entire genomes is very new. One of the pioneers in this area is Drew Endy, who has recently moved to M.I.T., and his whole-genome approach to evolution is a new trend in biology (see Chapter 9).

Calculating the Value of a Genome's Order Endy approached this experimentally difficult question of gene order by choosing a simple system (i.e., a model organism). Rather than selecting a species like yeast, which has about 6,200 genes, he decided to start smaller: a genome of only 56 genes. Endy chose the virus T7 as his model organism (Figure 8.28). T7's double-stranded DNA genome is only 39,937 bp long. It sounds so small that it is hard to imagine much complexity, but even this simple organism has some hidden secrets. Of the 59 proteins produced by T7's 56 genes, only 33 have known functions. That could present some problems with a simulated circuit. But computer models are not intended to be perfect; rather, they should incorporate current knowledge and allow predictions that can be tested experimentally to improve our understanding. Clearly

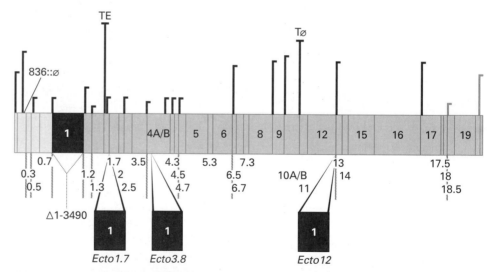

FIGURE 8.28 • **Wild-type T7 (T7⁺) genome contains 56 genes encoding 59 proteins.** Numbers represent coding regions (genes were numbered as space permitted). Vertical lines with half bars represent *E. coli* (black) and T7 (blue) promoters with bar height proportional to the strength of each promoter. *E. coli* RNAP (polymerase; TE) and T7 RNAP (Tø) terminators are shown as vertical lines with full bars. RNase III recognition sites are shown as vertical lines below the genome. All transcription goes from left to right. The positions of the *gene 1* (encoding the T7 RNA polymerase) in T7⁺ and in the three experimental strains constructed and characterized in the laboratory are shown below the genome and labeled *Ecto1.7*, *Ecto3.8*, and *Ecto12*; *wt gene 1* was deleted in the three Ecto-mutants.

there is room for improvement when only 56% of the proteins have known functions.

What Do We Know? When T7 infects an *E. coli* bacterium, about 100 T7 progeny will be released when the host cell lyses about 25 minutes after infection. Unlike viruses that inject their genomes quickly, it takes about 10 minutes for the T7 genome to enter its host. The first 850 bp are inserted quickly, and the rest is pulled into the host as it is being transcribed. Initially, the *E. coli* RNA polymerase binds to one or more of the 5 *E. coli* promoters in the T7 genome (spanning the first 15% of the genome in Figure 8.28). *E. coli* RNA polymerase pulls the T7 genome into the cell at about 45 bp per second. The first essential T7 gene to produce a protein is conveniently called *gene 1*. *Gene 1* encodes the T7 RNA polymerase, which takes over the role of transcription and genome movement into *E. coli*. T7 RNA polymerase uses 17 different promoter sequences (in the remaining 85% of the genome) and pulls the DNA in at a rate of 200 bp per second—a fourfold increase in genome movement. Therefore, it is easy to see why T7 would evolve to have *gene 1* as the first coding sequence in its genome.

Does *Gene 1* Have to be First? T7 is a clear example that gene order might matter even for a small genome. How much flexibility is there for *gene 1* position within the genome? Could it be second? third? Could the promoters be moved around and alter T7's evolutionary outcome? Endy started his computer simulation with the question about the optimum position for *gene 1* (Figure 8.29). Using in silico methods, Endy placed

gene 1 in every possible position (except as very first and very last) and calculated how long it would take T7 to double its numbers in a simulated flask of *E. coli*. For his simulation, he used this formula to determine the growth rate of T7:

$$\text{maximum doubling rate} = \mu_m = \max_t\{\log_2[Y(t)]/t\}$$

where $Y(t)$ is the computed number of intracellular progeny as a function of time (t).

μ_m defines the optimal time for phage-induced lysis in environments containing infinite uninfected hosts. For each of the 72 in silico *gene 1* positional mutants, μ_m was calculated and graphed. When he ran his simulation, Endy discovered some genotypes were better than the *wt* strain (see Figure 8.29). The *wt* genotype (triangle) has a doubling rate of about 0.35 per minute, but there are several genotypes with higher rates, some as much as 30% higher. This improvement on evolution flies in the face of evolutionary theory. Surely the T7$^+$ genome is not suboptimal in its evolutionary fitness! But remember, this was a computer simulation. How did Endy's simulations compare with reality?

Three **ectopic** (not in their normal genomic location) strains where copies of *gene 1* were placed into new locations were generated in the lab. *Ecto1.7* had *gene 1* inserted into a nonessential gene called *1.7*. *Ecto3.8* had *gene 1* inserted into the coding region of the nonessential gene *3.8*. In *Ecto12*, *gene 1* was inserted between a promoter and *gene 12*. Endy also constructed two control viruses in which a *gene 1*-sized piece of λ phage DNA was inserted into genes *1.7* and *3.8*, and a third control strain in which a T7 late promoter was inserted at base 836.

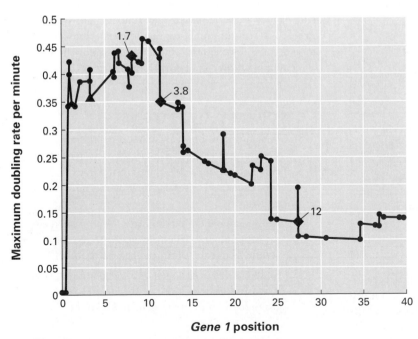

FIGURE 8.29 • Predicted consequences of altered *gene 1* position. Computed maximum phage doubling rate, μ_m, as *gene 1* was repositioned on the T7$^+$ genome. Black circles mark the 5' end of the in silico *ectopic* (or inserted) *gene 1* and the μ_m for 72 positional mutants. The blue diamonds indicate the computed μ_m for the ectopic *gene 1* strains that were constructed and characterized experimentally. The blue triangle indicates the μ_m computed for *wt gene 1* position.

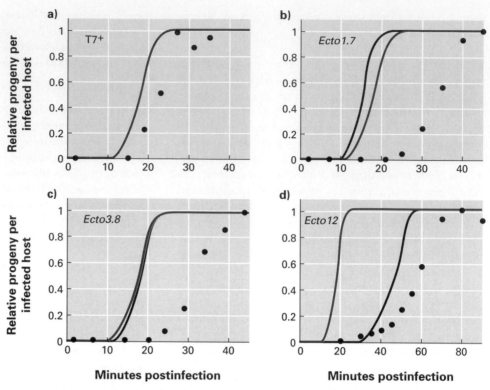

FIGURE 8.30 • **Computed (solid lines) and observed (black dots) intracellular one-step growth curves for T7⁺ and the ectopic *gene 1* strains.** Note the different time scale for *Ecto12*. **a)** *wt*, **b)** *Ecto1.7*, **c)** *Ecto3.8*, and **d)** *Ecto1.12*, are shown with T7⁺ growth curve (blue line) for reference.

DISCOVERY QUESTIONS

47. Think of reasons why the controls Endy used were not ideal. Design a different control for these experiments.

48. It is difficult to understand what would make a mutant strain more efficient, but easier to understand less efficient mutants. Explain why the mutations with *gene 1* positioned late in the genome would be the least efficient.

49. In the T7 computer simulation, identical viruses are programmed to behave identically. In reality, would you expect each virus to perform exactly the same? Do you think stochastic events might occur in natural settings? If so, what impact might this have when comparing computer growth curves and real growth curves?

Endy compared the simulated viruses to his three genetically engineered mutant strains of T7. Each real strain was added to a separate population of *E. coli* in a flask, and the real doubling rates for these T7 mutants were experimentally measured (Figure 8.30). The simulated growth curve for

T7⁺ was four minutes faster than the experimentally determined growth curve for T7⁺. Compare the slopes of computer growth rates and the real growth rates for T7, and you will see they are very similar. The simulated growth curve for *Ecto12* was ten minutes faster than the experimentally determined curve. Live *Ecto1.7* and *Ecto3.8* strains showed little agreement with predicted growth rates.

DISCOVERY QUESTIONS

50. Which part of the real curve differs the most from the simulated curve? Explain why this part of each curve might be so different, given what you know about stochastic protein kinetics and noisy toggle switches.

51. Reevaluate the data in Figure 8.29 after analyzing the experimental data from Figure 8.30. What can you say about the optimum gene order for T7?

52. Endy has been very open with his computer simulation and has put it online for anyone to use. Go to his T7web site to gain access to the T7web: The Bacteriophage T7 Simulation Online. Some genotypes are not viable. For example, if you put *gene 1* upstream of all promoters, then the virus is not viable since its RNA polymerase would never be produced.

   a. Try to create the most "evolutionarily fit" in silico T7 you can by altering only the position of *gene 1*. The fastest doubling rate is the measure we will use to determine fitness. Try at least five genotypes and record the one that produced the fastest doubling rate. You will probably want to consult a list of all the T7 genes, to learn what is known about each one.

   b. Alter the order of any gene(s) you want to maximize the doubling rate. Alternatively, you can try to create the least fit T7 virus. Try at least five genotypes and record the one that produced the most extreme value (fast or slow).

   c. At the bottom of the T7 web page you have the ability to alter characteristics of genes without altering their order. Try to produce the most fit T7 you can by altering any genetic trait you want, but not gene order. Record what you had to do to produce the "most fit" virus you can. Try at least five genotypes.

   d. Can you deduce any patterns to the fast- and slow-growing genotypes? If so, email Endy since he is continuing this research and is interested in any insights that might help.

In his paper, Endy simulated and measured plaque sizes, DNA replication, protein production, and lysis rates for his three real mutants. All this work (with web supplemental data available) indicated that he was successful in the scientific sense. He made a model, made predictions, tested them, and now he will refine his model based on new experimental evidence.

We studied Endy's work for two primary reasons. First, T7web allows you to experiment with genotypes of your choosing and begin your own in silico genomics research. Because the T7web is available online, you have a better understanding of how simulations are performed. Second, it points out how difficult understanding more complex genomes will be. Why? Chapter 9 will focus on one research project that attempted to simultaneously model yeast genomics, proteomics, and metabolomics.

Summary 8.2

Scaling up from simple circuits of a few genes to much more complex circuits has provided us with a few new insights. First, when analyzing interactions in larger circuits, we can discover new properties that are evident because of the more complex interactions. Single genes and small protein interaction circuits do not behave synergistically, and therefore, if we only study isolated proteins, we cannot fully understand how cells function. Second, we found feedback loops produce unexpected connections and consequences. Third, toggle switches can be composed of more than just one or two genes and can also include nonprotein components. We also saw how critical functions can utilize feedback loops to help flip a toggle switch from off to on. In the cancer circuits, we studied a larger vocabulary to describe more complex interactions. From these larger circuits, our fourth insight was to discern critical components (e.g., p53) and which processes were replicated by redundant pathways. Finally, we used computer simulations to model a whole genome to determine if gene order is critical. The model did not accurately predict experimental data, but it was still successful in that it highlighted the first few minutes of infection as the most critical step in modeling T7 infection and replication. In short, the postgenomic era must integrate computer simulations and classical molecular experiments in order to create more accurate models of genomes and proteomes.

Chapter 8 Conclusions

Understanding a genome requires more than just listing its genes and defining their roles. We need to understand how the proteins interact, how genes regulate each other's transcription, how positive and negative feedback loops produce synergistic properties that affect a cell's response to its changing needs. Toggle switches enable genomes to make "choices," and the inherent noise in toggle switches is a necessary part of how they function. Proteins exhibit stochastic behavior, so it may be impossible to produce a model that can predict with certainty how a particular cell will respond to a change in its environment. Complex circuits have unique properties, and as we construct increasingly complex circuit diagrams, we will discover more emergent properties that improve our understanding of cells and organisms. A "cell web" is composed of many interconnected components, and the ultimate challenge is to model the combined information of genome sequences, variations in the population, gene expression profiles, proteomics, metabolomics, and genomic circuits. These challenges are intimidating but exciting, too. In Chapter 9, we will use one case study to see what is possible now and where we want to go in the future.

References

Review Articles

Bonifer, Constanze. 2000. Developmental regulation of eukaryotic gene loci. *Trends in Genetics*. 16: 310–315.

Endy, Drew, and Roger Brent. 2001. Modelling cellular behavior. *Nature*. 409: 391–395.

Wray, G. A. 1998. Promoter logic. *Science*. 279: 1871–1872.

General Reference

Brent, Roger. 2000. Genomic Biology. <http://www.molsci.org/htdocs/publications/omicbio/brentcell.html>. Accessed 11 October 2000.

Cho, Raymond J., and Michael J. Campbell. 2000. Transcription, genomes, function. *Trends in Genetics.* 16: 409–415.

Clayton, David F. 2000. The genomic action potential. *Neurobiology of Learning and Memory.* 74: 185–216.

Davidson, Eric H. 2001. *Genomic Regulatory Systems: Development and Evolution.* San Diego: Academic Press.

Gladwell, Malcolm. 2000. *The Tipping Point: How Little Things Can Make a Big Difference.* Boston: Little, Brown and Co.

Hartwell, Leland H., John J. Hopfield, et al. 1999. From molecular to modular cell biology. *Nature.* 402 Supplement: C47–C52.

Legrain, Pierre, Jean-Luc Jestin, and Vincent Schächter. 2000. From the analysis of protein complexes to proteome-wide linkage maps. *Current Opinion in Biotechnology.* 11: 402–407.

Pirson, Isabelle, Nathalie Fortemaison, et al. 2000. The visual display of regulatory information and networks. *Trends in Cell Biology.* 10: 404–408.

Genetic Toggle Switches

Elowitz, Michael B., and Stanislas Leibler. 2000. A synthetic oscillatory network of transcriptional regulators. *Nature.* 403: 335–338.

Gardner, Timothy S., Charles R. Cantor, and James J. Collins. 2000. Construction of a genetic toggle switch in Escherichia coli. *Nature.* 403: 339–342.

Hasty, Jeff, Joel Pradines, et al. 2000. Noise-based switches and amplifiers for gene expression. *PNAS.* 97: 2075–2080.

McAdams, Harley, and Adam Arkin. 1999. It's a noisy business! Genetic regulation at the nanomolar scale. *Trends in Genetics.* 15: 65–69.

McAdams, Harley, and Adam Arkin. 1997. Stochastic mechanisms in gene expression. *PNAS.* 94: 814–819.

Ptashne, Mark. *A Genetic Switch: Gene Control and Phage* λ. 1987. Cambridge, Mass.: Cell Press and Blackwell Scientific Publications.

Interacting Pathways

Bhalla, Upinder S., and Ravi Iyengar. 1999. Emergent properties of networks of biological signaling pathways. *Science.* 283: 381–387.

Kohn, Kurt. 1999. Molecular interaction map of the mammalian cell cycle control and DNA repair systems. *Molecular Biology of the Cell.* 10: 2703–2734.

McAdams H. H., and L. Shapiro. 1995. Circuit simulation of genetic networks. *Science.* 269: 650–656.

Rzhetsky, Andrey, Tomohiro Koike, et al. 2000. A knowledge model for analysis and simulation of regulatory networks. *Bioinformatics.* 16 (12): 1120–1128.

Weng, Gezhi, Upinder S. Bhalla, and Ravi Iyengar. 1999. Complexity in biological signaling systems. *Science.* 284: 92–96.

Gene Order and T7

Dandekar, Thomas, Martijn Berend Snel, and Peer Bork. 1998. Conservation of gene order: A fingerprint of proteins that physically interact. *Trends in Biochemical Sciences.* 23: 324–328.

Endy, Drew. 2000. T7web: The Bacteriophage T7 Simulation Online. <http://virus.molsci.org/t7/sim/index.html>. Accessed 11 October 2000.

Endy, Drew, Lingchong You, et al. 2000. Computation, prediction, and experimental tests of fitness for bacteriophage T7 mutants with permuted genomes. *PNAS.* 97: 5375–5380.

E. coli Metabolic Simulations

Edwards, Jeremy S., and Bernhard O. Palsson. 2000. Metabolic flux balance analysis and the in silico analysis of Escherichia coli K-12 gene deletions. *BroMed Central Bioinformatics.* 1(1): 1. <http://biomedcentral.com/1471-2105/1/1>.

Edwards, J. S., and B. O. Palsson. 2000. The *Escherichia coli* MG1655 in silico metabolic genotype: Its definition, characteristics, and capabilities. *PNAS.* 97: 5528–5533.

Universal Networks

Jeong, H. B., R. Tombor, et al. 2000. The large-scale organization of metabolic networks. *Nature.* 407: 651–654.

Marcotte, Edward M., Matteo Pellegrin, et al. 1999. A combined algorithm for genome-wide prediction of protein function. *Nature.* 402: 83–86.

CHAPTER 9

Modeling

Whole-Genome

Circuits

Up to this point, we have examined the complexity and interactions within different subsets of genomics and proteomics. In this chapter, we begin to integrate those subsets into comprehensive and predictive models. The substantial difference between the field of genomics and traditional molecular biology is that genomics wants to understand all the molecules of life as they interact with each other. The methods developed over the last few years have provided us with new ways of looking within cells to see the connections made in biological circuits. On a weekly basis, leading scientific journals publish examples illustrating the rapidly evolving field of biology. Although numerous case studies exist, only one is discussed in this chapter: the modeling of the whole-genome response in yeast to a change in energy sources. If you read *Science* or *Nature,* you know that the field of genomics is gathering momentum and no textbook can compete with journals for current information. Therefore, use Chapter 9 as an *introduction* to discovering the new perspective of genomics.

9.1 Is Genomics a New Perspective?

There is a movement afoot in biology. Some call it **systems biology,** but it has many synonyms—genomics, circuits, intentional, modular, etc. Here is how the Institute for Systems Biology (ISB) defines the term.

> *Systems biology* is a unique approach to the study of genes and proteins which has only recently been made possible by rapid advances in computer technology. Unlike traditional science which examines single genes or proteins, systems biology studies the complex interaction of all levels of biological information: genomic DNA, mRNA, proteins, functional proteins, informational pathways and informational networks to understand how they work together. Complex systems give rise to emergent or systems properties—for the brain, these are memory, consciousness and the ability to learn. These systems properties cannot be understood by studying individual neurons; rather the whole system or subsystems must be analyzed. For the past 30 years, biologists have tended to study individual genes or proteins. Systems biology requires global technologies both to define the elements of the system and to follow the elements' behavior as the system carries out its functions. For example, if one is to understand how a car functions, biologists would have studied the individual parts in isolation—the transmission; the ignition; the brakes, etc. The systems approach defines all of the elements in a system (discovery science) and then studies how each behaves in relation to the others as the system is functioning. Ultimately, the systems approach requires a mathematical model which will both describe the nature of the system and its systems properties.

In his book *The Tipping Point,* Malcolm Gladwell dissects what is required for a small trend to spread to the rest of a culture. There are three critical components of this process: the people involved, the quality of the message, and the context within which the message is delivered. Gladwell is very clever in his exposition of the "tipping point" and gives many good examples. The question now is whether the newest fad in biology, systems biology, will reach some tipping point until "everyone is doing it." Let's look at the three critical components so you can decide if systems biology is the greatest thing since spliced genes.

The People Involved: Who Is Doing Systems Biology?

The list of names who are pushing for a systems approach reads like a who's who of biologists. Sydney Brenner, founder of the Molecular Science Institute (MSI), is responsible for the identification of mRNA (with François Jacob and Matthew Meselson), the idea that the genetic code consists of triplets (with Francis Crick), and the development of the nematode *Caenorhabditis elegans* as a model research organism. Most people would retire happily with that set of credentials, but Brenner is not resting on these laurels. MSI, located in Berkeley, California, includes on its board of trustees the 1999 Nobel laureate Gunter Blobel, who works at The Rockefeller University in New York, and Drew Endy, who produced the online T7 program that evaluated gene order (Chapter 8). MSI describes its goal as follows:

> One consequence of this work will be to bring into being an intentional biology. By this we mean a rational and, as possible, controlled human interaction with the living world. Such a transition requires an ability to observe what is happening in biological systems, an ability to understand what we observe, and an ability to affect biological systems based upon our understanding. An intentional biology will allow humans to leverage the existing molecular infrastructure and cellular architecture to produce food, energy, and materials with greater efficiency than is currently possible.

Other notable biologists have moved to areas of research that are complex and require new approaches. Francis Crick now studies consciousness of the human mind/brain, while James Watson is intrigued by "free will." To address the need for a new approach in biology, Cold Spring Harbor Laboratory (CSHL), where Watson is president, has created a new graduate school (Watson School of Biological Sciences) at a time when there is concern about producing too many PhDs for the market. Their mission includes this perspective:

> Advances in biology depend on multidisciplinary approaches, in which knowledge and technology from

diverse areas intersect to inspire new discoveries. Today, however, the breadth of accumulated knowledge about biology is immense—far more extensive than any individual can assimilate. The doctoral program of the Watson School of Biological Sciences at Cold Spring Harbor Laboratory will train biologists for this new era in biology.

At the end of his biography, *Time, Love, Memory,* Seymour Benzer talks with Crick about the future of biology and the need for a new level of understanding. With genome sequences becoming huge parts lists for very diverse biological systems, many are looking toward new areas of research. Benzer, who began the field of behavioral genetics, uses the model organism *Drosophila* to understand the brain and behavior. As Benzer continues to chip away at complex traits such as fly behavior, he knows many genes are involved. When it comes to human behavior, he is even more certain about the complexity. "So just having that (allele) doesn't mean you'll show that phenotype. Expression depends on a myriad of chemical reactions. And that's not generally understood. People think if you have the (allele), your fate is sealed." He is suggesting that we need to understand many genes, stochastic events, gene circuits, proteomics, single nucleotide polymorphisms (SNPs)—in short, systems biology.

The Quality of the Message: What Questions Do Systems Biologists Ask?

There are many people doing systems biology, including Lee Hartwell, who discovered many of the proteins involved in cell cycle control. In fact, his group isolated the original *cdc* (cell division cycle) described in every introductory biology textbook. Hartwell is the director of the Fred Hutchinson Cancer Research Center and another proponent of the systems approach, using it to examine the genetic variation in a population (see Chapter 3). In a 1999 commentary, Hartwell and his coauthors argued that a new approach is needed:

> We have recently become intrigued by a very fundamental difference between the artificial way we work on problems like these model systems and the reality of human biology. In models, we study the role of genes in a single genetic background. All individuals are genetically identical. However, in the human population as well as with most organisms living in nature, the genetic background is diverse. Natural genetic variation in the breeding population is often such that two or more forms (polymorphisms) are present at each gene such that most individuals are genetically individual. How do organisms accommodate this degree of variation? What variation contributes to the phenotypic diversity of the population? How do we think about genetic diversity in terms of molecular pathways? My lab is beginning to study the role of natural variation in the behavioral output of molecular pathways using cell division and cell mating as examples.

9.2 Can We Model Entire Eukaryotes with a Systems Approach?

Given all the holes in our knowledge, will we ever be able to model a complete cell? Trey Ideker, a graduate student at the University of Washington, Seattle, worked with Leroy Hood and others to use a systems approach to understand yeast. The ten-person team did not set out to understand every aspect of yeast's life, but they chose a well-defined system within metabolism—galactose utilization.

Yeast is normally grown in the lab with glucose as the energy source, but it can grow on galactose instead. Galactose is a six-carbon sugar with a very subtle but important distinction from glucose (Figure 9.1a, page 268). However, in order to metabolize galactose, the cell needs to induce some genes and repress others (Figure 9.1b). Galactose is given to the cells, but it is not able to enter the cytoplasm because it is too big and hydrophilic to cross the plasma membrane. So it must be imported by a galactose transport protein encoded by the gene *Gal2*. Genes *Gal1, Gal7, Gal10,* and *Gal5* encode metabolic enzymes that slowly convert galactose into glucose-6-phosphate, one of the basic molecules used in glycolysis and cellular respiration. Regulation of this galactose metabolism is controlled by a genome circuit with a toggle switch that is not bistable. *Gal4* encodes the transcription factor that activates all the genes listed above. However, in the absence of galactose, the Gal4 protein (Gal4p) is bound by the Gal80 protein (Gal80p), preventing Gal4p from initiating transcription. When galactose is present in the cytoplasm, it binds to Gal3p. When Gal3p has bound galactose, this complex binds to Gal80p, which must then let go of Gal4p. The inhibition of an inhibitor is analogous to a double negative in one sentence: "I will not not sleep through class" means you will sleep through class. Therefore, Gal3p-galactose inhibits Gal80p from inhibiting Gal4p. When not inhibited by Gal80p, Gal4p is able to bind to the promoters of *Gal* genes *2, 1, 7, 10,* and *5* to activate them, which allows the yeast cell to produce the proteins required to catabolize galactose. When galactose is totally consumed, the toggle switch will flip back to the default pathway of metabolizing glucose.

Unlike investigators using a traditional approach, Ideker and his team combined four steps in their analysis that are the hallmark of a systems biology approach.

1. Define all the genes in the genome, and the subset of genes, proteins, and other molecules (mostly sugars) constituting the galactose pathway. Build a model based on this information to make predictions.

METHODS
Respiration

STRUCTURES
Gal4p

LINKS
Seymour Benzer

Lee Hartwell

Fred Hutchinson Cancer Research Center

Trey Ideker

Leroy Hood

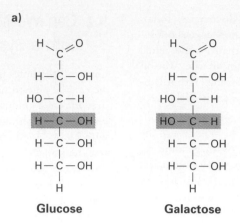

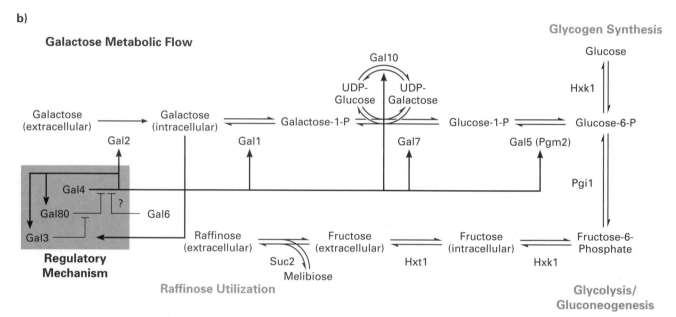

FIGURE 9.1 • **Galactose metabolism in yeast. a)** Line diagrams of glucose and galactose. Note the only difference is the position of one hydroxyl group. **b)** Metabolic circuit showing the galactose induction pathway. The toggle switch is controlled by the proteins located in the blue regulatory mechanism box. Gal6 is a repressor of the galactose induction pathway, but its mechanism is uncertain so it is shown with a question mark repressing Gal4.

<table>
<tr><td>**METHODS**</td></tr>
<tr><td>Knockout</td></tr>
<tr><td>Microarray</td></tr>
<tr><td>ICAT</td></tr>
</table>

2. Perturb each pathway component through a series of genetic (i.e., **knockout** deletions) and environmental (i.e., change the sugar food source) perturbations.

3. Utilize both gene **DNA microarray** and proteomics **isotope-coded affinity tags (ICAT)** data to create a more complete picture of the cells' response to experimentally introduced perturbations.

4. Refine the model when experimental data do not meet predictions.

With the complete genome sequence of *Saccharomyces cerevisiae* available, as well as years of biochemical data, step 1 was easily completed using a computer. Step 2 was also simple, since deletion strains are available from other researchers and altering the sugar is as simple as taking a spoonful of medicine. The grunt work came at step 3.

In step 3, they went to the lab to generate DNA microarray and ICAT data. They grew *wt* cells as well as knockout deletion strains (designated with the Greek letter delta, Δ) *gal2Δ*, *gal1Δ*, *gal5Δ*, *gal7Δ*, *gal10Δ*, *gal3Δ*, *gal4Δ*, *gal6Δ*, and *gal80Δ*. All strains were grown in the presence or absence (+/−) of galactose, and DNA microarrays were used to measure mRNA production. Each experiment was performed four times and the results averaged. The team compared some of their DNA microarray data with more traditional forms of mRNA analysis to ensure their data were accurate (Figure 9.2). The investigators wanted to get away from colorized microarray expression scales, so they

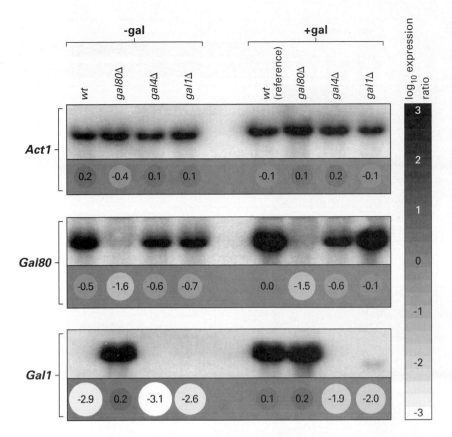

-gal **+gal**

wt | *gal80Δ* | *gal4Δ* | *gal1Δ* | *wt* (reference) | *gal80Δ* | *gal4Δ* | *gal1Δ*

log₁₀ expression ratio

Act1

0.2 | -0.4 | 0.1 | 0.1 | -0.1 | 0.1 | 0.2 | -0.1

Gal80

-0.5 | -1.6 | -0.6 | -0.7 | 0.0 | -1.5 | -0.6 | -0.1

Gal1

-2.9 | 0.2 | -3.1 | -2.6 | 0.1 | 0.2 | -1.9 | -2.0

FIGURE 9.2 • Comparison of Northern blot and microarray data. Total RNA from yeast growing in each of eight conditions (column headings) was probed with radiolabeled *Gal1, Gal80* cDNA, or *Act1* (actin gene as a control probe) using a Northern blot assay. Corresponding changes in gene expression measured with the yeast DNA microarray are depicted graphically beneath each RNA band. Change in gene expression was relative to *wt* yeast growing in galactose (column 5), with medium-gray representing no change (see color scale). Darker or lighter shades represent increasing or decreasing amounts of expression respectively, and spot size scales with the magnitude of change. The quantitative $\log_{10}$ change in expression level is annotated inside each spot.

created a black-and-white version. By eliminating colors, color-blind people are not at a disadvantage, and it is much easier to reproduce black-and-white than color images. In this black-and-white scale, no change is neutral gray, which blends into the background of the figure so you do not see any spot if mRNA production is identical between the two conditions.

As with all figures, examine the control data first. The expression of actin was relatively stable, as shown by **Northern blot** and DNA microarrays. Remember that microarray data represent the average of four experiments while the **Northern blot** data represent only one experiment. In the absence of galactose, *Gal80* mRNA is moderately reduced in three of the strains (average of $10^{-0.6}$ = fourfold repression; see Math Minute 4.1, page 112) but reduced 40-fold in *gal80Δ*. You would expect the *gal80Δ* strain to have a greater repression of *Gal80* mRNA than the other three strains, so this is a good control. However, when galactose was present, there is no change in the *wt* strain because this is the reference level of expression (i.e., self vs. self). In the absence of galactose, *Gal1* is not expressed ($10^{-2.9}$ or 794-fold repressed) as you would expect since it requires galactose for its induction. However, in the *gal80Δ* strain − galactose, *Gal1* was strongly induced, but compared to *wt* cells in the presence of galactose, it showed only a 1.6-fold induction ($10^{0.2}$). In the presence of galactose, *wt* and *gal80Δ* produced about the same amount of *Gal1* mRNA, as you would expect since *wt* should inhibit

Gal80p and the *gal80Δ* strain could never repress *Gal1*. However, in both *gal4Δ* and *gal1Δ* strains, *Gal1* mRNA is not produced ($10^{-2.0}$ = 100-fold repression) because the transcription factor (Gal4p) or the *Gal1* gene is absent.

METHODS

Northern Blot

Agreement between microarray and Northern blot data gave the investigators confidence to continue their microarray analysis (Figure 9.3). At this point, we will focus only on the top nine rows of data. It is always a good idea to begin with the control data, so we will examine the *wt* cells +/− galactose (far left column of data). In this column, the expression ratios for each gene were calculated as the amount of mRNA produced in the presence of galactose divided by the amount produced in the absence of galactose (*wt* + gal ÷ *wt* − gal). In the presence of galactose, *wt* cells strongly induce the expression of *Gal1, Gal7*, and *Gal10*. Induced at lower levels were *Gal2, Gal80*, and *Gal3*. Only *Gal4* was repressed, but only by a little. Column 2 is a control self vs. self experiment (*wt* + gal vs. *wt* + gal), and you can see that all the genes showed no change in expression levels. The first two columns verified that the method was working as expected and therefore we can turn our attention to the experimental columns.

The next nine columns of data compare the effect of galactose on the *galΔ* mutant strains vs. *wt* cells (+ gal; Δ vs. *wt*). As you would expect, there is a diagonal of white spots that indicate each *Gal* gene was repressed in the strain

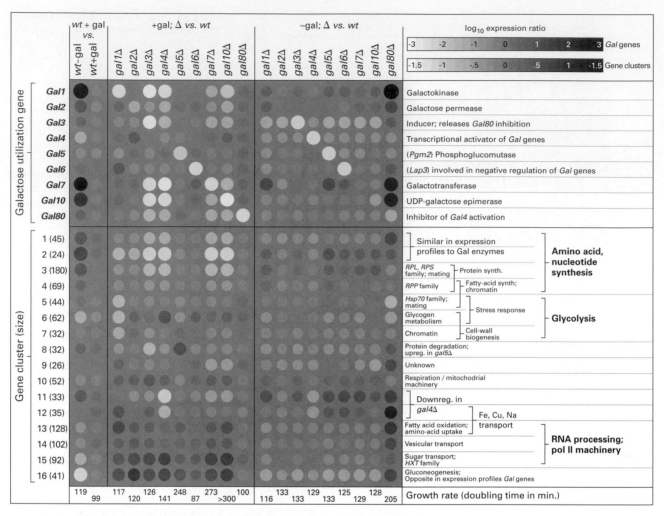

FIGURE 9.3 • **Galactose induction of yeast genes.** The top portion of the figure indicates the change in expression of a *Gal* gene (listed on the left side of each row) due to a particular perturbation (listed above each column). Medium-gray spots (the same color as the background) represent no change, darker spots represent increased mRNA production, and lighter spots indicate reduced mRNA production. The left half of the matrix shows expression changes for each deletion strain compared to *wt,* with both strains grown in the presence of galactose. The right half of the matrix compares mutants and *wt* cells in the absence of galactose. The far left two columns compare *wt* cells grown in +/− galactose. The bottom portion of the figure illustrates average expression profiles for genes in each of 16 clusters. Clusters contain genes involved in various categories, as noted on the far right. Listed at the very bottom are the growth rates for each strain and growth condition. A larger number of genes were analyzed for Figure 9.3 than for the published version.

where it was deleted when compared to *wt* cells. In the presence of galactose, *gal3Δ* and *gal4Δ* strains caused strong decreases of most, though not all, of the other *Gal* genes. This makes sense because Gal3p enables Gal4p to function by removing the inhibition caused by Gal80p (refer to Figure 9.1b). In *gal80Δ* strain, the *Gal* genes (except the deleted *Gal80* gene) were expressed slightly higher than in *wt* cells (shown as slightly darker gray spots), since the loss of the repressor Gal80p is functionally equivalent to galactose induction of the *Gal* genes. The only unexpected results in this panel were the column of lighter spots in the

gal7Δ and *gal10Δ* strains. Since these two genes are downstream of the toggle switch, it does not make sense that the absence of Gal7p or Gal10p would repress the other *Gal* genes. We will return to this issue.

In the absence of galactose (remaining nine columns), each *galΔ* mutant strain is compared to *wt* cells under identical growth conditions (−gal; Δ vs. *wt*). The most striking response is seen in the *gal80Δ* strain, which induced expression from many of the *Gal* genes because the repressor that normally inhibits their transcription had been deleted. Interestingly, the diagonal of white spots is not as obvious in the absence of galactose as it was in the presence of galactose, though *gal4Δ* is substantially repressed.

other. The top two clusters contain the *Gal* genes, while the bottom cluster contains genes involved in glycogen synthesis, which normally occurs only in the presence of excess glucose. The investigators chose a different way to display clustered gene expression profiles, illustrating yet again that visualization of data is an important component of the analysis.

DISCOVERY QUESTIONS

1. Some of the data in Figure 9.3 do not fit with traditional expectations. Look at the diagonal of spots in the absence of galactose. Why did some genes not show signs of repression when they were deleted?

2. In the presence of galactose, when Gal80p is inactivated by Gal3p, *wt* cells are still producing *Gal80* mRNA as indicated by the white spot for the *gal80Δ* strain. Why would *wt* cells continue to produce *Gal80* mRNA when the cells do not want to repress the galactose pathway?

Genomics versus Proteomics

Ideker's team included Ruedi Aebersold, who invented ICAT, the quantitative proteomics method (see pages 190–194). As a part of the systems biology approach, the members of ISB wanted to integrate ICAT and the DNA microarray data. They performed ICAT protein analysis on *wt* cells grown +/− galactose. The proteomics investigators were able to identify and quantify 289 proteins from the two proteomes. They plotted the change in protein levels in the presence of galactose vs. the absence of galactose; a positive ratio indicated the protein was more abundant in the presence of galactose (Figure 9.4). Changes in protein level are indicated by symbols further away from the horizontal line at zero. On the same graph, they used the DNA microarray data to plot the expression ratios for the 289 genes that encoded the

The bottom portion of Figure 9.3 shows the results of clustering 997 of the genes into 16 different clusters (see Math Minute 4.3, page 114). These are the only genes that had significantly altered expression in the experiments. Each cluster represents genes that were regulated similarly to each

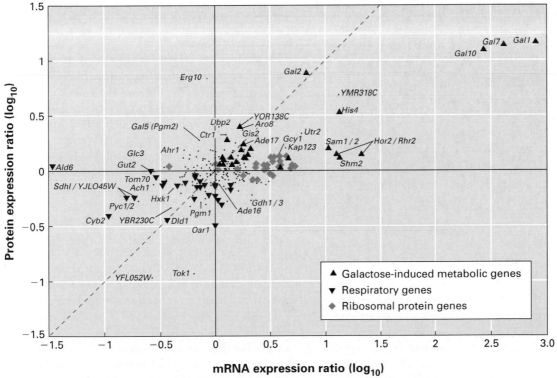

FIGURE 9.4 • Scatter plot of protein vs. mRNA expression for the 289 proteins and genes identified by both ICAT and DNA microarray methods. Blue triangles represent "galactose-induced genes"; gray diamonds represent ribosomal proteins; black inverted triangles denote components involved in cellular respiration. Dots represent other proteins and genes not in these three categories.

LINKS

Stan Fields

proteins measured by the ICAT method. Changes in mRNA production are indicated by symbols further away from the vertical line at zero. The diagonal dashed line indicates the location of genes/proteins whose protein and mRNA expression ratios were equal (i.e., both up or both down), as was the case for *Gal2* located in the top right quadrant.

Of the 289 quantified proteins, only 30 displayed substantial differences in abundance +/− galactose (i.e., they appear further away from the horizontal line). Of these 30 proteins, 15 showed no difference in mRNA levels using DNA microarray analysis (very near the vertical line). In other words, there were significant changes in protein levels even though there were no significant changes in mRNA levels. This is a classic case of **posttranscriptional control** of the **cell web.**

A lack of correlation between mRNA and protein levels has been documented before when studying single genes, but this is the first demonstration on a large scale, even though not all proteins could be measured. For the most part, metabolic and ribosomal components were up-regulated, while respiration components were down-regulated. Look at the three Gal data points in the top right quadrant that are located far to the right of the diagonal line. Their location indicates that the genes were induced more than the proteins accumulated in the cells. Once again, we see that *Gal7* and *Gal10* behaved in similar but unpredicted ways given the model in Figure 9.1b.

DISCOVERY QUESTIONS

3. What impact does the difference between protein and mRNA levels have on your evaluation of all microarray data?

4. Find the protein in Figure 9.4 that increased its concentration about 8-fold but had its mRNA reduced about 1.3-fold ($10^{-0.1}$). How could the mRNA be less abundant but the protein more abundant?

Building a Systems Model

Throughout Chapters 7 and 8, we saw that the real benefit of a model is not building a perfect one, but using it to make predictions, which can lead to new discoveries and improvements in the model. Ideker's group went back to the computer and used the research of Stan Fields (see pages 181–183) and his team to compile a list of 2,710 protein-protein interactions as well as 317 protein-DNA interactions. From this list of proteins, the investigators isolated the 1,012 genes (997 mRNAs and 15 proteins) whose expression levels were altered in the presence of galactose. These

1,012 genes/proteins were then analyzed by a computer program to overlay known interactions with expression changes for *gal4Δ* vs. *wt* + galactose. The final product was a large integrated circuit diagram (Figure 9.5 on your CD-ROM). This figure contains a lot of information, but parts have been enlarged for us (panel b). Notice the central role played by Gal4p (protein in the center of the circuit). Also, notice how many of the genes pointed to by *Gal4* have reduced mRNA expression (white), as one would expect if the transcription factor Gal4p was missing.

FIGURE 9.5 • Circuit diagram of 348 nodes and 362 interactions overlaid with mRNA expression data. Go to the CD-ROM to view this figure.

DISCOVERY QUESTIONS

5. Are the methods used to display this information (the gray spots, mostly black-and-white interaction network) the most effective way to communicate this information?

6. Study the expression changes in Figure 9.5b. Notice several of the genes are down-regulated (lighter spots) when *Gal4* was deleted. Develop a good hypothesis for why all the spots are not pure white if Gal4p normally interacts with the genes (arrows).

The investigators acknowledged that their circuit diagram is not a complete model since many of the proteins in the two proteomes (+/− galactose) were absent. They also pointed out predictions that were not supported by the data. The most striking example was that *gal7Δ* and *gal10Δ* strains cause a down regulation of many other genes. In fact, the *gal7Δ* and *gal10Δ* expression profiles in Figure 9.3 were very similar to *gal3Δ* and *gal4Δ* expression profiles. In Figure 9.1b, Gal7p and Gal10p have no apparent relationship to transcription and yet the data indicate they affect transcription. Are the data wrong or is the model wrong?

Typically, if good data conflict with your model, trust your data. The investigators believed their data so they reexamined their model. They recognized that what Gal7p and Gal10p had in common was **galactose-1-phosphate (Gal-1-P)** and that accumulations of Gal-1-P might be sensed by the cell and responded to by repressing the genes that helped produce Gal-1-P. If this were true, then you would predict that the transcription of *Gal5* would not be affected in *gal7Δ* and *gal10Δ* strains, which is consistent with the data in Figure 9.3. However, correlation alone is not satisfying, so they decided to test their revised model, with two types of experiments.

For their first experiment, the investigators used a genetic approach. They grew a double-deletion mutant *(gal1Δ gal10Δ)*, in the presence of galactose. For *gal10Δ* cells grown in galactose, the loss of Gal10p led to repression of *Gal 1, 2, 3, 7,* and *80* genes. However, in the double mutant, the loss of Gal1p would prevent the formation of Gal-1-P, and if their hypothesis was correct, then *Gal 2, 3, 7* and *80* should not be repressed. Microarrays were used to measure mRNA expression in the double mutants compared to *wt* + galactose. Their new prediction was correct; the other *Gal* genes were not repressed. Therefore, it appears that Gal-1-P can regulate the *Gal* genes in a new and unexpected way. There is one problem with this conclusion, however. Maybe the double mutation perturbed a different cellular circuit in addition to altering the production of Gal-1-P, and this different cellular circuit is the real cause, not Gal-1-P. To complement the double mutant experiment, the investigators used a biochemical experiment. They grew two strains in parallel, *gal7Δ* and *gal10Δ* in the absence of galactose. In the absence of galactose, the *Gal* genes were repressed. Then galactose was added to both strains for 20 minutes, enough time for all the *Gal* genes to be transcribed but not for Gal-1-P to accumulate. When microarrays were used to measure mRNA from these two mutant strains compared to *wt* cells + galactose, there was no repression in the *Gal* genes. Therefore, even in single mutation strains, the repression of *Gal* genes was not apparent within the first 20 minutes. This indicated that something other than mRNA (perhaps Gal-1-P) must accumulate before the *gal7Δ*- and *gal10Δ*-induced repression can occur.

DISCOVERY QUESTIONS

7. The experiments performed by Ideker's group to test their revised model used two different approaches to confirm their hypothesis. However, both experiments tested the hypothesis indirectly. What would you like to do to test more directly the role of Gal-1-P?

8. The experiment that looked for the repression of genes within 20 minutes has more than one interpretation. The investigators believe Gal-1-P affects other genes. Propose an alternative explanation.

9. Modify the circuit diagram in Figure 9.1b to illustrate the role Gal-1-P might play to regulate the expression of *Gal* genes.

Context of the Message

Although the initial galactose metabolism model was unable to fully explain all the data, the experimental design was a success. The investigators used existing data to generate a model, made predictions, tested those predictions with DNA microarrays and proteomic methods, found where the data did and did not support the model, refined the model,

made a new prediction, tested the new prediction with two additional experiments, and proposed a revised model. Some aspects of the data still cannot be explained by the data, which means new and creative experiments need to be designed to improve these weaknesses in the model. This **reiterative** approach will continue until the model gradually becomes better and better at predicting the ability of yeast to respond to changes in its environment.

Even with a lack of complete understanding of a relatively simple system, we gained some insight into emergent properties of an integrated biological circuit. First, it makes sense that a cell with too much Gal-1-P might want to slow down its production by repressing genes such as *Gal1, Gal10,* and *Gal7.* We saw in the learning circuitry (see pages 244–257) that feedback loops and subtle changes in the concentration of a few molecules are capable of regulating toggle switches. In the galactose circuit, there appears to be a toggle switch that can sense the accumulation of Gal-1-P to efficiently switch the circuit from on (make more Gal-1-P) to off (stop making Gal-1-P). When we examine the growth rate for the *gal7Δ* and *gal10Δ* strains + galactose, it appears that having too much Gal-1-P is not healthy (see Figure 9.3), since the doubling rates are very long (273 and > 300 minutes, respectively) compared to the other strains. Is too much of a good thing (Gal-1-P) bad for the cells? This scenario is further supported when you examine the growth rate for the *gal80Δ* strain minus galactose. In these cells, the repression system is lost, so the cells transcribe *Gal* genes even in the absence of galactose, which also produces the slowest growth rate of any strain in this condition (205 minutes).

A second emergent property became evident with this study. Gal4p appears to regulate many other genes in addition to the *Gal* genes. Initially, the investigators looked for Gal4p-binding DNA sequences upstream of genes that were in the same clusters (see Figure 9.3). Specifically, they looked at the 270 genes in clusters 1 and 2 (the *Gal* gene clusters), clusters 15 and 16 (genes whose profiles were the inverse of clusters 1 and 2), and clusters 11 and 12 (genes that were down-regulated in *gal4Δ* +/− galactose). Out of the possible 270 genes, 22 of them (8%) contained Gal4p binding sites, a significantly higher percentage than in all of the remaining clusters. Included in these 22 genes were *Pcl10* (cluster 1), *YMR318C* (from cluster 2), *YJL045W* (from cluster 15), and *YLR164W* and *Icl1* (both of cluster 16). *YMR318C* and *YJL045W* are open reading frames (ORFs) with no known function. So not only did we learn that Gal4p controls more genes than previously suspected, we also have good starting places to discern functions for two ORFs with unknown functions. This case study was a very good example of how a systems biology approach can lead to a deeper understanding than can be achieved when analyzing individual components in isolation. It is particularly interesting that Pcl10p is believed to play a role in repressing glycogen synthesis and Gal4p activated the *Pcl10*

LINKS

Al Gilman

AFCS

National Center
for Genome
Research

Integrated
Genomics

gene. Thus, Gal4p indirectly represses glycogen synthesis, an emergent property previously undetected. Remember how electrical engineers look at integrated circuits? They look for functional units (such as galactose utilization and glycogen synthesis) and connections between functional units (such as Gal4p). With a systems approach, we hope to find as many connections as possible in order to build a more complete model of a cell.

9.3 Will Systems Biology Go Systemic?

We began this chapter by asking whether systems biology is a fad that will come and go, or a new way to perform biological research for the next century. One component needed to establish a new trend is the people involved. Early in this chapter we listed some of the elite of biology who are moving toward a systems approach. The most profound manifestation of this movement would be the creation of new research institutions dedicated to systems biology. Leroy Hood and a number of prominent scientists left the comfort of their university setting and created the Institute for Systems Biology in Seattle. ISB members include biologists, chemists, computer scientists, physicists, and an astrophysicist who previously created complex computer models that simulate planetary formation. The ISB approach is exemplified by Ideker's galactose research, where they took a simple system and tried to understand it completely.

Nobel laureate Al Gilman at the University of Texas Southwestern Medical Center in Dallas has created the Alliance for Cell Signaling (AFCS). Gilman's approach differs from ISB's in two major ways. First, rather than create a physical institution de novo, Gilman used the Howard Hughes Medical Institute model where the institution is virtual, composed of people who retain their current jobs and institutional affiliations. In other words, he created a consortium of 52 investigators working at 21 different institutions rather than assemble 52 people under one roof. Second, AFCS's goal is different; they want to create a virtual cell by understanding the relationships that vary both temporally and spatially between sets of inputs and outputs in signaling cells. They will use multiple cell types and a wide range of methods to generate one integrated model. New models for research require new sources of funding. To launch AFCS, the National Institutes of Health awarded Gilman's team a "glue grant" of $25 million over five years to link the research efforts among different institutions. Gilman raised an additional $25 million from pharmaceutical companies and nonprofit research organizations.

Finally, there are a few other institutions whose missions are in line with a systems approach. The National Center for Genome Research is a nonprofit organization that develops computer tools that can be used by systems biologists. Integrated Genomics is a company that will create circuitry diagrams for the complete metabolism of your favorite organism. Of course, as a dot-com, they do this for profit and thus they expect the systems approach to be a growing business opportunity. Will they be the biological equivalent to Amazon.com? Or will they be closer to 8_track_tapes.com, to fade into oblivion in a couple of years?

DISCOVERY QUESTIONS

10. What technical hurdles must systems biology overcome to become a major force in biology?

11. Pretend you have several billion dollars in your savings account. Instead of buying tons of snacks and T-shirts, you want to contribute to some institutions that will be successful. Why would ISB or AFCS be a good place to donate money?

12. The most important step in any plan is to clearly identify your goals. If you were Hood or Gilman, how would you define your goal so that you would be able to know that you had successfully accomplished it? In other words, what would be a very good outcome for your new systems biology institution, and how would you know if you had achieved it?

Chapter 9 Conclusions

We began with an open-ended question—is systems biology a new perspective or just the latest advance in the continuum of biology? Is systems biology merely a faddish term that will fade quickly from popular usage? Or do systems biology, whole-genome research, biological circuits, etc., truly represent a paradigm change in furthering biological understanding? The case study in this chapter utilized many different forms of data discussed throughout the book. All aspects of life are important—sequence, variation, gene expression, proteomes, biological circuits. We can collect vast amounts of information, but can we process it? Can we build models that approach the complexity of a living yeast cell? Can the trillions of cells working cooperatively in your body be modeled in a comprehensive and realistic manner? Perhaps these goals will allow us to determine the success or shortcomings of systems biology. The pessimist may say humans will never know all there is to learn in biology, which is probably true. The optimist, however, may simply enjoy the process of attempting to learn everything and take pleasure from contributing to the cause.

References

Brent, Roger. 2000. Genomic Biology. <http://www.molsci.org/htdocs/publications/omicbio/brentcell.html>. Accessed 11 October 2000.

Cho, Raymond J., and Michael J. Campbell. 2000. Transcription, genomes, function. *Trends in Genetics.* 16: 409–415.

Clayton, David F. 2000. The genomic action potential. *Neurobiology of Learning and Memory.* 74: 185–216.

Davidson, Eric H. 2001. *Genomic Regulatory Systems: Development and Evolution.* San Diego: Academic Press.

Gladwell, Malcolm. 2000. *The Tipping Point: How Little Things Can Make a Big Difference.* Boston: Little, Brown and Co.

Hartwell, Leland H., John J. Hopfield, et al. 1999. From molecular to modular cell biology. *Nature.* 402 Supplement: C47–C52.

Ideker, Trey, Vesteinn Thorsson, et al. 2001. Integrated genomic and proteomic analyses of a systematically perturbed metabolic network. *Science.* 292: 929–934.

Legrain, Pierre, Jean-Luc Jestin, and Vincent Schächter. 2000. From the analysis of protein complexes to proteome-wide linkage maps. *Current Opinion in Biotechnology.* 11: 402–407.

Pirson, Isabelle, Nathalie Fortemaison, et al. 2000. The visual display of regulatory information and networks. *Trends in Cell Biology.* 10: 404–408.

UNIT FOUR

Transition from Genetics to Genomics: Medical Case Studies

CHAPTER 10

What's Wrong

with My Child?

Goals for Chapter 10

10.1 First Patients

Discover the complexity of apparently "simple" genetic diseases.

Utilize online bioinformatics tools to uncover new information.

Learn to visualize proteins in 3D.

10.2 The Next Steps in Understanding the Disease

Understand the value of a model research organism.

Recognize diagrams of interacting components as circuit diagrams.

In this chapter, we examine one genetic condition. Your job is to analyze the data as they are presented to you. Formulate rules; test them, look for exceptions that force you to modify them. Use this case study as a way to understand how genetics was used to understand the genomic mechanisms that cause this disease. Pay attention to the data that were collected and how they were analyzed as this case study follows the historical trail of many talented investigators. The case study is written in the style of a short mystery that you as the reader must solve. Put yourself in the place of the physician who tries to help the children in the story.

10.1 First Patients

Section 10.1 is broken into four phases of information. During each phase, you will be given some data and asked to search online sources for conclusions. This process is designed to help you strengthen your analytical skills and provide you with the perspective to appreciate how genomics is not just a collection of methods but has become an enhanced way of seeing life. So do not read ahead until you have answered the Discovery Questions as they appear in the text.

Phase I: Clinical Presentation

You are a family physician and a mother and her six-year-old boy come in to see you. (Since this is a fictitious case, no names will be used, though in reality, a good physician would refer to patients by their names.) On the chart, you see the reason for the office visit is "digestive problems." You have never seen this family before, and the paperwork indicates they do not have health insurance even though the mother has a full-time job at a small local company. When you enter the examination room, the mother is sitting in the chair and the boy is on the floor playing with some blocks. You greet them and ask the boy how he is feeling today, but you get no response. You turn to the mother since you figure that he is shy. She says that for a few months now, he has complained of a tummy ache but he never seems to be really sick.

You ask the boy to climb up on the examination table and remove his shirt. He starts to stir a bit, but then the mother gets out of her seat and helps him climb up. You get the sense that the boy is unable to stand and walk on his own. You inquire and find out that the boy could run and walk normally until he was about three and a half years old. You also notice that his calves seem larger and his thighs and arms thinner than you might expect for a boy of his size. You test his reflexes and they are nearly absent in the knee, diminished in the arms, but normal at the ankles. You listen to his heart and it seems abnormal as well. You ask if he attends school and the mother says that he does but he has to get special help with his writing. Being a parent yourself, you ask if he can write his name and the mother indicates that he has not learned all his letters yet. You know this is not normal for a six-year-old in first grade.

You decide to get some blood work done and take a small muscle biopsy. A couple of days later you get these results:

Red blood cells	Normal
White blood cells	Elevated
Glucose levels	Elevated
Creatine kinase	Extremely high
Glucose-6-phosphate dehydrogenase	Normal

The boy's muscle fibers did not look normal, with internal nuclei and varying fiber thickness (Figure 10.1). Scattered throughout the fibers were fat cells and connective tissue. Some nuclei are located away from the periphery.

a) Control Muscle Biopsy

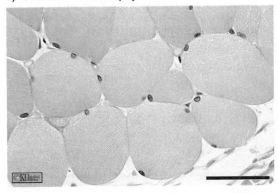

b) Patient's Muscle Biopsy

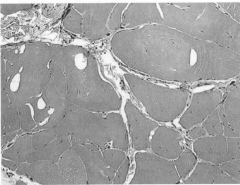

FIGURE 10.1 • Histology of *wt* and patient's skeletal muscle biopsy. a) Wild-type skeletal muscle histology. Notice that the fibers are regular in size, spacing, and "texture," with peripheral nuclei. Bar is 50 μm long. **b)** Patient's skeletal muscle biopsy. The white spaces show where fat had impregnated the muscle. Nuclei are not peripheral.

DISCOVERY QUESTIONS

1. Go to OMIM and see if you can diagnose this disease. Spend only about 5–10 minutes at this site.

2. Try NCBI Genes and Diseases and see if you can diagnose this disease. Again, spend only 5–10 minutes on this.

Phase II: Family Pedigree

Although you get the sense that this family has financial limitations, you call the mother and ask her to bring the boy back. She says she cannot pay and you "suddenly remember" a new policy that allows you to see him again as part of a free follow-up examination. Since you know she works, you suggest that she come by just before 5 P.M. when the doors will be locked.

When the mother and boy return, you meet them in your office rather than the examination room. You ask some general health questions and then begin to inquire about other family members. The boy has two older sisters and one younger brother. The girls exhibit no health problems out of the ordinary, and the younger brother is only 2 years old and has just barely begun to walk. The boy's older sister, age 24, is married, has one girl (age 2), and is expecting her next child in 7 months. You also ask about the extended family and learn that one of the boy's uncles died at age 20 from a crippling disease that reminded her of her son's illness. The mother also had two male cousins on her mother's side who died as young adults.

DISCOVERY QUESTIONS

3. Go to OMIM again and see if you can diagnose this disease. Spend only about 5–10 minutes at this site.

4. Try NCBI Genes and Diseases again and see if you can diagnose this disease. Spend no more than 5–10 minutes on this.

Phase III: Karyotyping and Linkage Analysis

You ask the mother if you could draw some blood from her and other family members. Also during the intervening week, your "new policy of free follow-up visits" has caught the attention of several neighbors of the original family. Your business is booming with people who have never seen a doctor before even though all the parents have full-time jobs. In this group, there is one girl who exhibits the same symptoms as the original boy. She is 11 years old, cannot walk on her

own, and has enlarged calves with normal ankle reflexes, but all other reflexes of the limbs are absent. She appears to have no intellectual deficits, unlike the original boy. This afflicted girl has three brothers, all of whom are healthy. When you collected pedigree information on her, no one else in a rather large extended family had any symptoms. You get permission to take blood from her and everyone in her family.

You are curious about the karyotypes of both patients as well as of their families. You also ask the lab to perform a restriction fragment length polymorphism (RFLP) and linkage analysis with special attention paid to the X chromosome (Figure 10.2).

From the boy's pedigree, you see the disease is recessive and several of the women in the pedigree are carriers. The next set of data you obtain are photographs of chromosomes taken from your two patients (Figure 10.3, page 282). The boy appears to have a deletion in his X chromosome; the girl has one typical X chromosome while the other one has recombined with one of her copies of chromosome number 6. At the top of the der(X) chromosome, the terminus is from chromosome 6 (6qter) and it continues down until band 21 on the long arm of chromosome 6 (6q21). Below this point, the remainder is X chromosome, beginning with band 21 on the short arm of X (Xp21). The der(6) chromosome contains the reciprocal DNA, as indicated in Figure 10.3e.

If this disease is recessive and the girl has only one mutated X chromosome, it does not make sense why she is suffering from the disease. She should be a healthy carrier just like the boy's mother. What's going on? Luckily, the pathologist who was working with you pointed out an unusual chromosome (Figure 10.4, page 283). Because the pathologist was curious, the karyotype was performed on 56 different cells taken from the affected girl. The wild-type (*wt*) X chromosome appeared to be replicating later than the mutant X chromosome in 55 out of 56 cells examined. In all human female cells, one X chromosome will be inactivated and the other one will remain active. In this girl, 55 out of 56 times, the mutant X chromosome was the active one and the *wt* X chromosome was the inactive one. For an unknown reason, her cells preferentially inactivated the *wt* X chromosome, so in about 98% of her cells, only genes on the mutated X chromosome could be expressed.

DISCOVERY QUESTIONS

5. What can you conclude about individual number 12 in the boy's extended family (see Figure 10.2a)? Identify its gender and probable disease status.

6. Go to OMIM and see if you can diagnose this disease with a quick search. Spend only about 5–10 minutes at this site.

a)

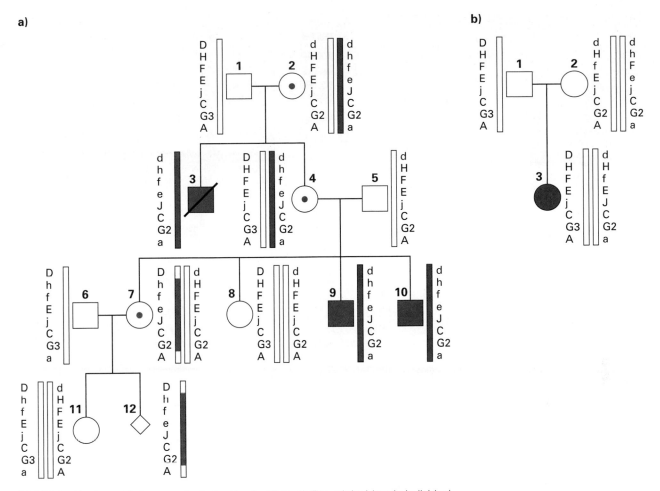

b)

FIGURE 10.2 • **Linkage analysis for the families. a)** The original boy is individual number 9. Dots indicate carrier status, and the bars represent X chromosomes with the RFLP markers as indicated. Blue on the chromosomes indicates regions carrying the disease allele. The capital letters indicate the most frequent markers in the population. **b)** Pedigree and linkage analysis information for the affected girl and her parents.

7. Go to NCBI Genes and Diseases and use the search window to see if you can diagnose the boy's disease in 5–10 minutes.

8. Does this girl represent an exception to the rule for X-linked diseases, or is she consistent with it? Explain your answer.

Phase IV: DNA Sequence Analysis

You decide that the definitive answer can only be found in cycle-sequencing DNA from the two patients. You get their permission to use the blood as a source of genomic DNA for sequencing a locus of interest and you obtain the deduced amino acid sequences shown in Table 10.1, page 283.

DISCOVERY QUESTIONS

9. Go to OMIM and see if you can diagnose this disease. Spend only about 5 minutes at this site.

10. Go to BLASTp (for protein) and see if you can diagnose this disease. To assist you, use this amino acid sequence from the locus of interest, which was conserved in all family members: NICKECPIIGFRYRSLKHFNYDICQSCFF. Save in electronic form the proteins you find with this search. We will come back to these hits.

11. Go to Protein Data Bank (PDB), enter the keyword "1dxx," and press "Find a Structure." This file con-

METHODS

Sequencing DNA

SEQUENCES

Locus of Interest

LINKS

BLASTp

PDB

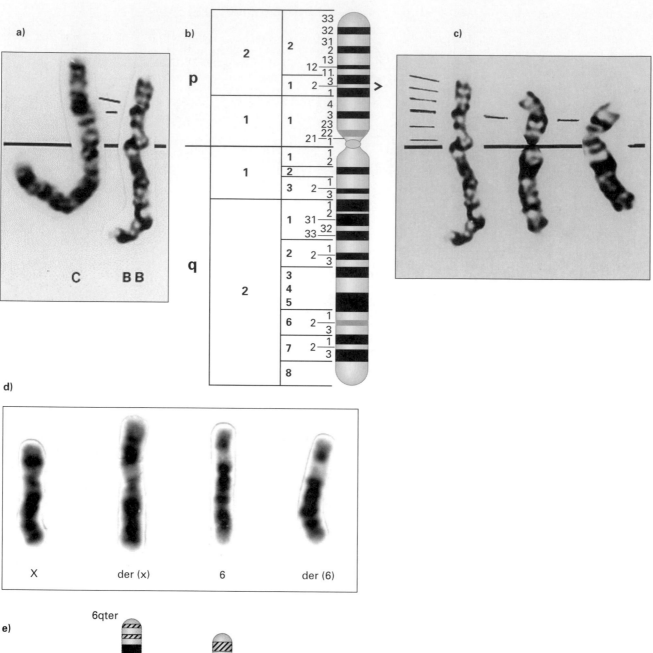

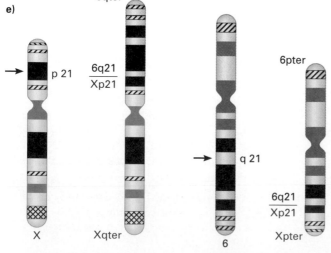

FIGURE 10.3 · **Karyotypes for the two patients.**
a) Metaphase X chromosome from control (C) and boy patient (BB). The thick horizontal bars mark the centromeres. The portion deleted from the boy's X chromosome is indicated by two thin lines. **b)** Ideogram of a *wt* X chromosome with the deleted region marked by ">." **c)** The BB chromosome (left) and two additional X chromosomes from the boy patient. **d)** The girl's *wt* X and 6 chromosomes and two that have recombined. The prefix "der" indicates the chromosomes are recombinant. **e)** Ideogram of the same four chromosomes in d). Arrows next to the *wt* chromosomes mark the site of recombination.

TABLE 10.1 • Deduced amino acid sequences for two patients and family members.

Boy:	1	MLWWEEVEDC	YEREDVQKKT	FTKWVNAQFS	KFGKQHIENL	FSDLQDGRRL	LDL**R**EGLTGQ	60
Boy's mother:	1	MLWWEEVEDC	YEREDVQKKT	FTKWVNAQFS	KFGKQHIENL	FSDLQDGRRL	LDL**R**EGLTGQ	60
	1	MLWWEEVEDC	YEREDVQKKT	FTKWVNAQFS	KFGKQHIENL	FSDLQDGRRL	LDLLEGLTGQ	60
Girl:	1	MLWWEEVEDC	YEREDVQKKT	FTKWVNAQFS	KFGKQHIENL	FSDLQDGRRL	LDLLEGLTGQ	60
	1	MLWWEEVEDC	YEREDVQKKT	FTKWVNAQFS	KFGKQHIENL	FSDLQDGRRL	LDL**R**EGLTGQ	60
Girl's mother:	1	MLWWEEVEDC	YEREDVQKKT	FTKWVNAQFS	KFGKQHIENL	FSDLQDGRRL	LDLLEGLTGQ	60
	1	MLWWEEVEDC	YEREDVQKKT	FTKWVNAQFS	KFGKQHIENL	FSDLQDGRRL	LDLLEGLTGQ	60
Girl's father:	1	MLWWEEVEDC	YEREDVQKKT	FTKWVNAQFS	KFGKQHIENL	FSDLQDGRRL	LDLLEGLTGQ	60

tains four copies of the amino portion of the full protein. Click on "View Structure" and view this fragment using the QuickPDB method (this file is a homotetramer). Find the location for the mutations found in the girl and the boy (see Table 10.1). Where within the protein do they lie? Look at the mutation's location in the secondary structure.

12. Find and highlight the sequence DVQKKTFTKW (amino acids 15–24). What kind of secondary structure does it help form? This will be useful later, so note its location.

Summary 10.1

You have progressed through about 40 years of biomedical research. Of course, it is easy to see the linear progress once the final answer is determined, but imagine what it was like for those who were trying to clone the Duchenne's muscular dystrophy (DMD) gene for the first time. This monumental task required many research groups learning from each other and building on the data. The big break came when an interdisciplinary team of hard-working and creative investigators devised a clever new experiment before anyone had a probe that could be used to clone the full-length cDNA and gene. Once they did identify the gene, you can imagine the collective groan that must have sounded when they realized the mRNA was over 14,000 nucleotides long and the gene was close to 1,000,000 bp long. At the time of this work, there was no such thing as automated DNA sequencing, so it all had to be done by hand. Using this method, one person might be able to read 500 bp a day. Furthermore, all results must be replicated since scientists cannot trust results produced only once, so they would need to sequence at least 28,000 bases in the cDNA. The full gene proved too large to sequence completely. The DNA sequence of the entire gene would have to wait until years later.

Although the complete gene was not sequenced initially, the work by Louis Kunkel and his many collaborators was groundbreaking, establishing the standards and procedures used to clone many other disease-causing genes. Kunkel's research had such a large impact that most other researchers have followed the naming convention established by Kunkel's lab—**dystrophin** is the gene/protein that causes muscular dystrophy. You will find that many other *wt* genes have been named after the mutant phenotype that led to their discovery.

METHODS
QuickPDB
LINKS
Louis Kunkel

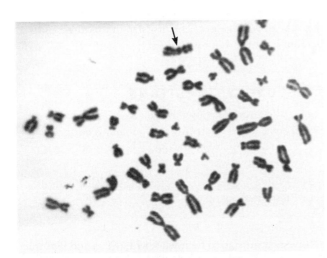

FIGURE 10.4 • Inactivated X chromosome. A full karyotype that is focused on a *wt* X chromosome of the girl patient (arrow). The X chromosome is delayed in its progression through mitosis.

DISCOVERY QUESTIONS

13. Another form of muscular dystrophy, Becker muscular dystrophy (BMD), affects primarily males. Whereas DMD is diagnosed on average by age 4.6 years, BMD patients do not show signs until they are in their twenties, develop less severe

symptoms, and live much longer. Assuming that these two diseases are caused by mutations in dystrophin, hypothesize a mechanism to explain the difference in clinical symptoms.

14. Another phenotype of DMD patients was mental retardation. Hypothesize how a "muscle" protein like dystrophin could cause brain deficiencies. This issue is addressed again beginning on page 293.

10.2 The Next Steps in Understanding the Disease

We Need an Animal Model System

In order to understand a disease, investigators need a lot of tissue on which to perform experiments. People are funny about donating their hearts and skeletal muscle for research purposes, so we have a problem. One solution would be to use animal models. For example, it would be nice if there were a strain of mice that had mutations in dystrophin that you could use to study muscular dystrophy. Nature has created many mutations that biologists have utilized for research.

Mutant mice called *mdx,* which develop DMD due to a spontaneous mutation, were isolated by biologists interested in muscle physiology. The mutant phenotype was observed long before dystrophin was cloned and sequenced. Having a mouse that develops muscular dystrophy has been a very helpful research tool. With this *model system* (the *mdx* mutant strain of mice), investigators could see what happens to a muscle as it develops the disease. They tested for the presence of dystrophin, as well as many other muscle-specific proteins, to see how one mutation might alter the expression of other proteins. Finally, we can test potential therapies and cures on mice that you know would otherwise develop muscular dystrophy. All these studies would be very difficult or impossible to perform without the mouse model system.

Much of what we know about dystrophin and muscular dystrophy was determined using *mdx* mice. For example, an antibody was made against the mouse dystrophin protein and used to immunologically label *wt* and *mdx* skeletal muscle. Investigators found that dystrophin is localized to the inside surface of the plasma membrane of skeletal muscles. They also found it is localized inside neurons of the brain (which helps to explain the mental retardation in some patients). In *mdx* mice, there was no dystrophin detected anywhere. This also helps explain why muscular dystrophy is a recessive disease, since it is the loss of function that causes the phenotype and not the creation of a "hyperactive" or overabundant protein as is the case with dominant diseases such as Huntington's disease.

What Was That Other Protein I Got Lots of Hits For?

When you did your sequence search of the dystrophin sequence (Discovery Question 10), you should have also gotten a lot of hits for a protein called utrophin. Go to the BLAST2 page and use these accession numbers to compare human dystrophin (P11532) with human utrophin (CAA48829). Be sure to select BLASTp and enter the accession numbers in the box to the right of the prompt "GI."

When you get your BLAST2 results, note the degree of sequence conservation; the statistics below the diagonal line are where you will find the percent amino acid identity and similarity (called positive) and the number of gaps needed to maximize alignment. Is this convergent evolution of two unrelated genes, or a pair of duplicated genes (**paralogs**) that have diverged but maintained some common sequence?

Find the first region from the BLAST2 result that is highly conserved between utrophin and dystrophin, DVQKKTFTKW. This is the region you found in Discovery Question 12. It is a part of an alpha helix in dystrophin. To determine what structure this region forms in utrophin, go to PDB and enter the word "utrophin." Choose the file that says "Actin Binding Region Of The Dystrophin Homologue Utrophin." This file contains only the amino portion of the full protein. View this structure using QuickPDB. Find the location for the amino acids DVQKKTFTKW. Use the secondary structure view to determine where this portion of utrophin is in relation to the patients' dystrophin mutation.

Does Utrophin Play a Role in Muscular Dystrophy, Too?

Based on their sequence and structural similarity, you would be insightful if you predicted that dystrophin and utrophin had similar functions. The question is whether utrophin is the cause of any diseases. The first step in this direction is to find out what the *wt* protein does.

DISCOVERY QUESTIONS

15. What experiment would you like to perform in order to see where utrophin protein is normally located? What tissue would you use for this experiment, and why?

16. Would you expect any diseases associated with the loss of utrophin?

Investigators started their work with the *wt* and *mdx* mice since they were easy sources of tissues. The first thing they wanted to know was where utrophin is located in *wt* mice. Their immunological localization images showed it inside skeletal muscles and clustered at the neuromuscular junction (Figure 10.5).

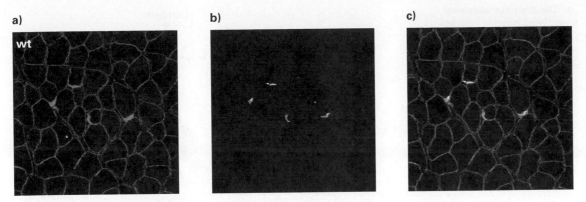

FIGURE 10.5 • **Immunofluorescent labeling of dystrophin and utrophin.** Wild-type adult mouse thigh muscle labeled for **a)** dystrophin, **b)** utrophin, and **c)** where they overlap is shown as bright white label.

Utrophin is expressed in fetal and regenerating muscle and in many other cell types during early development (utrophin got its name because it is similar to dys*trophin* in structure but it is *u*biquitously expressed. Further analysis revealed that the carboxyl terminus of utrophin is required for its localization to neuromuscular junctions of muscle cells. The amino terminus, as with dystrophin, interacts directly with the actin cytoskeleton.

When *mdx* mice were examined, utrophin expression was unaltered compared to *wt* mice. Utrophin was found throughout the muscle plasma membrane in small-caliber skeletal muscles (e.g., muscles that control eye movement) of adult *mdx* mice. Because these muscles show minimal pathologic changes in *mdx* mice, the findings raised the possibility that induction of utrophin might be a therapeutic approach to muscular dystrophy. To date, no clinical disorders relating to utrophin have been identified.

> **DISCOVERY QUESTION**
>
> 17. Why do you think there have been no diseases associated with a loss of function in utrophin?

What Does Dystrophin Do Anyway?

We have strayed a bit from the original protein, but that's how research goes sometimes; you follow leads as they come. Once antibodies to utrophin and dystrophin were available, a group at the University of Iowa asked a very simple question that had a profound impact on the field. Jim Ervasti and Kevin Campbell led a group that wanted to know if other proteins were physically attached to dystrophin. Up to this point, everyone had focused on dystrophin, which was responsible for the vast majority of cases diagnosed as DMD or BMD. But what did dystrophin do, and why was its loss so crippling? Based on its sequence, some features were known. For example, dystrophin is shaped like a dog bone with two blobs on either end of a

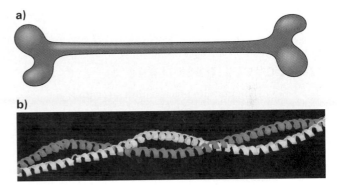

FIGURE 10.6 • **Proposed structure of dystrophin.**
a) Based on amino acid sequence, dystrophin was believed to be shaped like an elongated bone. **b)** The middle portion of dystrophin was predicted to form a coiled-coil as shown in this structure view of another protein with a similar motif. (PDB ID# 1C1G.)

long middle section (Figure 10.6a). The middle section of dystrophin had a sequence **motif** (or characteristic shape) called a coiled-coil. This redundant-sounding name is the same motif seen in myosin where two identical molecules (a **homodimer**) are wrapped around each other like snakes mating, and each molecule is an alpha helix (Figure 10.6b). Furthermore, it had been known for a while that the first 240 amino acids folded into a shape that was known to bind actin. This suggested dystrophin worked as a dimer and was anchored to the cytoskeleton via actin.

To answer their question, Ervasti and Campbell performed a classic molecular method called **immunoprecipitation.** The basic idea is to take the cells you are interested in (i.e., skeletal muscles) and create holes in the plasma membrane so that antibody-coated beads can get inside the cell. The antibodies will grab dystrophin and hold on very tightly. When the mixture is spun gently, the beads will settle to the bottom of the tube and all proteins stuck to the beads will be pulled down with them.

METHODS

Immuno-
precipitation

LINKS

Jim Ervasti

Kevin Campbell

FIGURE 10.7 • **Diagram summarizing dystrophin and associated protein interactions.** Notice that some labels do not name each molecule individually and only the group names are provided. The carboxyl (C) and amino (N) termini of dystrophin have been labeled. The branched beads represent glycosylation of the proteins.

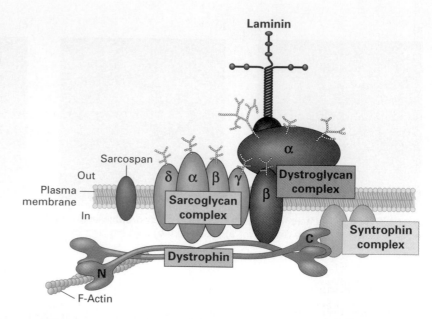

DISCOVERY QUESTIONS

18. Go to the immunoprecipitation animation and study the banding pattern produced for each lane. Explain the banding pattern for each lane.

19. Which lane or lanes best represents what you would see in an *mdx* mouse or a DMD patient?

Ervasti and Campbell discovered that in addition to actin, dystrophin is attached to a cluster of several more proteins. It took several years and a lot of work by the Iowa team (which included a growing number of investigators) as well as other labs around the world to identify these proteins. As you can imagine, most of this work was done in *mdx* mice first and later confirmed in humans. Eventually, the picture shown in Figure 10.7 emerged.

As you can see, dystrophin plays a critical role in connecting the cytoskeleton to the external skeleton (eventually tendons and bones). You can also see that the carboxyl terminus is linked to the dystroglycan complex, which is connected to sarcoglycan complex and laminin. Dystrophin is also linked to a pair of proteins called the syntrophin complex.

There are two ways to look at this portion of the muscle. One is the way most biologists think, in a 2D or 3D representation as seen in Figure 10.7. But equally valid is the way an electrical engineer would, as a **circuit diagram** (Figure 10.8). Imagine each protein as a node connected to others by arrows that represent wires. Using a circuit diagram, you can view the model in Figure 10.7 in a slightly different way, which may make some new properties of the system become more obvious.

When you look at this circuit diagram, a few interesting properties emerge. First, it is easier to see the link between the cytoskeleton and the extracellular skeleton. Also, it is more obvious that there are some places where many proteins work in concert to form a single node (e.g., sarcoglycan complex). In engineering terms, you might describe this as a **redundant** node. In circuit diagrams, redundancy means that several proteins have similar functions, so that if one is missing, the remaining complex may be sufficient to perform the task. We can also see places in the circuit where only one line connects two proteins in the pathway. Your understanding of electricity must tell you that if a lone wire were broken, the entire circuit would be broken. Therefore, redundancy creates a backup system for critical functions.

MATH MINUTE 10.1 • WHAT'S SPECIAL ABOUT THIS GRAPH?

Graph theory is the branch of mathematics that studies diagrams similar to those in Figure 10.8. In graph theory, the lines between nodes are called arcs or edges. A *directed* graph has arrows on the edges to indicate the direction of information flow between two nodes. In a graph like this one, in which the connections or relationships between nodes are important but there is no directional information, the arrows are normally left off and the graph is *undirected*.

Graph theorists study properties of graphs such as the degree of connectivity, typical path length from one node to another, and the existence of cycles (a path from a node back to itself without visiting any node more than once). One thing a graph theorist would notice about Figure 10.8 is that there are three edges going into dystrophin and the dystroglycan complex, while there are only one or two edges going into each of the remaining nodes. Therefore, dystrophin and the dystroglycan complex are critical nodes: their removal would cause the greatest disruption of the system represented by the graph.

We also consider the graph-theoretical view of biological systems in Chapters 7–9 and in Chapter 6 when we study proteomics.

DISCOVERY QUESTIONS

20. Use Figures 10.7 and 10.8 to answer these questions.
    a. Predict the phenotype if a person were homozygous for an actin deletion.
    b. Predict the phenotype if a person were homozygous for the deletion of the entire dystroglycan complex.
    c. Predict the phenotype if a person were homozygous for a deletion of only one protein from the dystroglycan complex.
    d. Predict the phenotype if a person were homozygous for a sarcospan deletion.

21. Given your predictions, explain why it makes sense that DMD and BMD were so easily identified as genetic diseases. Why was dystrophin initially identified as the only cause for muscular dystrophy?

22. Predict a molecular mechanism that could produce the differences seen in DMD and BMD.

23. Based on these diagrams and your predictions, do you think there are other human diseases that would present symptoms similar to those of muscular dystrophy?

Why Do DMD Patients' Muscles Deteriorate After the First Three Years?

Now that you have thought about a range of potential genetic causes for muscular dystrophy, you should be wondering how missing dystrophin causes children to develop their symptoms *after* a few years, and why the muscles *gradually* deteriorate. It seems that genetic diseases should begin at birth, yet DMD and BMD do not.

First, we should address the differences between BMD and DMD. If you look back to the peptides of the two afflicted children and their families (see Table 10.1), you will see the girl and boy had single missense mutations that created one amino acid substitution (**54L→R;** which means amino acid 54 is normally a leucine [L] but in the patients, it is an arginine [R]). This sort of mutation creates a drastic phenotype and DMD. However, mutation 168A→D (alanine to aspartic acid at amino acid 168) causes BMD. The difference lies in whether the dystrophin circuitry wire is cut completely (54L→R) or only part way (168A→D). In other words, DMD patients lack functional dystrophin. BMD patients have dystrophin that is partially active, and thus their symptoms begin later in life and are less severe.

The remaining question is why are there any symptoms at all. The working hypothesis is that when dystrophin is nonfunctional, the muscle contracts and causes the plasma

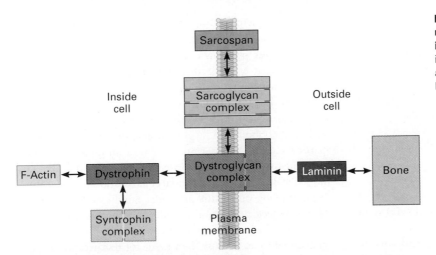

Inside cell

Outside cell

Plasma membrane

FIGURE 10.8 • Circuit diagram summarizing dystrophin and associated protein interactions. Dystrophin fits into the flow of information between the outside of the cell and the inside. The connection between laminin and bone is simplified here for clarity.

membrane to become damaged since its anchorage to the bone is intact but there is no coupling to the cytoskeleton. In *wt* muscles, the dystrophin connection serves to attenuate and redistribute the physical stress associated with contraction and relaxation. Therefore, the muscle contracts and the tendons pull on the bones, which causes the person to move. However, in DMD the two skeletons (intracellular and extracellular) are connected to each other only by the phospholipid bilayer of the muscle's plasma membrane. This would be analogous to making a bridge out of cement but not including any metal bars to provide the durability needed. Just as a poorly reinforced bridge would crumble, so too do dystrophic muscles. The late onset is due to the fact that children don't move around much during the first few months, so very little damage accumulates.

Although the exact mechanism of damage is uncertain, it seems logical to conclude that if the plasma membranes of muscles are damaged, the cells might die. The most popular idea is that a loss of integrity of the plasma membrane allows calcium to rush into the muscle and activate many proteases that only work in the presence of high calcium concentrations. As the muscle fibers began to die, there would be a loss of muscle mass and coordination, which is exactly what we see in DMD. In BMD patients, this damaging process is slowed down so the symptoms are less extreme.

The damaged membrane hypothesis works well for the muscle symptoms, but the mental retardation is still problematic. Neurons do not contract, so why are there mental deficiencies in some, but not all, patients? There are two parts to this answer. First is the percentage of people who develop the phenotype, which in genetic terms is referred to as **penetrance.** When *everyone* with a particular mutation gets the same phenotype, the disease is said to have complete penetrance. When few people exhibit the symptoms, the disease is said to have low penetrance. No one understands the reason for low penetrance for DMD's mental symptoms. However, it is easy to hypothesize why there might be some functional consequences. All cells, including neurons, have cytoskeletons. Every cell has to be able to bind to, and hold onto, its neighbors. Without this attachment, most cells enter apoptosis and die. Laminins are a common way to hold onto other cells, and so you might predict that analogs of the molecules in Figures 10.7 and 10.8 are also present in neurons. If the connection between the extracellular and intracellular worlds were broken, there might be some physical consequences for the cells, if not cell death.

Is it Possible to Have DMD and Be Wild-Type for Dystrophin?

About 20 years have passed since the first DMD boy entered your clinic. Sadly, he died due to heart failure; DMD is an incurable disease. It is late Friday afternoon on a dark and drizzly February day. It has been a long week, and you are ready for a break. At 4:55 P.M., a woman enters your clinic holding her son (age 4) and her 12-year-old daughter follows in a wheelchair. Neither child can walk, and you realize that your generosity continues to attract patients who are often neglected by the system.

In your examination, you notice that the children have hypertrophied calf muscles but their hips and shoulders look especially small. Their torsos also look noticeably reduced in musculature. You take a blood sample and tissue biopsies from all three and ask them to come back next Friday for the lab results. You send off the samples, requesting information about the levels of creatine kinase in the blood and asking for immunological detection of dystrophin in the muscle biopsies. Here is what you get back from the lab:

Boy and girl:
- elevated creatine kinase serum levels
- normal level and location of dystrophin in the muscle

Mother:
- normal creatine kinase serum levels
- normal level and location of dystrophin in the muscle

Next Friday, the family returns and you have to tell them you are not sure what's the matter with the children. You suspect that there is a single point mutation in the dystrophin gene, so you ask their permission to sequence their DNA. They agree and you request they return for one more free visit. When the sequence data arrive, you are truly baffled. All three individuals have *wt* dystrophin—not a single mutation anywhere. You are certain a mistake has been made, so you ask the lab to sequence a different locus that is highly polymorphic (the MHC locus) to confirm that the samples were contaminated. Surprisingly, the MHC sequence only confirms that the mother really is their mother and the children had the same father. There is no contamination of DNA, and they really are *wt* for dystrophin.

How Can They Have Muscular Dystrophy if Their Dystrophin Genes Are Normal?

You are completely stumped by this family. You had jumped to conclusions when you first examined the two children because of your previous experiences with DMD. However, your confidence has been shaken. You overlooked the most obvious feature—the mutation did not appear to be X-linked. Luckily, an old friend of yours from college does basic research on muscles. You decide to give your friend a call to chat about life in general and this troubling case in particular. The call goes on for almost an hour as you discover many of your current interests overlap. As a result, you both decide to establish a formal collaboration to understand what is going on in this particular family. You contact the family to seek permission for additional research with their tissue samples, explaining what you are doing and why. They are willing to let you use their tissue for research purposes.

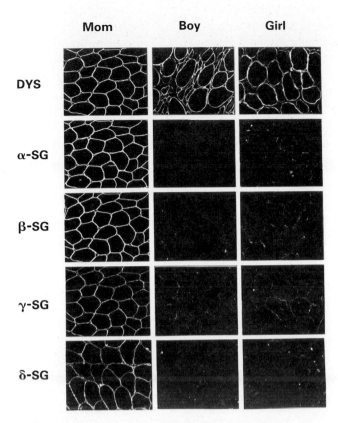

Mom   Boy   Girl

DYS

α-SG

β-SG

γ-SG

δ-SG

FIGURE 10.9 • Immunofluorescence labeling of biopsies from the mother and her two affected children. Each panel represents a different section of the thigh muscle labeled with a different antibody. The white label in these photos marks the presence of the indicated proteins (DYS: dystrophin; SG: sarcoglycan).

Being a Ph.D. researcher in the field of muscle physiology, your friend is more up to date with the most recent findings. As a result, you two decide to test their muscles with some new monoclonal antibodies that can detect specific subunits of the sarcoglycan complex. Your friend performs the immunofluorescence labeling experiments and sends you the results as an email attachment (Figure 10.9).

METHODS
Immunofluorescence

DISCOVERY QUESTIONS

24. Is this a dominant or recessive disease? What are the genotypes of the parents?

25. The father was not willing to participate due to a religious belief. Would having access to his tissue change your interpretation of these results?

26. Go to OMIM and find the chromosomal location for the four genes. Spell out the Greek letters when you search. Determine whether they are linked or unlinked.

27. Given their chromosomal location, do you think that all eight of the alleles were mutated in the two children, or is there another more likely answer? Refer to Figures 10.7 and 10.8.

When you look at the immunofluorescence photomicrograph in Figure 10.9, you are stunned by the clarity of the results. Although the father was not willing to participate in the study, you can make some inferences based on the available data. The children are suffering from a recessive disease that is not linked to the X chromosome. It appears that the heterozygous mother is phenotypically normal, though she must carry a mutated allele or alleles. Because each of these four genes is located on a different chromosome, it is highly unlikely that two children would both inherit the same eight alleles to produce the observed phenotype.

MATH MINUTE 10.2 • WHAT DO YOU MEAN BY HIGHLY UNLIKELY?

There are two simple explanatory models for the observations in Figure 10.9: (1) both children inherited 8 mutant alleles from four different loci, or (2) both children inherited 2 mutant alleles from a single locus, and that one gene affects expression of the sarcoglycan subunits. To help you choose the next step in your research, you calculate the probability that each of these situations occurs by chance, assuming both parents are carriers.

In the first model, since the 4 genes are on different chromosomes, the 16 inheritance events (8 alleles for each child) are mutually *independent*. Independent events have a special mathematical property: the probability that *all* the events occur is the product of the probabilities that the individual events occur. Because each allele is inherited with probability 1/2 and there are 16 inheritance events (8 alleles and 2 children), the probability that both children inherited 8 mutant alleles from 4 different loci is $(1/2)^{16}$ (approximately 0.000015).

In the second model, there are four independent inheritance events, leading to a probability of $(1/2)^4 = 0.0625$ that both children inherited 2 mutant alleles from a single locus. Therefore, although the first model remains a possibility and cannot be discarded yet, a single mutant gene is much more likely to be the cause of disease in this family.

METHODS

Western Blot

LINKS

NCBI

Your friend has been following Kevin Campbell's work and knows of some papers that are relevant. For example, Kathleen Holt in Campbell's lab did some experiments to study how the individual sarcoglycans assemble into a complex and insert into the plasma membrane. She found if any one member is missing, none of the sarcoglycans accumulate in the plasma membrane. Instead, they accumulate in a compartment that, to you, looks like Golgi (Figure 10.10).

In another paper, Rachelle Crosbie and Kevin Campbell led a team that studied γ-sarcoglycan in particular. The team looked at all four sarcoglycan subunits, as well as sarcospan. Although they screened more than 50 families with muscular dystrophies, only three had mutations in sarcospan (Figure 10.11), but these mutations did not seem to be the cause of the disease.

The more probable causes were mutations in their γ-sarcoglycan genes. In one particular patient Crosbie studied, the γ-sarcoglycan contained a frameshift mutation that produced a truncated γ-sarcoglycan protein (Figure 10.12). Interestingly, this patient's complete sarcoglycan complex and sarcospan were located in the plasma membrane, yet this patient still suffered from the disease.

All these data seem like too much detailed information for you. Unfortunately, you are conducting research on top of maintaining a full patient load. You decide the workload

is too much and ask your partners if you could take a one-month sabbatical to research this case more carefully. You are fortunate since your colleagues think your research is exciting and are willing to work extra hours to help you.

With your new freedom, you rush to the library to read about muscle proteins. You decide that you should sequence γ-sarcoglycan genes from the mother and her two children. The results show they have point mutations in one codon—TGT was changed to TAT, producing a 283C→Y amino acid substitution. From your reading, you realize this family has a very common mutation that may have originated in India 60 to 200 generations ago.

DISCOVERY QUESTIONS

28. Go to NCBI and select "protein" from the search menu at the top left corner. Enter the word "sarcoglycan" and search the protein database. How many different human sarcoglycan genes are there? You know there are at least four, but can you find any more?

29. What is the name of the disease associated with mutations in sarcoglycans? You might want to click on results from the question above, or search OMIM.

Now that you are on sabbatical, you read as many papers on the topic as you can find. In your reading, you realize that there are in fact five sarcoglycan genes in mammals, instead of four. The newest member in the family is ε-sarcoglycan, which has almost the same sequence as α-sarcoglycan. Being a student all over again, you draw out the sarcoglycan complex and realize that ε-sarcoglycan may be a redundant subunit that might act as a fail-safe for α-sarcoglycan—an interesting hypothesis you hope to pursue in the future. You also make a table of all the limb-girdle muscular dystrophy (LGMD) diseases and the causes of each form (Table 10.2, page 292).

You are not familiar with the proteins dysferlin and dystrobrevin, so you search OMIM to learn more. There are no mammalian orthologs of dysferlin but its sequence is similar to the spermatogenesis factor *fer-1* of the worm *Caenorhabditis elegans*. People from nine families, all of whom were homozygous for a frameshift mutation in dysferlin, have developed a form of dystrophy. The name "dysferlin" was derived from the role it plays in muscular dystrophy and its *C. elegans* ortholog. **Western blots** have demonstrated

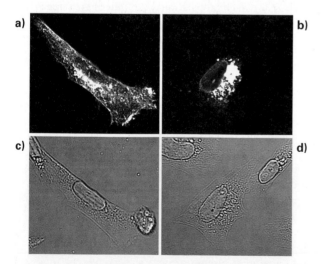

FIGURE 10.10 • Immunefluorescence localization of muted β-sarcoglycan. Immunofluorescence labeling of the sarcoglycan complex **a)** in a *wt* cell and **b)** in a cell where the β-sarcoglycan has been mutated. **c)** and **d)** Phase contrast images of the same cells let you see the full extent of the cells.

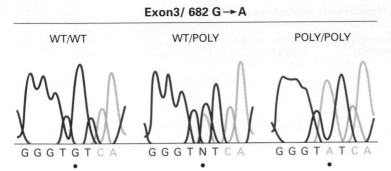

Exon3/ 682 G → A

WT/WT   WT/POLY   POLY/POLY

G G G T G T C A   G G G T N T C A   G G G T A T C A

FIGURE 10.11 • Sequence chromatograms showing homozygous *wt,* heterozygous, and homozygous mutant sequences for sarcospan. The mutation is highlighted with a dot below and is called "POLY" because the different base is a polymorphic site.

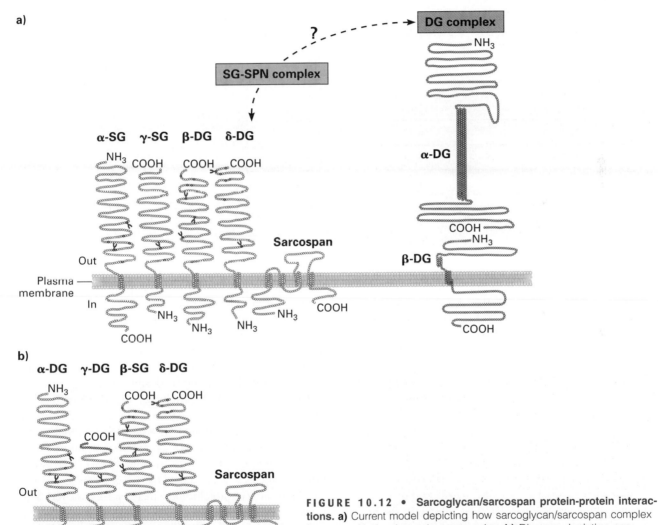

FIGURE 10.12 • **Sarcoglycan/sarcospan protein-protein interactions. a)** Current model depicting how sarcoglycan/sarcospan complex interacts with the dystroglycan complex. **b)** Diagram depicting one patient's sarcoglycan and sarcospan complex in the plasma membrane of muscle cells that are dystrophic. The small blue "Y"s represent glycosylation of the proteins (not drawn to scale for clarity).

dysferlin is 230 kDa in size and immunofluorescence experiments localized it to the muscle plasma membrane. Dysferlin is expressed at the earliest stages of human development when limbs start to form regional differentiation. The timing and role of dysferlin suggested it may contribute to the pattern of muscles affected in this form of muscular dystrophy—typically proximal or distal muscles.

OMIM describes the muscle plasma membrane proteins as seen in Figures 10.7 and 10.8, but it describes dystrobrevin localized near syntrophin in skeletal muscle

METHODS

Chromatograms

TABLE 10.2 • The dystrophin complex proteins, chromosomal location, and associated LGMD classifications.

Disease	Chromosomal Location	Protein
Autosomal Dominant (type 1)	—	—
1A	5q22.3–q31.3	unknown
1B	1q11–q21	unknown
1C	3p52	caveolin-3
1D	7q	unknown
Autosomal Recessive (type 2)	—	—
2A	15q15–21	calpain 3
2B	2p13	dysferlin
2C	13q12	γ-sarcoglycan
2D	17q12	α-sarcoglycan
2E	4q12	β-sarcoglycan
2F	5q31	δ-sarcoglycan
2G	7q11–12	telethonin
2H	9q31–33	unknown

Source: Modified from Betto, et al. *Italian Journal of Neurological Sciences.* 1999. 20: 375, Table 2.

LINKS

Réunion

(Figure 10.13, on pages 294 and 295). In patients with DMD and some forms of LGMD, dystrobrevin was almost absent from the muscles. Dystrobrevin amounts and cellular localization were normal in patients with other forms of LGMD where dystrophin and the rest of the dystrophin-associated protein complex are normally expressed. It appears that dystrobrevin deficiency is a common feature of dystrophies linked to dystrophin and the dystrophin-associated proteins. OMIM states "This was the first indication that a cytoplasmic component of the dystrophin-associated protein complex may be involved in the pathogenesis of limb-girdle muscular dystrophy." When you finish reading, you appreciate how much has been learned since you were in medical school. The information (the number of different proteins, their roles and subcellular localizations) seems complex and almost overwhelms you.

With all your newfound knowledge, you have another conversation with your collaborating friend. You ask why there was no antibody for ε-sarcoglycan in the analysis of the biopsies, and your friend confesses ignorance of this protein. (You smile with satisfaction for being up to speed and on par with your friend.) You continue to flex your intellectual muscles by discussing the in-frameshift deletion in laminin α2 that produced a mild form of congenital muscular dystrophy, reminiscent of the frameshift deletion in dystrophin that results in BMD.

By now you and your friend are deep into the causes of muscular dystrophies (note the term is now plural). You begin to focus on those proteins that seem less obvious. Why would proteins that play no obvious structural roles in the linkage between the cytoskeleton and the extracellular skeleton cause a muscular dystrophy? For example, caveolin-3 is a protein involved in endocytosis and not thought of as a skeletal protein in its function (see Figure 10.13b). Plus, LGMD1C (see Table 10.2) is a dominant disease that has no symptoms until later in life (20s–40s). Why? No one seems to know what dysferlin does (LGMD2B), and the real oddball is calpain 3. Calpains are a family of calcium-activated proteases and as such are thought to reside in the cytoplasm. Their job is to cut proteins into smaller pieces to either activate or inactivate them.

Your friend leans back and begins to reminisce about how the mutation linking calpain 3 to LGMD2A was discovered. Your friend was a postdoc in the lab that made this discovery, which focused on a tiny island in the Indian Ocean located several hundred kilometers east of Madagascar. Réunion was initially settled by a few people in the 17th century and has a very limited gene pool. In 1642, the island was explored for the first time by Europeans on the French ship "Le Saint-Louis." In 1644, the governor of Madagascar sent 12 mutineers into exile on the island. Two years later, they were found in excellent health by a delegation that expected to find them dead. From these genetic founders, the population has remained small, **consanguineous,** and a favorite spot for investigators in search of a good study population for genetic diseases.

Your friend describes how the shallow gene pool on Réunion permitted the team to isolate the genetic cause for LGMD2A, but there was something funny about the findings. Normally, one would expect everyone on this island who suffers from LGMD2A to share the same mutation since it has been genetically isolated for many generations—a **founder effect.** But instead of one mutation in calpain 3, the team found six different ones. Interestingly, the publication your friend coauthored also discussed the possibility that LGMD2A was **digenic,** or caused by two genes instead of just one gene. The authors also suggested that when mutated, the encoded protein may produce a *non*structural cause for LGMD2A. This hypothesis was radical since muscular dystrophies have "always" been caused by mutations in structural proteins.

You sit up abruptly and remember a paper you had read that discussed animal models of muscular dystrophies. In this paper, Volker Straub in Kevin Campbell's lab examined a strain of mice called *dy* which have a spontaneous mutation in the laminin α2 gene. What was striking was the observation that the plasma membranes in these mice seemed relatively intact. The authors concluded that there may be more than one pathway to produce muscular dystrophies, implying a nonstructural possibility.

DISCOVERY QUESTIONS

30. Do you think it is possible to have two separate pathways (structural and nonstructural) cause diseases with similar symptoms?

31. Using the circuit diagram in Figure 10.13c, make some predictions about nonstructural causes for muscular dystrophies. Choose any of the proteins discussed so far.

Where Is the Muscular Dystrophy Field Now?

You and your friend decide to cap off your sabbatical by going to an international meeting on muscular dystrophy. Your good luck holds out since the meeting is in Paris this year, where you have not been since your junior year abroad during college. In addition to the personal joy of returning to France, you are thrilled about the learning opportunity. It is a small meeting with only about 200 investigators, but they are the best in the world. You reserve your ticket and hotel, dust off your French-English dictionary, and do a PubMed search of the latest publications on muscular dystrophies.

DISCOVERY QUESTION

LINKS
PubMed
Lee Sweeney

32. Go to PubMed and search for "muscular dystroph*" (the asterisk is a wild card so you will get hits for "dystrophy," "dystrophies," and "dystrophin"). Look for papers that suggest nonstructural causes for muscular dystrophies.

Sixth International Conference on Molecular Causes for Muscular Dystrophies As you fly into Paris, you remind your college friend/collaborator that you went to France for your junior year. Your friend nods politely while silently trying to count the number of times you have said this. You rush off the plane and try out your rusty French as soon as possible and ask, "Where do the ducks get water?" After a puzzled look in response, you see the signs for the toilets and you say *merci* and move on. Next you ask "Where can I get the cousin in the kitchen?" More blank stares greet you. You begin to doubt your linguistic prowess, and decide to stick to your strength—muscular dystrophies. As you walk to get your luggage, you see several people laughing and asking in English, "Did you find the water for your ducks?"! You are thankful that the international meeting will be conducted in English.

The Meeting Begins The next morning, you head for the first big session where Lee Sweeney from the University of Pennsylvania School of Medicine will give a brief introduction to the session. In his remarks, you find a few interesting points. He argues that evolution suggests that most parts of dystrophin are important since the worm *C. elegans* contains all the same motifs in its dystrophin protein that humans do. He also discusses a knockout mouse in which the gene for syntrophin has been deleted. These mice do not have any pronounced dystrophies, but they do have altered cellular localization of the enzyme nitric oxide synthase (NOS) that produces nitric oxide (NO), as illustrated in Figure 10.13b. NO is a gas that is produced inside cells and is a part of the **signal transduction** (cellular communication) pathway in muscle cells. Recently, it had been shown that proper localization of NOS is necessary to increase blood flow during exercise to skeletal muscles in humans. Sweeney speculates that perhaps the mislocalization of NOS could cause dystrophic phenotypes in mice with mutant syntrophin or dystrophin that cannot bind syntrophin. As you listen to this, you realize that mutations in genes cannot be studied under constant laboratory conditions, as is often done. For example, you would like to study syntrophin-deficient mice with and without exercise regimens to see if there is a difference.

Structural Weaknesses The next speaker is Jim Ervasti, who discovered the glycoprotein complex (i.e., sarcoglycan complex and dystroglycan complex) associated with

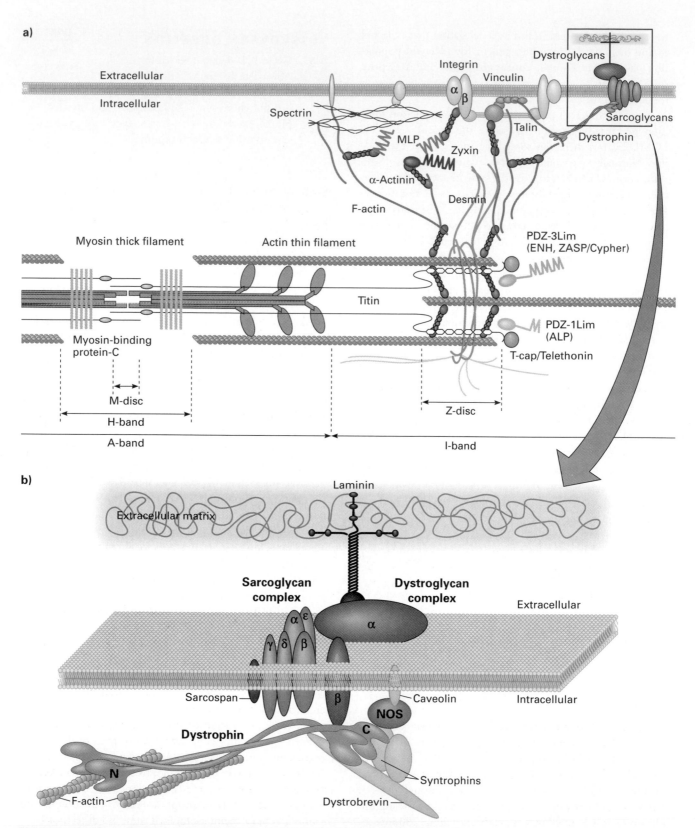

FIGURE 10.13 • **Working model of muscle molecules near dystrophin. a)** Schematic diagram of the major proteins involved in striated muscle function. **b)** Closer view of the dystrophin complex of proteins. **c)** Circuit diagram showing a different view of panel b. The blue arrows indicate additional connections to other parts of the cell web. Added to this diagram are two large circles that encompass two functional units of the circuitry—the structural unit (circled in blue) and the signaling unit (circled in gray).

c)

FIGURE 10.13 • Continued

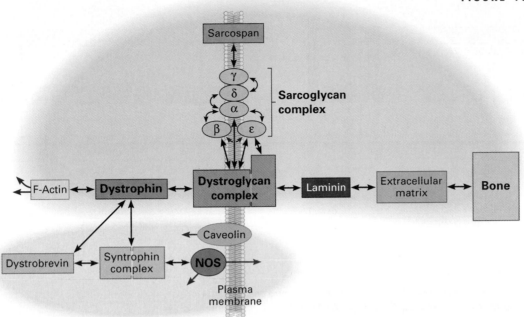

dystrophin. Ervasti presents some of his recent work examining the structural link between dystrophin and the underlying cytoskeleton. In his unique approach, Ervasti has peeled the plasma membrane off individual muscles and determined what proteins are still adhering to the cytoplasmic side of the removed membrane. He shows a series of beautiful immunofluorescence micrographs of peeled membranes from *wt* and *mdx* mice (Figure 10.14 on your CD-ROM). From your reading, you know that α-actinin is part of the cytoskeleton, and γ-actin can also form polymeric fibers like the more traditional α/β actin do in "normal" muscle actin fibers. Ervasti shows other slides that demonstrate that *mdx* mice express γ-actin, but it does not adhere to the plasma membrane.

FIGURE 10.14 • **Immunofluorescence detection of proteins adhering to plasma membrane.**
Go to the CD-ROM to view this figure.

You realize this is the first time anyone has studied the degree of adhesion between dystrophin and cytoskeletal proteins. But since all of this is news to you, you appreciate the circuit model Ervasti proposed to explain his data (Figure 10.15). Ervasti proposed that the lack of dystrophin produces a weak link between the plasma membrane and the cytoskeleton, and this weakness is revealed experimentally when the plasma membrane is peeled away. His presentation makes a good case for the structural basis for muscular dystrophies, but you begin to wonder about the other proteins in his circuit diagram. This diagram looks like an integrated circuit, which makes you wonder what roles the other proteins in the circuit might play in this pathology.

LINKS
Maria Rita
Passos-Bueno in
Sao Paulo

Yasuko Hagiwara

DISCOVERY QUESTIONS

33. Since Figure 10.15 is a circuit diagram, identify the three functional units in the model.

34. Using Ervasti's circuit diagrams, make a prediction of how there could be a signaling pathway that is also perturbed by the weak linkage between the plasma membrane and γ-actin. Even if you do not know what all these proteins do, use the circuitry to suggest possibilities.

The next talk is presented by Maria Rita Passos-Bueno, a respected researcher from Brazil. Passos-Bueno's presentation illustrates how her group discovered telethonin as the molecular cause of LGMD2G, the most recent addition to the growing list of proteins (Figure 10.13a and Table 10.2). The impact of her work is significant for the diagnosis and treatment of patients in Sao Paulo and all over the world. Her data are clear, but you are still troubled by something. Telethonin is very far away from the plasma membrane, so it is difficult to imagine how mutated telethonin could produce similar symptoms to those seen when a more peripheral protein is mutated—for example, caveolin-3 or one of the sarcoglycans.

As luck would have it, the next speaker is Yasuko Hagiwara from Tokyo, presenting his group's work with a knockout mouse that lacks caveolin-3. As you would expect, homozygous mutant mice produce on average 1.8 ± 0.5 **caveolae**/μm^2 (newly forming vesicles used for endocytosis) compared to 14.9 ± 1.6 in *wt* mice. Even the heterozygotes produce less than *wt* mice: 8.0 ± 0.7 (Figure 10.16, pag 296).

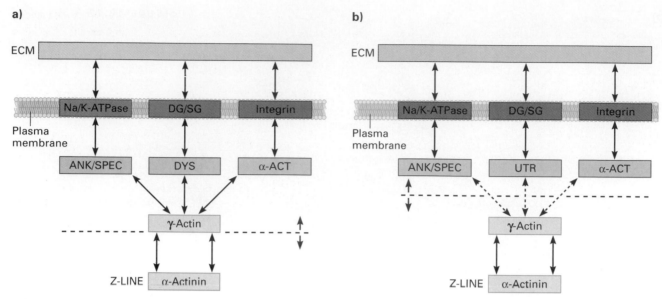

FIGURE 10.15 • **Circuit diagrams of proteins adhering to *wt* and *mdx* plasma membranes.** Circuit diagrams of **a)** *wt* and **b)** *mdx* linkage between the plasma membrane and cytoskeleton. The horizontal dashed line marks where the separation occurs in the experiments. Abbreviations: ANK/SPEC: ankyrin/spectrin cytoskeletal proteins; α-ACT: α-actinin cytoskeletal protein; IF: intermediate filament; DYS: dystrophin; DG/SG: dystroglycan/sarcoglycan; UTR: utrophin.

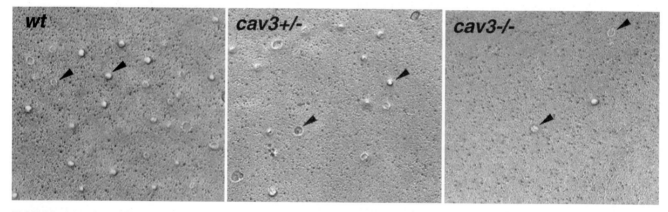

FIGURE 10.16 • **Electron micrograph of the cytoplasmic side of muscle plasma membranes from the indicated mice.** The bumps and craters are caveolae caught in the act of forming on the plasma membrane of muscle cells.

LINKS

Hiroyuki Sorimachi

Hagiwara's data are not surprising, but what is hard to understand is the next slide, which shows the muscle pathology of Cav 3$^{-/-}$ mice (Figure 10.17). During his talk, he says, "Dystrophin and its associated proteins were present at normal levels."

You want to follow up Hagiwara's talk with some questions such as, "What proteins did you include when you said 'dystrophin and its associated proteins were present at normal levels'?" but the organizers announce it is time for a break. You are relieved to stretch your legs when you notice a

small stampede to the refreshments table. Even your friend has sprinted away and left you alone. You stroll over and discover a lavish setting of pâté, *saucisson,* several cheeses, and other yummies you haven't tasted in years. You snap back to the meeting when the lights are lowered and you discover you are the only one still standing at the food table. You have not quite gotten the hang of international meetings yet.

Nonfunctional Mutations The next speaker is Hiroyuki Sorimachi from the University of Tokyo. He studies the role calpain 3 might play in muscular dystrophies. Sorimachi

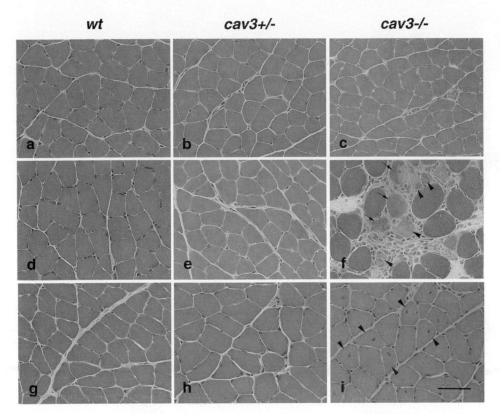

wt **cav3+/-** **cav3-/-**

FIGURE 10.17 • Effects of caveolin mutation on muscle development. Histology of the soleus muscles from **a)** to **c)** 6-week-old mice; **d)** to **f)** 8-week-old mice, and **g)** to **i)** 12-week-old mice from three strains of mice as indicated at the top of each column. Pathology is only apparent at 8 weeks of age and macrophages have invaded the tissue (arrowheads in panel f). By 12 weeks, the muscle has regenerated but many of the nuclei are abnormally located in the middle of the cells (arrowheads in i). Black bar in the lower right corner of i) equals 50 μm.

describes how his lab studied the biochemical properties of nine mutant forms of calpain isolated from LGMD2A patients. Almost all of them lost their proteolytic activities on a model substrate. However, some of them lost their **autocatalytic** capacity (a protein's ability to stimulate or act on itself) or could no longer bind to the structural protein titin (see Figure 10.13a). What is striking about his presentation is that he offers a nonstructural hypothesis. Maybe the enzymatic activity of calpain 3 is responsible for the pathology of muscular dystrophies. But which pathway was disrupted? Why would this lead to a delayed onset of symptoms? How could this lead to muscle atrophy? So many questions, and yet you know Hagiwara's hypothesis represents a step toward a more complete answer, even if it is not clear to you, or anyone, at this time. Science usually moves by slow sporadic progress, only rarely interrupted by sudden breakthroughs.

The next presentation summarizes a worldwide search for all the different mutations in calpain 3 that can lead to LGMD2A. You are surprised to learn that 97 different pathogenic mutations have been discovered scattered all over the gene (Figure 10.18a, page 298). Even more disturbing was the figure that illustrated the age of onset and the rate of disease progression as a function of genotype (Figure 10.18b–d). Once again, severity of the symptoms varies from person to person, which makes you question if this is "the" cause of LGMD2A.

DISCOVERY QUESTIONS

35. Formulate a hypothesis that can explain the data shown in Figure 10.18. Use the data in this case study to support your hypothesis.

36. If you could obtain tissue from any patient included in Figure 10.18, which ones would you like to study and why?

New Paradigms: Nonstructural Causes for Muscular Dystrophies As the next session begins, you are struck by the whispering in the audience. You ask your friend what's happening and are told that the final three presentations are very "hot" because they are controversial and provocative. You are not quite sure how a person's research can be characterized this way, so you are keen to hear a "hot" talk. Munekazu Shigekawa presents his findings, which suggest there is a bidirectional communication going on between the sarcoglycans and integrin. Although you have learned a lot during your sabbatical readings, you are not sure about integrin, though it sounds familiar. Your friend tells you rather tersely that integrins are plasma membrane proteins that allow one cell to hold onto another cell. "It is an adhesion protein!" Then you remember Ervasti had integrin in his circuit model (see Figure 10.15) that made you think about all the proteins with nonstructural connections.

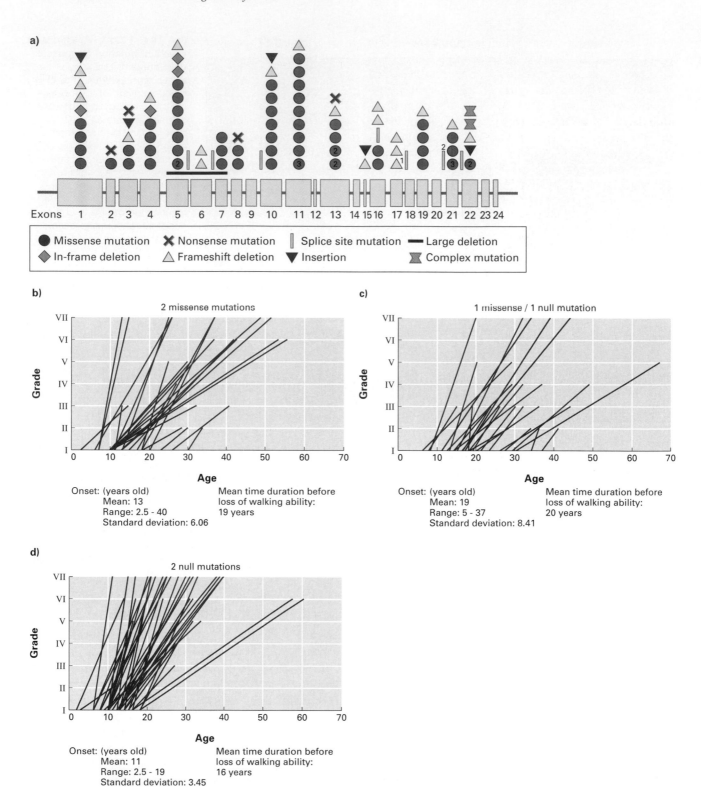

FIGURE 10.18 • **Different mutations in calpain and phenotype variations. a)** The 24 exons are indicated by open boxes with exon numbers below. The number of independent mutations is given either inside or above the symbol. **b)** to **d)** Progression-of-disease curves for LGMD2A patients. Functional stages are graded as I–VII. Lines are presented only for patients for whom there are at least two data points of disease progression, whereas the calculation of means and duration takes into account all available data.

Shigekawa shows how integrins are co-precipitated with dystrophin when an antidystrophin antibody is used in an immunoprecipitation experiment. The key point made during his presentation is that when muscle cells adhere to a substrate, α-sarcoglycan and γ-sarcoglycan become **phosphorylated.** Why? The most popular models do not indicate sarcoglycans are involved in adhesion. Shigekawa proposes that Ervasti's integrated circuitry (see Figure 10.15) transmits the adhesion status of a cell to other structural proteins such as the dystrophin complex. Why? Does the phosphorylation of sarcoglycans change the ability of the cell to respond to physical stress incurred during muscle contraction and relaxation? When this circuit is broken, does it create a signaling pathway that is misdirected? Are the structural proteins that normally protect cells less protective when they are not phosphorylated?

DISCOVERY QUESTION

37. Modify your earlier hypotheses (Discovery Questions 30 and 31) to explain how muscular dystrophies might be digenic in nature, or at least have two pathways as their molecular causes.

The next speaker is Mark Grady, whose introduction stimulates the audience to whisper with a nervous excitement. Grady begins his talk with a summary slide similar to Figure 10.13b. He plans to discuss the role dystrobrevin plays in muscular dystrophies. He has created a homozygous α-dystrobrevin (*adbn*) knockout mouse strain that he confirms with his next slide (Figure 10.19, page 300). He shows how α-dystrobrevin is expressed in skeletal muscle and brain, which reminds you that the mental retardation symptom has yet to be explained by anyone. Could dystrobrevin be the common link between the muscular and mental symptoms?

Grady then shows the pathology in skeletal and cardiac muscle (Figure 10.20, page 300), which you are very familiar with at the clinical level. You remember your first DMD patient died of heart failure.

It is clear from Grady's data that the three organs affected by muscular dystrophy are also affected in *adbn*$^{-/-}$ mice. But you begin to think that perhaps the dystroglycan/sarcoglycan complexes may be disrupted as well, which would support a model of structural causes and not signaling causes. The next slide stuns you and the rest of the audience, so there is a collective gasp of amazement (Figure 10.21, page 301). You marvel at the clarity of the data and the conclusions they provoke.

DISCOVERY QUESTIONS

38. What effect does a lack of dystrobrevin have on the dystroglycan complex?

39. Where is nonneural NOS (nNOS) normally located in *wt* muscle cells? Where is nNOS in *adbn*$^{-/-}$ mice? Where is nNOS in *mdx* mice?

40. Given that syntrophin binds to dystrobrevin, how can it be present in *adbn*$^{-/-}$ mice? Refer to Figure 10.13b.

METHODS
Knockout
STRUCTURES
cAMP
LINKS
Mark Grady

Your head is swimming, but Grady is not finished. He continues by showing data that indicate the plasma membrane of *adbn*$^{-/-}$ mice are not as weakened as in *mdx* mice. He also shows data that reveal that cytosolic levels of nNOS are normal, so the only perturbation is the plasma membrane localization of nNOS in *adbn*$^{-/-}$ mice. He cannot explain why nNOS fails to localize to the plasma membrane, especially since syntrophin appears to be in its proper location.

Grady's next slide silences the audience with its implications. He reminds everyone that nNOS plays a signaling role in many cells, and is required to relax the smooth muscles in blood vessels when muscle exercise is vigorous. Nitric oxide stimulates the production of cyclic guanylic acid (cGMP), similar to cAMP, but the base is guanine not adenine). In *mdx* mice, nNOS is undetectable near the plasma membrane, cGMP is undetectable, and the blood vessels constrict rather than dilate. Grady wanted to know whether cGMP concentrations could be elevated by stimulation in *adbn*$^{-/-}$ mice as is the case in *wt* mice (Figure 10.22, page 302).

DISCOVERY QUESTIONS

41. What normally happens to cGMP levels in *wt* muscles that have been stimulated to contract? What happens to cGMP levels in *adbn*$^{-/-}$ mice? What happens to cGMP levels in nNOS$^{-/-}$ mice?

42. Do you think NO is required for the production of cGMP? Support your answer with data from Figure 10.22.

43. Can you formulate a model that would explain why the loss of nNOS at the plasma membrane of a muscle cell could lead to muscular dystrophy? A correct answer gets you a nomination for a Nobel Prize!

Skeletal muscle **Brain**

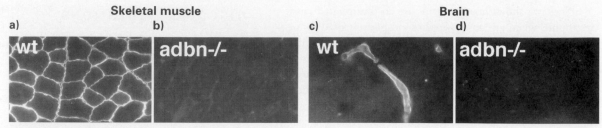

FIGURE 10.19 • **Location of α-dystrobrevin in *wt* and mutant mice.** Antibody labeling of *adbn* in skeletal muscle (a and b) and brain tissue (c and d) in: **a)** and **c)** *wt*, **b)** and **d)** dystrobrevin knockout mice.

FIGURE 10.20 • **Pathology of *adbn⁻/⁻* muscle. a)** and **b)** Sections of skeletal muscle from *wt* and *adbn⁻/⁻* mice. Small areas of necrosis and centrally nucleated fibers are seen in *adbn⁻/⁻* muscle. **c)** and **d)** Sections of skeletal muscle labeled with an antibody to embryonic and fetal myosin heavy chains. The positive fibers in *adbn⁻/⁻* muscle indicate actively regenerating muscle fibers. **e)** and **f)** Sections of cardiac muscle, showing dystrophic areas in *adbn⁻/⁻* tissue.

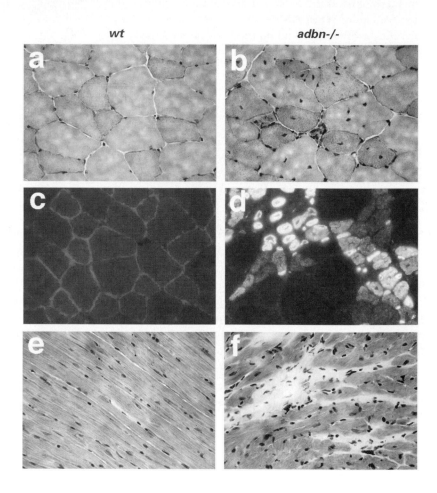

wt *adbn-/-*

Grady concludes by stating he does not know if the dystroglycan complex is destabilized even though it is located along the plasma membrane. During the question and answer session, Ervasti suggests that his new method could answer Grady's question, and they establish a collaboration on the spot. Another audience member asks whether the nNOS might be stretch activated, in which case structural damage may exacerbate the cGMP signaling defect. Grady does not know but says that the question is a good one.

Your head is about to explode from information overload when Kathy Wilson from Johns Hopkins is introduced. Her talk focuses on a form of muscular dystrophy you have never heard of before, Emery-Dreifuss muscular dystrophy. Your brain has neared saturation, and it is hard for you to fully appreciate what she is saying. But the general story is about proteins called lamins that interact in the inner nuclear membrane. Somehow, mutations in lamin can lead to sudden death due to irregular rhythms in the heart. Wilson suggests two models. One is more structural and affects the pacemaker cells of the heart. The other suggests that a lack of lamins leads to gene expression irregularities, which you know is another form of signaling. She also notes that during **apoptosis** (genetically determined cell suicide), lamins are

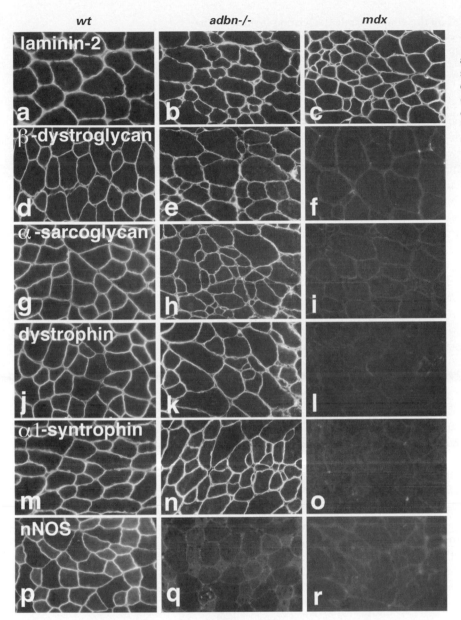

FIGURE 10.21 • **Immunofluorescence labeling of skeletal muscle from *wt*, *adbn*$^{-/-}$ and *mdx* mice. a)** to **c)** Levels of laminin-α2 were similar in all three genotypes. **d)** to **o)** Levels of the DGC proteins β-dystroglycan, α-sarcoglycan, dystrophin, and α1-syntrophin were markedly reduced in *mdx* muscle but normal in *adbn*$^{-/-}$ muscle. **p)** to **r)** In contrast, levels of nNOS were greatly reduced in both *mdx* and *adbn*$^{-/-}$ muscle.

MATH MINUTE 10.3 • IS cGMP PRODUCTION ELEVATED?

What you want to learn from Figure 10.22 is whether, in each strain of mice, the cGMP levels in unstimulated and stimulated muscles are the same or different. Since both quantities vary from mouse to mouse, it is the population means (averages of the two quantities over all mice) that are compared. You can never know exactly what the population means are (you can't test every mouse in the world), but their values can be estimated by measuring the two quantities in samples of several individuals. In Figure 10.22, error bars help depict population mean estimates. The following example illustrates how error bars are produced from sample data.

Example: Suppose Grady measured cGMP levels in unstimulated muscles to be 2.7, 3.2, 6.0, and 5.4 in a sample of *wt* mice. The *sample size (n)* is 4. The *sample mean* (or average) of these measurements is 4.3. The *sample standard deviation (s)* is a measure of "spread," or variation from the sample mean. In this example, *s* is approximately 1.6 (Math Minute 2.1, page 33, explains how to compute *s*). The *standard* *error of the mean* (s.e.m.) is estimated by the sample standard deviation divided by the square root of the sample size. The s.e.m. is about 0.8, which is how far the error bar extends above and below the sample mean of 4.3 in Figure 10.22.

Notice that the error bar does not tell you anything about the range of the data (2.7–6.0). The error bar extends from 3.5 to 5.1, which does not include some of the observed values. Rather, the sample mean ± s.e.m. error bar marks the boundaries of a *confidence interval,* a range in which you can be 60% sure the population mean lies. The *confidence level* of 60% is obtained from a table of probabilities specifically for $n = 4$.

A rule of thumb test for whether two population means are the same or different is based on the error bars for the two samples. If the bars overlap, the means are likely the same. If they don't overlap, the means are likely different. To be more precise, and quantify the likelihood associated with various degrees of overlap, the *t*-test is necessary.

The ***t*-test** addresses the question of whether the population mean of the difference between cGMP levels in unstimulated and stimulated muscles in a particular genotype is zero. Therefore, you must subtract the unstimulated muscle cGMP level from the stimulated muscle cGMP level for each mouse of this genotype, and find the sample mean and s.e.m. of the resulting numbers. If the population mean of the difference is zero, the sample mean of the difference should be "small." How small depends on the sample size and s.e.m.

The result of the *t*-test is a ***p*-value:** the probability of seeing a sample mean of the difference as large or larger than the observed value, under the assumption that the population mean of the difference is zero. The *p*-value is obtained by dividing the sample mean of the difference by the s.e.m. and looking up the result in a table of probabilities. A small *p*-value (typically 0.1 or less) indicates the zero population mean hypothesis should be rejected; a larger *p*-value indicates a lack of evidence against the zero mean hypothesis.

The *p*-values above the bars in Figure 10.22 imply (as expected) that cGMP is elevated by stimulation in *wt* mice and not in nNOS$^{-/-}$ mice. The borderline *p*-value in *adbn*$^{-/-}$ mice leaves room for doubt whether cGMP is elevated in this strain: there is a greater than 1 in 10 chance of seeing cGMP levels like those observed in the sample *adbn*$^{-/-}$ mice even if cGMP is not elevated.

LINKS

Jacques Beckmann

cleaved by proteases. You wonder if lamins have anything to do with calpain 3 causing muscular dystrophy but do not bother to ask since you are not sure you could absorb any answer she might offer.

Final Presentation The final speaker is Jacques Beckmann from Evry, France. Beckmann gives a wonderful historical summary of the research in muscular dystrophy and LGMD in particular, which was first diagnosed as a unique disease in 1884. He gradually works his way toward the most recent findings and some of the unanswered questions. For example, does the lack of calpain 3 create an enzymatic or a structural problem? It has been shown that calpain 3 binds to titin, and perhaps this is the structural cause of LGMD2A. No one knows for sure. He wants to know why there is variation in severity of symptoms and suggests that the particular combination of mutations found in a person's two alleles may affect the phenotype. Beckmann speculates that other possibilities include linked genes, unlinked genes, or nongenetic factors. This brings him to the original study that

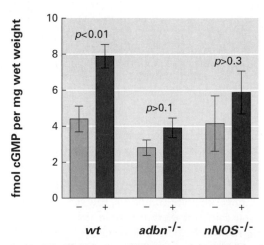

FIGURE 10.22 • cGMP levels in control and mutant muscle. Amounts of cGMP in isolated extensor digitorum muscles from unstimulated (−) or electrically stimulated (+; 30 Hz for 15 s) *wt* ($n = 4$), *adbn*$^{-/-}$ ($n = 6$), or nNOS$^{-/-}$ ($n = 6$) mice. Bar graphs show means ± s.e.m. The significance of differences between stimulated and unstimulated muscles was assessed by the *t*-test.

identified calpain 3 as the genetic cause for LGMD2A. Beckmann cites your friend's research as seminal in our understanding. Beckmann says the results from Réunion were groundbreaking because nonstructural roles were suggested for the first time, as well as the real possibility of a second genetic locus. Of course, your friend is trying to keep a poker face, but you can see the radiating pride.

As you know, genetically isolated and homogeneous populations make ideal places to find genes that cause diseases. On Réunion, the expectation was to find a single founder allele that had propagated throughout the island and caused LGMD2A. However, the "Réunion Paradox" in this homogeneous population (as well as other isolated populations in the world) is based on the observation that there are many different mutant alleles that seem to cause LGMD2A. Beckmann proposes a simple solution. The founder effect on Réunion is not the calpain 3 gene but a second genetic locus that has not been discovered yet. The existence of a second locus would explain the Réunion Paradox as well as the wide variation in severity and rate of disease progression seen in LGMD2A, as well as many other forms of muscular dystrophy. Beckmann supports his model with a lot of good logic and unexplained observations. He also cites studies by Richard Rozmahel et al. showing in mice that the very well documented case of cystic fibrosis can vary in severity due to the influence of other genes.

In closing, Beckmann leaves the audience with a proposal in which at least two genes are responsible for muscular dystrophies. Beckmann says, "Thus, the 'one mutation—several diseases' concept could well be explained by a digenic inheritance model." The simplistic view of "one gene—one function—one phenotype" may well be the exception rather than the rule, even for a single gene. To understand diseases at the genomic level will require understanding of **epistatic** interaction and an improvement in the detection of subtle phenotypes. We need to understand a protein's role in genomic circuits and identify all the proteins with which it interacts. In short, we need to think of cells as complex ecosystems with many interconnected components.

He continues by citing the many accomplishments of the reductionist approach to biology. However, as we learn more, the clear distinctions between monogenic and multilocus traits may prove to be too simplistic and arbitrary. Beckmann predicts that traits are best described as a continuum with most traits best described as at least digenic or conditionally monogenic. He closes his talk with a final caution to those who like to categorize biological traits. "Furthermore, within specific genes, there may be alleles that behave as typical monogenic characters, while the expression of other allelic variants may require the interaction with specific genetic—or even environmental—contexts." In other words, our cells are ecosystems within the larger ecosystem and every component is capable of direct or indirect communication.

Summary 10.2: Your Final Thoughts

On the plane home, you are haunted by Beckmann's remarks. Is any disease monogenic? If a biological "rule" has exceptions, does the rule still hold or does it need to be revised? Was Mendel wrong? Are there any traits that can be explained by a single locus? Has all the molecular genetic work performed by studying genes in isolation or artificial settings been a wasted effort? What about all the genomes being sequenced—is there any information in there that will help us understand complex genetics, or is that too a waste of time and money?

On the TV screen in front of you, you watch the silhouette of a plane gradually move over the Atlantic Ocean toward home. The plane seems so directed and certain of its destiny, and you feel so lost.

DISCOVERY QUESTIONS

44. Is muscular dystrophy one disease with many causes? Are muscular dystrophies many different diseases with many related/interconnected causes? Are muscular dystrophies many unrelated diseases with convergent symptoms? Explain your answer.

45. Is mental retardation seen in some muscular dystrophy patients related to the molecular causes of the muscle pathology, or are they mechanistically unrelated?

Chapter 10 Conclusions

This case study was designed to confront a common misconception. As stated by Beckmann, "'one gene—one function—one phenotype' is far from representative of the whole story." Mendel made significant discoveries about the inheritance of traits, but not all traits are as simple as pea color and shape. As you study genomics and its related fields, keep an open mind and look for connections between proteins and functional units of cellular functions. One of the major goals of genomics is to build more complex models and to study proteins in their normal setting—inside functioning cells. Complex interactions inside cells are similar to those studied in ecology. The term "food web" refers to the interrelated nature of all the members in an ecosystem. Perhaps we need to think of cells as ecosystems and the proteins as components of "cell webs." Regardless of the terminology, the complexity of cells is an integral aspect of how proteins function. *Discovering Genomics, Proteomics, and Bioinformatics* will lead you through the challenge Beckmann made—to reconsider cells and their web of interrelated proteins.

References

Phase I

Francke, Uta, Hans D. Ochs, et al. 1985. Minor Xp21 chromosome deletion in a male associated with expression of Duchenne muscular dystrophy, chronic granulomatous disease, retinitis pigmentosa, and McLeod syndrome. *American Journal of Human Genetics*. 37: 250–267.

Zatz, Mayana, Angela M. Vianna-Morgante, et al. 1981. Translocation (X;6) in a female with Duchenne muscular dystrophy: Implications for the localization of the DMD locus. *Journal of Medical Genetics*. 18: 442–447.

Sources for pathology images.
> <http://www.bplab.com.my/telepath98/03953_muscular_dystrophy/p03953-1.htm>.
> <http://www.bplab.com.my/telepath98/03953_muscular_dystrophy/p03953-n1.jpg>.
> <http://www.bplab.com.my/telepath98/03953_muscular_dystrophy/p03953-b1.jpg>.
> <http://www.kumc.edu/instruction/medicine/anatomy/histoweb/muscular/muscular .htm>.
> <http://www.kumc.edu/instruction/medicine/anatomy/histoweb/muscular/muscle04 .htm>.

Phases II and III

Bakker, E., M. H. Hofker, et al. 1985. Prenatal diagnosis and carrier detection of Duchenne muscular dystrophy with closely linked RFLPs. *Lancet*. 1(8430): 655–658.

Francke, U., H. D. Ochs, et al. 1985. Minor Xp21 chromosome deletion in a male associated with expression of Duchenne muscular dystrophy, chronic granulomatous disease, retinitis pigmentosa, and McLeod syndrome. *American Journal of Human Genetics* 37(2): 250–267.

Verellen-Dumoulin, C., M. Freund, et al. 1984. Expression of an X-linked muscular dystrophy in a female due to translocation involving Xp21 and non-random inactivation of the normal X chromosome. *Human Genetics*. 67(1): 115–119.

Zatz, M., A. M. Vianna-Morgante, et al. 1981. Translocation (X;6) in a female with Duchenne muscular dystrophy: Implications for the localisation of the DMD locus. *Journal of Medical Genetics*. 18(6): 442–447.

Phase IV

Koenig, M., E. P. Hoffman, et al. 1987. Complete cloning of the Duchenne muscular dystrophy (DMD) cDNA and preliminary genomic organization of the DMD gene in normal and affected individuals. *Cell*. 50(3): 509–517.

Kunkel, L. M., et al. 1986. Analysis of deletions in DNA from patients with Becker and Duchenne muscular dystrophy. *Nature*. 322: 73–77.

Monaco, A. P., C. J. Bertelson, et al. 1985. Detection of deletions spanning the Duchenne muscular dystrophy locus using a tightly linked DNA segment. *Nature*. 316: 842–845.

Monaco, A. P., R. L. Neve, et al. 1986. Isolation of candidate cDNAs for portions of the Duchenne muscular dystrophy gene. *Nature*. 323: 646–650.

The Next Steps in Understanding the Disease

Betto, R., D. Biral, and D. Sandona. 1999. Functional roles of dystrophin and of associated proteins. New insights for the sarcoglycans. *Italian Journal of Neurological Sciences*. 20(6): 371–379.

Campbell, Kevin. *Molecular Studies of Muscular Dystrophy.* <http://www.physiology.uiowa.edu/campbell/Netscape%20Site/DGCResearch.htm>. Accessed 7 December 2000.

Campbell, K. P. 1995. Three muscular dystrophies: Loss of cytoskeleton-extracellular matrix linkage. *Cell*. 80(5): 675–679.

Crosbie, Rachelle H., Connie S. Lebakken, et al. 1999. Membrane targeting and stabilization of sarcospan is mediated by the sarcoglycan subcomplex. *Journal of Cell Biology*. 145: 153–165.

Crosbie, R. H., L. E. Lim, et al. 2000. Molecular and genetic characterization of sarcospan: Insights into sarcoglycan-sarcospan interactions. *Human Molecular Genetics*. 9(13): 2019–2027.

Jim Ervasti web site. <http://www.physiology.wisc.edu/www/ervasti.html>.

Ervasti, J. M., K. Ohlendieck, et al. 1990. Deficiency of a glycoprotein component of the dystrophin complex in dystrophic muscle. *Nature*. 345: 315–319.

Holt, K. H., and K. P. Campbell. 1998. Assembly of the sarcoglycan complex: Insights for muscular dystrophy. *Journal of Biological Chemistry*. 273(52): 34667–23670.

Disease information web sites
DMD and BMD: <http://www.ncbi.nlm.nih.gov/htbin-post/Omim/dispmim?310200>.

Utrophin: <http://www.ncbi.nlm.nih.gov/htbin-post/Omim/dispmim?128240>.

Where Is the Muscular Dystrophy Field Now?

Beckmann, J. S., I. Richard, et al. Identification of muscle-specific calpain and b-sarcoglycan genes in progressive autosomal recessive muscular dystrophies. *Neuromuscular Disorders*. 1996; 6: 455–462.

Beckmann, Jacques S. 1999. Disease taxonomy—Monogenic muscular dystrophy. *British Medical Bulletin*. 55(2): 340–357.

Chien, Kenneth R. 2000. Genomic circuits and the integrative biology of cardiac diseases. *Nature*. 407: 227–232.

Grady, R. Mark, Robert W. Grange, et al. 1999. Role for α-dystrobrevin in the pathogenesis of dystrophin-dependent muscular dystrophies. *Nature Cell Biology*. 1: 215–220.

Hack, Andrew A., Chantal T. Ly, et al. 1998. γ-sarcoglycan deficiency leads to muscle membrane defects and apoptosis independent of dystrophin. *The Journal of Cell Biology*. 142(5): 1279–1287.

Hack, A. A., L. Cordier, et al. 1999. Muscle degeneration without mechanical injury in sarcoglycan deficiency. *PNAS USA*. 96: 10723–10728.

Hagiwara, Yasuko, Toshikuni Sasaoka, et al. 2000. Caveolin-3 deficiency causes muscle degeneration in mice. *Human Molecular Genetics*. 9(20): 3047–3054.

Jung, Daniel, Franck Duclos, et al. 1996. Characterization of d-sarcoglycan, a novel component of the oligomeric sarcoglycan complex involved in limb-girdle muscular dystrophy. *Journal of Biological Chemistry*. 271(50): 32321–32329.

LGMD <http://www.ncbi.nlm.nih.gov/htbin-post/ Omim/dispmim?253700>.

Map of Réunion—similar version: <http://www.maptown.com/reunionmaps.html>.

McNally, Elizabeth M., Chantal T. Ly, and Louis M. Kunkel. 1998. Human ε-sarcoglycan is highly related to α-sarcoglycan (adhalin), the limb girdle muscular dystrophy 2D gene. *FEBS Letters*. 422: 27–32.

Minetti, C., F. Sotgia, et al. 1998. Mutations in the caveolin-3 gene cause autosomal dominant limb girdle muscular dystrophy. *Nature Genetics.* 18: 365–8.

Moreira, Eloisa S., Tim J. Wiltshire, et al. 2000. Limb-girdle muscular dystrophy type 2G is caused by mutations in the gene encoding the sarcomeric protein telethonin. *Nature Genetics.* 24: 163–166.

Ono, Yasuko, Hiroko Shimada, et al. 1998. Functional defects of a muscle-specific calpain, p94, caused by mutations associated with limb-girdle muscular dystrophy type 2A. *Journal of Biological Chemistry.* 273: 17073–17078.

Richard, I., O. Broux, et al. 1995. A novel mechanism leading to muscular dystrophy: Mutations in calpain 3 cause limb girdle muscular dystrophy type 2A. *Cell.* 81: 27–40.

Richard, I., C. Roudaut, et al. 1999. Calpainopathy: A survey of mutations and polymorphisms. *American Journal of Human Genetics.* 64(6): 1524–1540.

Rozmahel, Richard, Michael Wilschanski, et al. 1996. Modulation of disease severity in cystic fibrosis transmembrane conductance regulator deficient mice by a secondary genetic factor. *Nature Genetics.* 12(3): 280–287.

Rybakova, Inna N., Jitandrakumar R. Patel, and James M. Ervasti. 2000. The dystrophin complex forms a mechanically strong link between the sarcolemma and costameric actin. *Journal of Cell Biology.* 158(5): 1209–1214.

Sweeney, H. Lee, and Elisabeth R. Barton. 2000. The dystrophin-associated glycoprotein complex: What parts can you do without? *PNAS USA.* 97: 13464–13466.

Yoshida, Tomokazu, Yan Pan, et al. 1998. Bidirectional signaling between sarcoglycans and the integrin adhesion system in cultured L6 myocytes. *The Journal of Biological Chemistry.* 27(3): 1583–1590.

Wilson, Katherine L. 2000. The nuclear envelope, muscular dystrophy, and gene expression. *Trends in Cell Biology.* 10: 125–129.

Why Can't I Just Take a Pill to Lose Weight?

Goals for Chapter 11

Hungry for Knowledge

Discover the complexity of a "simple" genetic trait.

Extract useful information from databases.

Perceive proteins in 3D.

Improve data interpretation skills.

Analyze circuit diagrams for new insights.

In this chapter, we use a combination of a case study and short story approach to introduce you to the genomic mind-set. Most students assume that each gene encodes a single protein and each protein has one function that affects one phenotype. This Mendelian outlook works well when you are learning the fundamental rules of inheritance, but is too simplistic to fully understand more complex traits such as body weight. Each person has a genetic and behavioral component to his or her weight. This chapter examines the genetic and behavioral control of weight and what you can and cannot do to regulate it. While you read the story, put yourself in the lead role, and keep an open mind as the data are presented.

Hungry for Knowledge

Saturday, 21 October. 7:30 A.M.

It's a cool fall Saturday morning, and you want to sleep late. You love this time of the week when school pressures are temporarily lifted and you can do what you want. Just as you adjust your pillow for another segment of sleep, the phone rings. You pick it up and hear the chipper voice of your grandma Pauline. She lived with you when you were a child and is like a second mother. Pauline came to America from Germany between the two world wars. She arrived at Ellis Island in 1924 speaking no English, but she was well schooled and very smart. You answer the phone with a groggy voice that sounds very sleepy and somewhat angry for being disturbed. Your grandma responds with one of her classic phrases and remnants of her German heritage, "Only vimps sleep in!" This makes you smile and clear your throat so you sound wide awake, though she knows better.

Grandma is calling because of an article she just read in the Science section of the *New York Times* describing a protein that can make mice lose weight. She wants to know what protein this is so she can add it to her diet. Since you are a biology major, Grandma likes to ask you about all kinds of things. "Why are my plants turning yellow?" was last month's question. Today, she asks a question that you cannot answer right away (certainly not before breakfast), so you promise you will investigate the matter and call her back.

Library Opens at 8:30 A.M. on Saturdays

You don't often visit the library on Saturday morning and wait for it to open, but for Grandma you make an exception. You would do anything for her, including Saturday morning library research. You go to PubMed and do a search for "obesity gene" and get a lot of hits. Luckily, you scan the titles and quickly find a word that reappears often—**leptin.**

You decide to start at the beginning and find the first paper that talks about leptin. It is a 1994 paper written by

Jeffrey Friedman's lab at Rockefeller University. You know this is a big-splash paper since it made the cover shot of the journal *Nature*. You begin your reading with a news summary in the same issue about the leptin discovery. You learn:

LINKS

PubMed

Jeffrey Friedman

ob

Jackson Labs

1. A mutant strain of mice isolated in the 1950s was called *ob* for *ob*ese. The obese phenotype is only observed in homozygous mice. These mice are sterile and often get sick. (To find out what it would take to purchase a pair of these mice for research purposes, browse Jackson Labs in Maine and search for "leptin").

2. Localized damage to the brain region called the hypothalamus also causes an obese phenotype (Figure 11.1)

3. Surgical removal of fat from an individual produces increased food consumption and fat storage to replace what was removed. This led to the coining of the term "**lipostat,**" meaning a *lip*id (or fat) regulator similar to the way a thermo*stat* regulates temperature.

4. If fat is removed from one wild-type (*wt*) animal and put into another *wt* individual, the recipient will lose the weight and revert to its original size.

5. If a homozygous *ob* mouse *(ob/ob)* is surgically joined to a *wt* mouse so that they cocirculate blood, the *ob* mouse will eat less and lose weight. The *wt* mouse is unaffected.

6. Proteins can be extracted from fat in *wt* mice and injected into an *ob* mouse. The *ob* mouse will eat less and lose weight.

7. There is another mouse strain called *db* (for *d*iabetic) that has the obese phenotype when homozygous for the *db* mutation.

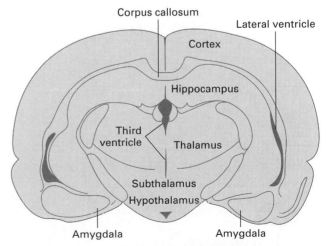

FIGURE 11.1 • Cross-sectional view of a rat brain. The hypothalamus is located near the bottom of the brain, in about the middle from front to back.

307

METHODS
Chromosomal Walk
Northern Blot

LINKS
Human-Mouse

8. When an *ob/ob* mouse is surgically joined to a *db/db* mouse, the *ob* mouse loses weight but the *db* mouse does not.

9. When a *db* mouse is joined to a *wt* mouse, the *wt* mouse will refuse to eat and starve to death.

10. If you inject a *db* mouse with the *wt* fat protein extract, it does not lose any weight.

DISCOVERY QUESTION

1. What conclusions can you make from the ten observations given above?

Building a Model for Weight Homeostasis

After reading these facts in the library, you list your conclusions. Then you construct a very simple model so you can keep track of what's going on (Figure 11.2). You know Grandma will want to know everything and she likes pictures.

Preliminary Conclusions

1. *ob* and *db* are recessive mutations, so these mice fail to make something produced in *wt* mice.

2. The brain has a role in maintaining weight.

3. Lipostat: An individual has an "ideal fat content" where this level is maintained even if fat cells are removed or added surgically. Lipostats explain why most people who go on diets or have liposuction return to their original weight. For example, Grandma has tried every fad diet, but always returns to her normally plump size.

4. Because joining an *ob* mouse with a *wt* mouse **complements** the mutation (makes the *ob* mouse normal size), there must be something in *wt* blood that can cause an animal to lose weight.

5. *db* and *ob* are not alleles of each other because when surgically joined to a *wt* mouse, the *ob* mouse loses weight but the *db* mouse does not.

6. *db* mice produce something (the same as *wt*?) that causes weight loss in *ob* mice. However, the blood of *ob* mice does not contain something that can complement the *db* mutation.

7. There is a protein in fat cells that causes an individual to lose weight. Is this the same protein produced by *db* but lacking in *ob* mice?

8. If there is a protein that controls weight, it does not work in *db* mice. Therefore, the *db* mutation might lack the receptor for this protein.

9. If *db* mice can cause *wt* mice to starve, then the *db* mouse produces an excess amount of this weight-controlling protein and it works "too well" in *wt* mice.

Cloning the Leptin Gene

Since the phenotype for *ob* has been known for a long time, Friedman's group started with some classical breeding experiments to map the genetic locus for *ob*. It is located on mouse chromosome 6, between loci *Pax4* and *Ptn*. The investigators used probes to *Pax4* and started a chromosomal walk toward *Ptn*.

DISCOVERY QUESTIONS

2. Go to the Human-Mouse Homology Map, which aligns the mouse and human chromosomes so you can choose one species' chromosome and find out where the orthologs are in the other species. Choose "Mouse" and chromosome "6" from the menus at the top and hit "Go."
   a. Find *Ptn,* which has a map location of 13.5 cM, then look up a few lines to *Pax4* that has a map location of 9 cM. Did you see the leptin gene (called *Lep*) in between these two loci? On which chromosome is the human leptin gene located?
   b. Click on the small yellow dot next to the human *LEP* gene. You will see an alignment of the mouse and human leptin mRNA sequences. Are these two highly conserved or not? What percentage of the nucleotides are identical? Would you predict protein sequences to be more or less identical than nucleotide sequences?

The Friedman lab isolated a portion of genomic DNA they called 2G7 that was used as a probe for a **Northern blot** using mRNA from many tissues in *wt* mice (Figure 11.3). They also probed the same blot with a probe specific for actin, and a band was seen in each lane.

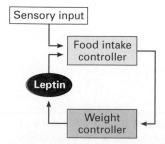

FIGURE 11.2 • Your initial model for weight control. Sensory input includes both environmental and physical stimuli.

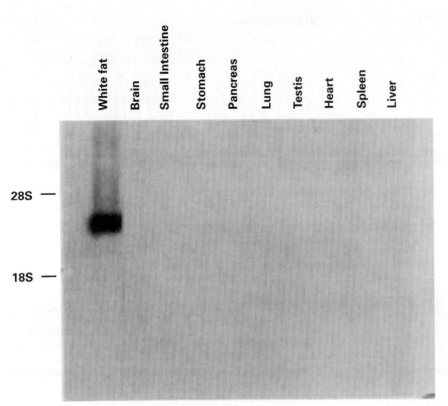

FIGURE 11.3 • **Northern blot using 2G7 DNA as probe.** RNA was isolated from *wt* mouse tissues as indicated. 28S and 18S are ribosomal RNA used as molecular weight markers.

DISCOVERY QUESTIONS

3. What was the purpose of the actin probe?

4. What can you conclude about 2G7 DNA? Does 2G7 encode the protein extracted from fat that can cause mice to lose weight? Support your conclusion with data.

Since there are two strains of *ob* mice, Friedman's group wanted to see if 2G7 was expressed in either strain (Figure 11.4, page 310). Given that *ob* is a recessive mutation, they did not expect to see the leptin gene expressed in these mice. Fat from several strains were tested: SM/Ckc - +Dac +/+, which is a lean mouse, as indicated by the homozygous *ob* +/+ genotype; SM/Ckc - +Dac ob²ᴶ/ob²ᴶ, which is an obese mouse homozygous for the *ob* allele called 2J; C5BL/6J +/+, which is a lean mouse and the *wt* parental stock for the original *ob* strain found in the 1950s; and C57BL/6J *ob/ob*, which is the original *ob* mutant strain.

DISCOVERY QUESTIONS

5. Which strains transcribe 2G7 DNA? Do you think 2G7 is the *ob* gene? Support your conclusion with data.

6. Since Figure 11.4 is a Northern blot, it measures transcripts only. What kind of data is needed to see if functional leptin protein is produced in these mice?

7. Search Genbank with the accession number NM_008493. How long is the mRNA, and how long is the protein? Save the protein sequence for Discovery Question 9.

8. What kind of posttranslational processing would need to happen to the leptin protein in order for it to be secreted into the serum?

9. Go to PDB and enter leptin's ID (1AX8). Click on "View Structure" and then select "QuickPDB." What is the first amino acid in the protein structure vs. the sequence from Discovery Question 7? Rotate the protein and notice it has two separate parts. Click on 3I and 24I. This should highlight the two ends of the separate alpha helix. Begin clicking on amino acids to the right of number 24 until you highlight more of the structure. How many amino acids are missing, and how can this be?

10. Go to the Leptin Structures page to see more of leptin's structure. This web site adds visual information that may provide insights into the questions above.

METHODS
QuickPDB

STRUCTURES
Leptin

LINKS
Genbank
PDB

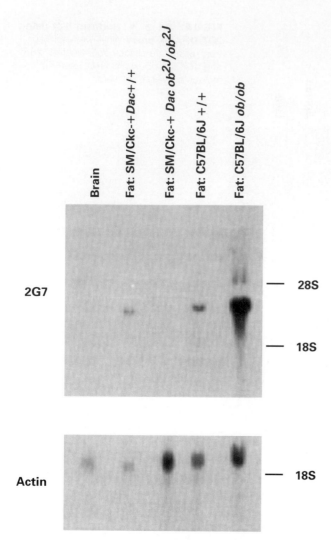

FIGURE 11.4 • **Northern blot of fat RNA from various strains of mice.** The probes were either 2G7 (see Figure 11.3) or actin cDNA. 28S and 18S are ribosomal RNA bands used as molecular weights. RNA was extracted from fat or brain as indicated. The source of the brain mRNA was a *wt* mouse.

METHODS

DNA Sequence
Southern Blot

When the investigators compared the DNA sequence of *wt* mice with *ob* mice, they found a significant point mutation (Figure 11.5). Friedman's group understood the mechanism for two different alleles of *ob*. SM/Ckc-+Dac ob^{2J}/ob^{2J} does not make any mRNA (see Figure 4.4) and C57BL/6J *ob/ob* makes truncated versions (see Figure 11.5). Next they wanted to determine if the leptin gene was conserved in other species. They performed a form of **Southern blot** that is jokingly called a zooblot, since the DNA from many different animals is probed on the same blot (Figure 11.6).

DISCOVERY QUESTIONS

11. Describe the mutation found in C57BL/6J mice.

a) C57BL/6J *wt*

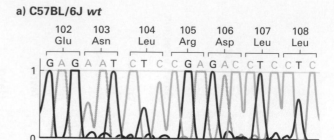

b) C57BL/6J *ob/ob*

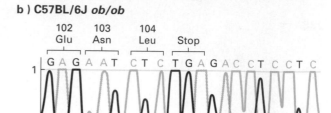

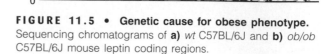

FIGURE 11.5 • **Genetic cause for obese phenotype.** Sequencing chromatograms of **a)** *wt* C57BL/6J and **b)** *ob/ob* C57BL/6J mouse leptin coding regions.

12. How does this mutation help you understand the Northern blot data from Figure 11.4? What can you conclude about the ob^{2J} mice?

13. If you made an antibody against leptin that bound upstream of leucine number 104, would you see any protein in C57BL/6J *ob* serum? Explain your answer.

14. How widespread is the leptin gene? Can you conclude whether the cat or *Drosophila* genomes contain leptin genes? Explain your answer.

15. Has Friedman demonstrated that 2G7 encodes the leptin protein?

Functional Tests for Leptin

As you read these papers in the library, you are uncertain whether Friedman's group had cloned the leptin gene or a different gene by mistake. The definitive proof must always be a functional test. If they could make the 2G7-encoded protein in bacteria, inject this into *ob* mice, and see a weight loss, that would be a good test. That's what they did (Figure 11.7, page 312).

DISCOVERY QUESTIONS

16. Does 2G7 encode leptin?

17. What can you conclude about the *db* mice?

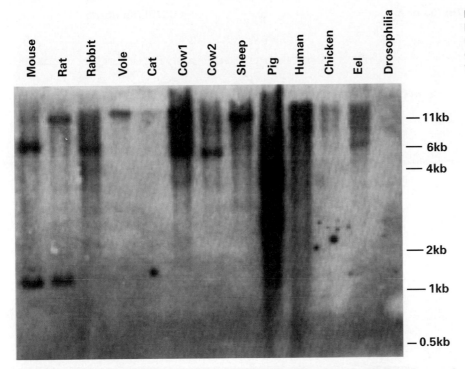

FIGURE 11.6 • Conservation of leptin gene in animals. Southern blot using genomic DNA from a wide range of species, as indicated. The probe was made from *ob* cDNA.

18. Predict what would happen to *ob* mice if they were taken off the leptin treatment.

Time to Visit Grandma

You have spent Saturday morning reading about leptin and how it controls weight gain and loss. You have a couple of figures that you photocopied for Grandma so she can see the data and your revised model that incorporates what you have learned this morning (Figure 11.8, page 313). You are excited to go over to her house for two reasons. First, you are happy to help her understand something since she played a major role in your upbringing. Second, you *know* she is going to make lunch for you and she is a great cook.

As you walk over to her house, you see this headline in a local paper:

> Fat Gene Found: Amgen Pays $20 Million
> for Rights to Protein to Cure Obesity

You chuckle rather confidently as you buy a copy. Now you have the definitive answer for her. All she will have to do is take a leptin pill and she will be able to lose weight.

You explain the data and your models to Grandma, and from her tone, you can tell she is impressed. All you can think about now are the wonderful smells coming out of the kitchen. You start to rise to get ready to eat, but she says she has a few questions that should be easy for you to answer.

1. How does leptin regulate weight homeostasis?
2. What is wrong with *db* mice since injected leptin did not cause them to lose weight?
3. Are all obese people either *ob* or *db* mutants?
4. Why are *ob* mice sterile? Why do they get sick?
5. What is the link between obesity and diabetes? Does leptin control this, too?
6. Can leptin pills or injections cure all cases of obesity?
7. What foods are rich in leptin, and can I eat these to control my weight?
8. What role does the brain play in weight homeostasis?
9. Is there anything I can do to adjust my lipostat?
10. What would you like to have for dinner tonight when you return with all these answers?

Grandma Gives You Homework!

The lunch Grandma fed you was delicious, though for the first time you were aware of the high fat content. All this research into the circuitry of weight homeostasis has made you think more about your own weight. Since Grandma is rather large, and you realize that obesity has a genetic component, you wonder what role diet has, if any. But as you were eating your second helping of dessert, you realized that she was asking too many questions and you would have to spend more of your Saturday in the library. . . . You made a tactical mistake. Rather than searching out the most recent information, you stopped after

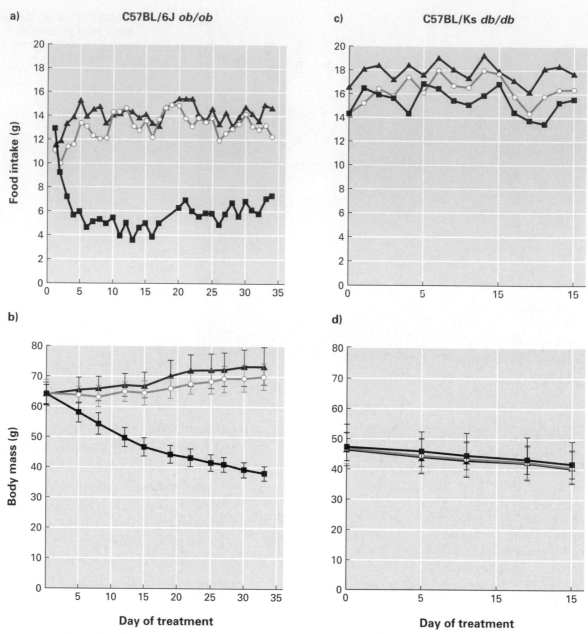

FIGURE 11.7 • **Effect of leptin on *ob* and *db* mice. a)** and **b)** Ten homozygous *ob* mice, and **c)** and **d)** 10 homozygous *db* mice received daily intraperitoneal injections of 2G7-encoded protein (filled squares) or saline (open circles), or no injections (filled triangles).

reading the original leptin cloning paper. Although it gave you the foundation you needed, it did not answer all the questions Grandma had. So you decided to answer her questions in order.

How Does Leptin Regulate Weight Homeostasis? The term leptin comes from the Greek word *leptos,* which means thin. Associating leptin with the word "thin" will help you remember how leptin works. Each person has a lipostat set for his or her own ideal percentage of body fat (Figure 11.9, page 314). Leptin is made by fat cells. The more fat you have, the more leptin you make. When a person gains weight

through increased fat storage, he or she will produce more leptin, resulting in lowered appetite and increased metabolism so that less fat will be created. If the genetic lipostat is unaltered, temporary weight loss is just that, temporary. When the amount of fat is reduced (through dietary restriction, surgical removal, or starvation), leptin production is reduced, which results in increased appetite and fat storage.

What Is Wrong with *db* Mice Since Leptin Injections Did Not Cause Them to Lose Weight? *db* stands for the *diabetes* phenotype, and these obese mice have been used as a model

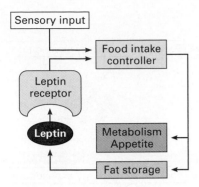

FIGURE 11.8 • Revised model for weight control.
Notice that the leptin receptor has been added. In addition, the controller now produces two separate signals: one for metabolism/appetite and the other for weight gained through fat production.

for diabetes research for many years. The molecular cause for the *db* mutation was identified in 1995 by a group from Millennium Pharmaceuticals Inc. Since it had been documented that *ob* and *db* mice had almost identical phenotypes, it was believed that *db* was probably the leptin receptor. In 1996, the same investigators confirmed that *db* mice have defective leptin receptors. This explains why *db* mice do not lose weight when injected with leptin.

The leptin receptor is produced in three forms that are all derived from a single gene—a long, a short, and a very short form. The long form has a long cytoplasmic domain that allows it to initiate signal transduction when leptin binds. The short form lacks the cytoplasmic domain but is still an integral membrane protein. This form is expressed mostly in the choroid plexis of the brain and is believed to act as a transporter protein that allows leptin to cross the blood-brain barrier. This would explain how leptin can be made in fat cells and found in the cerebrospinal fluid. The very short form lacks the **transmembrane domain** and therefore is secreted into the blood. The very short form is produced by pregnant women, allowing them to increase their fat stores to help ensure a full-term pregnancy. By secreting a soluble receptor, the woman's cells would sense less leptin and thus increase food intake and store more fat (more about this later).

19. Go to the Human-Mouse Homology Map, select mouse chromosome 4, and look for the leptin receptor *(Lepr)* near the bottom of the map (at 46.7 cM).
    a. On which chromosome is the human leptin receptor located?
    b. Perform a BLAST2 with the mouse *Lepr* (accession number P48356) and the human

LEPR (accession number P48357). What percentage of the receptor sequence is shared between human and mouse? Do you think human and mouse leptin could work in both organisms?

20. How could three different forms of one protein be derived from a single gene?

Are All Obese People Either *ob* or *db* Mutants? One set of human twins was identified that failed to make leptin and thus had the *ob* genotype. However, they did not suffer from hypothermia or diabetes the way *ob* mice do. These twins were successfully treated with leptin injections. From 1995 to 2000, over a dozen key components to the neurocircuitry underlying appetite control were discovered. Therefore, the genetic causes for obesity are numerous. Five percent of obese children have mutations in the melanocortin circuit, making it the most common set of genetic defects associated with obesity. (The brain circuitry is discussed later.)

Most obese people are resistant to leptin to varying degrees. In addition, serum leptin levels vary among people who have the same percentages of body fat. These data indicate that a variety of factors, genetic and environmental, probably contribute to people's sensitivity to leptin. For example, it is known that a high-fat diet can cause mice to become insensitive to leptin and therefore obese. It has also been shown that the lipostat can be lowered by exercise. Exercise and a high-fat diet are the two most important environmental factors that can alter a lipostat's set point. Leptin levels change during the day, and in most people there is a significant rise in serum leptin at midnight and 4 A.M. When the leptin cycle is abnormal, it may contribute to a behavior pattern that leads to obesity due to an inadequate suppression of nighttime appetite.

DISCOVERY QUESTIONS

21. How could the twins have the *ob* genotype (mutant leptin gene) but not have all the same symptoms found in *ob* mice? What kind of information about this particular human allele of leptin would you want in order to understand this apparent exception to the rule?

22. How can obese people with the same percentage of body fat have different levels of leptin in their blood?

Why Are *ob* Mice Sterile? Why Do They Get Sick? These two questions are specific, but they also ask a larger and more general question. Does leptin have more than just one function beyond simply controlling the amount of fat in an

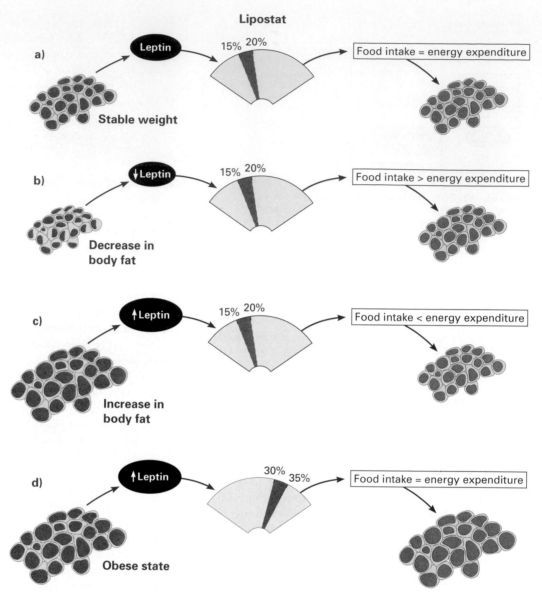

FIGURE 11.9 • **Lipostat controls body fat content. a)** Leptin affects the lipostat (scale) in a feedback loop regulating fat mass. At an individual's stable weight (shown as 15–20% fat for a nonobese subject), the amount of circulating leptin elicits a state in which food intake equals energy expenditure. Lipostats respond **b)** to a loss in body fat, and **c)** an increase of fat. **d)** A lipostat in an obese person has a higher set point for body fat (30–35% fat).

individual? It is probably the hardest to answer because leptin's different functions are still being discovered. However, you will answer the two specific questions first and then list some of the other physiological processes affected by leptin.

Why Sterility? Just prior to puberty, the amount of leptin in a person's serum increases. It is not known if this plays a causal role or not, but it does support the hypothesis that an individual is not reproductively ready until he or she has achieved a certain body mass, including fat. Children who have a higher proportion of body fat tend to enter puberty at an earlier age, which also correlates with the recent trend (over the last 200 years) of an

increase in the average weight of children and a decrease in the average age for entering puberty. In order for menstruation to occur, a girl's percent body fat needs to be approximately 24% or greater.

Women who are not pregnant have about 40% higher serum levels of leptin than men do. Falling leptin levels due to starvation results in a decrease in estradiol levels and amenorrhea (loss of menstrual cycle). Amenorrhea is commonly seen in women with anorexia nervosa or in female athletes who exercise strenuously over many days. A requirement for body fat is a part of the evolutionary selection pressure on women to have enough energy stores to be able to carry a pregnancy to term, and may also explain why, on average, women have more subcutaneous fat than men.

Male and female *ob* mice are sterile, but *ob* mice treated with leptin become fertile. Leptin allowed the maturation of the hypothalamic-pituitary-gonad circuit which corrected their sterile phenotype (Figure 11.10).

DISCOVERY QUESTION

23. If *ob* mice are sterile, how could researchers breed *ob* mice starting in the 1950s, before leptin was discovered? (Hint: They did not give the breeding mice any special diet or injections.)

In mice, leptin is not required for any stage of reproduction in females after mating (i.e., it is needed only for sexual maturation). Surprisingly, leptin treatment of *ob* females during pregnancy did not reduce food intake. This apparent contradiction has not been resolved fully, but we do have some suggestive evidence.

Pregnant women secrete a variant form of the leptin receptor, one that has no transmembrane domain. This tailless form of the leptin receptor is called a **soluble receptor.** By secreting a soluble leptin receptor into the serum, a pregnant woman **chelates** leptin (binds to and thus causes the ligand to be removed from circulation). By having less leptin in her circulation, her body responds by eating more and converting her energy into fat. This explains why so many women gain weight during pregnancy, which would have been selected for during evolution when a steady supply of food was not assured. It has been hypothesized that women who have difficulty losing weight after giving birth may also have difficulty turning off the expression of the soluble leptin receptor.

DISCOVERY QUESTION

24. How could anyone produce a *soluble* receptor if there is only one gene for the receptor?

Why Sick? This question is not easy to answer, but here are three relevant facts.

- Leptin's structure is similar to that of interleukin-6 (a well-known cytokine secreted by white blood cells), which stimulates proliferation and blocks apoptosis in T cells, leukemia cells, and hematopoietic (blood cell formation) progenitors. There is also a relationship between serum leptin levels and inflammatory cytokines such as tumor necrosis factor-α in patients with chronic heart failure.

- There are leptin receptors on a subset of your white blood cells.

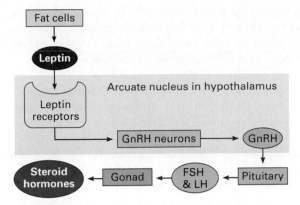

FIGURE 11.10 • **Circuit diagram illustrating the role fat accumulation plays in sexual maturity.** Abbreviations: GnRH is gonadtrophin-releasing hormone; FSH is follicle-stimulating hormone; and LH is leuteinizing hormone.

- People with infections and inflammations produce more leptin.

It is easy to imagine that insensitivity to leptin or abnormal production of leptin might lead to a compromised immune system. Hyperleptinemia (high levels of leptin in a person's serum) has been observed in 90% of obese individuals; the remaining 10% have normal leptin levels.

Other Functions? Again, you cannot answer this with any certainty, but it appears that leptin is a very busy protein.

1. Leptin receptors are present on the pancreatic islet cells, liver, kidney, lung, vascular endothelial cells, and skeletal muscles.

2. Leptin plays a role in angiogenesis (blood vessel formation).

3. *ob* and *db* mice have increases in bone formation, though leptin does not stimulate signaling in osteoblasts (bone-forming cells).

4. Leptin increases sympathetic nerve activity to tissues not normally thought of as thermogenic, including the kidney, hind limbs, and adrenal glands. Chronic administration of leptin to the central nervous system increases arterial pressure and heart rate. Therefore, leptin probably plays a role in maintenance of arterial pressure and may be involved in hypertension (high blood pressure), which can lead to problems such as heart attacks and strokes.

5. Obesity is a major risk factor for obstructive sleep apnea syndrome. Subjects treated with nasal continuous positive airway pressure (blowing air up nostrils all night to prevent snoring and apnea) for six months demonstrated significant reduction in serum leptin levels, which one would expect to increase fat production.

What Is the Link Between Obesity and Diabetes? Pancreatic islet cells die when they accumulate excess fat in their cytoplasm. This occurs frequently in obese individuals, which explains part of the link between obesity and diabetes. There might be other, more direct roles, but this was the only clear one you found during your day of reading.

Can Leptin Pills or Injections Cure All Cases of Obesity?

Leptin has been used to cure a few people who fail to produce it, but only a small percentage of those who are obese. In a News Feature in *Nature,* you read "In October [1999], came the news that [leptin] had performed poorly in its first clinical trial." But the October 1999 *JAMA* publication cited by the *Nature* News Feature stated, "A dose-response relationship with weight and fat loss was observed with subcutaneous recombinant leptin injections both in lean and obese subjects" (Figure 11.11).

DISCOVERY QUESTIONS

25. Do the data support the authors' claim or the *Nature* News Feature? Explain your decision with data from Figure 11.11.

26. In the *JAMA* paper that claimed the positive results for leptin, it is clearly indicated that 5 of the 11 authors are employees of Amgen. Remember, Amgen paid $20 million for the commercial rights to leptin. How does this affect your view of the different perspectives (*JAMA* vs. *Nature*) of the therapeutic potential for leptin?

27. What additional information would you like to know about the leptin study and the individuals who participated in the trial?

There are five critical points in the fat homeostatic circuitry that are available to medical intervention to alter body weight (Table 11.1).

The bottom line is that intervention to control weight is fighting millions of years (remember mammals, birds, and eels all have leptin genes; see Figure 11.6) of evolutionary selection that has resulted in a lipostat. Drugs that control only intake or only metabolism are doomed to failure since obesity is controlled by a feedback mechanism (see Chapter 8). For example, altering food intake will reduce fat, lower leptin, and result in increased fat storage and lowered metabolism.

What Foods Are Rich in Leptin, and Can Grandma Eat These to Control Her Weight?

Leptin is a protein, and our digestive tracts are designed to break down proteins. So eating leptin would not cause anyone to lose weight. But to answer her question more directly, high-fat meats contain a lot of leptin since it is produced in fat cells. But high-fat diets lead to leptin resistance, though the mechanism is uncertain. It is known that "westernization" has created obesity in many nonwestern peoples. It is believed that a change in diet (increased fat) and reduced activity levels are key factors in westernization obesity.

What Role Does the Brain Play in Weight Homeostasis?

Deciding whether to eat or not is a consequence of a complex biochemical mental process. There must be integrated circuitry that can process the genetic (e.g., leptin circuitry), environmental (e.g., smells, time of day, room location), emotional (e.g., distress or worry makes some people want to eat), and physical input (e.g., full or empty stomach). Understanding this circuitry will provide insights into many other behavioral circuitries as well.

Leptin is a link between the nervous system and adipose tissue. It is the central player in the control of **satiety** and energy metabolism. Leptin also plays a key role in sexual maturity, as you have already outlined. Although leptin has an impact on both energy and reproduction, the pathways for each are distinct. For example, if the MC4 (melanocortin-4) receptor is knocked out in mice, or transgenic mice are created that **overexpress** agouti-related protein (AgRP), the mice gain

FIGURE 11.11 • **Pattern of weight change over 24 weeks in obese people who received recombinant human leptin.** The amount of leptin given to people depended on their weight as indicated in the figure's key. The number of people was not constant over the course of the study. Error bars indicate s.e.m.; thin horizontal line indicates no change.

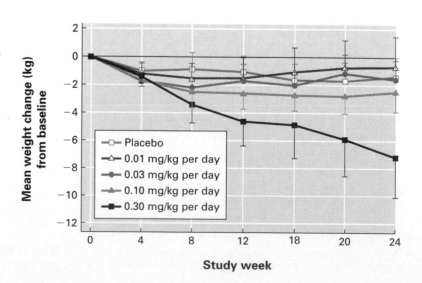

TABLE 11.1 • **Five critical points in the fat homeostatic circuitry.**

Category	Action	Negative Side Effects
1. Reduce food intake by reducing appetite	• *Sibutramine* reduces serotonin and dopamine reuptake from synaptic junctions. • *Desensitization* of serotonin and dopamine receptors by high levels of agonist (ligand that activates the receptor) • *Stimulation* of dopamine receptors (D1/D5 and D2/D3/D4) reduces food intake. These receptors also influence preference for food and the "enjoyment" factor associated with it.	Dry mouth, insomnia, asthenia, increase in blood pressure and heart rate Leads to an increase in seizures in mice
2. Blocking nutrient absorption in the gut	• *Orlistat* inhibits pancreatic lipase and thus lipid uptake in the gut.	Fecal fat loss that can lead to loose stools and diarrhea, absorption of fat-soluble vitamins such as E and β-carotene
3. Increasing thermogenesis	• *Ephedrine* and *caffeine* increase oxygen consumption (i.e., metabolism and heat production). • *Dinitrophenol* uncouples oxidative phosphorylation from ATP production.	Increased heart rate and sense of palpitations Cataracts, neuropathy
4. Blocking adipose synthesis		
5. Modulating the lipostat to reset the "optimum" percentage body fat		

weight but are not sterile. This indicates that the same ligand (leptin) binds to portions of the brain that initiate at least two separate circuits.

In January 2001, a new hormone called resistin was discovered. It is produced by fat cells and appears to be involved in the link between obesity and diabetes. Fat homeostasis will not be understood completely for many years. Below is a partial list of neurological players in this very complex integrated circuitry controlled by leptin.

Appetite Suppression
• Pro-opiomelanocortin (POMC) is the precursor for α-melanocyte-stimulating hormone (α-MSH)
• α-MSH binds to MC4 receptor located in the hypothalamus. α-MSH belongs to the melanocortin family of peptides that regulate pigment formation, food intake, fat storage, immune function, and nervous system function. Initial interest by pharmaceutical companies in α-MSH was for cosmetic tanning.
• Cocaine- and amphetamine-regulated transcript (CART)
• Ciliary neurotrophic factor (CNTF) binds to receptors in the hypothalamus and produces a leptin-like weight response when injected into mice.
• MC4: **knockout** mice eat more; binds the ligand α-MSH
• Opioid receptors μ and κ stimulation reduces appetite; this helps explain why drug addicts are often very skinny.
• Serotonin (also called 5-HT) binding to its receptor blocks appetite but may increase seizure susceptibility.

Stimulate Appetite

• Agouti is normally expressed only in hair follicle cells and stimulates pigment formation (dark colors). Certain mutations lead to agouti production in every cell. Agouti is an antagonist to MC4, which means it binds to MC4 but blocks its signaling ability; thus agouti causes an increase in appetite.
• AgRP is an antagonist to melanocortin-3 (MC3) and MC4 receptors, which leads to reduced activity, increased fat production (MC3), and increased appetite (MC4)
• Neuropeptide Y (NPY) is overproduced in hypothalamus of *ob* mice. *ob* mice with knocked out NPY are less obese because of a reduced appetite and increased energy expenditure.

Regulators of Energy Metabolism
• Perosixome proliferator-activated receptor-γ (PPAR-γ)
• PPAR-γ co-activator-1 (PGC-1)
• MC3: Loss of MC3 reduces activity level and increases fat production.

The complete neurological pathways are not known, but we can summarize what happens when a person rapidly loses fat or accumulates fat (Figure 11.12, page 318).

Is There Anything I Can Do to Adjust My Lipostat? There are only two ways a lipostat can be adjusted. As noted above, a high-fat diet can override the lipostat and result in an obese phenotype even if the individual has a lean lipostat genotype.

a) What happens when you lose fat through "fad" diet: **b) What happens when you eat too much at one meal:**

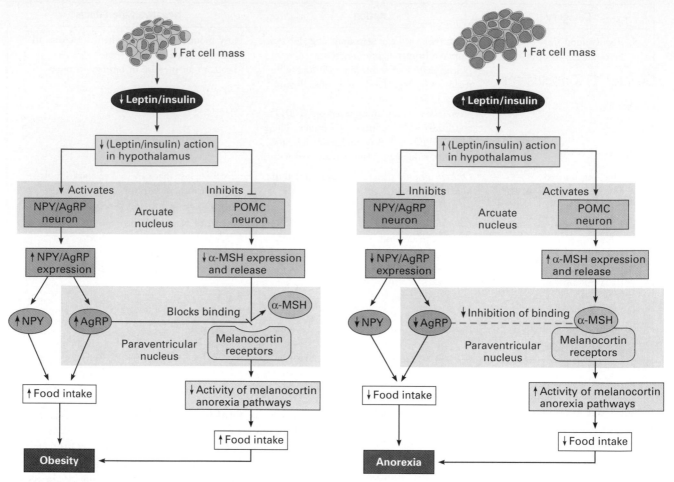

FIGURE 11.12 • Lipostat response to rapid changes in body fat. a) When fat is lost through surgery or "fad" diet, the body responds by producing less leptin. The outcome of reduced fat is increased food intake and fat storage. **b)** When you eat too much at a holiday meal, you accumulate body fat, which produces more leptin. The result of increased fat storage is reduced food intake, technically called "anorexia" but not to be confused with the pathological condition anorexia nervosa. Eventually, the body fat will be reduced to its preholiday level.

The only other way that has been shown to reduce the set point of a lipostat is to exercise. For unknown reasons, exercise reduces the set point for lipostats and can result in the conversion of an obese phenotype to a lean phenotype. The mechanism is not understood, but it is more than simply burning more energy. Sustained aerobic exercise alters the food intake and fat storage ratios so that the net result is a loss in body fat.

What Would You Like to Have for Dinner Tonight When You Come Back with All These Answers? You know that no one can make *lenza und spetzla* like she does, so you request this treat. She smiles and is happy to spend her afternoon making dinner for you. She tells you about a recent article she read in the *New York Times Magazine* (24 December 2000). Lisa Belkin had written a provocative story about why girls are entering puberty sooner than they used to. Now the average age

is 1–2 years lower than textbooks say. Theories range from obesity to sexual images in the media, pheromones, diet, and the relatedness of males living in the same house. Grandma thinks back to her childhood when children were not exposed to such variables and cannot imagine the answer will be simple or singular. "Humans are the most complex people on earth," she proclaims. She loves to invent sayings that she hopes will catch on, but you're pretty sure this one won't leave the room.

As you kiss her goodbye and give her a big hug, you remember one of your childhood wishes. When you were six years old, you wished aloud that someday you could put your arms all the way around Grandma. Now that you are full grown, your hands still cannot touch when you hug her. You realize that her lipostat is set rather high and/or her fatty diet has probably made her insensitive to leptin so her body inappropriately monitors how much fat is present. It makes you a little sad as you walk away. But then you hear her phoning her best friend, bragging about her new insights and why she can never lose weight. She brags about you, and you realize this was a great way to spend your Saturday after all.

Chapter 11 Conclusions

The obesity case study presents you with increasingly complex information. What started as a simple question quickly evolved into an intricate genomic circuit where components interacted with multiple partners and produced varied outcomes. Gregor Mendel was right: for each locus we have two alleles, and you can make predictions about the inheritance patterns of any gene. Mendel chose his traits carefully because they were simple and not parts of complex circuits that produced confounding outcomes. Simple genetics cannot explain all aspects of physical traits, and the growing field of genomics is demonstrating how few traits are simple.

Perhaps one of the most fundamental traits for any animal is the ability to utilize energy sources efficiently. When mammals evolved, there must have been a high premium for the ability to survive periods of starvation. Accumulation of fat allows us to save energy for later, which could be tied to reproductive maturity and immune capacity. Genomes can evolve over time but they rarely create new genes from scratch. Typically we see gene duplication and/or the coopting of one protein for multiple functions. In your study of genomics, you will need to keep an open mind for unexpected outcomes. Data come in many forms and can accumulate rapidly. As a genomic scientist, you will need to evaluate the validity and meaning of large data sets. This book will help you use bioinformatics tools to mine genomic and proteomic data that need to be pieced together in order to understand the complete story. Don't settle for simple answers if the data indicate a more complex model is more accurate.

References

Boston, B. A., K. M. Blaydon, et al. 1997. Independent and additive effects of central POMC and leptin pathways on murine obesity. *Science.* 278: 1641–1644.

Bray, G. A., and L. A. Tartaglia. 2000. Medicinal strategies in the treatment of obesity. *Nature.* 404: 672–677.

Campfield, L. Arthur, Françoise J. Smith, et al. 1995. Recombinant mouse *ob* protein: Evidence for a peripheral signal linking adiposity and central neural networks. *Science.* 269: 546–549.

Chehab, Farid F. 2000. Leptin as a regulator of adipose mass and reproduction. *Trends Pharmacological Sciences.* 21: 309–314.

Chen, A. S., D. J. Marsh, et al. 2000. Inactivation of the mouse melanocortin-3 receptor results in increased fat mass and reduced lean body mass. *Nature Genetics.* 26: 97–102.

Chen, H., O. Charlat, et al. 1996. Evidence that the diabetes gene encodes the leptin receptor: Identification of a mutation in the leptin receptor gene in db/db mice. *Cell.* 84: 491–495.

Chicurel, Marina. 2000. Whatever happened to leptin? *Nature.* 404: 538–540.

Chua, Streamson C., Iakovos K. Koutras, et al. 1997. Fine structure of the murine leptin receptor gene: Splice site suppression is required to form two alternatively spliced transcripts. *Genomics.* 45: 264–270.

Doehner, Wolfram, and Stefan D. Anker. 2000. The significance of Leptin in human: Do we know it yet? *International Journal of Cardiology.* 76: 122–124.

Fantuzzi, G., and R. Faggioni. 2000. Leptin in the regulation of immunity, inflammation, and hematopoiesis. *Journal of Leukocyte Biology.* 68: 437–446.

Fried, S. K., M. R. Ricci, et al. 2000. Regulation of leptin production in humans. *Journal of Nutrition.* 130(12 Suppl): 3127S–3131S.

Friedman, Jeffrey M. 2000. Obesity in the new millennium. *Nature.* 404: 632–634.

Halaas, Jeffrey L., Ketan S. Gajiwala, et al. 1995. Weight-reducing effects of the plasma protein encoded by the *obese* gene. *Science.* 269: 543–546.

Haynes, W. G. 2000. Interaction between leptin and sympathetic nervous system in hypertension. *Current Hypertension Reports.* 2(3): 311–318.

Heymsfield, S. B., A. S. Greenberg, et al. 1999. Recombinant leptin for weight loss in obese and lean adults: A randomized, controlled, dose-escalation trial. *JAMA.* 282: 1568–1575.

Marik, Paul E. 2000. Leptin, obesity, and obstructive sleep apnea. *Chest.* 118: 569–571.

Mantzoros, C. S. 2000. Role of leptin in reproduction. *Annals of the New York Academy of Sciences.* 900: 174–183.

Morgan, H. D., G. E. Heidi, et al. 1999. Epigenetic inheritance at the agouti locus in the mouse. *Nature Genetics.* 23: 314–318.

Ollmann, Michael M., Brent D. Wilson, et al. 1997. Antagonism of central melanocortin receptors in vitro and in vivo by agouti-related protein. *Science.* 278: 135–138.

OMIM: Leptin. <http://www.ncbi.nlm.nih.gov:80/entrez/dispomim.cgi?id=164160>. Accessed 21 December 2000.

Pelleymounter, Mary Ann, Mary Jane Cullen, et al. 1995. Effects of the *obese* gene product on body weight regulation in *ob/ob* mice. *Science.* 269: 540–543.

Rink, Timothy J. 1994. News and views: In search of a satiety factor. *Nature.* 372: 406–407.

Stephenson, Joan. 1999. Knockout science: Chubby mice provide new insights into obesity. *JAMA.* 282: 1507–1508.

Steppan, Claire M., Shannon T. Balley, et al. 2001. The hormone resistin links obesity to diabetes. *Nature.* 409: 307–312.

Schwartz, M. W., S. C. Woods, et al. 2000. Central nervous system control of food intake. *Nature.* 404: 661–671.

Zhang, Yiying, Ricardo Proenca, et al. 1994. Positional cloning of the mouse *obese* gene and its human homolog. *Nature.* 372: 425–432.

Why Can't We Cure More Diseases?

Goals for Chapter 12

How to Develop a New Medication

Understand requirements for developing a new medication.

Evaluate the potential for drugs to produce unexpected consequences.

Appreciate the challenge inherent in drug development.

In Chapters 10 and 11, we discovered the complexity associated with muscular dystrophy and obesity. The purpose of these first two case studies was to illustrate a key point. You need to think of cell and molecular biology the way an ecologist thinks of ecosystems. You have heard of the food web; now you need to imagine a "**cell web.**" Everything inside a cell is connected to something else. If you pluck one string of the cell web (e.g., mutate one gene), the entire cell reverberates and reacts to this perturbation. The pregenomic approach of reductionism is no longer the only one to consider.

We have learned a lot by dissecting the genome and understanding what a single gene does in isolation. The "one-gene-at-a-time" approach, though still very useful and beneficial, has its limitations. An ecologist would never study the food choice of a koala without also considering its environment. It makes no sense for a mammal to eat only one type of leaf until that you understand that koalas are surrounded by eucalyptus trees with very few competitors for these leaves. Why then should we be restricted to studying how one protein works in a microfuge tube? We must consider proteins in the context of their environment, which leads us to genomic medicine and potential treatments.

This chapter uses several case studies to examine why curing a disease by medical intervention sounds easier than it really is. We begin with a general overview of how to approach the task of developing a new medication and then apply what we have learned to specific examples. The examples range from making a better aspirin to curing muscular dystrophy by gene therapy. Although these two examples are extremes on a continuum of medical treatments, the difficulty of developing cures is universal. Can we introduce a drug into the cell web without perturbing the entire network?

How to Develop a New Medication

Define the Problem and Devise a Solution

We have studied two genetically influenced conditions that have multiple causes and effects. Although the problems seem complex, surely we can devise some solutions. In this chapter, imagine yourself as a biotechnology engineer who will design the "ultimate cure." You have unlimited resources (money, time, personnel, and a factory with unlimited space that generates nonpolluting electricity with wind and solar generators, etc.). So go crazy with creativity. Throughout this chapter we focus on different aspects of genomic medical therapy. For example, the first focus is on location: concentrate on *where* you want your medicine to go. Later, we focus our attention on other aspects, so keep tailoring your ideas to the appropriate focus.

Focus 1: Location, Location, Location

When you buy a home or rent an apartment, the most important consideration is location. Every town has some vacancies at any given moment. If the inexpensive places are an hour's drive from your work, you may consider this an unacceptable compromise in your pursuit of the right blend of cost and benefit. "The price is right, but losing two hours of my day is too much." Likewise, in genomic medicine, you have to address the location problem first. Where in the body do you need to send your "cure"? If the problem occurs in the brain, then a skin ointment may not be a good idea. You might think that simple injections of the desired protein would work, but remember two things: (1) foreign proteins do not last long in your bloodstream so injections would need to occur at least once a day, which is not the "ultimate cure" we were hoping for; and (2) proteins cannot cross the plasma membrane to enter cells. As you design a cure, think about the location of the problem in muscular dystrophy and obesity. Where is the problem?

> ### DISCOVERY QUESTIONS
>
> 1. Choose either obesity or muscular dystrophy, and define the location of the problem. Stick to the level of cells and organs for now. You probably will need to create an outline with each Roman numeral being a different molecular cause for the disease.
>
> 2. Once you have the cellular location defined, list the proteins known to play a role in the disease.
>
> 3. For each protein you listed, describe the subcellular location (e.g., plasma membrane, endoplasmic reticulum [ER], cytoplasm, blood stream, etc.).
>
> 4. Finally, determine if you need to eliminate the defective protein or add a functional copy.

Focus 2: Delivery Vehicles

Once you have chosen a location, you have to select the right vehicle to deliver your cure. For example, if you delivered pizzas for a living and you got an order that said "blue boat moored 100 meters south of lighthouse," you would need a boat and not a car. In the human body, there are three main delivery vehicles: viruses, protein carriers, and DNA.

Viral Vehicles You know that viruses contain nucleic acids (DNA or RNA) surrounded by a protein structure. When a virus binds to its target cell, it injects its genome into the new host cell and forces it to make viral proteins. However, any given virus can infect only a narrow range of species and cell types. For example, the virus that gives you a cold binds to epithelial cells in your respiratory passages but not to your skin or the epithelium that lines your blood vessels. Likewise, human immunodeficiency virus (HIV) infects only certain white blood cells. Viruses have evolved to be specific, and there is a virus for every cell type. So, if you could figure out which cells to target, you should be able to find a virus that infects this cell type.

There are a few common viruses currently in use for carrying medications. Adenoviruses, adeno-associated viruses, lentiviruses, herpesviruses, and retroviruses are among the most popular. Each of these has its advantages and disadvantages; nature has not yet produced the perfect virus to be used as a vehicle for genomic cures. So we are left with the same types of choices illustrated by the apartment that was inexpensive but located an hour away from your work.

DISCOVERY QUESTIONS

5. What structures of the virus determine the specificity of the viral infection? Could this be bioengineered to allow you to change the specificity?

6. What are the negative consequences to infecting with a virus? What does your immune system do to virally infected cells? What would this response do to your virally delivered cure?

7. Is there a way to create viruses that do not stimulate an immune response? What properties would such a virus need to have in order to hide from the immune system?

8. Devise a method that uses a virus that could infect any cell and yet retains the specificity you wanted. You might want to utilize surgery in your design.

Two strategies are currently in vogue due to the properties of the viruses. The first uses an **in vitro** (originally in glass petri dishes, but now everyone uses plastic) approach, which means the infection takes place in petri dishes. Cells are removed from the patient, infected in vitro, and then put back into the patient. This approach has the advantage of delivering high doses of a virus but not stimulating the immune system. However, it means the affected cells must be removable from the patient. This would *not* be a good way to affect the hypothalamus or all skeletal muscles, but might work well on a short-term basis for heart, lung, or liver cells.

DISCOVERY QUESTIONS

9. Why would you want to use the patient's cells and not a cell line that has the cure gene already built in?

10. For some diseases, you might not have to infect the cells responsible for the disease. For example, the inability to produce insulin causes diabetes. If you could get the cold virus to "convince" your lung cells to secrete insulin, you might be able to treat diabetes without having to infect pancreatic cells (the site of normal insulin production). Would muscular dystrophy or obesity be treatable in this "indirect" approach? Explain your answer.

The second approach leaves the diseased cells in the patient and infects them **in vivo** (in the living host). The in vivo approach requires the virus to be very specific, nonharmful, and invisible to the immune system. On the positive side, this approach would be appropriate for cells that are difficult to grow in vitro and successfully replaced into the patient. This issue of in vivo infection is addressed again near the end of this chapter.

Protein Carriers Viruses have obvious problems, so you might wonder why they would ever be used. First, they can be produced very cheaply and on a large scale. Second, they are good at producing a lot of protein inside cells for a long time. So, economically, viruses are good solutions. But there are alternatives. The traditional one is called a **liposome**, which consists of an artificial membrane (i.e., made in a microfuge tube and not by a living cell) stuffed with any cargo you want. A liposome can be used to deliver proteins or nucleic acids to a cell. However, liposomes are not specific so this process must be done in vitro to selected cells, in vivo to all cells, or by injection to a small localized set of cells (e.g., skeletal muscles).

The newest method is très chic but not well understood. Steven Dowdy found that some proteins can cross phospholipid bilayers unimpeded, another biological rule (proteins cannot cross membranes) with exceptions. These proteins all contain a protein-transduction domain (PTD), which means about 10–16 amino acids (a **domain**) are sufficient to move entire proteins across membranes (transduction). What is amazing is that even large iron beads and large proteins can be imported into cells with PTDs (Figure 12.1). This is a nonspecific method and would work on any cell type, which has advantages and disadvantages.

Nucleic Acids Surprisingly, muscle cells (cardiac and skeletal) can take up DNA without any special coating or delivery vehicle. So to treat such cells, you could simply inject your therapeutic DNA into muscles and they will take up the DNA. Even more surprising, the DNA can move to the nucleus and become expressed as protein. The physiology of

a)

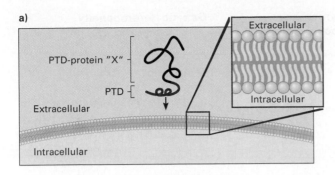

b)

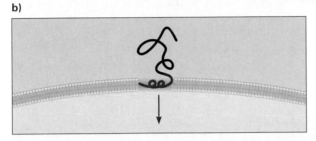

c)

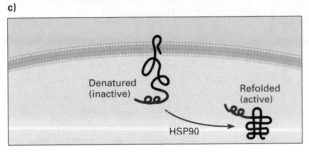

FIGURE 12.1 • Cellular internalization of a larger protein fused onto a protein transduction domain. a) The positively charged PTD makes contact with the negatively charged outer membrane. **b)** The protein translocates through the membrane in an unfolded state. **c)** Once inside the cell, members of the HSP90 protein family refold the larger protein into an active conformation.

skeletal muscles makes them especially amenable to DNA cures. Skeletal muscles are formed when individual cells fuse during development to form multinucleated giant cells that can be as long as the distance between two joints. So if therapeutic DNA enters a skeletal muscle, it needs to enter only a small number of nuclei to affect a large muscle cell. The mechanism for this DNA-uptake is unknown and has not been observed in nonmuscle cells.

DISCOVERY QUESTIONS

11. Choose either muscular dystrophy or obesity and design the vehicle you think would be best for delivering a miracle cure to the right cells and subcellular location.

12. Once you have designed your potential Nobel prize–winning ultimate cure, you need to test it out. What series of steps would you want to go through before you try it out on a real person? Outline your protocol leading up to the first human clinical trial.

13. Given that viral vehicles are often species specific, what affect would this have on your protocol leading up to human clinical trials?

LINKS

Genetic Causes

William Evans

Mary Relling

Focus 3: Specificity—"If It Ain't Broke, Don't Fix It"

Now that you know how to deliver your ultimate cure to the right location, you need to think about specificity. Specificity has two meanings: (1) hit only the specific protein you want to target and do not disturb the other proteins that are working fine; (2) given that each person might have different alleles for a particular disease gene, how can you be sure your cure will work on their specific allele? This question is the heart of a new field called **pharmacogenomics** that focuses on the genetic polymorphisms that translate into inherited differences in drug effects (see Chapter 3).

With the advent of molecular methods, pharmacogenomics has tried to understand the molecular causes behind clinical observations first documented in the 1950s. All individuals with the same disease will not respond equally to the same treatment. Although there are many nongenetic reasons as well, sometimes the inability to be treated is seen within families and has genetic causes, as explained by William Evans and Mary Relling (see page 89). Proteins that process the drugs into active forms or destroy the drug at a certain rate can lead to one form of drug responsiveness polymorphism. The second form is when the binding site for the active drug cannot bind properly and so is ineffective. These two classes of polymorphisms can produce at least nine different genotypes in a population (Figure 12.2, page 324). Genetic diversity in the population is one reason why treatments such as "the patch" to stop smoking are not 100% effective for everyone.

As shown in Figure 12.2, different genotypes will respond differently to the same medication. The first three graphs show how some will metabolize the drug quickly, others slowly. The difference in metabolic rates will affect the amount of active drug in a person's circulation. The second locus is the receptor, and the next three graphs depict the rate at which the drug binds to a person's receptor. Some alleles bind the medication better than others and thus can produce a stronger effective dose even at identical drug concentrations.

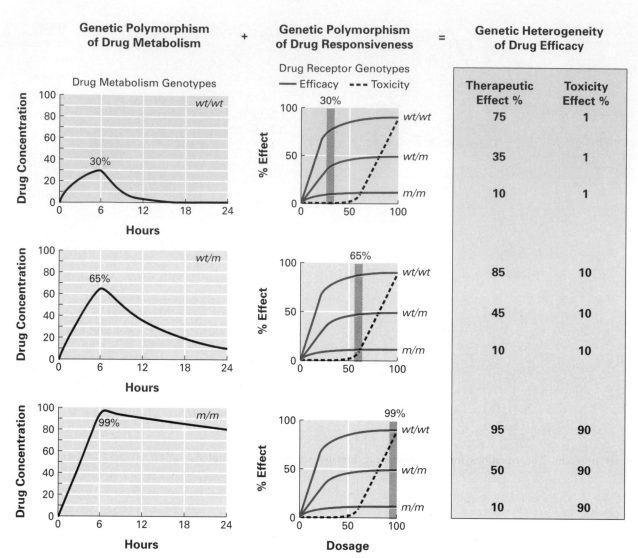

FIGURE 12.2 • Two loci that can affect a person's response to medications. In the left column, three graphs illustrate the amount of active medication present in a person's circulation, depending on their genotype. The second column illustrates the effectiveness of three different concentrations of medication due to three genotypes for the drug receptor. Superimposed in the second column is a dotted line that shows the toxicity of the drug. The final table summarizes the consequences of medication for each of the nine genotypes.

MATH MINUTE 12.1 • WHAT'S THE RIGHT DOSE?

One of the goals of pharmacogenomics is to find appropriate drugs and dosages for individuals of different genotypes. The hypothetical drug described in Figure 12.2 has minimal therapeutic effect on patients with *m/m* receptor genotypes, regardless of dose. However, patients with *wt/wt* or *wt/m* receptor genotypes can benefit from the drug, provided the dose is nontoxic. For concentrations greater than 50%, toxicity increases rapidly, while therapeutic effect increases very slowly. Therefore, drug concentration should probably be kept near 50% for all patients. Depending on factors such as disease severity, drug cost, and patient risk tolerance, lower concentrations might be desirable.

Using the data in Figure 12.2 and a mathematical model of drug metabolism, you can determine an appropriate dose for each drug metabolism genotype to achieve 50%

peak concentration. A rough model of drug concentration in the bloodstream is based on what happens to the administered drug dose in an interval of time, say one hour. (A finer scale model of drug concentration can be built using differential equations, in which the time interval is infinitesimal.)

For example, suppose 100 mg of the drug is administered orally to a patient with *wt/wt* metabolism genotype. The drug is absorbed into the bloodstream at a constant rate of 16 mg per hour, causing an increase in overall concentration over the first several hours. In what follows, we simplistically assume that the 16 mg enters the bloodstream all at once, at the beginning of the hour. We further assume that metabolizing proteins inactivate the drug at a rate proportional to the amount of active drug in the bloodstream at the beginning of the hour. Using an inactivation proportionality constant of 0.32, the following equations describe the amount of drug in the stomach at the end of hour n (S_n) and the amount of active drug in the bloodstream at the end of hour n (B_n):

$$S_{n+1} = \max(S_n - 16, 0)$$
$$B_{n+1} = (B_n + S_n - S_{n+1}) - 0.32 \times (B_n - S_n + S_{n+1}),$$

where the initial value of B is 0, and the initial value of S is 100. In the second equation, the quantity in parentheses is the amount of drug in the bloodstream at the beginning of hour $n+1$, since B_n is the amount still active at the end of hour n and $S_n - S_{n+1}$ is the amount absorbed into the bloodstream at the beginning of hour $n+1$. Thus the second equation says that the amount of active drug in the bloodstream at the end of an hour is the amount at the beginning of the hour, minus 32% of that amount. The equation can be simplified algebraically by factoring out the quantity in parentheses. By entering these equations into a spreadsheet program or graphing calculator, you can verify that they generate the drug concentration curve for the *wt/wt* drug metabolism genotype (see Figure 12.2a).

If you would like to explore these equations further, open the file drugmodel.xls in a spreadsheet program. The equations for the *wt/wt* drug metabolism model are implemented in this file. Using the same 16 mg/hr absorption rate, find an inactivation proportionality constant that generates the drug concentration curve for the *wt/m* metabolism genotype (see Figure 12.2b). Now, using this inactivation proportionality constant, adjust the initial dose down from 100 until the peak concentration is roughly 50%. This dosage would be more appropriate for patients with the *wt/m* metabolism genotype.

DISCOVERY QUESTIONS

14. Here is another example of genotypes affecting the outcome of a drug treatment. Go to Minoxidil Part 1 and Part 2. Propose a model to explain why only about half of the men (women's study is still underway as of 2001) who use Rogaine (its technical name is minoxidil) have more hair than they started with.

15. What modification could you make to improve Rogaine's success rate?

Hard Facts Another issue related to specificity is diagnosis. How would you know which particular mutation has led to the disease? As we saw in muscular dystrophy and obesity, these diseases are not monogenic. You would want to determine which gene is mutated in each patient. This can be done by one of the many companies that employ small armies to

sequence DNA, so don't worry about that. . . . Or should you? Figure 12.3, on page 326, shows the results from three years of genotyping data for cystic fibrosis. In July 2000, Elisabeth Dequeker and Jean-Jacques Cassiman published the results from a multiyear study. In 1996, 1997, and 1998, these two investigators surveyed 136, 145, and 159 laboratories that perform genotyping services in Europe. They sent six known cystic fibrosis DNA clones carrying common mutations and asked that the samples be processed using routine procedures. Only 48% of the 114 labs that participated in all three years made zero mistakes. In addition, some of the kits used for detection have difficulty detecting particular mutations such as the amino acid substitutions **551G→D** and **553R→X** (where X represents any amino acid). Although the quality of genotyping is improving with time, it does raise the possibility that you might try to cure for the wrong mutation!

LINKS
Minoxidil Part 1
Minoxidil Part 2
Elisabeth Dequeker
Jean-Jacques Cassiman

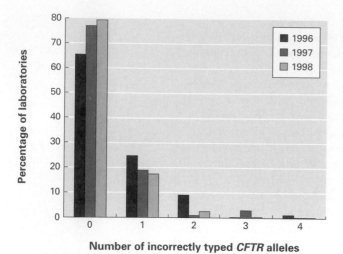

FIGURE 12.3 • **The percentage of laboratories that produced incorrect genotype determinations.** Interestingly, 10% of the mistakes were due to reporting mistakes despite the correct sequence data. 1996 is blue; 1997 is dark gray; 1998 is light gray.

SEQUENCES

Drugmodel.xls

STRUCTURES

Aspirin

Cox Enzyme

Acetylated

LINKS

400 B.C.

How Many Drugs Does It Take to Cure a Disease? We now have hit an interesting problem for drug development. If there are at least nine genotypes for each disease treatment (see Figure 12.2), does that mean that each disease needs at least nine different drugs? What effect will these different genotypes have on drug development and governmental approval? Each drug approved by the U.S. Food and Drug Administration has to be effective against the disease *and* safe to take. So does that mean that each drug will need to be tested in each of the nine genotypes? If so, that would require each of the nine drugs to be subjected to nine genotype-specific trials instead of the one genotype-blind trial currently used! In our search for a simple cure, we have gone from developing one drug for one disease to be tested in one clinical trial to developing nine drugs, each of which needs to be tested in nine clinical trials—an increase of 81-fold! And this number is an underestimate since we have assumed only two alleles at each of only two loci. What if more than one locus were required to process the drug? What if there were more than two alleles in the population? The potential expense for genotype-specific clinical testing may make pharmaceutical companies wish pharmacogenomics had stayed in the 1950s mentality—sometimes it works, sometimes it doesn't.

DISCOVERY QUESTION

16. Here is Discovery Question 12 again, though you might have a different answer this time. Once you have designed your potential Nobel prize–winning ultimate cure, you will need to test it out. What series of steps would you want to go through before you tried it out on a real person? Outline your protocol leading up to the first human clinical trial.

Eye of Newt . . . ?

Now that you have thought about where to send your drug and how specific it needs to be, it is time to decide what type of compound to make (not a small detail). Two types of drugs are available to you: chemical compounds and DNA molecules. The first can be represented by aspirin—a compound produced by pharmaceutical companies for over 100 years and by people since 400 B.C. This type of drug is not a protein but a small molecule that has physical properties that are optimized for its effect. For example, when you swallow a pill, it should be resistant to destruction in the stomach, readily taken up by cells, and have an appropriate half-life (not too long or too short).

That's Not Helpin' My Headache! Aspirin is a great example of a small and simple wonder drug. It is a very small molecule, which means it is easy to synthesize in vitro, which cuts down on cost. Also, its simple structure means that it is stable at room temperature (good for shipping and storage) and it has a good half-life inside humans. What could be better? Well, that's not an easy question to answer, but a lot of chemists are busy trying to build a better aspirin.

Aspirin works because it binds to an enzyme called **cyclooxygenase** (**Cox**) and then adds an acetyl group $(CH_3-\overset{\overset{\textstyle O}{\|}}{C}-)$ to one serine amino acid out of about 600 amino acids in the entire Cox enzyme. When Cox is acetylated by aspirin, it is irreversibly inactivated because the modified serine is in the active site, which the normal substrate, arachidonic acid, can no longer reach. When functioning normally, Cox converts arachidonic acid into prostaglandins, which are lipid-signaling molecules that trigger the sensation of pain, inflammation, and fever. So if you want to stop the pain, inflammation, or fever, just take aspirin. But there is a complex twist to this simple story.

Humans have two genes that encode slightly different cyclooxygenases called Cox-1 and Cox-2. Cox-1 is expressed in every cell in your body, with the highest levels in the stomach and kidneys. The prostaglandins produced here are needed to keep the cells working properly. Cox-2 is expressed primarily in the brain, lung, kidneys, and white blood cells when stimulated by cellular damage or infection. What we *want* to happen when we take aspirin is to block the activity of Cox-2 but leave Cox-1 unaffected. Unfortunately, aspirin has about 100 times higher affinity for Cox-1 than Cox-2. This explains why some people have stomach problems if they take too much aspirin—their stomach-friendly Cox-1 is preferentially inhibited and the stomach cells lose the beneficial prostaglandins produced by Cox-1.

METHODS
Knockout
LINKS
Genetech
Immunex
Disease Genes

DISCOVERY QUESTIONS

17. Why does aspirin prefer to bind to Cox-1 over Cox-2? What is the structural reason for this difference? (These questions are meant to be general and not require sequence information.)

18. In general terms, what would you want to produce in order to change the affinity of your "new and improved" aspirin so that it prefers Cox-2 over Cox-1?

19. What effect might allelic variations in Cox-1 and Cox-2 have on the effectiveness of new "super-aspirins"?

Some Assembly Required . . . Modifying the structure of aspirin so that it preferentially binds Cox-2 but still inhibits it is like assembling a child's Christmas present late on December 24th. It sounds easy, but it's not. Many chemists and biologists have spent years designing the "perfect" super-aspirin that will remove your headache but protect your stomach. Several companies have succeeded to varying degrees and are testing their new products now. However, there are some potential problems that need to be addressed.

Genetically engineered **knockout** (both alleles of a gene are deleted) mice were produced that lacked any Cox-2 enzyme. Individuals lacking Cox-2 enzyme are biochemically equivalent to individuals subjected to long-term exposure to the new generation of super-aspirins. Sadly, these mice developed severe kidney problems and were susceptible to peritonitis (inflammation of the peritoneum, the membrane that lines the wall of the abdomen and covers the organs). Female mice showed reproductive problems in ovulation, fertilization, and implantation. Plus, the area of highest Cox-2 activity is the brain, so there may be mental effects not detected in mice that would be unacceptable for humans. The good news was that loss of Cox-2 activity decreased the rate of colon cancer in these mice.

Conclusions? The bottom line in the aspirin case study is that what sounds simple at first may not be simple to implement. There are new forms of aspirin, but each will have new side effects that may outweigh the potential benefits of getting rid of a headache or reducing a fever. In case you think this is an isolated example, let's look at another situation that was even more difficult to detect.

A protein called TRAIL is being developed by two of the biotech giants—Genentech and Immunex. From 1996 to mid-2000, TRAIL was becoming very popular in cancer re-search labs because it appeared to kill many cancerous cells while leaving the noncancerous cells unharmed. Think of it: a single drug that could treat many forms of cancer. It is what we all want to see developed, but a good scientist should not jump to conclusions simply because it would be the best thing since . . . aspirin. It turns out that TRAIL does not harm healthy liver cells in mice or monkeys, but it kills *human* liver cells. Once again we see the benefit for requiring that new drugs undergo long periods of testing. Given the diversity in the human gene pool, extreme care must be taken since no drug can produce a uniform response in every person—not even simple ones like aspirin.

Don't Treat the Symptom, Treat the Cause

Of course, the sexiest solution to genetic diseases is **gene therapy** (replacing the defective gene with a functional one). When the human genome project was being proposed, many said it would change medicine as we know it. Although public health policies that guaranteed prenatal care and immunizations would be more cost-effective solutions to many health problems, such mundane solutions are not exciting enough to garner billions of dollars from the public and private sector. So, why not fix the problem at its source? Fix the mutated gene and be done with it!

The first genetic disease to be figured out was sickle cell, though the gene was not cloned until much later. The first "disease-causing gene" to be discovered was the dystrophin gene (see Chapter 10). The list of "disease genes" grows every day. But knowing the genetic cause does not mean that we can generate a genomic cure by simply replacing the inappropriate nucleotides. This has been the dream of many investigators, and after many trials, there has been only limited success.

DISCOVERY QUESTIONS

20. Let's assume you could accurately diagnose which gene is mutated. How could you introduce a *wt* sequence into the nucleus of a cell? Can you insert the therapeutic gene anywhere in the genome, or does it need to go in a certain locus?

21. Do you need to replace every copy of the gene in all cells or only a subset? Explain your answer.

22. What sort of control do you need over the replacement gene's transcription?

23. If the patient never produced the protein before, how do you know that the patient won't produce an immune response to the new protein since it may be seen as foreign?

LINKS

Xiao Xiao

Paper in *PNAS*

Alain Fisher

IHGT

Good Questions . . . The Discovery Questions above are not trivial, and if you don't have all the answers, well . . . join the club. If these questions were easy, gene therapy would be routine. But every disease requires a unique combination of answers. For example, there are dominant and recessive forms of limb-girdle muscular dystrophy (LGMD, see Chapter 10). Since a recessive disease is caused by the lack of a functional protein (only homozygous mutants have the disease) and a dominant disease is caused by an overactive protein (heterozygotes have the disease), inserting a single *wt* allele may cure a recessive disease but not a dominant disease. With obesity, the cause may be neuronal (located in a few nondividing cells) or due to the adipocytes (many mitotically active cells). Obesity may be due to a missing secreted molecule (leptin) or its receptor. Each of these causes for the same symptoms would require different strategies for gene therapy.

Since dystrophin was one of the earliest disease genes cloned, it was also one of the earliest candidates for gene therapy. But the gene itself presented problems. First, it is a huge gene—approximately three million bp long. That is too big to fit inside any viral vector or liposome. In fact, it is too big to pipet! Unfortunately, the 14,000 bp cDNA was also too big for easy manipulations. In December 2000, Xiao Xiao from the University of Pittsburgh published a paper in the PNAS (the Proceedings of the National Academy of Sciences USA) describing how they used a "minidystrophin gene" that was only about 4,000 bp long. This minigene is small enough to fit inside an adeno-associated virus that cannot replicate in human cells and does not elicit an immune response. But here is the obvious question: "How can a truncated version of dystrophin complement a mutant allele that is itself a truncated version of dystrophin?" It all depends where the truncation occurs (Figure 12.4). The *wt* protein has two functional domains on either end and a long rodlike middle section. Various mutant alleles encode for proteins that lack either one terminus or the other. The minigene created in Xiao's lab contained both terminal domains and a shortened version of the center piece. When this minigene

was given to *mdx* mice, the mice were cured. *mdx* mice lack a dystrophin gene, so this minigene therapy worked for the most prevalent form of muscular dystrophy.

Kevin Campbell's group in Iowa has also used gene therapy to treat hamsters that had LGMD2F. The Iowa group injected an adenovirus containing a *wt* allele of δ-sarcoglycan into skeletal muscles and corrected the mutant phenotype. So it appears that gene therapy may soon be an experimental option for certain patients enrolled in clinical trials. In fact, gene therapy has worked for at least one person who had severe combined immune deficiency (SCID, often called "the boy in the bubble disease"). SCID patients have almost no immune system because they lack a functional copy of an enzyme that metabolizes adenosine. Beginning in September 1990, a four-year-old girl with SCID was given the missing gene. Her immune system has improved, though she is also injected with the enzyme to bolster the amount produced by her therapeutic gene. In April 2000, a French group led by Alain Fischer treated two infants with SCID who lacked a different protein, and for ten months (time from treatment to publication date) the children have been able to produce normal levels of white blood cells. In this case, a retrovirus was used to infect bone marrow stem cells for three days in vitro before being reinjected into the patients. Sadly, many others who have also undergone SCID gene therapy have not been as fortunate.

It is always great to hear stories with happy endings, but science and medical research rarely progress without some sad stories, too. In September 2000, Jesse Gelsinger, an 18-year-old man with a liver enzyme deficiency, was killed by his experimental gene therapy treatment. Investigators at the University of Pennsylvania's Institute for Human Gene Therapy (IHGT) were performing an experimental procedure when he was given too many adenoviruses. Gelsinger died four days after being injected with the potential cure—a tragic outcome for everyone involved, especially the Gelsinger family.

So, what should be done now? Should we prohibit all gene therapy? Should it only be done when no other therapy exists? Should we demand that more research be done in the area of "traditional drugs" as illustrated with the aspirin story? Is gene therapy an inappropriate use of limited dollars available for health care? Should we insist that more cost-effective medical treatments be performed universally before very expensive treatments for relatively rare genetic diseases? Should we insist that people who want these expensive treatments share more of the expense? Should experimental treatments be reserved for the rich who can afford to pay for it themselves? There are no easy answers, but one thing is certain. These questions will need to be discussed openly by people like you who understand the complexity of the science and can communicate with those who do not. To address these ethical issues, we need a national dialogue, not small discussions restricted to academic institutions and in-

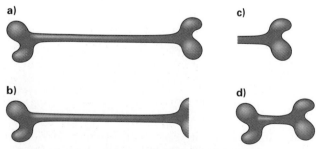

FIGURE 12.4 • Three truncated versions of dystrophin. a) Wild-type protein. **b)** Carboxyl-terminus truncation. **c)** Lacking most of the amino-terminus. **d)** Minidystrophin protein produced for gene therapy.

surance companies (see Chapter 3). Below is a series of questions without obvious answers. Consider them in light of what you know about developing new medications.

1. In October 2000, Tejvir Khurana from the University of Pennsylvania reported that in rats, 14 genes are expressed in skeletal muscles that control eye movement but not in skeletal muscles of the limb. Because patients with muscular dystrophy do not have any problems with the muscles that control eye movement, he speculated that one possible treatment for muscular dystrophy would be to "convert limb muscles to eye muscles by up-regulating the genes expressed in eye but not limb muscles."
   a. Based on what you know of muscular dystrophies, critique this proposal.
   b. Based on what you know about the complexity of "cell webs," critique this proposal.

2. Gene therapy sounds like the ideal cure for genetic diseases, even though there are many technical difficulties. Assume that a given gene could be delivered to the desired cells. Hypothesize reasons why gene therapy may not work in some patients.

3. National health policies cannot be based on individual cases, no matter how moving they are. Health policy must be based on the greatest good for the most people. List the pros and cons of gene therapy as if you were commissioned by the president. When you are done, decide whether or not to continue federal funding for gene therapy research.

4. There are evolutionary implications for gene therapy, too. For example, many people today wear glasses. However, 150,000 years ago, hominids did not need glasses because those that did were eaten by animals and did not live long enough to reproduce. Today, the selection pressures have changed so that the gene pool is loaded with "bad eyesight" alleles.

   If gene therapy is conducted on people with genetic diseases, should we as a society require them to be sterilized so they cannot pass on their disease alleles? If we do not, won't the gene pool become loaded with more disease alleles and thus require more and more gene therapy?

5. Since gene therapy may lead to increased need for more gene therapy, should we require all gene therapy to be performed on fetuses so that the resulting individual will not have the disease and neither will his or her children?

6. What about complex traits like intelligence, beauty, honesty? Do you think these have any genetic basis? If so, could they be "treated" with gene therapy?

7. Now that we have raised many real-world ethical questions, step back into your hypothetical vacuum where you have unlimited resources. Choose either muscular dystrophy or obesity and design the ultimate cure. Of course, most of you will not be able to come up with an idea that

is worth some money, but decide the major issues such as location, specificity, vehicle, and drug or gene therapy.

LINKS
Tejvir Khurana

Chapter 12 Conclusions

Now that you have finished this chapter on genomic medicine, you realize that it was not very genomic in the sense that it only focused on a few genes. However, genomics is more than just massive data sets and sequences of DNA. It is also a mind-set, a new perspective. Genomics requires molecular biologists to think differently. We need to think of cells as integrated circuits with massive complexity and interconnections. Imagine a world in which every single person had a cell phone. How quickly would news spread around the world if we all had phones and unlimited free minutes? That is what each of your proteins has—unlimited free minutes. They talk to their neighbors all the time, who talk to their neighbors, etc. Some proteins communicate with proteins located in different cells, and so the message spreads from one cell web to another, and soon the whole organism knows what has just happened in the little toe of your left foot.

The main purpose of Chapters 10–12 was to illustrate that Gregor Mendel was right and wrong at the same time. He correctly deduced how traits are passed on from one generation to the next. But in the real world, outside the garden walls of an abbey, life is much more complicated. Cell webs, where everything is connected to something else, provide us with a better model. Traits (e.g., muscular dystrophy, obesity, personality, intelligence, sexual orientation, etc.) have multiple inputs and connections. Now you are in the right frame of mind to start thinking on the genomic scale. This book will expose you to a wide range of topics, methods, and data. Use the tools provided to discover the hidden information in genomes.

References

Anderson, W. French. 2000. The best of times, the worst of times. *Science.* 288: 627.

Belkin, Lisa. 2000. The making of an 8-year-old woman. *New York Times Magazine.* 24 December. 38–43.

Black, Harvey. 2000. Seeing a solution. *The Scientist.* 14(22): 18.

Campbell, Kevin. 2000. *Molecular Studies of Muscular Dystrophy.* <www.physiology.uiowa.edu/campbell/Netscape%20Site/DGCResearch.htm>. Accessed 19 September 2000.

Cavazzana-Calvo, Marina, Salima Hacein-Bey, et al. 2000. Gene therapy of human severe combined immunodeficiency (SCID)-X1 disease. *Science.* 288: 669–672.

Feng, L., W. Sun, et al. 1993. Cloning two isoforms of rat cyclooxygenase: Differential regulation of their expression. *Archives of Biochemical Biophysics.* 307(2): 361–368.

Gura, Trisha. 2000. Caution raised about possible new drug. *Science.* 288: 786.

Kalgutkar, Amit S., Brenda C. Crews, et al. 1998. Aspirin-like molecules that covalently inactivate cyclooxygenase-2. *Science.* 280: 1268–1270.

Kay, Mark A., Joseph C. Glorioso, and Luigi Naldini. 2001. Viral vectors for gene therapy: The art of turning infectious agents into vehicles of therapeutics. *Nature Medicine.* 7: 33–40.

Loll, P. J., D. Picot, and R. M. Garavito. 1995. The structural basis of aspirin activity inferred from the crystal structure of inactivated prostaglandin H2 synthase. *Nature Structural Biology.* 2(8): 637–643.

Luong, C., A. Miller, et al. 1996. Flexibility of the NSAID binding site in the structure of human cyclooxygenase-2. *Nature Structural Biology.* 3(11): 927–933.

Marshall, Eliot. 1999. Gene therapy death prompts review of adenovirus vector. *Science.* 286: 2244.

Morham, S. G., R. Langenbach, et al. 1995. Prostaglandin synthase 2 gene disruption causes severe renal pathology in the mouse. *Cell.* 83(3): 473–482.

OMIM: ROSTAGLANDIN-ENDOPEROXIDE SYNTHASE 2; PTGS2 2000. <http://www.ncbi.nlm.nih.gov:80/entrez/dispomim.cgi?id=600262>. Accessed 5 January 2001.

Pennisi, Elizabeth. 1998. Building a better aspirin. *Science.* 280:1191–1192.

Service, Robert. 1996. Closing in on a stomach-sparing aspirin substitute. *Science.* 273: 1660.

Steppan, Claire M., Shannon T. Bailey, et al. 2001. The hormone resistin links obesity to diabetes. *Nature.* 409: 307–312.

Wang, Bing, Juan Li, et al. 2000. Adeno-associated virus vector carrying human minidystrophin genes effectively ameliorates muscular dystrophy in mdx mouse model. *PNAS USA.* 97:13714–13719.

GLOSSARY

Symbolic Notation

60-mer oligonucleotide 60 bases long. Any number can be used and the suffix -mer attached to indicate the length of the oligonucleotide.

123L→R a symbol using the pattern of **number:letter:letter** indicates that the first amino acid chime (L for leucine) at amino acid position 123 in the protein has been changed (to R for arginine).

Δ*tup1* the Greek letter delta (Δ) before or after a gene name is used to symbolize a deleted gene. In this case, the gene *Tup1* has been deleted. Italics is used in yeast to denote the gene is being described.

p53 a lower case "p" preceding a number indicates a protein of that molecular weight. For example, the **tumor suppressor** p53 has a molecular weight of 53,000 Daltons.

Tup1p when a gene name (e.g., *Tup1*) is followed by a lowercase "p," the *protein* encoded by *Tup1* is being discussed. This symbolic notation can be use for any gene/protein combination and unlike the gene, the protein is not italicized.

A

accession number identification number given to every DNA and protein sequence submitted to NCBI or equivalent database. For example, the human leptin receptor SwissProt accession number is P48357.

aliquot can be used as a noun or verb. When a liquid volume is divided into portions and placed in microfuge tubes, these *aliquots* of liquid are *aliquotted* into different microfuge tubes.

alkaline phosphatase enzyme that cleaves phosphate from its substrate at alkaline pH. Commonly used as a reporter gene and to label probes for blots of all kinds.

aneuploidy abnormal chromosomal number. For example, a loss of chromosome 4 and an extra copy of chromosome 21 could both be described as aneuploidy. Aneuploidy is also applied to portions of chromosomes.

annotated a gene is annotated when it has been recognized from a large segment of genome sequence and often we know something about its cellular role. Genomes can also be described as annotated once they have been analyzed for gene content.

antagonistic pleiotropy a theory of evolution that states for every beneficial gain in function, there is a compensatory loss in another function. For each genetic change, there are many effects (pleiotropy), some of which are beneficial while others are detrimental (antagonistic) in other environments.

anthrax *(Bacillus anthracis)* rod-shaped **Eubacteria** that can infect skin or lungs; may be used as a biological weapon.

antibody microarray one example of a **protein microarray** where antibodies are spotted onto glass to determine the nature and amount of **antigens** in a solution.

antigen any molecule that stimulates an immune response in the form of an antibody. Pollen and vaccines are both antigens.

antisense technology molecular method that uses a nucleic acid sequence complementary to an mRNA so that the two bind and the mRNA is effectively neutralized.

apoptosis a normal function for many cells, apoptosis is a genetically encoded sequence of cellular actions that leads to the cell's death; often referred to as "programmed cell death."

aquaporin protein that forms a channel (pore) across a membrane to allow the flow of water *(aqua)* molecules.

Arabidopsis a flowering plant about 15 cm tall that reproduces very quickly; sometimes referred to as the plant equivalent of the fruit fly. First plant to have its genome fully sequenced.

Archaea one of the three domains of life (along with **Eubacteria** and **Eukaryotes**). This domain includes prokaryotes that typically live in extreme environments or utilize atypical sources of energy.

archenteron during **gastrulation**, a portion of the embryo **invaginates** to form a pocket inside the embryo. This pocket will become the gut and is called an archenteron prior to gut formation.

array an orderly pattern of objects. In genomic studies, there are *micro*arrays and *macro*arrays. Microarrays are small spots of DNA or protein, and the identity of the spotted material is known. Macroarrays are bacterial, yeast, or similar colonies on plates used to determine functional consequences of genomic manipulations.

autocatalytic an enzyme capable of stimulating its own (auto) activity (catalyst).

autophosphorylate an enzyme capable of phosphorylating itself.

avidin a protein found in egg whites and bacteria (e.g., *Streptomyces*) that binds with very high affinity to the vitamin **biotin** (also called vitamin H). In the lab, avidin is used to purify biotin-tagged molecules from complex mixtures as was done in the **ICAT** method.

B

bacterial artificial chromosome (BAC) cloning vector replicated in bacteria that can hold an insert of about 150,000 base pairs. BACs were used extensively in the public **HGP.**

bar codes analogous to the black-and-white striped bar codes on packages. In genomic studies, unique nucleotide sequences are used to replace each gene in the yeast genome. There are two halves to the bar codes: an **upstream** UPTAG and a **downstream** DOWNTAG.

bimodal toggle switch a protein/DNA component that allows a circuit to accept input from two different sources to produce one of two outcomes. This switch is not stable.

biotin *See* **avidin.**

binomial coefficients the subject of Math Minutes 6.1 and 8.1; a counting formula, used to represent the number of ways a sample of k cells can be selected from a population of n cells.

bioinformatics a field of study that extracts biological information from large data sets such as sequences, protein interactions, microarrays, etc. This field also includes the area of data visualization.

biological process coined by **Gene Ontology** to describe broad cellular outcomes, such as mitosis or energy production, that are accomplished by ordered assemblies of molecules.

bistable circuit similar to **bistable toggle switch,** but describes the entire circuit and not just the switching mechanism.

bistable toggle switch a protein/DNA component of a circuit that chooses between two different outcomes. Once a choice is made, the switch maintains the outcome until the circuit receives new stimulus to change.

BLAST the protein and nucleic acid sequence search engine developed at NCBI that allows you to search sequence databases. BLASTn searches for nucleotide sequences; BLASTp searches for amino acid sequences; BLAST2 compares two sequences.

blastula an early embryonic stage of development where the embryo consists of a hollow ball of cells prior to **gastrulation.** It is from this stage that embryonic stem cells are isolated.

bootstrap analysis a statistical technique used to assess the reliability of the branching structure in phylogenetic trees.

Brownian motion random movement of particles or molecules that is powered by kinetic energy of the solution.

C

Caenorhabditis elegans nematode worm about the size of an eyelash and the first animal to have its genome sequenced completely. This model organism, pioneered by Sydney Brenner in the 1970s, is ideal for studying neurons; was instrumental in the discovery of **apoptosis.**

capillary electrophoresis a method that allows small amounts of compounds to be separated for further analysis. This method is at the heart of automated DNA sequencing and proteomics.

CAT *See* **chloramphenicol acetyl transferase.**

caveolae indentations on the inner surface of plasma membranes that are used for endocytosis.

CD *See* **conserved domain.**

cDNA *See* **complementary DNA.**

CDS *See* **coding sequence.**

cellular component a term coined by **Gene Ontology** to describe subcellular structures, locations, and macromolecular complexes such as nucleus, telomere, and mitotic spindles.

cell web a term coined in this book to describe the subcellular ecosystem of interacting components. Analogous to "food web," which indicates when one component of an interacting system is perturbed, the entire system is affected.

checkpoint a pause during a vital and complex cell process to make sure events have progressed properly before proceeding.

chelate to bind and remove from solution. For example, EDTA chelates divalent cations such as Mg^{2+}. Receptors can chelate ligands as well.

chloramphenicol acetyl transferase (CAT) a bacterial antibiotic resistance enzyme used as a reporter gene to assay the ability of cis-regulatory elements to promote transcription.

chromatogram (chromat) a four-colored graph produced from nonradioactive **dideoxy sequencing methods** including cycle sequencing.

CIM *See* **clustered image map.**

circuit a term borrowed from electrical engineering that indicates several interconnected proteins or genes. Circuits can describe connections needed to regulate a single gene or entire genomes.

circuit diagram a term borrowed from engineering that illustrates the interconnections among proteins or genes. Circuit diagrams allow us to visualize critical components and connections that are otherwise difficult to perceive.

cis-regulatory elements the portion of a gene that controls transcription of the same gene. The term describes the diverse DNA sequences that are bound by transcription factors and repressors and is inclusive of promoter and enhancer sequences.

clone noun or verb. A clone is any molecule/cell/organism present in more than one identical copy. To clone something means to produce more than one copy of the original molecule/cell/organism.

clustered image map (CIM) a complex integration of more than one data set. In this book, a CIM was used to integrate DNA microarray data with the effects of chemotherapy drugs on cells.

Clusters of Orthologous Groups (COG) NCBI compilation of evolutionarily related gene sequences from several microbial genomes. This site allows you to search by gene or cellular role and produces **dendrograms** to show sequence similarities.

coding sequence (CDS) abbreviation used at NCBI to indicate which bases constitute the **open reading frame.**

COG *See* **Clusters of Orthologous Groups.**

compartmentalization a way to maximize efficiency of cellular processes that require multiple proteins to work in concert (e.g., organelles).

complement when one allele is able to encode a functional product and compensate for a nonfunctional gene.

complementary DNA (cDNA) created by investigators by incubating mRNA, reverse transcriptase, and **dNTPs**. Initially, the cDNA is single stranded and can be used as probes for **DNA microarrays**. Often a second strand of cDNA is produced to form **dsDNA**.

consanguineous synonym for incest; breeding among closely related individuals that can accumulate deleterious alleles in the offspring and thus more genetic diseases.

conserved domain (CD) a **domain** that has been retained during evolution presumably due to its essential role within the protein's structure. Conserved domain searches are a part of the **BLAST** search.

constitutive always active; constitutive promoters or genes are active at all times.

contiguous (contigs) overlapping DNA segments that as a collection form a longer and gapless segment of DNA.

cooperation through communication a principle that enhances efficiency of a complex cellular process. Proteins often signal each other as the process continues so the cell can achieve the intended outcome (e.g., **checkpoints**).

correlation coefficient a measure of the corresponding between two sets of data.

Cox *See* cyclooxygenase.

Cy3 and Cy5 fluorescent green and red dyes, respectively, that are commonly used for **microarray** experiments.

cycle sequencing PCR-based method for sequencing DNA that uses fluorescent dyes and automated DNA sequencers.

cyclooxygenase (Cox) Cox converts arachidonic acid into prostaglandins, which are lipid-signaling molecules that trigger the sensation of pain, inflammation, and fever. Humans have two Cox genes called Cox-1 and Cox-2. Cox-1 is expressed in every cell, with the highest levels in the stomach and kidneys. Cox-2 is expressed primarily in the brain, lung, kidney, and white blood cells.

cytogenetic markers banding patterns on metaphase chromosomes used as physical landmarks along chromosomes, used to map the location of genes.

D

ddNTPs *See* **dideoxyribonucleotide triphosphates.**

degrade destroy a molecule such as a protein or mRNA. The implication is that the molecule is destroyed by a mechanism or protein and not simple entropy.

dendrogram a branching diagram that shows the relative sequence similarity between many different proteins or genes. Typically, horizontal lines indicate the degree of differences in sequences, but vertical lines are used for clarity to separate the branches. A scale bar should be included with each dendrogram.

deoxyribonucleotide triphosphates (dNTPs) these are the normal DNA monomers of dGTP, dCTP, dATP, and dTTP.

dideoxy [sequencing] method invented by Fred Sanger, a method for sequencing DNA that utilizes **ddNTPs.**

dideoxyribonucleotide triphosphates (ddNTPs) a modified **dNTP** that lacks the 3' hydroxyl group. ddNTPs are used in the **dideoxy sequencing method** developed by Fred **Sanger**. There are four ddNTPs: ddGTP, ddCTP, ddATP, and ddTTP.

digenic a trait is described as digenic if two genes collectively encode the proteins that produce the trait. *Polygenic* is a related term that indicates more than two genes contributing to the phenotype. Hair and eye colors are examples of polygenic traits.

discovery science perhaps the oldest form of science, when a person performs an experiment to see "What if . . . ?" Discovery science is a departure from hypothesis testing, which has been the standard for many years in molecular biology.

DNA chips *See* **DNA microarrays.**

DNA microarrays or **DNA chips** synonyms for gene sequences spotted on glass slides used to measure simultaneously the level of transcription of many genes.

dNTPs *See* **deoxyribonucleotide triphosphates.**

domain (1) the highest level of taxonomic organization. All life is divided into three domains: **Archaea, Eubacteria,** and **Eukaryotes.** (2) a region within a protein that has a particular shape and function (see **motif**).

double-stranded DNA (dsDNA) formed when two DNA strands bind to each other.

double-stranded RNA (dsRNA) Formed when two complementary RNA sequences bind to each other. Transfer RNAs contain sections of dsRNA as do **external guide sequences.**

downstream a relative direction for DNA sequence. Since DNA is usually written with the 5' end to the left, downstream would be to the right of a reference point. For example, the start codon is downstream of the promoter.

draft sequence a description of the degree of confidence in a DNA sequence. When the DNA has been sequenced only four times, it is described as a "draft sequence," compared to a "**finished sequence,**" which has been sequenced about eight times. With a draft sequence, about 95% of the genes should be identifiable.

drug signature genes genes that are either **induced** or **repressed** when a particular drug is administered to cells and usually identified by **DNA microarrays.**

dsDNA *See* double-stranded DNA.

dsRNA *See* double-stranded RNA.

dystrophin gene/protein that causes muscular dystrophy when mutated. Dystrophin helps connect the contractile proteins inside muscles to the extracellular matrix. A mouse model of muscular dystrophy has a mutated dystrophin gene, called *mdx.*

E

EC numbers *See* **Enzyme Commission numbers.**

ectoderm one of the three layers of cells in a developing embryo. Ectoderm cells on the outer layer will form the epidermis and nervous system.

ectopic a gene inserted in an unnatural location. For example, in mutant strains of T7, ectopic copies of *gene 1* were inserted throughout the genome to test the positional effects of gene locations.

EGS *See* **External Guide Sequence.**

ELSI *See* **Ethical Legal and Social Issues.**

endoderm one of the three layers of cells in a developing embryo. Endoderm cells in the innermost layer will form the lining of the digestive system and connected organs such as liver and pancreas.

endosymbiotic symbionts are two organisms who live together so each benefits from the other. Endosymbionts live inside the host organism while both continue to benefit.

Endo16 sea urchin gene expressed early in development in endothelial cells that eventually become a part of the gut of the larva.

environmental stress response (ESR) coined as a result of **DNA microarray** experiments, ESR indicates a collection of genes regulated by sudden changes in the local environment (intracellular or extracellular).

Enzyme Commission (EC) numbers a systematic way to identify every enzyme regardless of species of origin or language used by investigator. For example, cytoplasmic IDH has an EC number of 1.1.1.42.

epigenetic regulation control of gene activity without altering DNA sequence. One example of epigenetic regulation is **imprinting,** which is affected in part by **methyltransferases** adding a methyl group ($-CH_3$) to cytosine bases in DNA.

epistatic Gene A is said to be epistatic to gene B if an allele of gene A masks the encoded effects of gene B.

epitope tag epitopes are portions of larger molecules to which an antibody binds. Epitope tags are added onto different molecules to provide an antibody binding site.

ESI *See* **mass spectrometry.**

ESR *See* **environmental stress response.**

ESTs *See* **expressed sequence tags.**

Ethical Legal and Social Issues (ELSI) a division of the **Human Genome Project,** ELSI was intended to fund research and educational efforts about the ethical impact of sequencing the human genome.

Eubacteria one of the three **domains** of life; includes all the gram-positive and gram-negative bacteria such as *E. coli, M. tuberculosis,* and *Salmonella.*

Eukaryotes one of the three **domains** of life; includes all plants, animals, fungi, and protists whose cells contain membrane-bound organelles such as nuclei, endoplasmic reticulum, Golgi body, etc.

E-value *See* **expect value.**

exons parts of RNA molecule spliced together to form mRNA.

expect value (E-value) when performing a **BLAST** search, you will obtain an E-value for each sequence that is retrieved. An E-value can be thought of as the probability that two sequences are similar to each other by chance.

Therefore, E-values are best when they are small (e.g., 1×10^{-12}) compared to larger E-values (e.g., 0.06).

expressed sequence tags (ESTs) short DNA sequences obtained from either the 5' or 3' ends of **cDNAs.** An EST database has been established to help determine coding sequences within genomes, alternative splicing, and **single nucleotide polymorphisms.**

External Guide Sequence (EGS) short RNA sequences that bind to mRNAs to form a 3D shape that resembles tRNA precursors. RNaseP recognizes this pre-tRNA 3D shape and cleaves the mRNA, thus inactivating it.

extremophiles organisms that live in extreme environments, members of the domain **Archaea.** For example, Archaea can live in boiling water, extremely acidic water, or at the bottom of the ocean near hot vents.

F

fate map indicates the long-term outcome for each cell in an embryo. For example, a sea urchin fate map tells you what will happen to a particular cell once its development has locked it into a set pathway.

features in the context of **DNA microarrays,** each spot on the microarray can be called a feature. Spots and features are synonyms.

feedback loop biochemical term to describe how the products of a circuit pathway can either enhance (positive feedback) or repress (negative feedback) the circuit.

finished sequence DNA has been sequenced at least eight times and there are no gaps, and contains no more than 1 error in 10,000 base pairs (see **draft sequence**).

founder effect a population genetics term that explains a lack of genetic diversity in a population. For example, when a small number of adults land on a small island, all subsequent children typically contain no more variability than was present in the original inhabitants.

frozen genome description of the genome sequence from a particular date that does not change over time. It was necessary to create a frozen genome to annotate the draft human genome since it would be impossible to analyze the sequence if it changed every day as **draft sequence** became **finished sequence.**

G

galactose 1-phosphate (Gal-1-P) an intermediate metabolite as galactose is converted into glucose for ATP production.

gas chromatography a method for separating and identifying many different types of molecules.

gastrula an embryo in the process of **gastrulation.**

gastrulation a process of embryonic development when part of a **blastula invaginates** and forms a second layer of embryonic cells.

GCAT *See* **Genome Consortium for Active Teaching.**

GenBank developed and housed at **NCBI,** GenBank is the U.S. repository for all DNA and protein sequences.

Gene Ontology a collaborative effort of investigators to unify and standardize terms associated with the role a gene or protein plays in an organism. Represented model organisms include fruit fly *(Drosophila melanogaster),* yeast *(Saccharomyces cerevisiae* and *Schizosaccharomyces pombe),* mouse *(Mus musculus),* plant *(Arabidopsis thaliana),* worm *(C. elegans),* rat *(Rattus norvegicus),* and slime mold *(Dictyostelium discoideum).*

gene therapy correcting a defective gene by inserting a *wt* allele. Gene therapy can only work to correct recessive alleles unless the defective dominant allele is replaced by a knockout deletion.

genetic applet a term coined to indicate that simple toggle switches are capable of storing digital information (on or off).

genetic determinism the idea that all human traits are encoded in DNA. Examples of genetic determinism can be found in popular media stories that tout the discovery of a "smart gene" or "worry gene."

genetically modified organism (GMO) in contrast to **gene therapy,** GMO has a *transgene* inserted into the genomes in every cell in its body, as will all its offspring.

Genome Consortium for Active Teaching (GCAT) a nonprofit organization dedicated to bringing genomic methods into the undergraduate curriculum.

genomic representations a subset of the genome created by randomly choosing restriction fragments. Especially useful when the genome has not been sequenced but you want to survey genome-scale phenomena.

genomics a vague term that encompasses the study of **reference genome** sequences, variations within a species' genome, **DNA microarrays, circuits,** and **systems biology.** Some people include wider areas such as **proteomics, metabolomics,** etc., under the genomics umbrella. Due to its recent and changing definition, the term does not have a universally accepted meaning.

GMO *See* **genetically modified organism.**

gram-negative bacteria a subset of **Eubacteria** that do not stain with a dye called Gram stain (e.g., *E. coli).*

guilt by association an inference made from **DNA microarray** data. If Gene A is induced and repressed at the same time as Gene B, then guilt by association predicts Genes A and B perform similar functions.

H

hemagglutinin (HA) a protein encoded by the human influenza virus and often used as an **epitope tag.**

haplotype (derived from haploid genotype) a collection of alleles in one individual that are located on one chromosome. Alleles within a haplotype often are inherited as a single unit from one generation to the next. In **SNP** studies, haplotypes refer to a group of genomic variations found repeatedly in many people within a population.

heat denaturation use of heat to deform the 3D shape of proteins or nucleic acids (e.g., turning dsDNA into ssDNA).

heterodimer protein made of two subunits. *Hetero* denotes different subunits, and *dimer* means it is composed of two subunits. There are also heterotrimers (three subunits), heterotetramers (four subunits), etc.

HGP *See* **Human Genome Project.**

hierarchical clustering a method for organizing large numbers of genes, tumors, or other objects into dendograms.

high-throughput methods that produce large volumes of data and can process many samples quickly. Robots and computerized data collection are common themes in high-throughput methods.

hippocampus the location near the center of the brain where learning takes place and memories are stored. A critical layer of the hippocampus is the CA1 layer. See Brain Anatomy web page for details.

hominid living or extinct human/human-like animals. Examples include the famous fossils called Lucy and Peking Man, as well as broader categories such as *Homo erectus, Neanderthal,* or *Homo sapiens.*

homology a term with two different meanings that are often confused. Initially, homology referred to two sequences (DNA or amino acid) that were similar due to evolutionary relatedness. The newer and less specific meaning is simply two sequences that are similar. One other usage refers to homologous chromosomes, the pair of chromosomes in diploid organisms.

homologous recombination molecular method that targets a single gene for modification. Genes can be **knocked out** by this method, or more subtle changes can be created. Homologous recombination works better in some organisms (e.g., yeast) than others (e.g., humans).

homotetramer proteins composed of four identical subunits. *Homo* denotes identical subunits, and *tetramer* means it is composed of four subunits. There are also homotrimers (three subunits), **homodimers** (two subunits), etc.

horizontal transfer the process of passing genes from one species to another without sexual reproduction. The mechanism is unknown, though **pathogens** and **transposons** are often suspected as the cause.

Human Genome Project (HGP) the multinational, public domain DNA sequencing consortium composed of academic labs in the U.S., Europe, and Asia. Funding for this work came from government agencies and private philanthropic organizations.

hybridize process of incubating a probe with its target as well as the process of the probe binding to its target. The probe can be a protein (e.g., an antibody) binding to another protein or a nucleic acid (e.g., **ssDNA**) binding to its complementary sequence.

hydropathy plot (Kyte-Doolittle) a computer-generated graph that uses the amino acid sequence to predict whether or not a protein will span a membrane.

I

ICAT *See* **Isotope-Coded Affinity Tags.**
IDH *See* **isocitrate dehydrogenase.**
immunoglobulin synonym for antibody.

immunoprecipitation method that uses an antibody to extract your favorite protein from a mixture of many proteins. Often, proteins that are bound to your favorite protein are co-purified along with your protein.

imprinting process through which mammalian paternal and maternal alleles can be treated differently, depending on which parent contributed them to the offspring.

induced a gene with increased transcription. Typically refers to the switch from none to some transcription, but it could also refer to a switch from low to high transcription.

in silico experimental process performed on a computer and not by bench research.

integrated circuits when a particular process requires more than one individual **circuit** of genes or proteins, it can be described as an integrated circuit. The distinction between a circuit and an integrated circuit is subjective and therefore indicates only relative complexity.

intergenic sequence DNA sequence between two genes, sometimes referred to as "junk DNA."

intrinsic gene subset coined as a part of a **DNA microarray** study of breast cancer, these genes are expressed in similar ways within a patient's tumor but not in tumors from other patients.

intron portion of the gene and initial RNA transcript that will be excised and not included in mRNA; usually begin with the sequence GT and end with AG.

invaginate when a mass of **blastula** cells migrate to form a pocket inside the larger embryo. Invaginating cells of a **gastrula** look as if someone is pushing a finger into a balloon.

in vitro experimental process performed in a tube or petri dish and not in a living cell or organism. Literally translated as "in glass."

in vivo a term that means the experimental process was performed in live cells or organism, compared to **in vitro** or **in silico.**

isocitrate dehydrogenase (IDH) ubiquitous metabolic isozymes encoded by five genes in yeast and other eukaryotes. IDH converts isocitrate into α-ketoglutarate and requires a coenzyme (either NAD^+ or $NADP^+$) and a divalent cation (either Mn^{2+} or Mg^{2+}).

isoelectric point (pI) when the net charge of a protein is zero, the pH of the local environment will be equivalent to the isoelectric point of the protein.

isoforms or **isozymes** these closely related terms refer to two versions of highly similar proteins. Isozymes is specifically used for enzymes, while isoforms can be used for any protein.

Isotope-Coded Affinity Tags (ICAT) proteomic method that identifies proteins in a mixture and determines relative amounts for each protein from two protein mixtures.

isozymes *See* **isoforms.**

iteration when a process is repeated in an attempt to reach the ideal outcome, each repetition is called an iteration. Each iteration is slightly different from the previous one since we learn from the first and improve the second iteration.

J

Joint Genome Institute (JGI) Department of Energy–funded organization that uses **high-throughput** methods and computational analysis to understand basic biology. Included in the JGI are the Human Genome Project, Microbial Genome Projects, fish, sea urchin, etc.

"junk DNA" an outdated term that indicates how little people knew about genomes. It was intended to recognize that about 98% of the human genome does not contain any genes and therefore had no function.

K

knockout a molecular method to target a single allele for deletion and replacement with DNA of your choice using **homologous recombination.** Through selective breeding, homozygous individuals can be produced.

knockout mouse a mouse that has had one or more (typically just one) gene deleted (**knocked out**).

Kyte-Doolittle plot *See* **hydropathy plot.**

L

leprosy a disfiguring disease caused by the bacterium *Mycobacterium leprae*. They infect white blood cells called **macrophages,** which normally engulf and kill pathogenic bacteria. Later, Schwann cells surrounding the nerves become infected, which leads to the loss of myelin, nerve damage, loss of sensation, and eventually

loss of extremities due to reduced blood circulation.

leptin mammalian protein hormone (encoded by the *ob* gene) produced by fat cells that regulate fat homeostasis (the **lipostat**). If you produce too much leptin, you will lose fat, but if you don't produce enough, you will store more fat. In addition to fat homeostasis, leptin also influences sexual reproduction, immune and brain functions.

linkage when two *genes* are located near each other on the same chromosome; quantified by the frequency of recombination between two loci and measured in map units or centiMorgans, where 1 map unit equals 1% recombination frequency.

linkage disequilibrium linkage refers to *genes*, while linkage disequilibrium focuses on *alleles*. Linkage disequilibrium occurs when a set of alleles on one copy of a chromosome (a **haplotype**) stay associated with each other at a higher frequency than would be expected if recombination were completely random. When haplotypes are retained for many generations, it is assumed there is a selective advantage in not separating these particular alleles.

linker a short segment of **dsDNA** that can be ligated onto a second fragment of DNA to facilitate the cloning of that fragment. Linkers contain a restriction site, so they can be digested to produce the desired sticky ends for ligation.

liposomes spheres of lipids created **in vitro** to carry DNA or other biological reagent inside eukaryotic cells.

lipostat a term to describe the **integrated circuit** that uses **leptin** and other molecules to regulate fat homeostasis. Each person's lipostat is set for a percentage of body fat, and this set point can be increased by eating a high-fat diet or decreased by exercising.

long-term potentiation (LTP) when a neuron is stimulated and the consequence of the stimulation is maintained after the original stimulus is gone. This is the central component to memory and learning, controlled in large part in the **hippocampus** area of the brain. See Brain Anatomy web page for details.

lysogenic bacterial viruses such as lambda (λ) insert their genomes into the chromosome of the host bacterium and remain dormant, or lysogenic.

lytic one of the two choices bacterial viruses such as lambda (λ) can make that result in the rupture of the infected bacterium and the release of many new viruses.

M

macroarrays individual spots in macroarrays are composed of colonies of cells such as yeast and permit **high-throughput** screening of whole cell phenotypes.

macrophage one type of white blood cell that engulfs **pathogens** and presents fragments of the destroyed pathogen to other white blood cells for subsequent immune response.

MALDI *See* **mass spectrometry.**

mass spectrometry (MS) a technique that allows investigators to separate proteins based on their **mass to charge ratio (m/z).** The m/z for each protein allows them to be identified and quantified from complex mixtures. This **proteomics** tool is often used in pairs and called tandem mass spectrometry (MS/MS). Protein samples are first ionized and then inserted into MS/MS by either laser-based methods (MALDI and SELDI) or an electrospray method (ESI).

mass to charge ratio *See* **mass spectrometry.**

megabase (Mb) 1,000 kilobases or 1 million bases of DNA.

mesoderm one of the three layers of cells in a developing embryo. The middle layer of mesoderm cells will form many tissues such as heart, kidney, and reproductive organs as well as blood, bone, muscles, and tendons.

metabolome term coined to encompass the entire metabolic content of a cell or organism.

methylome term coined to refer to the methylation state of a genome (methylcytosine bases in the chromosome).

microarray *See* **DNA microarray.**

microsatellite a short segment of DNA (2 to 50 bases) repeated multiple times. Microsatellites vary in length and base composition, which makes them useful tools for distinguishing members of a population. For example, one allele may contain GCGCGC, while another might contain GCGCGCGCGC.

modular when a larger process or structure is composed of individual units that can perform functions, the larger process or structure can be described as modular or composed of modules.

molecular function coined by **Gene Ontology,** describes tasks performed by individual gene products such as transcription factor and calcium transportation.

molecular weights molecules of known mass used to determine the mass of unknowns. For example, protein molecular weight standards or markers are measured in Daltons (Da), and DNA/RNA are measured in bp or bases.

monomers individual units that can be polymerized. For example, ATP is an RNA monomer.

most recent common ancestor (MRCA) a description of a prehistoric population or species that gave rise to two or more species. Parents are the MRCA for siblings; the MRCA for humans and gorillas lived about 1.2 million years ago.

motif a sequence of amino acids or nucleotides that performs a particular role and is often conserved in other species or molecules.

MRCA *See* **most recent common ancestor.**

MS *See* **mass spectrometry.**

multiplex when a series of reagents are mixed in a single tube so more than one outcome will be produced simultaneously; multiplex PCR produces several different-sized bands of DNA that can be detected.

mutation accumulation an evolutionary theory to explain why selective advantages in one aspect often are accompanied by loss of advantages in others. Random mutations accumulate gradually and can become fixed by genetic drift in genes that are not subject to selection pressure. Adaptation to one environment and loss of adaptation to another are caused by distinct and unrelated variations in the genome.

myelin the fatty layer of insulation that surrounds nerves outside the CNS. Myelin is produced by Schwann cells.

m/z *See* **mass spectrometry.**

N

Na/K ATPase enzyme found in the plasma membrane of all animal cells. It pumps three sodium ions out of the cell and two potassium ions into the cell. Moving these ions consumes one ATP and produces the membrane potential used in muscle and nerve depolarization.

National Center for Biotechnology Information (NCBI) a federally funded part of the U.S. National Library of Medicine, NCBI is the home of **GenBank, BLAST, COG,** and many other genomic databases and computational tools.

National Cancer Institute (NCI) federally funded part of the U.S. **National Institutes of Health,** the NCI focuses on basic and applied research to treat and prevent cancer. Home for the Cancer Genome Anatomy Project, which intends to catalog the gene activity found in every type of cancer and compare this to the gene activity to *wt* tissues.

National Institutes of Health (NIH) federally funded center for biomedical research and conduit for funding at institutions in the U.S. The NIH has several campuses located in many states, all of which are dedicated to understanding and improving human health.

NCBI *See* **National Center for Biotechnology Information.**

NCI *See* **National Cancer Institute.**

NCI60 a panel of 60 human cell lines that represent the major forms of cancer. New drugs being developed for chemotherapy are tested first on the NCI60 to determine their effects.

ncRNAs *See* **noncoding RNAs.**

NIH *See* **National Institutes of Health.**

noise in the context of genomic **circuits,** inputs and outputs that are not identical each time. A genomic circuit needs to tolerate a level of noise not typical of real electronic circuits.

nonannotated open reading frame (NORFs) an open reading frame that was considered not to be a real gene when the genome was **annotated.**

noncoding RNAs (ncRNAs) DNA that is transcribed but not translated. The best known examples are rRNA and tRNA.

NORFs *See* **nonannotated open reading frame.**

normal distribution the subject of Math Minute 8.1, normal distribution indicates data have an average value around which all the individuals are clustered. A graph of normal distribution results in the classic "bell curve."

Northern blot named in reaction to **Southern blot,** RNA is separated according to size in an agarose gel, blotted onto a membrane, and then probed to detect specific sequences. Northern blots are used to determine which cells express the gene of interest, the size, and relative abundance of the mRNA.

O

ob *See* **leptin.**

Occam's Razor a guiding principle; when deciding which explanation to accept, always start with the simplest one.

oligonucleotide microarray *See* **DNA microarray.**

oligonucleotides (oligos) ssDNA polymers of unspecified length. The oligo sequence is determined by the investigator and synthesized **in vitro.** Oligos are used to probe blots, prime sequencing reactions, and PCR, as well for spotting on **DNA microarrays.**

OMIM *See* **Online Mendelian Inheritance in Man.**

oncogenes mutant dominant alleles of vital genes (e.g., *Ras*). Oncogenes (acting like the accelerator on a car) force the cell cycle to speed up, while **tumor suppressors** are similar to brakes trying to slow down the cell cycle.

Online Mendelian Inheritance in Man (OMIM) a comprehensive web site that catalogs all the genetic and molecular information related to human diseases (not just male diseases).

open reading frame (ORF) a portion of a cDNA or gene that begins with the start codon and ends with the stop codon. Synonym for **coding sequence (CDS)** on **GenBank** results.

ORF *See* **open reading frame.**

orthologs two genes in different species that are evolutionarily related. For example, the mouse and human **leptin** genes are orthologous because they evolved from a common ancestral leptin gene.

overexpress when genes are bioengineered to produce excessive amounts of protein, they overexpress their encoded proteins.

P

p53 a protein of 53 kDa molecular weight and a very important **tumor suppressor.**

pair-fed when one mouse is treated with **leptin** injections and reduces its food intake, the amount of food consumed by a control mouse is restricted to equal that consumed by the leptin-injected mouse. The control mouse is called a pair-fed mouse since the amount it was fed was paired with a treated mouse.

paralogs two genes within the same species are called paralogs if one evolved from the other. For example, yeast has three *IDP* genes that encode different **isoforms** of IDH. These three could be called paralogous genes if two of them evolved in yeast from one ancestral gene.

pathogen bacterium that can harm its host.

PCR *See* **polymerase chain reaction.**

PDB *See* **Protein Data Bank.**

penetrance the percentage of individuals who develop the phenotype for a particular genotype. If ten people are homozygous at one locus but only eight of them exhibit the phenotype, then this gene has 80% penetrance.

phagosomes internal organelles of white blood cells that engulf and kill **pathogens.**

pharmacogenetics study of the relationship between particular genes or alleles and the effectiveness of medications.

pharmacogenomics very similar to **pharmacogenetics,** pharmacogenomics attempts to study genome-wide influences on the efficacy of medications.

phosphorylate (phosphoprotein) when an enzyme adds a phosphate to another protein, the enzyme is called a kinase and the substrate protein becomes a phosphoprotein.

phylogenetic tree graphic way to illustrate the evolutionary relatedness of genes, proteins, individuals, strains, or species.

phytoplankton microscopic organisms that live in water and photosynthesize.

pI *See* **isoelectric point.**

pleiotropic one gene affects several seemingly unrelated phenotypes.

pointillism a school of art mastered by Georges Seurat in the 19th century and Chuck Close in the 20th. **DNA microarrays** are a collection of spots that can be examined as a whole or as a series of dots similar to the art work of pointillists.

polymerase chain reaction (PCR) molecular method that allows you to mass-produce any segment of DNA as long as you have two **oligos** that **hybridize** to the two strands of target DNA with their 3′ ends pointed toward each other.

postgenomic era describes the time in biology after entire genomes have become sequenced routinely. The postgenomic era began around the year 2000, though no single event signaled its beginning.

posttranscriptional control after a gene has been transcribed, the RNA can be modified and regulated, processes that constitute posttranscriptional control.

pRb *See* **retinoblastoma protein.**

prions proteins that have two shapes, one benign and the other contagious, which

leads to the conversion of the benign shape to the contagious shape. Contagious prion proteins can spread from one organism to anther and cause neurological diseases such as scrapie in sheep, mad cow disease, and Creutzfeldt-Jakob Disease (CJD) in humans.

prostaglandins produced when **cyclooxygenases** cleave **arachadonic acid.** Prostaglandins are lipid-signaling molecules that trigger the sensation of pain, inflammation, and fever.

protease enzyme **degrades** proteins into smaller pieces. Two examples are cathepsins found in **phagosomes** and trypsin produced by the pancreas to help digest food.

Protein Data Bank (PDB) database of every protein for which the 3D structure is known; it also contains a few nonprotein structures.

protein microarrays proteomic method similar to **DNA microarrays** in size and scale; proteins spotted onto glass and are used to determine protein interaction or to identify and quantify molecules found in various solutions. One example is an **antibody microarray.**

proteome the complete collection of proteins in a cell/tissue/organism at a particular time. Unlike genomes, which are stable over the lifetime of the organism, proteomes change rapidly as each cell responds to its changing environment and produces new proteins and at different amounts. You have one body-wide proteome, about 200 tissue proteomes, and about a trillion individual cell proteomes.

proteomics the study of proteomes that includes determining the 3D shapes of proteins, their roles inside cells, the molecules with which they interact, and defining which proteins are present and how much of each is present at a given time.

pseudogenes segments of DNA that resemble genes by their sequence of bases but are nonfunctional. Pseudogenes often have **transposons** inserted in them, or they may have other mutations that led to their inability to encode a functional protein.

PubMed an extensive database of biomedical literature hosted by **NCBI** that is searchable. You can subscribe to PubCrawler and automatically search PubMed and receive email results on a schedule of your choosing.

p-**value** probability associated with a statistical test of the difference between populations. Populations are considered significantly different if the associated *p*-value is small (typically 0.1 or smaller).

Q

quantitative trait loci (QTL) genes that encode phenotypes that can be measured on a scale, such as autism or skin tone.

R

rate constants a characteristic of enzymes that catalyze a reaction at a consistent pace. For example, the rate constant for **IDH** would indicate how many moles of isocitrate would be consumed per unit time.

real-time PCR (RT-PCR) a molecular method that is sometimes confused with reverse transcriptase PCR. Real-time PCR uses the specificity of PCR to measure the number of template molecules in your starting material and is being developed to detect biological weapons.

reductionist a person who dissects a complex system into increasingly smaller parts in order to understand it. For example, to understand how a watch works, you might take it apart to determine its components and deduce how they work together to form a watch.

redundant in the context of **genomic circuits,** redundant means that a critical process can be performed by more than one gene or individual pathway. For example, there are three redundant **IDH** genes that utilize $NADP^+$ and consume isocitrate to produce NADPH and α-ketoglutarate.

reference genome or **reference sequence** genome that was sequenced first for a species and thus represents a standard but not necessarily "normal" example. The term "reference" implies that variations exist within the population, but the reference is used as a common point for comparisons.

reference sequence *See* **reference genome.**

regional organization of components a principle that allows complex systems to behave more efficiently. When many proteins are required to perform a task, the task can be accomplished more rapidly if the necessary proteins are located near each other.

reiterative a process can be described as reiterative if you keep trying to improve the quality with each successive attempt. For example, models that describe how genes are regulated are reiterative because each version of the model is built upon more information so the model gradually approaches the truth.

reliability an engineering term used to measure the probability that the desired outcome will be accomplished. The reliability of a biological process is higher if **redundant** genes ensure there are multiple ways to accomplish the task.

reporter gene gene from another species that produces an mRNA or protein that can be detected easily. Reporter genes are often used to study the capacity of promoters or **cis-regulatory elements**. CAT, GFP, and *lacZ* are three common examples.

repressed when the level of transcription is reduced, a gene has been repressed. Repressor proteins bind to **cis-regulatory elements** and block or hinder transcription.

retinoblastoma protein (pRb) a very important **tumor suppressor**.

ribozymes RNA enzymes. Contrary to the initial rule "all enzymes are proteins," some enzymatic reactions are performed by RNA.

RIKEN Japanese Institute of Physical and Chemical Research; a genome center that hosts many databases and funds research.

RNA interference (RNAi) also referred to as siRNA (short inhibitory RNA); short **dsRNA** capable of inactivating genes by blocking the production of the encoded proteins. First discovered in the worm *C. elegans,* RNAi works in a wide range of species.

RNase P an RNA-cleaving enzyme that is required to produce mature tRNA molecules from precursors. **External guide sequences** form tRNA-like shapes when they bind to mRNAs, and **RNase P** cleaves the mRNA and thus prevents its translation.

robust an engineering term that indicates the ability to function in less than optimum conditions (i.e., **noise**). Methods such as **Southern blots** and **DNA sequencing** can be described as robust because they work properly even when investigators unintentionally alter the reaction conditions. Tolerant biological processes can also be described as robust.

RT-PCR *See* **real-time PCR.**

S

Sanger method *See* **dideoxy [sequencing] method.**

satiety the satisfied feeling of no hunger.

second messengers a vague term that refers to the way a cell relays information intracellularly. cAMP and calcium ions are examples of second messengers since they are produced in response to a primary message such as the binding of a ligand to its receptor.

sequence-tagged site (STS) unique locus in a **genome** defined by PCR primers that amplify a single locus. STSs were used to map the human genome, but are still useful as markers defining chromosomal positions.

serotypes analogous to genotypes, serotypes describe pathogens based on the ability of antibodies to bind to different subsets. For example, *Neisseria meningitides* exists in several serotypes.

shotgun sequencing a strategy for sequencing whole genomes, it was pioneered by the for-profit company Celera. Genomes are cut into very small pieces, cloned into plasmids, sequenced, and then assembled into whole chromosomes or genomes. This method is faster than hierarchical shotgun sequencing but more prone to assembly errors.

signal sequence hydrophobic in nature, the first 20 amino acids of proteins synthesized that pause translation until the ribosome docks with the rough endoplasmic reticulum (ER).

signal transduction conveyance of information from the outside to the inside of a cell. When a ligand binds to its receptor, this information is conveyed to the rest of the cell through a complex pathway of signal transduction that involves **second messengers.**

signature genes often cited in **DNA microarray** experiments; a collection of genes that are characteristic for a particular sample. For example, you would expect all green leaves to express a set of signature genes necessary to conduct photosynthesis (see Math Minute 5.1 for details).

single nucleotide polymorphisms (SNPs) very similar to point mutations except SNPs are considered to represent the genetic variation present in *wt* genotypes. By definition, SNPs differ from the **reference sequence** of a species.

small nuclear RNAs (snRNAs) example of a ncRNA.

small nucleolar RNAs (snoRNAs) example of a **ncRNA.**

SNP minor alleles SNPs that are less common than most other SNPs at the same position.

SNPs *See* **single nucleotide polymorphisms.**

soluble receptor alternative form of a receptor that is not anchored to a membrane. Soluble receptors can be bioengineered or naturally occurring (e.g., **leptin** receptor in pregnant women).

somatic hypermutation immunology term that in part refers to the ability of B cells to alter the DNA of antibody-encoding genes in order to produce antibodies with improved binding capacity. This takes place in the germinal centers of lymph nodes.

Southern blot named to honor its inventor, Dr. Ed Southern. A classic molecular method that allows the investigator to separate DNA by size in a gel, transfer the DNA to specialized paper (called a membrane), and hybridize the DNA with a particular probe.

stable isotopes differing from radioactive isotopes such as ^{35}S and ^{32}P, stable isotopes are atoms that do not emit radiation but have heavier or lighter masses based on the number of neutrons they carry. They are used in **proteomic** methods such as **ICAT.**

stationary phase a description of cells growing in culture that have slowed down their growth rate. Stationary cells are not increasing in number, but they are not dead.

stochastic not exactly the same as random, stochastic refers to genes or proteins that can produce widely variable outcomes.

stomata plural of stoma, the openings in leaves that permit gas exchange between the environment and cells within the leaves.

Strongylocentrotus purpuratus the Latin name for the sea urchin model organism; its genome is completely sequenced. Expresses the model gene *Endo16* during development.

structural proteomics a discipline within proteomics that focuses on the 3D shape of proteins.

STS *See* **sequence-tagged site.**

Sup35 a nontoxic yeast protein involved in termination of translation; has **prion**-like properties. It can assume two shapes, one of which is able to convert other Sup35 proteins to the contagious shape that no longer terminates translation.

Swiss-Prot a proteomics database based in Switzerland.

synergy the total output is greater than the sum of its parts. For example, sperm and egg synergistically work together to produce a new individual, not just a diploid cell.

synteny a term that has experienced a gradual change in its meaning, but current usage refers to multiple genetic loci from different species located on a chromosomal region of common evolutionary ancestry. Many mouse and human loci are syntenic.

systems biology coined to denote the new perspective for research in the **postgenomic era**. Systems biology studies whole cells/tissues/organisms not by a traditional **reductionist**'s approach but by holistic means in a **reiterative** attempt to model the complete cell/ tissue/organism.

T

tandem MS *See* **mass spectrometry**.

TATA box a portion of the eukaryote promoter 25 bases upstream of the start transcription site; contains the sequence TATA.

TB *See* **tuberculosis**.

teratogen a substance such as lithium or alcohol that causes developmental abnormalities.

The Institute for Genomics Research (TIGR) Maryland-based genomics and proteomics nonprofit organization that produces public domain genome sequences and analysis software.

The SNP Consortium (TSC) central repository of human SNP data hosted at Cold Spring Harbor Laboratories, NY.

threshold stimulation referring to a critical point that determines which direction a toggle switch will go. Similar to the action potential of neurons where depolarization above the critical point leads to one of two options (i.e., action potential or no response).

TIGR *See* **The Institute for Genomics Research**.

transcriptome coined to describe the complete RNA content of a cell/tissue/ organism and often measured by **DNA microarrays**.

transmembrane domain the portion of a protein that spans a phospholipids

bilayer, typically about 20 amino acids long and predominantly hydrophobic.

transposons sometimes referred to as "jumping genes," segments of DNA that can move from one place in a genome to another. Transposons have moved throughout the human genome and constitute a significant percentage of our genome. In yeast, the transposon Tn3 was bioengineered to be used as a genomics/proteomics research tool called mTn.

TSC *See* **The SNP Consortium**.

t-test a statistical test used to assess whether two quantities are significantly different.

tuberculosis (TB) the **pathogenic** bacterium *Mycobacterium tuberculosis* causes TB, which is a debilitating and potentially lethal respiratory infection that can be antibiotic resistant and thus difficult to treat.

tumor suppressor protein slows down the progression of the cell cycle and prevents cancers. Typically, tumor suppressors work at **checkpoints** to ensure the cell is functioning properly before permitting the cell cycle or mitosis to continue. p53 and **pRB** are two examples.

Tup1p the yeast protein encoded by the gene *Tup1* that functions as a transcription repressor.

two-dimensional (2D) gel electrophoresis proteomics method that separates proteins based on the **isoelectric point** (first dimension) and **molecular weight** (second dimension). Often spots from 2D gels are excised and sequenced to identify them. ExPASy (Expert Protein Analysis System) has a good database of 2D gels.

U

upstream a relative direction for nucleic acids often used to describe the location of a promoter relative to the start transcription site. For example, the start codon is upstream of the stop codon.

untranslated region (UTR) a portion of the mRNA that is not utilized to produce protein and is located **upstream** of the start codon (5' UTR) and **downstream** of the stop codon (3' UTR).

V

virulent **pathogen** or parasite that has the potential to do serious harm to its host.

W

Western blot proteins that have been separated by size using SDS-PAGE, transferred to a special type of paper (called a membrane), and probed with an antibody. Western blots are used to determine molecular weight, tissue distribution, and relative amount of the protein of interest.

wild-type (wt) allele, genotype, or phenotype that is considered to be the standard for a given strain or species. Wild-type alleles encode functional proteins and produce typical phenotypes. *Italics* is used if referring to genes or alleles.

Y

Y2H *See* **yeast two-hybrid**.

YAC *See* **yeast artificial chromosome**.

Yap1p a yeast transcription factor encoded by the *Yap1* gene.

yeast artificial chromosome (YAC) a cloning vector that replicates in yeast and can contain inserts about 1 Mb in size.

yeast two-hybrid (Y2H) proteomics method to detect protein-protein interactions. Variations of this method have been produced for mammalian and bacterial cells as well. The protein of interest is used as a *bait* to "fish out" proteins that bind to it (called *prey*).

yoctomoles 1 ymol = 10^{-24} moles = 0.6 molecules. A quantity recently coined due to increased sensitivity in proteomics.

Z

zeptomoles 1 zmol = 10^{-21} mol = 602 molecules. A quantity recently coined due to increased sensitivity in proteomics.

CREDITS

This section constitutes a continuation of the copyright page.

PHOTO CREDITS

CHAPTER 1 **Fig. 1.15** Reprinted with permission from Paul B. Vrana, Xiao-Juan Guan, Robert S. Ingram, and Shirley M. Tilghman, "Genomic imprinting is disrupted in interspecific Peromysus hybrids." 1998. *Nature Genetics.* 20: 362, fig.1a.

CHAPTER 2 **Fig. 2.16a–c** Reprinted with permission from Mark A. Burns, Brian N. Johnson, Sundaresh N. Brahmasandra, Kalyan Handique, James R. Webster, Madhavi Krishnan, Timothy S. Sammarco, Piu M. Man, Darren Jones, Dylan Heldsinger, Carlos H. Mastrangelo, and David T. Burke. "An integrated nanoliter DNA analysis device." 1998. *Science.* 282: 485–486, figs. 1, 4. © 1998 by The American Association for the Advancement of Science. All rights reserved; **Fig. 2.17a–b** Reprinted with permission from Phillip Belgrader, William Bennett, Dean Hadley, James Richards, Paul Stratton, Raymond Mariella, Jr., Fred Milanovich. "PCR detection of bacteria in seven minutes." 1999. *Science.* 284: 449, fig. 1. © 2001 by The American Association for the Advancement of Science. All rights reserved; **Fig. 2.19a–b** Reprinted with permission from Russell H. Vreeland, William D. Rosenzweig, and Dennis W. Powers. "Isolation of a 250 million-year-old halotolerant bacterium from a primary salt crystal." 2000. *Nature.* 407: 897, fig. 1. © 2000 Macmillan Publishers Ltd.; **Fig. 2.21** and **2.22a–b** Reprinted with permission from Wilmar L. Salo, Arthur C. Aufderheide, Jane Buikstra, and Todd A. Holcomb. "Identification of *Mycobacterium tuberculosis* DNA in a pre-Columbian Peruvian mummy." 1994. *Proceedings of the National Academy of Sciences.* 91: 2092–2093, figs. 1, 2, 4. © 1994 National Academy of Sciences, U.S.A.; **Fig. 2.23** Reprinted with permission from Andreas G. Nerlich, Christian J. Haas, Albert Zink, Ulrike Szeimies, and Hjalmar G. Hagedorn. "Molecular evidence for tuberculosis in an ancient Egyptian mummy." 1997. *The Lancet.* 350: 1404. Reprinted with permission from Elsevier Science; **Fig. 2.24** Kathleen D. Eisenach, M. Donald Cave, Joseph H. Bates, and Jack T. Crawford. "Polymerase chain reaction amplification of a repetitive DNA sequence specific for *Mycobacterium tuberculosis*." 1990. *Journal of Infectious Diseases.* 161: 979, fig 3a. ©1990 University of Chicago Press; **Fig. 2.25a–b** Courtesy of Laura Richman.

CHAPTER 3 **Fig. 3.06a–b** Courtesy of Ginger Armbrust.

CHAPTER 4 **Fig. 4.25a–c** Reprinted with permission from Timothy R. Hughes, Christopher J. Roberts, Hongyue Dai, Allan R. Jones, Michael R.

Meyer, David Slade, Julja Burchard, Sally Dow, Teresa R. Ward, Matthew J. Kidd, Stephen H. Friend, and Matthew J. Marton. "Widespread aneuploidy revealed by DNA microarray expression profiling." 2000. *Nature Genetics.* 25: 335, fig. 3.

CHAPTER 5 **Fig. 5.10a–c** Reprinted with permission from Robert Lucito, Joseph West, Andrew Reiner, Joan Alexander, Diane Esposito, Bhubaneswar Mishra, Scott Powers, Larry Norton, and Michael Wigler. "Detecting gene copy number fluctuations in tumor cells by microarray analysis of genomic representations." 2000. *Genome Research.* 10: 1730, figs. 3a–b, 4. Cold Spring Harbor Lab Press; **Fig. 5.24** Reprinted with permission from Alexander Soukas, Paul Cohen, Nicholas D. Socci, and Jeffrey M. Friedman. "Leptin-specific patterns of gene expression in white adipose tissue." 2000. *Genes and Development.* 14: 966, fig. 1a. Cold Spring Harbor Lab Press.

CHAPTER 6 **Fig. 6.02a–f, 6.03a–b,** and **6.06a–g** Reprinted with permission from Petra Ross-Macdonald, Paulo S. R. Coelho, Terry Roemer, Seema Agarwal, Anuj Kumar, Ronald Jansen, Kei-Hoi Cheung, Amy Sheehan, Dawn Symoniatis, Lara Umansky, Matthew Heidtman, F. Kenneth Nelson, Hiroshi Iwasaki, Karl Hager, Mark Gerstein, Perry Miller, G. Shirleen Roeder, and Michael Snyder. "Large-scale analysis of the yeast genome by transposon tagging and gene disruption.' 1999. *Nature.* 402: 415–416, figs. 3a, 3b, 5. ©1999 Macmillan Publishers Ltd.; **Fig. 6.13a–d** Heather L. True and Susan L. Lindquist. "A yeast prion provides a mechanism for genetic variation and phenotypic diversity." 2000. *Nature.* 407: 480, figs. 2c, 3a–c. ©2000 Macmillan Publishers Ltd.; **Fig. 6.18e** The 1999 SWISS-2DPAGE Database Update. © Swiss Institute of Bioinformatics, www.expasy.org; **Fig. 6.18f–g** Reprinted with permission from "The Yeast SWISS-2DPAGE database." 1996. *Electrophoresis.* 17: 556–565; **Fig. 6.21, 6.22,** and **6.23** Reprinted with permission from Jérome Garin, Roberto Diez, Sylvie Kieffer, Jean-François Dermine, Sophie Ducios, Etienne Gagnon, Remy Sadoul, Christiane Rondeau, and Michel Desjardins. "The phagosome proteome insight into phagosome functions." 2001. *Journal of Cell Biology.* 152(1): 167–174, figs. 1, 4, 6. By copyright permission of The Rockefeller University Press; **Fig. 6.36c** Reprinted with permission from Heng Zhu, James F. Klemic, Swan Chang, Paul Bertone, Antonio Casamayor, Kathryn G. Klemic, David Smith, Mark Gerstein, Mark A. Reed, and Michael Snyder. 2000. "Analysis of yeast protein kinases using protein chips." *Nature Genetics.* 26: 285, Fig. 2b; **Fig. 6.37c–f** Reprinted with permission from Heng Zhu, James F. Klemic, Swan Chang, Paul Bertone, Antonio Casamayor, Kathryn G. Klemic, David Smith, Mark Gerstein, Mark A. Reed, and

341

Michael Snyder. "Analysis of yeast protein kinases using protein chips." 2000. *Nature Genetics.* 26: 285, adapted from fig. 3.

CHAPTER 7 **Fig. 7.04a–f** Reprinted with permission from Andrew Ransick, Susan Ernst, Roy J. Britten and Eric H. Davidson. "Whole mount in situ hybridization shows *Endo 16* to be a marker for the vegetal plate territory in sea urchin embryos." 1993. *Mechanisms of Development.* 42: 119. fig. 1. Reprinted with permission from Elsevier Science; **Fig. 7.07** Adapted from Chiou-Hwa Yuh and Eric H. Davidson. "Modular *cis*-regulatory organization of *Endo16,* a gut-specific gene of the sea urchin embryo." 1996. *Development.* 122: 1076, fig. 2. © 1996. The Company of Biologists Ltd.

CHAPTER 8 **Fig. 8.06a–c** Reprinted with permission from Michael B. Elowitz and Stanislas Leibler. "A synthetic oscillatory network of transcriptional regulators." 2000. *Nature.* 403: 336. fig. 2. © 2000 Macmillan Publishers Ltd.

CHAPTER 9 **Fig. 9.02** and **9.03** Reprinted with permission from Trey Ideker, Vesteinn Thorsson, Jeffrey A. Ranish, Rowan Christmas, Jeremy Buhler, Jimmy K. Eng, Roger Bumgarner, David R. Goodlett, Ruedi Aebersold, and Leroy Hood. "Integrated genomic and proteomic analyses of a systematically perturbed metabolic network." 2001. *Science.* 292, no. 5518: 929–930, supplemental fig. 2 and fig. 2. Copyright © 2001 by The American Association for the Advancement of Science. All rights reserved.

CHAPTER 10 **Fig. 10.01a** Courtesy of The University of Kansas Medical Center; **Fig. 10.01b** Courtesy of BP Healthcare (BPHealthcare.com); **Fig. 10.03a–c** Courtesy of Uta Francke; **Fig. 10.03d–e** and **10.04** Adapted from Mayana Zatz, Angelina M. Vianna-Morgante, Patricia Campos and A. J. Diament. "Translocation (X;6) in a female with Duchenne muscular dystrophy implications for the localization of the DMD locus." *Journal of Medical Genetics.* 18: 443–444, fig. 2, 3, 4; **Fig. 10.05a–c** Reprinted with permission from Rachelle H. Crosbie, Connie S. Lebakken, Kathleen H. Holt, David P. Venzke, Volker Straub, Jane C. Lee, R. Mark Grady, Jeffery S. Chamberlain, Joshua R. Sanes, and Kevin P. Campbell. "Membrane targeting and stabilization of sarcospan is mediated by the sarcoglycan subcomplex." 1999. *Journal of Cell Biology.* 145 (1): 159, adapted from fig. 4. By copyright permission of The Rockefeller University Press; **Fig. 10.06b** Courtesy of Protein Data Bank; **Fig. 10.09** Adapted from Daniel Jung, Franck Duclos, Barbara Apostol, Volker Straub, Jane C. Lee, Valérie Allamand, David P. Venzke, Yoshihide Sunada, Carolyn R. Moomaw, John D. McPherson, and Kevin P. Campbell. "Characterization of ∂-sarcoglycan, a novel component of the oligomeric sarcoglycan complex involved in limbgirdle muscular dystrophy." 1996. *Journal of Biological Chemistry.* 271(50): 32328, fig. 6; **Fig. 10.10a–c** Reprinted with permission from Kathleen H. Holt and Kevin P. Campbell. "Assembly of the sarcoglycan complex." 1998. *Journal of Biological Chemistry.* 273 (52): 34669, fig. 4c; **Fig. 10.16a–c** and **10.17a–i** Reprinted with permission from Yasuko Hagiwara, Toshikuni Sasaoka, Kenji Araishi, Michihiro Imamura, Hiroshi Yorifuji, Ikuya Nonaka, Eijiro Ozawa, and Tateki Kikuchi. "Caveolin-3 defiency causes muscle degeneration in mice." 2000. *Human Molecular Genetics.* 9(20): 3050–3052, figs. 3, 5b; **Fig. 10.19a–d, 10.20a–f,** and **10.21a–r** Reprinted with permission from R. Mark Grady, Robert W. Grange, Kim S. Lau, Margaret M. Maimone, Mia C. Nichol, James T. Stull, and Joshua R. Sanes. "Role for œ-dystrobrevin in the pathogenesis of dystrophin-dependent muscular dystrophies." 1999. *Nature Cell Biology.* 1: 216–218, adapted from fig. 2d, 3 and 5.

CHAPTER 11 **Fig. 11.03, 11.04,** and **11.06** Reprinted with permission from Yiying Zhang, Ricardo Proenca, Margherita Maffel, Marisa Barone, Lori Leopold, and Jeffrey M. Friedman. "Positional cloning of the mouse obese gene and its human homologue." 1994. *Nature.* 372: 427–430, figs. 2a, 3a, 6a. © 1994 Macmillan Publishers Ltd.

ILLUSTRATION CREDITS

CHAPTER 1 **Fig. 1.4a–b** Adapted from Banfi et al. 2000. *Science.* 287: 139, fig.3B; **Fig. 1.5a–b** Reprinted by permission from *Nature.* 409: 878 (left)

and 897 (right). Copyright 2001, Macmillan Magazines Ltd.; **Table 1.1** Reprinted by permission from *Nature.* 409: 896. Copyright 2001, Macmillan Magazines Ltd.; **Fig. 1.6a–b** Reprinted by permission from *Nature.* 409: 896. Copyright 2001, Macmillan Magazines Ltd.; **Fig. 1.7** Courtesy of Michael Green; **Fig. 1.8** Reprinted by permission from *Nature.* 409: 902. Copyright 2001, Macmillan Magazines Ltd.; **Fig. 1.11** Courtesy of Shirley Tilghman.

CHAPTER 2 **Fig. 2.1a–b, 2.2a–d,** and **2.3a–b** Reprinted by permission from Horiike et al. *Nature Cell Biology.* 3: 211; **Fig. 2.4** Adapted from Horiike et al. *Nature Cell Biology.* 3: 213, fig. 2; **Fig. 2.6** Adapted from Shingenobu et al. 2000. *Nature.* 407: 84, fig. 3; **Table 2.1** Courtesy of Stewart Cole; **Fig. 2.9** Reprinted by permission from Cole et al. 2001. *Nature.* 409: 1009, fig. 3; **Fig. 2.11** Reprinted by permission from Madsen et al. 2001. *Nature.* 409: 611, fig. 1b; **Fig. 2.12** Reprinted by permission from Ingman et al. 2000. *Nature.* 408: 709, fig. 2; **Fig. 2.13** Reprinted by permission from Ingman et al. 2000. *Nature.* 408: 653, fig. 1; **Fig. 2.14** Adapted from Kopp et al. 1990. *Science.* 280: 1046, fig. 1; **Fig. 2.15a–b** Adapted from Burns et al. 1998. *Science.* 282:485–486, figs. 1, 4; **Fig. 2.17** Reprinted with permission from Belgrader et al. 1999. *Science.* 284: 450, fig. 2; **Fig. 2.18a–b** Reprinted by permission from Vreeland et al. 2000. *Nature.* 407: 897, fig. 1; **Fig. 2.19** Reprinted by permission from Vreeland et al. 2000. *Nature.* 407: 899, fig. 2; **Fig. 2.25** Reprinted with permission from Richman et al. 1999. *Science.* 283: 1174, fig. 4B; **Fig. 2.26a–b** Courtesy of Robert Lanciotti; **Fig. 2.27** Reprinted with permission from Lanciotti et al. 1999. *Science.* 286: 1174, fig. 2; **Fig. 2.28** Reprinted with permission from Pizza et al. 1999. *Science.* 287: 1818, fig. 2; **Fig 2.29** Reprinted with permission from Pizza et al. 1999. *Science.* 287: 1818, fig. 3; **Fig 2.30a–b**<http://gsbs.utmb.edu/microbook.ch002.htm>; **Fig 2.30c** Adapted from Pizza et al. 1999. *Science.* 287: 1818, fig. 2; **Fig. 2.31a–b** Adapted from Onishi et al. 1996. *Science.* 274: 981, fig. 3; **Fig. 2.32a** Reprinted with permission from Ma et al. 1998. *Antisense and Nucleic Acid Drug Development.* 8: 416, fig. 1a; **Fig. 2.32b** Reprinted with permission from Werner et al. 1999. *Antisense and Nucleic Acid Drug Development.* 9: 84, fig. 2a; **Fig. 2.33a** Reprinted with permission from Ma et al. 1998. *Antisense and Nucleic Acid Drug Development.* 8: 416, fig. 1b; **Fig. 2.33b** Reprinted with permission from Werner et al. 1999. *Antisense and Nucleic Acid Drug Development.* 9: 84, fig. 2b; **Fig. 2.34** Reprinted with permission from Ma et al. 1998. *Antisense and Nucleic Acid Drug Development.* 8: 423, fig. 6.

CHAPTER 3 **Fig. 3.1** Courtesy of Kennie Comstock; **Table 3.1** Courtesy of Kennie Comstock; **Fig. 3.4** Courtesy of Ginger Armbrust; **Fig. 3.5** Adapted from Rynearson and Armbrust. 2000. *Limnology and Oceanograpy.* 45(6): 1335, fig. 4; **Fig. 3.6a** Adapted from Rynearson and Armbrust. 2000. *Limnology and Oceanograpy.* 45(6): 1333, fig. 2; **Fig. 3.7** Adapted from Rynearson and Armbrust. 2000. *Limnology and Oceanograpy.* 45(6): 1335, fig. 3. **Fig. 3.8** Adapted from Rynearson and Armbrust. 2000. *Limnology and Oceanograpy.* 45(6): 1336, fig. 6; **Fig. 3.12a–b** Reprinted with permission from Evans and Relling. 1999. *Science.* 286: 488, fig. 2; **Fig. 3.13** Adapted from Walker et al. 2000. *Nature.* 405: 296, fig. 1a; **Fig. 3.14** Adapted from Walker et al. 2000. *Nature.* 405: 296, fig. 1b; **Fig. 3.17** Adapted from Evans et al. 2001. *British Medical Journal.* 322: 1052, fig. 1.

CHAPTER 4 **Fig. 4.5** Reprinted by permission from Chu et al. 1998. *Science.* 282: 701, fig. 3; **Fig. 4.6a–b** Reprinted by permission from Chu et al. 1998. *Science.* 282: 701, fig. 3; **Fig. 4.9** Reprinted by permission from Chu et al. 1998. *Science.* 409: 1009, fig. 3; **Fig. 4.10** Adapted from DeRisi et al. 1997. *Science.* 282: 485–486, fig. 1, fig. 4; **Fig. 4.11** Reprinted with permission from DeRisi et al. 1997. *Science.* 278: 681, fig. 1; **Fig. 4.12** Reprinted with permission from DeRisi et al. 1997. *Science.* 278: 682, fig. 2; **Fig. 4.13** Reprinted with permission from DeRisi et al. 1997. *Science.* 278: 684, fig. 4; **Fig. 4.14** Adapted from DeRisi et al. 1997. *Science.* 278: 685, fig. 2c, and e; **Fig. 4.15** Adapted from DeRisi et al. 1997. *Science.* 278: 685, fig. 5b; **Fig. 4.16** Adapted from DeRisi et al. 1997. *Science.* 278: 682, fig. 2; **Fig. 4.17** Reprinted with permission from DeRisi et al. 1997.

Science. 278: 683, fig. 3; **Fig. 4.18** Reprinted with permission from Gasch et al. 2000. 11: 4245, fig. 1; **Fig. 4.19** Reprinted with permission from Gasch et al. 2000. *Molecular Biology of the Cell.* 11: 4247, fig. 3; **Fig. 4.20** Reprinted with permission from Gasch et al. 2000. *Molecular Biology of the Cell.* 11: 4246, fig. 2; **Fig. 4.21** Adapted from Gasch et al. 2000. *Molecular Biology of the Cell.* 11: 4250, fig. 5; **Fig. 4.22** Adapted from Gasch et al. 2000. *Molecular Biology of the Cell.* 11: 4253, fig. 8a; **Fig. 4.23** Adapted from Gasch et al. 2000. *Molecular Biology of the Cell.* 11: 4251, fig. 6a–b; **Fig. 4.24a–c** Reprinted by permission from Hughes et al. 2000. *Nature Genetics.* 25: 333, fig. 4; **Fig. 4.25** Reprinted by permission from Hughes et al. 2000. *Nature Genetics.* 25: 335, fig. 3; **Fig. 4.26** Adapted from Hughes et al. 2000. *Nature Genetics.* 25: 335, fig. 3; **Fig. 4.27** Adapted from Shoemaker et al. 2001. *Nature.* 409: 923, fig. 2; **Fig. 4.28a–c** Adapted from Shoemaker et al. 2001. *Nature.* 409: 924, fig. 2; **Fig. 4.29a–e** Reprinted with permission from Shoemaker et al. 2001. *Nature.* 409: 923, fig. 2; **Fig. 4.30** Reprinted with permission from Shoemaker et al. 2001. *Nature.* 409: 925, fig. 3; **Fig. 4.31a–d** Reprinted with permission from Shoemaker et al. 2001. *Nature.* 409: 926, fig. 4.

CHAPTER 5 **Fig. 5.1** Reprinted with permission from Alizaden et al. 2000. *Nature.* 403: 505, fig. 1; **Fig. 5.2** Reprinted with permission from Alizadeh et al. 2000. *Nature.* 403: 506, fig. 2; **Fig. 5.3a–d** Reprinted with permission from Alizadeh et al. 2000. *Nature.* 403: 507, fig. 3; **Fig. 5.4a–b** Reprinted with permission from Alizadeh et al. 2000. *Nature.* 403: 509, fig. 5a–b; **Fig. 5.5** Reprinted with permission from Alizadeh et al. 2000. *Nature.* 403: 509, fig. 5c; **Fig. 5.6a** Reprinted with permission from Perou et al. 2000. *Nature.* 406: 748, fig.1c–j; **Fig. 5.6b** Reprinted with permission from Perou et al. 2000. *Nature.* 406: 748, fig.1a; **Fig. 5.7a–e** Reprinted with permission from Perou et al. 2000. *Nature.* 406: 750, fig. 3a, c–f, scale bar; **Fig. 5.8a–c** Adapted from Lucito et al. 2000. *Genome Research.* 10: 1727, fig. 1; **Fig. 5.9** Reprinted with permission from Lucito et al. 2000. *Genome Research.* 10: 1729, fig. 2; **Fig. 5.10a–c** Reprinted with permission from Lucito et al. 2000. *Genome Research.* 10: 1730, fig. 3a–b, fig. 4; **Fig. 5.11** Reprinted with permission from Behr et al. 1999. *Science.* 284: 1521, fig. 1a; **Fig. 5.12** Reprinted with permission from Behr et al. 1999. *Science.* 284: 1521, fig. 1b; **Fig. 5.13** Reprinted with permission from Behr et al. 1999. *Science.* 284: 1521, fig. 1c; **Fig. 5.14** Reprinted with permission from Behr et al. 1999. *Science.* 284: 1522, fig. 2; **Fig. 5.15** Adapted from Marton et al. 1998. *Nature Medicine.* 4(11): 1294, fig. 1; **Fig. 5.16a–c** Adapted from Marton et al. 1998. *Nature Medicine.* 4(11): 1295, fig. 2b,c,d; **Fig. 5.17a–b** Reprinted with permission from Marton et al. 1998. 4(11): 1295, Table 1; 1297, Table 2; **Fig. 5.18a–b** Reprinted with permission from Marton et al. 1998. *Nature Medicine.* 4(11): 1297, fig. 5a–b; **Fig. 5.19** Reprinted with permission from Marton et al. 1998. *Nature Medicine.* 4(11): 1298, fig. 6; **Fig. 5.20** Reprinted with permission from Ross et al. 2000. *Nature Genetics.* 24: 228, fig. 1a; **Fig. 5.21** Reprinted with permission from Ross et al. 2000. *Nature Genetics.* 24: 231, fig. 3c; **Fig. 5.22a–c** Reprinted with permission from Scherf et al. 2000. *Nature Genetics.* 24: 241, fig. 4; **Fig. 5.23** Reprinted with permission from Scherf et al. 2000. *Nature Genetics.* 24: 221, fig. 5; **Fig. 5.24** Reprinted with permission from Soukas et al. 2000. *Genes and Development.* 14: 966, fig. 1a; **Fig. 5.25** Adapted from Soukas et al. 2000. *Genes and Development.* 14: 967, fig. 2a; **Fig. 5.26** Adapted from Soukas et al. 2000. *Genes and Development.* 14: 968, fig. 3a,h,l,m; **Fig. 5.27** Adapted from Soukas et al. 2000. *Genes and Development.* 14: 970, fig. 4a; **Fig. 5.28** Reprinted with permission from Soukas et al. 2000. *Genes and Development.* 14: 970, fig. 4b; **Fig. 5.29** Adapted from Soukas et al. 2000. *Genes and Development.* 14: 972, fig. 5; **Fig. 5.30** Adapted from Soukas et al. 2000. *Genes and Development.* 14: 974, fig. 7b.

CHAPTER 6 **Fig. 6.1** Reprinted with permission from Ross-Macdonald et al. 1999. *Nature.* 402: 413, fig. 1; **Fig. 6.2a–f** Reprinted with permission from Ross-Macdonald et al. 1999. *Nature.* 402: 415, fig. 3a; **Fig. 6.3a–b** Reprinted with permission from Ross-Macdonald et al. 1999. *Nature.* 402: 415, fig. 3b; **Fig. 6.5** Reprinted with permission from Ross-Macdonald et al. 1999. *Nature.* 402: 415, fig. 4; **Fig. 6.6a–g**

Reprinted with permission from Ross-Macdonald et al. 1999. *Nature.* 402: 465, fig. 5; **Fig. 6.7** Reprinted with permission from Winzeler et al. 1999. *Science.* 285: 903, fig. 2a; **Fig. 6.8a–b** Reprinted with permission from Winzeler et al. 1999. *Science.* 285: 903, fig. 2b, d; **Fig. 6.9** Reprinted with permission from Winzeler et al. 1999. *Science.* 285: 902, fig.1a; **Fig. 6.10** Reprinted with permission from Christiendat et al. 2000. *Nature.* 7(10): 415, fig. 2; **Fig. 6.11a–c** Reprinted with permission from Murata et al. 2000. *Nature.* 407: 604, fig. 5a–c; **Fig. 6.12a–c** Reprinted with permission from Nienaber et al. 2000. *Nature Biotechnology.* 18: 1105, fig. 1; **Fig. 6.13** Reprinted with permission from True and Lindquist et al. 2000. *Nature.* 407: 480, fig. 3a–c; **Fig. 6.14** Courtesy of Stan Field; **Fig. 6.15** Reprinted with permission from Uetz et al. 2000. *Nature.* 623: 480, fig. 3 b; **Fig. 6.16** Reprinted with permission from Marcotte et al. 1999. *Nature.* 402: 84, fig. 3; **Fig. 6.17** Reprinted with permission from Schwikowski et al. 2000. *Nature Biotechnology.* 18: 1260, fig. 3; **Fig. 6.18a–d, 6.19a–d,** and **6.20a–b** Courtesy of Stan Field; **Fig. 6.21** Reprinted with permission from Garin et al. 2001. *Journal of Cell Biology.* 152(1): 167, fig. 1; **Fig. 6.22** Reprinted with permission from Garin et al. 2001. *Journal of Cell Biology.* 152(1): 174, rows 3–9; **Fig. 6.23** Reprinted with permission from Garin et al. 2001. *Journal of Cell Biology.* 152(1): 176, fig. 6; **Fig. 6.24** Adapted from Oda et al. 1999. *PNAS.* 96: 6592, fig. 3; **Fig. 6.25** Adapted from Oda et al. 1999. *PNAS.* 96: 6592, fig. 4; **Fig. 6.26** Adapted from Oda et al. 1999. *PNAS.* 96: 6595, fig. 5; **Fig. 6.27** Adapted from Giygi et al. 2000. *Current Opinion in Biotechnology.* 11: 399. fig. 3a; **Fig. 6.28** Adapted from Giygi et al. 2000. *Current Opinion in Biotechnology.* 11: 399. fig. 3b; **Fig. 6.29a–b** Adapted from Giygi et al. 2000. *Current Opinion in Biotechnology.* 11: 399. fig. 3b; **Table 6.3** Reprinted with permission from Giygi et al. 1999. *Nature Biotechnology.* 17: 997. Table 2; **Fig. 6.30** Reprinted with permission from Giygi et al. 1999. *Nature Biotechnology.* 17: 998, fig. 5a; **Fig. 6.31a–b** Reprinted with permission from Giygi et al. 1999. *Nature Biotechnology.* 17: 998, fig. 5b; **Fig. 6.32a–f** Adapted from Haab et al. 2001. *Genome Biology.* 2(2): 4.3, fig. 1; **Fig. 6.33** Reprinted with permission from Haab et al. 2001. *Genome Biology.* 2(2): 4.8, fig. **Fig. 6.34a–d** Reprinted with permission from MacBeath and Schreibner. 2000. *Science.* 289: 1761, fig. 1,2; **Fig. 6.35a–d** Reprinted with permission from MacBeath and Schreibner. 2000. *Science.* 289: 1761, fig. 5; **Fig. 6.36a–b** Adapted from Zhu et al. 2000. *Nature Genetics.* 26: 285, fig. 2a–b; **Fig. 6.37a–f** Adapted from Zhu et al. 2000. *Nature Genetics.* 26: 285, fig. 2a; **Fig. 6.38** Reprinted with permission from Craig et al. 1996. *Journal of the American Chemical Society.* 118: 5249. figs. 4–5; **Fig. 6.39** Reprinted with permission from Craig et al. 1996. *Journal of the American Chemical Society.* 118(22): 5249. fig. 6: 5251 fig. 7; **Fig. 6.40** Reprinted with permission from Zhang et al. 2000. *Analytical Chemistry.* 72(2): 319, fig. 2; 320. fig. 3; **Fig. 6.41** Reprinted with permission from Fiehn et al. 2000. *Nature Biotechnology.* 18: 1159, fig. 3.

CHAPTER 7 **Fig. 7.2** Adapted from Eric Davidson. 1999. *Current Opinion in Genetics and Development.* 9: 534, fig. 2a; **Fig. 7.3** Adapted from Davidson. 1999. *Current Opinion in Genetics and Development.* 9: 534, fig. 2a; **Fig. 7.5** Adapted from Yuh and Davidson. 1996. *Development.* 122: 1072, fig. 1a; **Fig. 7.8** Adapted from Yuh and Davidson. 1996. *Development.* 122: 1072, fig. 1b; **Table 7.2** Reprinted with permission from Yuh and Davidson. 1996. *Development.* 122: 1072, fig. 1b; **Fig. 7.9** Adapted from Yuh and Davidson. 1996. *Development.* 122: 1072, fig. 1b; **Fig. 7.10** Adapted from Yuh and Davidson. 1996. *Development.* 122: 1078, fig. 4; **Fig. 7.11** Reprinted with permission from Yuh and Davidson. 1996. *Development.* 122: 1079, fig. 5; **Fig. 7.12a–b** Reprinted with permission from Yuh and Davidson. 1996. *Development.* 122: 1077, fig. 3a–b; **Fig. 7.13a–c** Reprinted with permission from Yuh and Davidson. 1996. *Development.* 122: 1077, fig. 3d; **Fig. 7.14** Adapted from Yuh and Davidson. 1996. *Development.* 122: 1080, fig. 6a; **Fig. 7.15** Adapted from Yuh and Davidson. 1996. *Development.* 122: 1080, fig. 6b; **Fig. 7.16** Reprinted with permission from Yuh el al. 1998. *Science.* 279: 1897, fig. 1c; **Fig. 7.17** Adapted from Yuh el al. 1998. *Science.* 279: 1897, fig. 1a–b; **Fig. 7.18a–d** Adapted from Yuh el al. 1998. *Science.* 279: 1898, fig. 2; **Fig. 7.19a–c** Adapted from Yuh el al. 1998. *Science.* 279: 1899, fig. 3; **Fig. 7.20** Adapted from Yuh el al. 1998. *Science.* 279: 1900, fig. 4; **Fig. 7.21** Adapted from Yuh el al. 1998. *Science.* 279: 1900, fig. 5; **Fig. 7.22** Adapted from Davidson. 1999. *Current Opinion in Genetics and Development.* 9: 534, fig. 2d; **Fig. 7.23** Courtesy of Eric Davidson.

CHAPTER 8 **Fig. 8.1** Reprinted with permission from McAdams and Arkin. 1999. *Trends in Genetics.* 15(2): 66, fig. 1; **Fig. 8.2** Reprinted with permission from McAdams and Arkin. 1999. *Trends in Genetics.* 15(2): 67, fig. 2; **Fig. 8.3** Adapted from Gardner et al. 2000. *Nature.* 403: 339, fig. 1; **Fig. 8.4** Reprinted with permission from Gardner et al. 2000. *Nature.* 403: 339, fig. 4b; **Fig. 8.5** Reprinted with permission from Elowitz and Leibler. 2000. *Nature.* 403: 336, fig. 1a; **Fig. 8.6** Adapted from Elowitz and Leibler. 2000. *Nature.* 403: 336, fig. 2; **Fig. 8.7a–c** Reprinted with permission from Elowitz and Leibler. 2000. *Nature.* 403: 337, fig. 3a–c; **Fig. 8.8** Adapted from McAdams and Arkin. 1999. *Trends in Genetics.* 15(2): 68, fig. 3a; **Fig. 8.9** Reprinted with permission from McAdams and Arkin. 1999. *Trends in Genetics.* 15(2): 68, fig. 3b; **Fig. 8.10** Reprinted with permission from McAdams and Arkin. 1999. *Trends in Genetics.* 15(2): 68, fig. 3c; **Fig. 8.11** Reprinted with permission from McAdams and Arkin. 1999. *Trends in Genetics.* 15(2): 68, fig. 3d; **Fig. 8.12** Reprinted with permission from McAdams and Arkin. 1999. *Trends in Genetics.* 15(2): 68, fig. 3e; **Fig. 8.13** Adapted from McAdams and Arkin. 1999. *Trends in Genetics.* 15(2): 68, fig. 3e; **Fig. 8.14a–c** Reprinted with permission from Weng et al. 1999. *Science.* 284: 93, fig. 1; **Fig. 8.15a–o** Reprinted with permission from Bhalla and Iyengar. 1999. *Science.* 283: 383, fig. 1; **Fig. 8.16** Reprinted with permission from Bhalla and Iyengar. 1999. *Science.* 283: 383, fig. 2a; **Fig. 8.17a–b** Reprinted with permission from Bhalla and Iyengar. 1999. *Science.* 283: 383, fig. 2a and 2c; **Fig. 8.18** Reprinted with permission from Bhalla and Iyengar. 1999. *Science.* 283: 383, fig. 2d; **Fig. 8.19** Reprinted with permission from Bhalla and Iyengar. 1999. *Science.* 283: 383, fig. 2e; **Fig. 8.20a–b** Reprinted with permission from Bhalla and Iyengar. 1999. *Science.* 283: 383, fig. 2g, 2h; **Fig. 8.21** Reprinted with permission from Bhalla and Iyengar. 1999. *Science.* 283: 384, fig. 3a; **Fig. 8.22a–d** Reprinted with permission from Bhalla and Iyengar. 1999. *Science.* 283: 384, fig. 3b–e; **Fig. 8.23** Reprinted with permission from Bhalla and Iyengar. 1999. *Science.* 283: 385, fig. 4a; **Fig. 8.24a–b** Reprinted with permission from Bhalla and Iyengar. 1999. *Science.* 283: 385, fig. 4b–c; **Fig. 8.25a–d** Reprinted with permission from Bhalla and Iyengar. 1999. *Science.* 283: 385, fig. 4d–g; **Fig. 8.26** Reprinted with permission from Kohn. 1999. *Molecular Biology of the Cell.* 10: 2705, fig. 1; **Fig. 8.27** Reprinted with permission from Kohn. 1999. *Molecular Biology of the Cell.* 10: 2705, fig. 1; **Fig. 8.28** Reprinted with permission from Endy et al. 2000. *PNAS.* 97(10): 5377, fig. 1b; **Fig. 8.29** Reprinted with permission from Endy et al. 2000. *PNAS.* 97(10): 5377, fig. 1a; **Fig. 8.30a–d** Reprinted with permission from Endy et al. 2000. *PNAS.* 97(10): 5377, fig. 2a–d.

CHAPTER 9 **Fig. 9.1a** and **9.2** Adapted from Ideker et al. 2001. *Science.* 292: 929, fig. 1; **Fig. 9.3** Reprinted with permission from Ideker et al. 2001. *Science.* 292: 930, fig. 2; **Fig. 9.4** Reprinted with permission from Ideker et al. 2001. *Science.* 292: 931, fig. 3; **Fig. 9.5** Adapted from Ideker et al. 2001. *Science.* 292: 932, fig. 4.

CHAPTER 10 **Fig. 10.2** Adapted from Bakker et al. 1985. *The Lancet.* Saturday 23 March: 657, fig. 2; **Fig. 10.7** Courtesy of Jim Ervasti; **Fig. 10.11** Reprinted with permission from Crosbie et al. 2000. *Human Molecular Genetics.* 9(13): 2023. fig. 2d; **Fig. 10.12a** Reprinted with permission from Crosbie et al. 1999. *Journal of Cell Biology.* 145(1): 163, fig. 10; **Fig. 10.12b** Reprinted with permission from Crosbie et al. 2000. *Human Molecular Genetics.* 9(13): 2024, fig. 4; **Fig. 10.13a** Adapted from Chien. 2000. *Nature.* 405: 229, fig. 2; **Fig. 10.13b** Reprinted with permission from Sweeney and Barton. 2000. *PNAS.* 97(25): 13465, fig. 1; **Fig. 10.14a–b** Reprinted with permission from Rybakora. Et al. 2000. *Journal of Cell Biology.* 150(5): 1212, fig. 4; **Fig. 10.18a** Adapted from Richard et al. 1999. *American Journal of Human Genetics.* 64: 1531, fig. 1; **Fig. 10.19b–d** Adapted from Richard et al. 1999. *American Journal of Human Genetics.* 64: 1534, fig. 2; **Fig. 10.22** Reprinted with permission from Grady et al. 1999. *Nature Cell Biology.* 1: 219, fig. 7.

CHAPTER 11 **Fig. 11.5** Reprinted with permission from Zhang et al. 1994. *Nature.* 372: 429, fig. 4b; **Fig. 11.7a–d** Reprinted with permission from Pelleymounter. 1995. *Science.* 269: 544, fig. 2a–d; **Fig. 11.8** Reprinted with permission from Friedman. 1994. *Nature.* 404: 632, fig. 1; **Fig. 11.11** Reprinted with permission from Heymsfield et al. 1999. JAMA: 1572, fig. 2; **Fig. 11.12a–b** Reprinted with permission from Schwartz et al. 2000. *Nature.* 404: 664, fig. 21.

CHAPTER 12 **Fig. 12.1a–c** Reprinted with permission from Schwartz et al. 2000. *Trends in Cell Biology.* 10: 291, fig.1; **Fig. 12.2** Reprinted with permission from Evans and Relling. 1999. *Science.* 286: 487, fig. 1; **Fig. 12.3** Adapted from Dequeker and Cassiman. 2000. *Nature Genetics.* 25: 259, fig. 1.

INDEX

Page references followed by *f* indicate an illustration; followed by *t* indicates a table; followed by *b* indicates a box.